MW01323280

IMAGE PROCESSING FOR REMOTE SENSING

IMAGE PROCESSING FOR REMOTE SENSING

EDITED BY

C. H. CHEN

CRC Press is an imprint of the
Taylor & Francis Group, an **informa** business

The material was previously published in *Signal and Image Processing for Remote Sensing* © Taylor and Francis 2006.

CRC Press
Taylor & Francis Group
6000 Broken Sound Parkway NW, Suite 300
Boca Raton, FL 33487-2742

Printed in the United States of America on acid-free paper
10 9 8 7 6 5 4 3 2 1

International Standard Book Number-13: 978-1-4200-6664-7 (Hardcover)

Library of Congress Cataloging-in-Publication Data

Image processing for remote sensing / [edited by] C.H. Chen.
p. cm.
Includes bibliographical references and index.
ISBN-13: 978-1-4200-6664-7
ISBN-10: 1-4200-6664-1
1. Remote sensing--Data processing. 2. Image processing. I. Chen, C.H. (Chi-hau), 1937- II. Title.

G70.4.I44 2008
621.36'78--dc22 2007030188

Visit the Taylor & Francis Web site at
http://www.taylorandfrancis.com

and the CRC Press Web site at
http://www.crcpress.com

Preface

This volume is a spin-off edition derived from *Signal and Image Processing for Remote Sensing*. It presents more advanced topics of image processing in remote sensing than similar books in the area. The topics of image modeling, statistical image classifiers, change detection, independent component analysis, vertex component analysis, image fusion for better classification or segmentation, 2-D time series modeling, neural network classifications, etc. are examined in this volume. Some unique topics like accuracy assessment and information-theoretic measure of multiband images are presented. An emphasis is placed on the issues with synthetic aperture radar (SAR) images in many chapters. Continued development on imaging sensors always presents new opportunities and challenges on image processing for remote sensing. The hyperspectral imaging sensor is a good example here. We believe this volume not only presents the most up-to-date developments of image processing for remote sensing but also suggests to readers the many challenging problems ahead for further study.

Original Preface from *Signal and Image Processing for Remote Sensing*
Both signal processing and image processing have been playing increasingly important roles in remote sensing. While most data from satellites are in image forms and thus image processing has been used most often, signal processing can contribute significantly in extracting information from the remotely sensed waveforms or time series data. In contrast to other books in this field which deal almost exclusively with the image processing for remote sensing, this book provides a good balance between the roles of signal processing and image processing in remote sensing. The book covers mainly methodologies of signal processing and image processing in remote sensing. Emphasis is thus placed on the mathematical techniques which we believe will be less changed as compared to sensor, software and hardware technologies. Furthermore, the term "remote sensing" is not limited to the problems with data from satellite sensors. Other sensors which acquire data remotely are also considered. Thus another unique feature of the book is the coverage of a broader scope of the remote sensing information processing problems than any other book in the area.

The book is divided into two parts [now published as separate volumes under the following titles]. Part I, *Signal Processing for Remote Sensing*, has 12 chapters and Part II [comprising the present volume], *Image Processing for Remote Sensing*, has 16 chapters. The chapters are written by leaders in the field. We are very fortunate, for example, to have Dr. Norden Huang, inventor of the Huang–Hilbert transform, along with Dr. Steven Long, to write a chapter on the application of the transform to remote sensing problem, and Dr. Enders A. Robinson, who has made many major contributions to geophysical signal processing for over half a century, to write a chapter on the basic problem of constructing seismic images by ray tracing.

In Part I, following Chapter 1 by Drs. Long and Huang, and my short Chapter 2 on the roles of statistical pattern recognition and statistical signal processing in remote sensing, we start from a very low end of the electromagnetic spectrum. Chapter 3 considers the classification of infrasound at a frequency range of 0.001 Hz to 10 Hz by using a parallel bank neural network classifier and a 11-step feature selection process. The >90% correct classification rate is impressive for this kind of remote sensing data. Chapter 4 through

Chapter 6 deal with seismic signal processing. Chapter 4 provides excellent physical insights on the steps for construction of digital seismic images. Even though the seismic image is an image, this chapter is placed in Part I as seismic signals start as waveforms. Chapter 5 considers the singular value decomposition of a matrix data set from scalar-sensors arrays, which is followed by independent component analysis (ICA) step to relax the unjustified orthogonality constraint for the propagation vectors by imposing a stronger constraint of fourth-order independence of the estimated waves. With an initial focus of the use of ICA in seismic data and inspired by Dr. Robinson's lecture on seismic deconvolution at the 4th International Symposium, 2002, on Computer Aided Seismic Analysis and Discrimination, Mr. Zhenhai Wang has examined approaches beyond ICA for improving seismic images. Chapter 6 is an effort to show that factor analysis, as an alternative to stacking, can play a useful role in removing some unwanted components in the data and thereby enhancing the subsurface structure as shown in the seismic images. Chapter 7 on Kalman filtering for improving detection of landmines using electromagnetic signals, which experience severe interference, is another remote sensing problem of higher interest in recent years. Chapter 8 is a representative time series analysis problem on using meteorological and remote sensing indices to monitor vegetation moisture dynamics. Chapter 9 actually deals with the image data for digital elevation model but is placed in Part I mainly because the prediction error (PE) filter is originated from the geophysical signal processing. The PE filter allows us to interpolate the missing parts of an image. The only chapter that deals with the sonar data is Chapter 10, which shows that a simple blind source separation algorithm based on the second-order statistics can be very effective to remove reverberations in active sonar data. Chapter 11 and Chapter 12 are excellent examples of using neural networks for retrieval of physical parameters from the remote sensing data. Chapter 12 further provides a link between signal and image processing as the principal component analysis and image sharpening tools employed are exactly what are needed in Part II.

With a focus on image processing of remote sensing images, Part II begins with Chapter 13 [Chapter 1 of the present volume] that is concerned with the physics and mathematical algorithms for determining the ocean surface parameters from synthetic aperture radar (SAR) images. Mathematically Markov random field (MRF) is one of the most useful models for the rich contextual information in an image. Chapter 14 [now Chapter 2] provides a comprehensive treatment of MRF-based remote sensing image classification. Besides an overview of previous work, the chapter describes the methodological issues involved and presents results of the application of the technique to the classification of real (both single-date and multitemporal) remote sensing images. Although there are many studies on using an ensemble of classifiers to improve the overall classification performance, the random forest machine learning method for classification of hyperspectral and multisource data as presented in Chapter 15 [now Chapter 3] is an excellent example of using new statistical approaches for improved classification with the remote sensing data. Chapter 16 [now Chapter 4] presents another machine learning method, AdaBoost, to obtain robustness property in the classifier. The chapter further considers the relations among the contextual classifier, MRF-based methods, and spatial boosting. The following two chapters are concerned with different aspects of the change detection problem. Change detection is a uniquely important problem in remote sensing as the images acquired at different times over the same geographical area can be used in the areas of environmental monitoring, damage management, and so on. After discussing change detection methods for multitemporal SAR images, Chapter 17 [now Chapter 5] examines an adaptive scale–driven technique for change detection in medium resolution SAR data. Chapter 18 [now Chapter 6] evaluates the Wiener filter-based method,

Mahalanobis distance, and subspace projection methods of change detection, with the change detection performance illustrated by receiver operating characteristics (ROC) curves. In recent years, ICA and related approaches have presented many new potentials in remote sensing information processing. A challenging task underlying many hyperspectral imagery applications is decomposing a mixed pixel into a collection of reflectance spectra, called endmember signatures, and the corresponding abundance fractions. Chapter 19 [now Chapter 7] presents a new method for unsupervised endmember extraction called vertex component analysis (VCA). The VCA algorithms presented have better or comparable performance as compared to two other techniques but require less computational complexity. Other useful ICA applications in remote sensing include feature extraction, and speckle reduction of SAR images. Chapter 20 [now Chapter 8] presents two different methods of SAR image speckle reduction using ICA, both making use of the FastICA algorithm. In two-dimensional time series modeling, Chapter 21 [now Chapter 9] makes use of a fractionally integrated autoregressive moving average (FARIMA) analysis to model the mean radial power spectral density of the sea SAR imagery. Long-range dependence models are used in addition to the fractional sea surface models for the simulation of the sea SAR image spectra at different sea states, with and without oil slicks at low computational cost.

Returning to the image classification problem, Chapter 22 [now Chapter 10] deals with the topics of pixel classification using Bayes classifier, region segmentation guided by morphology and split-and-merge algorithm, region feature extraction, and region classification.

Chapter 23 [now Chapter 11] provides a tutorial presentation of different issues of data fusion for remote sensing applications. Data fusion can improve classification and for the decision level fusion strategies, four multisensor classifiers are presented. Beyond the currently popular transform techniques, Chapter 24 [now Chapter 12] demonstrates that Hermite transform can be very useful for noise reduction and image fusion in remote sensing. The Hermite transform is an image representation model that mimics some of the important properties of human visual perception, namely local orientation analysis and the Gaussian derivative model of early vision. Chapter 25 [now Chapter 13] is another chapter that demonstrates the importance of image fusion to improving sea ice classification performance, using backpropagation trained neural network and linear discrimination analysis and texture features. Chapter 26 [now Chapter 14] is on the issue of accuracy assessment for which the Bradley–Terry model is adopted. Chapter 27 [now Chapter 15] is on land map classification using support vector machine, which has been increasingly popular as an effective classifier. The land map classification classifies the surface of the Earth into categories such as water area, forests, factories or cities. Finally, with lossless data compression in mind, Chapter 28 [now Chapter 16] focuses on information-theoretic measure of the quality of multi-band remotely sensed digital images. The procedure relies on the estimation of parameters of the noise model. Results on image sequences acquired by AVIRIS and ASTER imaging sensors offer an estimation of the information contents of each spectral band.

With rapid technological advances in both sensor and processing technologies, a book of this nature can only capture certain amount of current progress and results. However, if past experience offers any indication, the numerous mathematical techniques presented will give this volume a long lasting value.

The sister volumes of this book are the other two books edited by myself. One is *Information Processing for Remote Sensing* and the other is *Frontiers of Remote Sensing Information Processing*, both published by World Scientific in 1999 and 2003, respectively. I am grateful to all contributors of this volume for their important contribution and, in particular, to Dr. J.S. Lee, S. Serpico, L. Bruzzone and S. Omatu for chapter contributions to all three volumes. Readers are advised to go over all three volumes for a more complete information on signal and image processing for remote sensing.

C. H. Chen

Editor

Chi Hau Chen was born on December 22nd, 1937. He received his Ph.D. in electrical engineering from Purdue University in 1965, M.S.E.E. degree from the University of Tennessee, Knoxville, in 1962, and B.S.E.E. degree from the National Taiwan University in 1959.

He is currently chancellor professor of electrical and computer engineering at the University of Massachusetts, Dartmouth, where he has taught since 1968. His research areas are in statistical pattern recognition and signal/image processing with applications to remote sensing, geophysical, underwater acoustics, and nondestructive testing problems, as well as computer vision for video surveillance, time series analysis, and neural networks.

Dr. Chen has published 25 books in his area of research. He is the editor of *Digital Waveform Processing and Recognition* (CRC Press, 1982) and *Signal Processing Handbook* (Marcel Dekker, 1988). He is the chief editor of *Handbook of Pattern Recognition and Computer Vision*, volumes 1, 2, and 3 (World Scientific Publishing, 1993, 1999, and 2005, respectively). He is the editor of *Fuzzy Logic and Neural Network Handbook* (McGraw-Hill, 1966). In the area of remote sensing, he is the editor of *Information Processing for Remote Sensing* and *Frontiers of Remote Sensing Information Processing* (World Scientific Publishing, 1999 and 2003, respectively).

He served as the associate editor of the *IEEE Transactions on Acoustics Speech and Signal Processing* for 4 years, *IEEE Transactions on Geoscience and Remote Sensing* for 15 years, and since 1986 he has been the associate editor of the *International Journal of Pattern Recognition and Artificial Intelligence.*

Dr. Chen has been a fellow of the Institutue of Electrical and Electronic Engineers (IEEE) since 1988, a life fellow of the IEEE since 2003, and a fellow of the International Association of Pattern Recognition (IAPR) since 1996.

Contributors

Bruno Aiazzi Institute of Applied Physics, National Research Council, Florence, Italy

Selim Aksoy Bilkent University, Ankara, Turkey

V.Yu. Alexandrov Nansen International Environmental and Remote Sensing Center, St. Petersburg, Russia

Luciano Alparone Department of Electronics and Telecommunications, University of Florence, Florence, Italy

Stefano Baronti Institute of Applied Physics, National Research Council, Florence, Italy

Jon Atli Benediktsson Department of Electrical and Computer Engineering, University of Iceland, Reykjavik, Iceland

Fabrizio Berizzi Department of Information Engineering, University of Pisa, Pisa, Italy

Massimo Bertacca ISL-ALTRAN, Analysis and Simulation Group—Radar Systems Analysis and Signal Processing, Pisa, Italy

L.P. Bobylev Nansen International Environmental and Remote Sensing Center, St. Petersburg, Russia

A.V. Bogdanov Institute for Neuroinformatich, Bochum, Germany

Francesca Bovolo Department of Information and Communication Technology, University of Trento, Trento, Italy

Lorenzo Bruzzone Department of Information and Communication Technology, University of Trento, Trento, Italy

Chi Hau Chen Department of Electrical and Computer Engineering, University of Massachusetts Dartmouth, North Dartmouth, Massachusetts

Salim Chitroub Signal and Image Processing Laboratory, Department of Telecommunication, Algiers, Algeria

José M.B. Dias Department of Electrical and Computer Engineering, Instituto Superior Técnico, Av. Rovisco Pais, Lisbon, Portugal

Shinto Eguchi Institute of Statistical Mathematics, Tokyo, Japan

Boris Escalante-Ramírez School of Engineering, National Autonomous University of Mexico, Mexico City, Mexico

Toru Fujinaka Osaka Prefecture University, Osaka, Japan

Gerrit Gort Department of Biometris, Wageningen University, The Netherlands

Sveinn R. Joelsson Department of Electrical and Computer Engineering, University of Iceland, Reykjavik, Iceland

O.M. Johannessen Nansen Environmental and Remote Sensing Center, Bergen, Norway

Dayalan Kasilingam Department of Electrical and Computer Engineering, University of Massachusetts Dartmouth, North Dartmouth, Massachusetts

Heesung Kwon U.S. Army Research Laboratory, Adelphi, Maryland

Jong-Sen Lee Remote Sensing Division, Naval Research Laboratory, Washington, D.C.

Alejandra A. López-Caloca Center for Geography and Geomatics Research, Mexico City, Mexico

Arko Lucieer Centre for Spatial Information Science (CenSIS), University of Tasmania, Australia

Enzo Dalle Mese Department of Information Engineering, University of Pisa, Pisa, Italy

Gabriele Moser Department of Biophysical and Electronic Engineering, University of Genoa, Genoa, Italy

José M.P. Nascimento Instituto Superior, de Eugenharia de Lisbon, Lisbon, Portugal

Nasser Nasrabadi U.S. Army Research Laboratory, Adelphi, Maryland

Ryuei Nishii Faculty of Mathematics, Kyusyu University, Fukuoka, Japan

Sigeru Omatu Osaka Prefecture University, Osaka, Japan

S. Sandven Nansen Environmental and Remote Sensing Center, Bergen, Norway

Dale L. Schuler Remote Sensing Division, Naval Research Laboratory, Washington, D.C.

Massimo Selva Institute of Applied Physics, National Research Council, Florence, Italy

Sebastiano B. Serpico Department of Biophysical and Electronic Engineering, University of Genoa, Genoa, Italy

Anne H.S. Solberg Department of Informatics, University of Oslo and Norwegian Computing Center, Oslo, Norway

Alfred Stein International Institute for Geo-Information Science and Earth Observation, Enschede, The Netherlands

Johannes R. Sveinsson Department of Electrical and Computer Engineering, University of Iceland, Reykjavik, Iceland

Maria Tates U.S. Army Research Laboratory, Adelphi, Maryland, and Morgan State University, Baltimore, Maryland

Xianju Wang Department of Electrical and Computer Engineering, University of Massachusetts Dartmouth, North Dartmouth, Massachusetts

Carl White Morgan State University, Baltimore, Maryland

Michifumi Yoshioka Osaka Prefecture University, Osaka, Japan

Contents

1

Polarimetric SAR Techniques for Remote Sensing of the Ocean Surface

Dale L. Schuler, Jong-Sen Lee, and Dayalan Kasilingam

CONTENTS

1.1 Introduction

Selected methods that use synthetic aperture radar (SAR) image data to remotely sense ocean surfaces are described in this chapter. Fully polarimetric SAR radars provide much more usable information than conventional single-polarization radars. Algorithms, presented here, to measure directional wave spectra, wave slopes, wave–current interactions, and current-driven surface features use this additional information.

Polarimetric techniques that measure directional wave slopes and spectra with data collected from a single aircraft, or satellite, collection pass are described here. Conventional single-polarization backscatter cross-section measurements require two orthogonal passes and a complex SAR modulation transfer function (MTF) to determine vector slopes and directional wave spectra.

The algorithm to measure wave spectra is described in Section 1.2. In the azimuth (flight) direction, wave-induced perturbations of the polarimetric orientation angle are used to sense the azimuth component of the wave slopes. In the orthogonal range direction, a technique involving an alpha parameter from the well-known Cloude–Pottier entropy/anisotropy/averaged alpha ($H/A/\bar{\alpha}$) polarimetric decomposition theorem is used to measure the range slope component. Both measurement types are highly sensitive to ocean wave slopes and are directional. Together, they form a means of using polarimetric SAR image data to make complete directional measurements of ocean wave slopes and wave slope spectra.

NASA Jet Propulsion Laboratory airborne SAR (AIRSAR) P-, L-, and C-band data obtained during flights over the coastal areas of California are used as wave-field examples. Wave parameters measured using the polarimetric methods are compared with those obtained using *in situ* NOAA National Data Buoy Center (NDBC) buoy products.

In a second topic (Section 1.3), polarization orientation angles are used to remotely sense ocean wave slope distribution changes caused by ocean wave–current interactions. The wave–current features studied include surface manifestations of ocean internal waves and wave interactions with current fronts.

A model [1], developed at the Naval Research Laboratory (NRL), is used to determine the parametric dependencies of the orientation angle on internal wave current, wind-wave direction, and wind-wave speed. An empirical relation is cited to relate orientation angle perturbations to the underlying parametric dependencies [1].

A third topic (Section 1.4) deals with the detection and classification of biogenic slick fields. Various techniques, using the Cloude–Pottier decomposition and Wishart classifier, are used to classify the slicks. An application utilizing current-driven ocean features, marked by slick patterns, is used to map spiral eddies. Finally, a related technique, using the polarimetric orientation angle, is used to segment slick fields from ocean wave slopes.

1.2 Measurement of Directional Slopes and Wave Spectra

1.2.1 Single Polarization versus Fully Polarimetric SAR Techniques

SAR systems conventionally use backscatter intensity-based algorithms [2] to measure physical ocean wave parameters. SAR instruments, operating at a single-polarization, measure wave-induced backscatter cross section, or sigma-0, modulations that can be developed into estimates of surface wave slopes or wave spectra. These measurements, however, require a parametrically complex MTF to relate the SAR backscatter measurements to the physical ocean wave properties [3].

Section 1.2.3 through Section 1.2.6 outline a means of using fully polarimetric SAR (POLSAR) data with algorithms [4] to measure ocean wave slopes. In the Fourier-transform domain, this orthogonal slope information is used to estimate a complete directional ocean wave slope spectrum. A parametrically simple measurement of the slope is made by using POLSAR-based algorithms.

Modulations of the polarization orientation angle, θ, are largely caused by waves traveling in the azimuth direction. The modulations are, to a lesser extent, also affected by range traveling waves. A method, originally used in topographic measurements [5], has been applied to the ocean and used to measure wave slopes. The method measures vector components of ocean wave slopes and wave spectra. Slopes smaller than 1° are measurable for ocean surfaces using this method.

An eigenvector or eigenvalue decomposition average parameter $\bar{\alpha}$, described in Ref. [6], is used to measure wave slopes in the orthogonal range direction. Waves in the range direction cause modulation of the local incidence angle ϕ, which, in turn, changes the value of $\bar{\alpha}$. The alpha parameter is "roll-invariant." This means that it is not affected by slopes in the azimuth direction. Likewise, for ocean wave measurements, the orientation angle θ parameter is largely insensitive to slopes in the range direction. An algorithm employing both ($\bar{\alpha}$, θ) is, therefore, capable of measuring slopes in any direction. The ability to measure a physical parameter in two orthogonal directions within an individual resolution cell is rare. Microwave instruments, generally, must have a two-dimensional (2D) imaging or scanning capability to obtain information in two orthogonal directions.

Motion-induced nonlinear "velocity-bunching" effects still present difficulties for wave measurements in the azimuth direction using POLSAR data. These difficulties are dealt with by using the same proven algorithms [3,7] that reduce nonlinearities for single-polarization SAR measurements.

1.2.2 Single-Polarization SAR Measurements of Ocean Surface Properties

SAR systems have previously been used for imaging ocean features such as surface waves, shallow-water bathymetry, internal waves, current boundaries, slicks, and ship wakes [8]. In all of these applications, the modulation of the SAR image intensity by the ocean feature makes the feature visible in the image [9]. When imaging ocean surface waves, the main modulation mechanisms have been identified as tilt modulation, hydrodynamic modulation, and velocity bunching [2]. Tilt modulation is due to changes in the local incidence angle caused by the surface wave slopes [10]. Tilt modulation is strongest for waves traveling in the range direction. Hydrodynamic modulation is due to the hydrodynamic interactions between the long-scale surface waves and the short-scale surface (Bragg) waves that contribute most of the backscatter at moderate incidence angles [11]. Velocity bunching is a modulation process that is unique to SAR imaging systems [12]. It is a result of the azimuth shifting of scatterers in the image plane, owing to the motion of the scattering surface. Velocity bunching is the highest for azimuth traveling waves.

In the past, considerable effort had gone into retrieving quantitative surface wave information from SAR images of ocean surface waves [13]. Data from satellite SAR missions, such as ERS 1 and 2 and RADARSAT 1 and 2, had been used to estimate surface wave spectra from SAR image information. Generally, wave height and wave slope spectra are used as quantitative overall descriptors of the ocean surface wave properties [14]. Over the years, several different techniques have been developed for retrieving wave spectra from SAR image spectra [7,15,16]. Linear techniques, such as those having a linear MTF, are used to relate the wave spectrum to the image spectrum. Individual MTFs are derived for the three primary modulation mechanisms. A transformation based on the MTF is used to retrieve the wave spectrum from the SAR image spectrum. Since the technique is linear, it does not account for any nonlinear processes in the modulation mechanisms. It has been shown that SAR image modulation is nonlinear under certain ocean surface conditions. As the sea state increases, the degree of nonlinear behavior generally increases. Under these conditions, the linear methods do not provide accurate quantitative estimates of the wave spectra [15]. Thus, the linear transfer function method has limited utility and can be used as a qualitative indicator. More accurate estimates of wave spectra require the use of nonlinear inversion techniques [15].

Several nonlinear inversion techniques have been developed for retrieving wave spectra from SAR image spectra. Most of these techniques are based on a technique developed in Ref. [7]. The original method used an iterative technique to estimate the wave spectrum from the image spectrum. Initial estimates are obtained using a linear transfer function similar to the one used in Ref. [15]. These estimates are used as inputs in the forward SAR imaging model, and the revised image spectrum is used to iteratively correct the previous estimate of the wave spectra. The accuracy of this technique is dependent on the specific SAR imaging model. Improvements to this technique [17] have incorporated closed-form descriptions of the nonlinear transfer function, which relates the wave spectrum to the SAR image spectrum. However, this transfer function also has to be evaluated iteratively. Further improvements to this method have been suggested in Refs. [3,18]. In this method, a cross-spectrum is generated between different looks of the same ocean wave scene. The primary advantage of this method is that it resolves the 180° ambiguity [3,18] of the wave direction. This method also reduces the effects of speckle in the SAR spectrum. Methods that incorporate additional a posteriori information about the wave field, which improves the accuracy of these nonlinear methods, have also been developed in recent years [19].

In all of the slope-retrieval methods, the one nonlinear mechanism that may completely destroy wave structure is velocity bunching [3,7]. Velocity bunching is a result of moving scatterers on the ocean surface either bunching or dilating in the SAR image domain. The shifting of the scatterers in the azimuth direction may, in extreme conditions, result in the destruction of the wave structure in the SAR image.

SAR imaging simulations were performed at different range-to-velocity (R/V) ratios to study the effect of velocity bunching on the slope-retrieval algorithms. When the (R/V) ratio is artificially increased to large values, the effects of velocity bunching are expected to destroy the wave structure in the slope estimates. Simulations of the imaging process for a wide range of radar-viewing conditions indicate that the slope structure is preserved in the presence of moderate velocity-bunching modulation. It can be argued that for velocity bunching to affect the slope estimates, the (R/V) ratio has to be significantly larger than 100 s. The two data sets discussed here are designated "Gualala River" and "San Francisco." The Gualala river data set has the longest waves and it also produces the best results. The R/V ratio for the AIRSAR missions was 59 s (Gualala) and 55 s (San Francisco). These values suggest that the effects of velocity bunching are present, but are not sufficiently strong to significantly affect the slope-retrieval process. However, for spaceborne SAR imaging applications, where the (R/V) ratio may be greater than 100 s, the effects of velocity bunching may limit the utility of all methods, especially in high sea states.

1.2.3 Measurement of Ocean Wave Slopes Using Polarimetric SAR Data

In this section, the techniques that were developed for the measurement of ocean surface slopes and wave spectra using the capabilities of fully polarimetric radars are discussed. Wave-induced perturbations of the polarization orientation angle are used to directly measure slopes for azimuth traveling waves. This technique is accurate for scattering from surface resolution cells where the sea return can be represented as a two-scale Bragg-scattering process.

1.2.3.1 Orientation Angle Measurement of Azimuth Slopes

It has been shown [5] that by measuring the orientation angle shift in the polarization signature, one can determine the effects of the azimuth surface tilts. In particular, the shift in the orientation angle is related to the azimuth surface tilt, the local incidence angle, and, to a lesser degree, the range tilt. This relationship is derived [20] and independently verified [6] as

$$\tan\theta = \frac{\tan\omega}{\sin\phi - \tan\gamma\cos\phi} \tag{1.1}$$

where θ, $\tan\omega$, $\tan\gamma$, and ϕ are the shifts in the orientation angle, the azimuth slope, the ground range slope, and the radar look angle, respectively. According to Equation 1.1, the azimuth tilts may be estimated from the shift in the orientation angle, if the look angle and range tilt are known.

The orthogonal range slope $\tan\gamma$ can be estimated using the value of the local incidence angle associated with the alpha parameter for each pixel. The azimuth slope $\tan\omega$ and the range slope $\tan\gamma$ provide complete slope information for each image pixel.

For the ocean surface at scales of the size of the AIRSAR resolution cell (6.6 m $\times$ 8.2 m), the averaged tilt angles are small and the denominator in Equation 1.1 may be approximated by $\sin\phi$ for a wide range of look angles, $\cos\phi$, and ground range slope, $\tan\gamma$, values. Under this approximation, the ocean azimuth slope, $\tan\omega$, is written as

$$\tan\omega \cong (\sin\phi)\cdot\tan\theta \tag{1.2}$$

The above equation is important because it provides a direct link between polarimetric SAR measurable parameters and physical slopes on the ocean surface. This estimation of ocean slopes relies only on (1) the knowledge of the radar look angle (generally known from the SAR viewing geometry) and (2) the measurement of the wave-perturbed orientation angle. In ocean areas where the average scattering mechanism is predominantly tilted-Bragg scatter, the orientation angle can be measured accurately for angular changes $<1^\circ$, as demonstrated in Ref. [20].

POLSAR data can be represented by the scattering matrix for single-look complex data and by the Stokes matrix, the covariance matrix, or the coherency matrix for multi-look data. An orientation angle shift causes rotation of all these matrices about the line of sight. Several methods have been developed to estimate the azimuth slope–induced orientation angles for terrain and ocean applications. The "polarization signature maximum" method and the "circular polarization" method have proven to be the two most effective methods. Complete details of these methods and the relation of the orientation angle to orthogonal slopes and radar parameters are given [21,22].

1.2.3.2 Orientation Angle Measurement Using the Circular-Pol Algorithm

Image processing was done with both the polarization signature maximum and the circular polarization algorithms. The results indicate that for ocean images a significant improvement in wave visibility is achieved when a circular polarization algorithm is chosen. In addition to this improvement, the circular polarization algorithm is computationally more efficient. Therefore, the circular polarization algorithm method was chosen to estimate orientation angles. The most sensitive circular polarization estimator [21], which involves RR (right-hand transmit, right-hand receive) and LL (left-hand transmit, left-hand receive) terms, is

$$\theta = [\mathrm{Arg}(\langle S_{\mathrm{RR}} S_{\mathrm{LL}}^{*} \rangle) + \pi]/4 \tag{1.3}$$

A linear-pol basis has a similar transmit-and-receive convention, but the terms (HH, VV, HV, VH) involve horizontal (H) and vertical (V) transmitted, or received, components. The known relations between a circular-pol basis and a linear-pol basis are

$$\begin{aligned} S_{\mathrm{RR}} &= (S_{\mathrm{HH}} - S_{\mathrm{VV}} + i2S_{\mathrm{HV}})/2 \\ S_{\mathrm{LL}} &= (S_{\mathrm{VV}} - S_{\mathrm{HH}} + i2S_{\mathrm{HV}})/2 \end{aligned} \tag{1.4}$$

Using the above equation, the Arg term of Equation 1.3 can be written as

$$\theta = \mathrm{Arg}(\langle S_{\mathrm{RR}} S_{\mathrm{LL}}^{*} \rangle) = \tan^{-1}\left(\frac{-4\mathrm{Re}\left(\langle (S_{\mathrm{HH}} - S_{\mathrm{VV}}) S_{\mathrm{HV}}^{*} \rangle \right)}{-\langle |S_{\mathrm{HH}} - S_{\mathrm{VV}}|^{2} \rangle + 4 \langle |S_{\mathrm{HV}}|^{2} \rangle} \right) \tag{1.5}$$

The above equation gives the orientation angle, θ, in terms of three of the terms of the linear-pol coherency matrix. This algorithm has been proven to be successful in Ref. [21]. An example of the accuracy is cited from related earlier studies involving wave–current interactions [1]. In these studies, it has been shown that small wave slope asymmetries could be accurately detected as changes in the orientation angle. These small asymmetries had been predicted by theory [1] and their detection indicates the sensitivity of the circular-pol orientation angle measurement.

1.2.4 Ocean Wave Spectra Measured Using Orientation Angles

NASA/JPL/AIRSAR data were taken (1994) at L-band imaging a northern California coastal area near the town of Gualala (Mendocino County) and the Gualala River. This data set was used to determine if the azimuth component of an ocean wave spectrum could be measured using orientation angle modulation. The radar resolution cell had dimensions of 6.6 m (range direction) and 8.2 m (azimuth direction), and 3 × 3 boxcar averaging was done to the data inputted into the orientation angle algorithms.

Figure 1.1 is an L-band, VV-pol, pseudo color-coded image of a northern California coastal area and the selected measurement study site. A wave system with an estimated dominant wavelength of 157 m is propagating through the site with a wind-wave direction of 306° (estimates from wave spectra, Figure 1.4). The scattering geometry for a single average tilt radar resolution cell is shown in Figure 1.2. Modulations in the polarization orientation angle induced by azimuth traveling ocean waves in the study area are shown in Figure 1.3a and a histogram of the orientation angles is given in Figure 1.3b. An orientation angle spectrum versus wave number for azimuth direction waves propagating in the study area is given in Figure 1.4. The white rings correspond to ocean wavelengths of 50, 100, 150, and 200 m. The dominant 157-m wave is propagating at a heading of 306°. Figure 1.5a and Figure 1.5b give plots of spectral intensity versus wave number (a) for wave-induced orientation angle modulations and (b) for single-polarization (VV-pol)-intensity modulations. The plots are of wave spectra taken in the direction that maximizes the dominant wave peak. The orientation angle–dominant wave peak

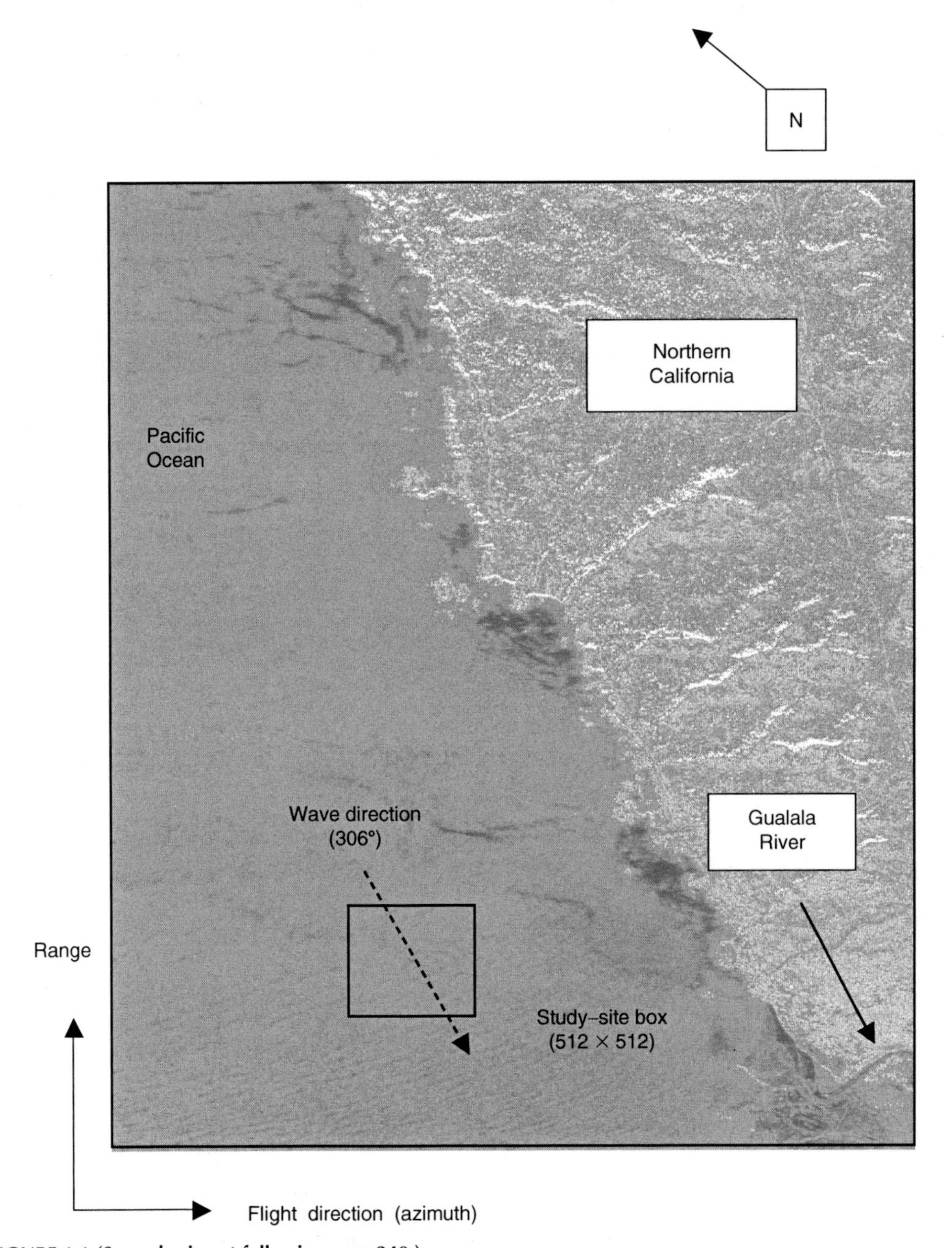

FIGURE 1.1 (See color insert following page 240.)
An L-band, VV-pol, AIRSAR image, of northern California coastal waters (Gualala River dataset), showing ocean waves propagating through a study-site box.

(Figure 1.5a) has a significantly higher signal and background ratio than the conventional intensity-based VV-pol-dominant wave peak (Figure 1.5b).

Finally, the orientation angles measured within the study sites were converted into azimuth direction slopes using an average incidence angle and Equation 1.2. From the estimates of these values, the ocean rms azimuth slopes were computed. These values are given in Table 1.1.

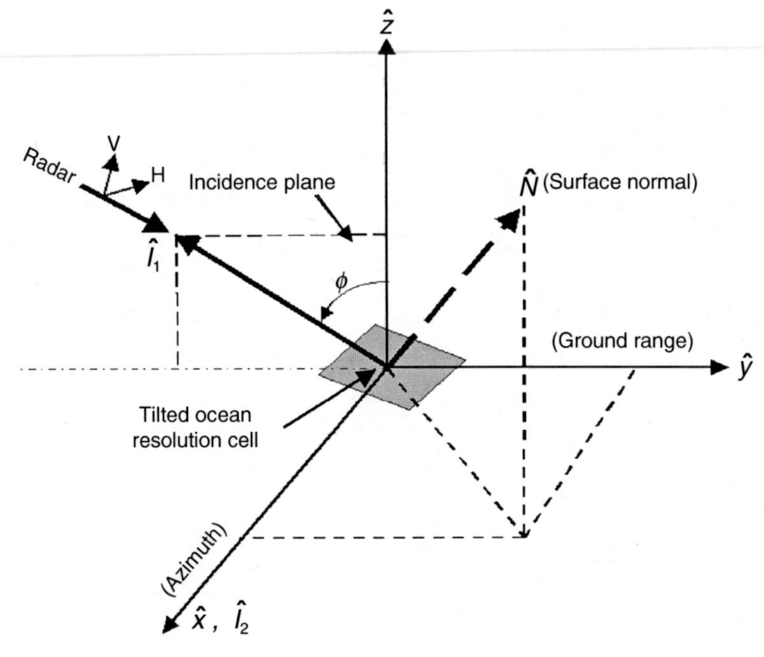

FIGURE 1.2
Geometry of scattering from a single, tilted, resolution cell. In the Gualala River dataset, the resolution cell has dimensions 6.6 m (range direction) and 8.2 m (azimuth direction).

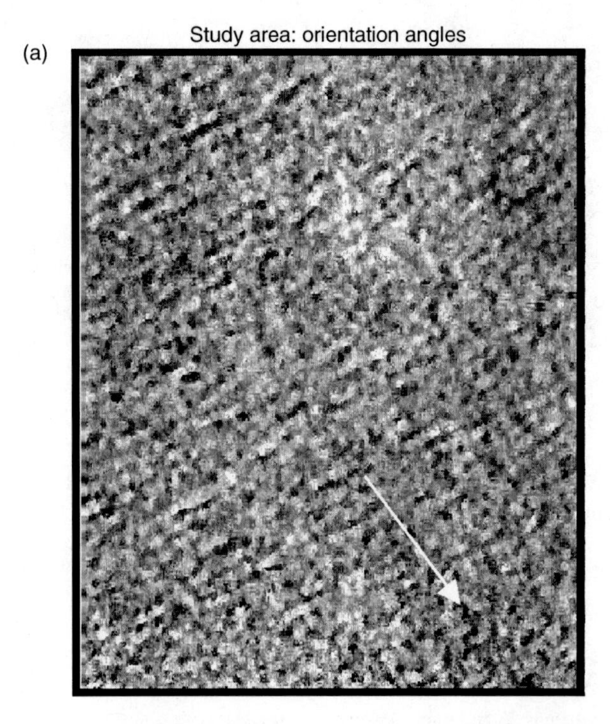

FIGURE 1.3
(a) Image of modulations in the orientation angle, θ, in the study site and (b) a histogram of the distribution of study-site θ values.

(*continued*)

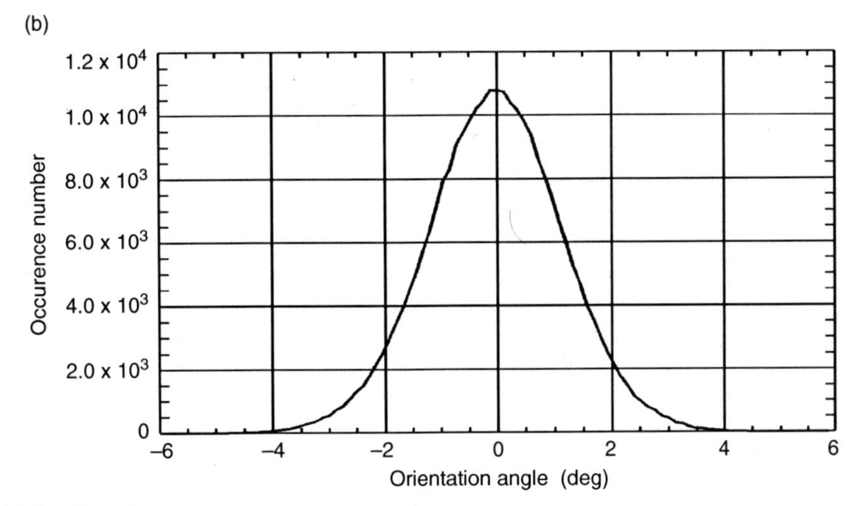

FIGURE 1.3 (continued)

1.2.5 Two-Scale Ocean-Scattering Model: Effect on the Orientation Angle Measurement

In Section 1.2.3.2, Equation 1.5 is given for the orientation angle. This equation gives the orientation angle θ as a function of three terms from the polarimetric coherency matrix T. Scattering has only been considered as occurring from a slightly rough, tilted surface equal to or greater than the size of the radar resolution cell (see Figure 1.2). The surface is planar and has a single tilt θ_s. This section examines the effects of having a distribution of azimuth tilts, $p(\varphi)$, within the resolution cell, rather than a single averaged tilt.

For single-look or multi-look processed data, the coherency matrix is defined as

$$T = \langle kk^{*T} \rangle = \frac{1}{2}\begin{bmatrix} \langle |S_{HH}+S_{VV}|^2 \rangle & \langle (S_{HH}+S_{VV})(S_{HH}-S_{VV})^* \rangle & 2\langle (S_{HH}+S_{VV})S_{HV}^* \rangle \\ \langle (S_{HH}-S_{VV})(S_{HH}+S_{VV})^* \rangle & \langle |S_{HH}-S_{VV}|^2 \rangle & 2\langle (S_{HH}-S_{VV})S_{HV}^* \rangle \\ 2\langle S_{HV}(S_{HH}+S_{VV})^* \rangle & 2\langle S_{HV}(S_{HH}-S_{VV})^* \rangle & 4\langle |S_{HV}|^2 \rangle \end{bmatrix} \tag{1.6}$$

$$\text{with } k = \frac{1}{\sqrt{2}}\begin{bmatrix} S_{HH}+S_{VV} \\ S_{HH}-S_{VV} \\ 2S_{HV} \end{bmatrix}$$

We now follow the approach of Cloude and Pottier [23]. The composite surface consists of flat, slightly rough "facets," which are tilted in the azimuth direction with a distribution of tilts, $p(\varphi)$:

$$p(\varphi) = \begin{cases} \frac{1}{2\beta} & |\varphi - s| \le \beta \\ 0 & \text{otherwise} \end{cases} \tag{1.7}$$

where s is the average surface azimuth tilt of the entire radar resolution cell.

The effect on T of having both (1) a mean bias azimuthal tilt θ_s and (2) a distribution of azimuthal tilts β has been calculated in Ref. [22] as

$$T = \begin{bmatrix} A & B\sin c(2\beta)\cos 2\theta_s & -B\sin c(2\beta)\sin 2\theta_s \\ B^*\sin c(2\beta)\cos 2\theta_s & 2C(\sin^2 2\theta_s + \sin c(4\beta)\cos 4\theta_s) & C(1-2\sin c(4\beta))\sin 4\theta_s \\ -B^*\sin c(2\beta)\sin 2\theta_s & C(1-2\sin c(4\beta))\sin 4\theta_s & 2C(\cos^2 2\theta_s - \sin c(4\beta)\cos 4\theta_s) \end{bmatrix} \tag{1.8}$$

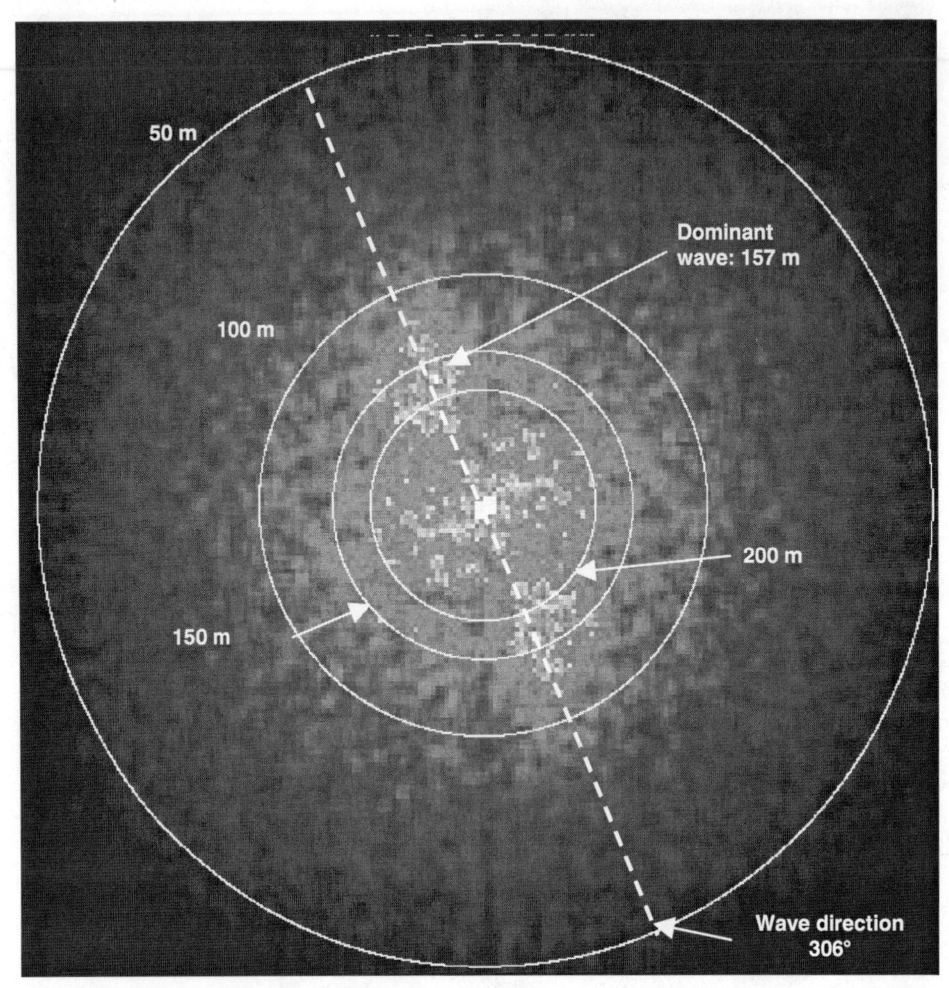

FIGURE 1.4 (See color insert following page 240.)
Orientation angle spectra versus wave number for azimuth direction waves propagating through the study site. The white rings correspond to 50, 100, 150, and 200 m. The dominant wave, of wavelength 157 m, is propagating at a heading of 306°.

where $\sin c(x)$ is $= \sin(x)/x$ and

$$A = |S_{HH} + S_{VV}|^2, \quad B = (S_{HH} + S_{VV})(S_{HH}^* - S_{VV}^*), \quad C = 0.5|S_{HH} - S_{VV}|^2$$

Equation 1.8 reveals the changes due to the tilt distribution (β) and bias (θ_s) that occur in all terms, except in the term $A = |S_{HH} + S_{VV}|^2$, which is roll-invariant. In the corresponding expression for the orientation angle, all of the other terms, except the denominator term $\langle |S_{HH} - S_{VV}|^2 \rangle$, are modified

$$\theta = \text{Arg}(\langle S_{RR} S_{LL}^* \rangle) = \tan^{-1}\left(\frac{-4\text{Re}\left(\langle (S_{HH} - S_{VV}) S_{HV}^* \rangle\right)}{-\langle |S_{HH} - S_{VV}|^2 \rangle + 4\langle |S_{HV}|^2 \rangle} \right) \tag{1.9}$$

From Equation 1.8 and Equation 1.9, it can be determined that the exact estimation of the orientation angle θ becomes more difficult as the distribution of ocean tilts β becomes stronger and wider because the ocean surface becomes progressively rougher.

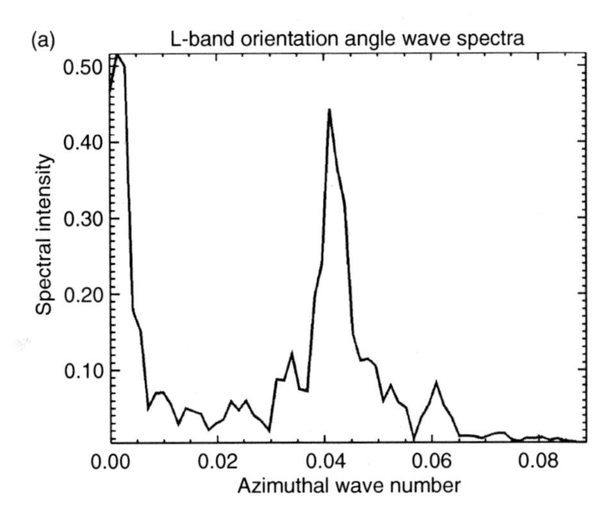

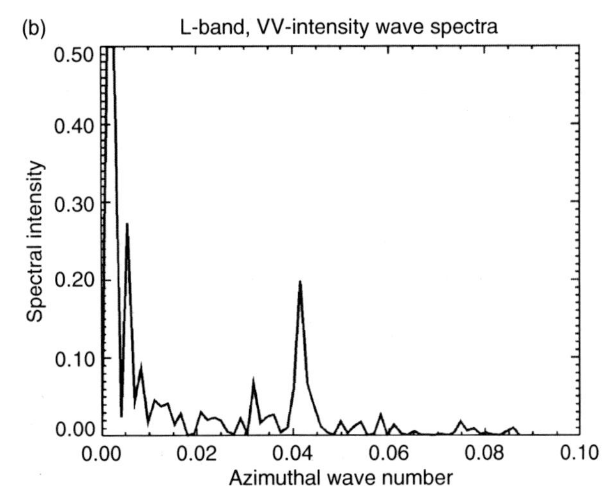

FIGURE 1.5
Plots of spectral intensity versus wave number (a) for wave-induced orientation angle modulations and (b) for VV-pol intensity modulations. The plots are taken in the propagation direction of the dominant wave (306°).

1.2.6 Alpha Parameter Measurement of Range Slopes

A second measurement technique is needed to remotely sense waves that have significant propagation direction components in the range direction. The technique must be more sensitive than current intensity-based techniques that depend on tilt and hydrodynamic modulations. Physically based POLSAR measurements of ocean slopes in the range direction are achieved using a technique involving the "alpha" parameter of the Cloude–Pottier polarimetric decomposition theorem [23].

1.2.6.1 Cloude–Pottier Decomposition Theorem and the Alpha Parameter

The Cloude–Pottier entropy, anisotropy, and the alpha polarization decomposition theorem [23] introduce a new parameterization of the eigenvectors of the 3×3 averaged coherency matrix $\langle |T| \rangle$ in the form

TABLE 1.1

Northern California: Gualala Coastal Results

Parameter	*In Situ* Measurement Instrument: Bodega Bay, CA 3-m Discus Buoy 46013	*In Situ* Measurement Instrument: Point Arena, CA Wind Station	Orientation Angle Method	Alpha Angle Method
Dominant wave period (s)	10.0	N/A	10.03 From dominant wave number	10.2 From dominant wave number
Dominant wavelength (m)	156 From period, depth	N/A	157 From wave spectra	162 From wave spectra
Dominant wave direction (°)	320 Est. from wind direction	284 Est. from wind direction	306 From wave spectra	306 From wave spectra
rms slopes azimuth direction (°)	N/A	N/A	1.58	N/A
rms slopes range direction (°)	N/A	N/A	N/A	1.36
Estimate of wave height (m)	2.4 Significant wave height	N/A	2.16 Est. from rms slope, wave number	1.92 Est. from rms slope, wave number

Date: 7/15/94; data start time (UTC): 20:04:44 (BB, PA), 20:02:98 (AIRSAR); wind speed: 1.0 m/s (BB), 2.9 m/s (PA) Mean = 1.95 m/s; wind direction: 320° (BB), 284° (PA), mean = 302°; Buoy: "Bodega Bay" (46013) = BB; location: 38.23 N 123.33 W; water depth: 122.5 m; wind station: "Point Arena" (PTAC – 1) = PA; location: 38.96° N, 123.74 W; study-site location: 38°39.6′ N, 123°35.8′ W.

$$\langle|T|\rangle = [U_3]\cdot\begin{bmatrix}\lambda_1 & 0 & 0\\ 0 & \lambda_2 & 0\\ 0 & 0 & \lambda_3\end{bmatrix}\cdot[U_3]^{*\mathrm{T}} \tag{1.10}$$

where

$$[U_3] = \mathrm{e}^{\mathrm{j}\phi}\begin{bmatrix}\cos\alpha_1 & \cos\alpha_2\mathrm{e}^{\mathrm{j}\phi_2} & \cos\alpha_3\mathrm{e}^{\mathrm{j}\phi_3}\\ \sin\alpha_1\cos\beta_1\mathrm{e}^{\mathrm{j}\delta_1} & \sin\alpha_2\cos\beta_2\mathrm{e}^{\mathrm{j}\delta_2} & \sin\alpha_3\cos\beta_3\mathrm{e}^{\mathrm{j}\delta_3}\\ \sin\alpha_1\sin\beta_1\mathrm{e}^{\mathrm{j}\gamma_1} & \sin\alpha_2\sin\beta_2\mathrm{e}^{\mathrm{j}\gamma_2} & \sin\alpha_3\sin\beta_3\mathrm{e}^{\mathrm{j}\gamma_3}\end{bmatrix} \tag{1.11}$$

The average estimate of the alpha parameter is

$$\bar{\alpha} = P_1\alpha_1 + P_2\alpha_2 + P_3\alpha_3 \tag{1.12}$$

where

$$P_i = \frac{\lambda_i}{\sum_{j=1}^{j=3}\lambda_j}. \tag{1.13}$$

The individual alphas are for the three eigenvectors and the Ps are the probabilities defined with respect to the eigenvalues. In this method, the average alpha is used and is, for simplicity, defined as $\bar{\alpha} \equiv \alpha$. For the ocean backscatter, the contributions to the average alpha are dominated, however, by the first eigenvalue or eigenvector term.

The alpha parameter, developed from the Cloude–Pottier polarimetric scattering decomposition theorem [23], has desirable directional measurement properties. It is (1) roll-invariant in the azimuth direction and (2) in the range direction, it is highly sensitive to wave-induced modulations of ϕ in the local incidence angle ϕ. Thus, the alpha parameter is well suited for measuring wave components traveling in the range direction, and discriminates against wave components traveling in the azimuth direction.

1.2.6.2 Alpha Parameter Sensitivity to Range Traveling Waves

The alpha angle sensitivity to range traveling waves may be estimated using the small perturbation scattering model (SPSM) as a basis. For flat, slightly rough scattering areas that can be characterized by Bragg scattering, the scattering matrix has the form

$$S = \begin{bmatrix} S_{HH} & 0 \\ 0 & S_{VV} \end{bmatrix} \tag{1.14}$$

Bragg-scattering coefficients S_{VV} and S_{HH} are given by

$$S_{HH} = \frac{\cos\phi_i - \sqrt{\varepsilon_r - \sin^2\phi_i}}{\cos\phi_i + \sqrt{\varepsilon_r - \sin^2\phi_i}} \quad \text{and} \quad S_{VV} = \frac{(\varepsilon_r - 1)(\sin^2\phi_i - \varepsilon_r(1 + \sin^2\phi_i))}{\left(\varepsilon_r\cos\phi_i + \sqrt{\varepsilon_r - \sin^2\phi_i}\right)^2} \tag{1.15}$$

The alpha angle is defined such that the eigenvectors of the coherency matrix T are parameterized by a vector k as

$$k = \begin{bmatrix} \cos\alpha \\ \sin\alpha\cos\beta e^{j\delta} \\ \sin\alpha\sin\beta e^{j\gamma} \end{bmatrix} \tag{1.16}$$

For Bragg scattering, one may assume that there is only one dominant eigenvector (depolarization is negligible) and the eigenvector is given by

$$k = \begin{bmatrix} S_{VV} + S_{HH} \\ S_{VV} - S_{HH} \\ 0 \end{bmatrix} \tag{1.17}$$

Since there is only one dominant eigenvector, for Bragg scattering, $\alpha = \alpha_1$. For a horizontal, slightly rough resolution cell, the orientation angle $\beta = 0$, and δ may be set to zero. With these constraints, comparing Equation 1.16 and Equation 1.17 yields

$$\tan\alpha = \frac{S_{VV} - S_{HH}}{S_{VV} + S_{HH}} \tag{1.18}$$

For $\varepsilon \to \infty$,

$$S_{VV} = 1 + \sin^2\phi_i \quad \text{and} \quad S_{HH} = \cos^2\phi_i \tag{1.19}$$

which yields

$$\tan\alpha = \sin^2\phi_i \tag{1.20}$$

Figure 1.6 shows the alpha angle as a function of incidence angle for $\varepsilon \to \infty$ (blue) and for (red) $\varepsilon = 80\text{–}70j$, which is a representative dielectric constant of sea water. The sensitivity to alpha values to incidence angle changes (this is effectively the polarimetric

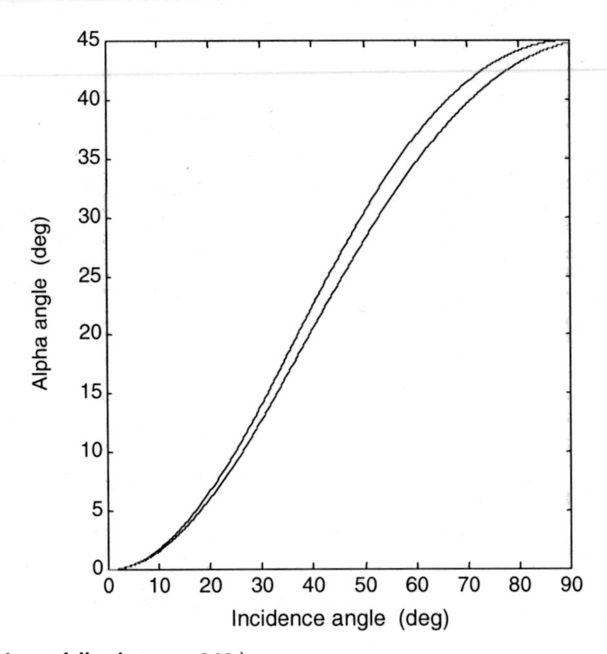

FIGURE 1.6 (See color insert following page 240.)
Small perturbation model dependence of alpha on the incidence angle. The red curve is for a dielectric constant representative of sea water (80–70j) and the blue curve is for a perfectly conducting surface.

MTF) as the range slope estimation is dependent on the derivative of α with respect to ϕ_i. For $\varepsilon \to \infty$

$$\frac{\Delta\alpha}{\Delta\phi_i} = \frac{\sin 2\phi_i}{1 + \sin^4 \phi_i} \tag{1.21}$$

Figure 1.7 shows this curve (blue) and the exact curve for $\varepsilon = 80\text{–}70j$ (red). Note that for the typical AIRSAR range of incidence angles (20–60°), across the swath, the effective MTF is high (>0.5).

1.2.6.3 Alpha Parameter Measurement of Range Slopes and Wave Spectra

Model studies [6] result in an estimate of what the parametric relation α versus the incidence angle ϕ should be for an assumed Bragg-scatter model. The sensitivity (i.e., the slope of the curve of $\alpha(\phi)$) was large enough (Figure 1.6) to warrant investigation using real POLSAR ocean backscatter data.

In Figure 1.8a, a curve of α versus the incidence angle ϕ is given for a strip of Gualala data in the range direction that has been averaged 10 pixels in the azimuth direction. This curve shows a high sensitivity for the slope of $\alpha(\phi)$. Figure 1.8b gives a histogram of the frequency of occurrence of the alpha values.

The curve of Figure 1.8a was smoothed by utilizing a least-square fit of the $\alpha(\phi)$ data to a third-order polynomial function. This closely fitting curve was used to transform the α values into corresponding incidence angle ϕ perturbations. Pottier [6] used a model-based approach and fitted a third-order polynomial to the $\alpha(\phi)$ (red curve) of Figure 1.6 instead of using the smoothed, actual, image $\alpha(\phi)$ data. A distribution of ϕ values has been made and the rms range slope value has been determined. The rms range slope values for the data sets are given in Table 1.1 and Table 1.2.

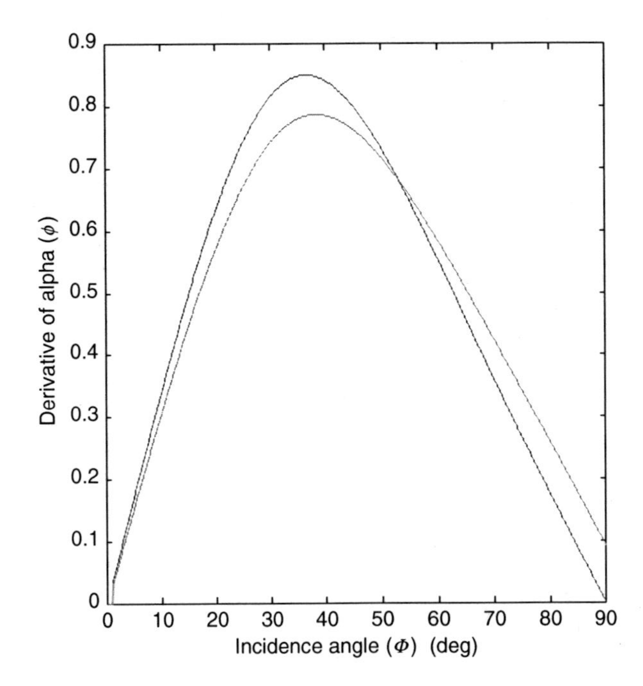

FIGURE 1.7 (See color insert following page 240.) Derivative of alpha with respect to the incidence angle. The red curve is for a sea water dielectric and the blue curve is for a perfectly conducting surface.

Finally, to measure an alpha wave spectrum, an image of the study area is formed with the mean of $\alpha(\phi)$ removed line by line in the range direction. An FFT of the study area results in the wave spectrum that is shown in Figure 1.9. The spectrum of Figure 1.9 is an alpha spectrum in the range direction. It can be converted to a range direction wave slope spectrum by transforming the slope values obtained from the smoothed alpha, $\alpha(\phi)$, values.

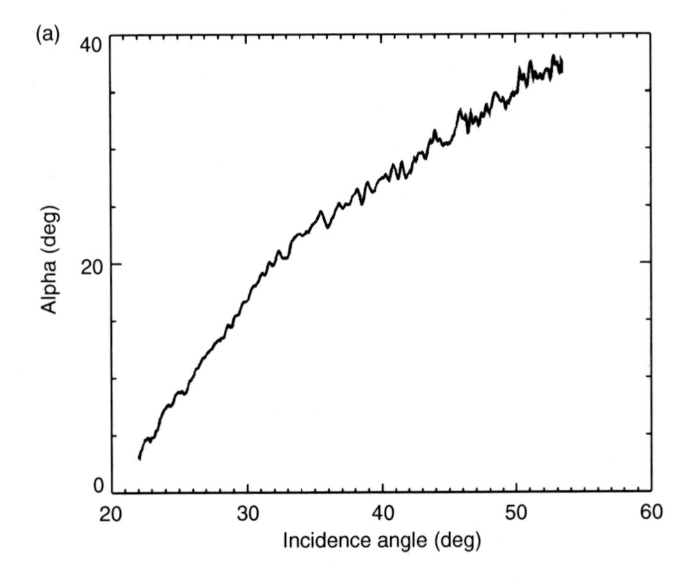

FIGURE 1.8
Empirical determination of the (a) sensitivity of the alpha parameter to the radar incidence angle (for Gualala River data) and (b) a histogram of the alpha values occurring within the study site.

(continued)

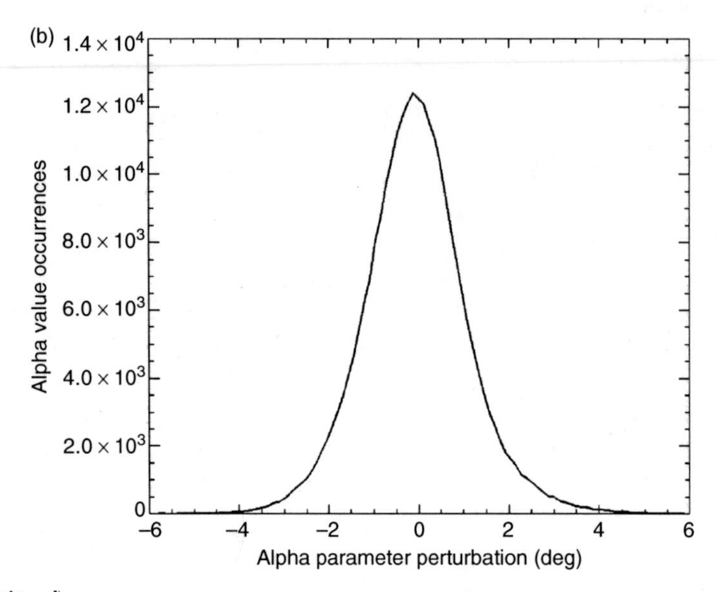

FIGURE 1.8 (continued)
(b) a histogram of the alpha values occurring within the study site.

1.2.7 Measured Wave Properties and Comparisons with Buoy Data

The ocean wave properties estimated from the L- and P-band SAR data sets and the algorithms are the (1) dominant wavelength, (2) dominant wave direction, (3) rms slopes (azimuth and range), and (4) average dominant wave height. The NOAA NDBC buoys provided data on the (1) dominant wave period, (2) wind speed and direction, (3) significant wave height, and (4) wave classification (swell and wind waves).

Both the Gualala and the San Francisco data sets involved waves classified as swell. Estimates of the average wave period can be determined either from buoy data or from the SAR-determined dominant wave number and water depth (see Equation 1.22).

The dominant wavelength and direction are obtained from the wave spectra (see Figure 1.4 and Figure 1.9). The rms slopes in the azimuth direction are determined from the distribution of orientation angles converted to slope angles using Equation 1.2. The rms slopes in the range direction are determined by the distribution of alpha angles converted to slope angles using values of the smoothed curve fitted to the data of Figure 1.8a.

Finally, an estimate of the average wave height, H_d, of the dominant wave was made using the peak-to-trough rms slope in the propagation direction S_{rms} and the dominant wavelength λ_d. The estimated average dominant wave height was then determined from $\tan(S_{rms}) = H_d/(\lambda_d/2)$. This average dominant wave height estimate was compared with the (related) significant wave height provided by the NDBC buoy. The results of the measurement comparisons are given in Table 1.1 and Table 1.2.

1.2.7.1 Coastal Wave Measurements: Gualala River Study Site

For the Gualala River data set, parameters were calculated to characterize ocean waves present in the study area. Table 1.1 gives a summary of the ocean parameters that were determined using the data set as well as wind conditions and air and sea temperatures at the nearby NDBC buoy ("Bodega Bay") and wind station ("Point Arena") sites. The most important measured SAR parameters were rms wave slopes (azimuth and range directions), rms wave height, dominant wave period, and dominant wavelength. These quantities were estimated using the full-polarization data and the NDBC buoy data.

TABLE 1.2

Open Ocean: Pacific Swell Results

Parameter	*In Situ* Measurement Instrument		Orientation Angle Method	Alpha Angle Method
	San Francisco, CA, 3 m Discus Buoy 46026	Half Moon Bay, CA, 3 m Discus Buoy 46012		
Dominant wave period (s)	15.7	15.7	15.17 From dominant wave number	15.23 From dominant wave number
Dominant wavelength (m)	376 From period, depth	364 From period, depth	359 From wave spectra	362 From wave spectra
Dominant wave direction (°)	289 Est. from wind direction	280 Est. from wind direction	265 From wave spectra	265 From wave spectra
rms slopes azimuth direction (°)	N/A	N/A	0.92	N/A
rms slopes range direction (°)	N/A	N/A	N/A	0.86
Estimate of wave height (m)	3.10 Significant wave height	2.80 Significant wave height	2.88 Est. from rms slope, wave number	2.72 Est. from rms slope, wave number

Date: 7/17/88; data start time (UTC): 00:45:26 (Buoys SF, HMB), 00:52:28 (AIRSAR); wind speed: 8.1 m/s (SF), 5.0 m/s (HMB), mean = 6.55 m/s; wind direction: 289° (SF), 280° (HMB), mean = 284.5°; Buoys: "San Francisco" (46026) = SF; location: 37.75 N 122.82 W; water depth: 52.1 m; "Half Moon Bay" (46012) = HMB; location: 37.36 N 122.88 W; water depth: 87.8 m.

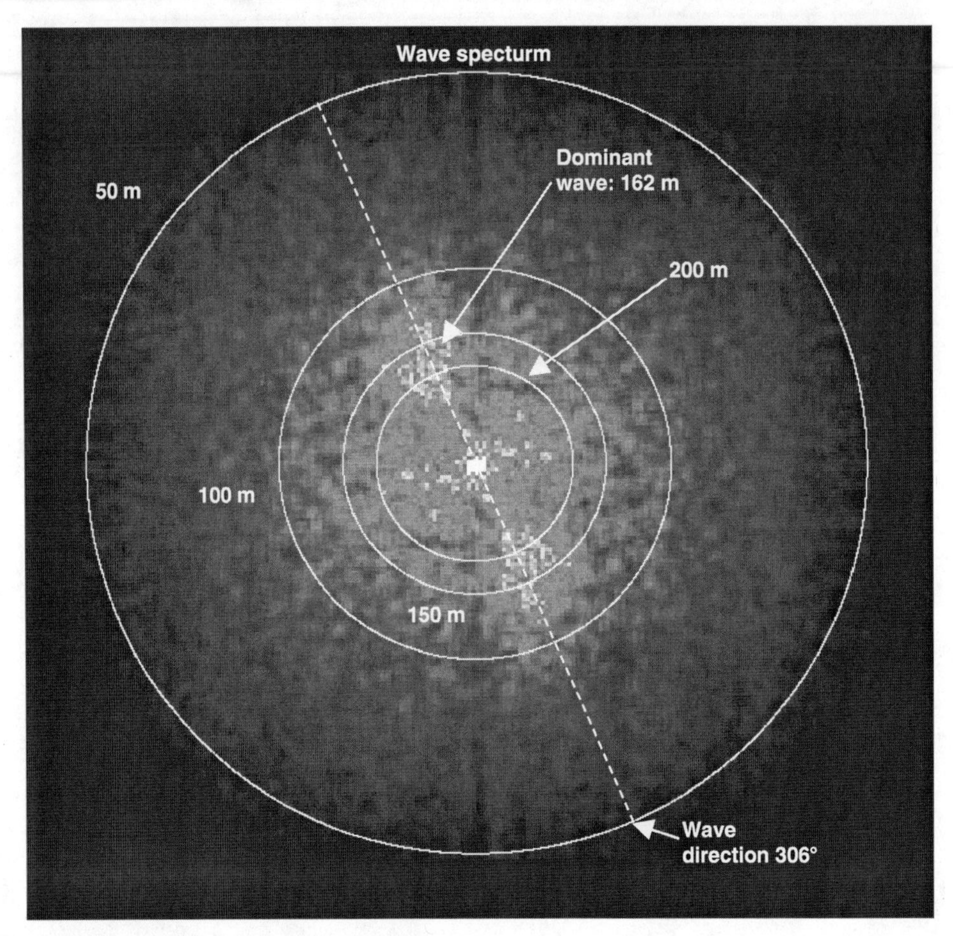

FIGURE 1.9 (See color insert following page 240.)
Spectrum of waves in the range direction using the alpha parameter from the Cloude–Pottier decomposition method. Wave direction is 306° and dominant wavelength is 162 m.

The dominant wave at the Bodega Bay buoy during the measurement period is classified as a long wavelength swell. The contribution from wind wave systems or other swell components is small relative to the single dominant wave system. Using the surface gravity wave dispersion relation, one can calculate the dominant wavelength at this buoy location where the water depth is 122.5 m. The dispersion relation for surface water waves at finite depth is

$$\omega_{\mathrm{W}}^2 = gk_{\mathrm{W}} \tan\mathrm{h}(kH) \tag{1.22}$$

where ω_{W} is the wave frequency, k_{W} is the wave number $(2\pi/\lambda)$, and H is the water depth. The calculated value for λ is given in Table 1.1.

A spectral profile similar to Figure 1.5a was developed for the alpha parameter technique and a dominant wave was measured having a wavelength of 156 m and a propagation direction of 306°. Estimates of the ocean parameters obtained using the orientation angle and alpha angle algorithms are summarized in Table 1.1.

1.2.7.2 Open-Ocean Measurements: San Francisco Study Site

AIRSAR P-band image data were obtained for an ocean swell traveling in the azimuth direction. The location of this image was to the west of San Francisco Bay. It is a valuable

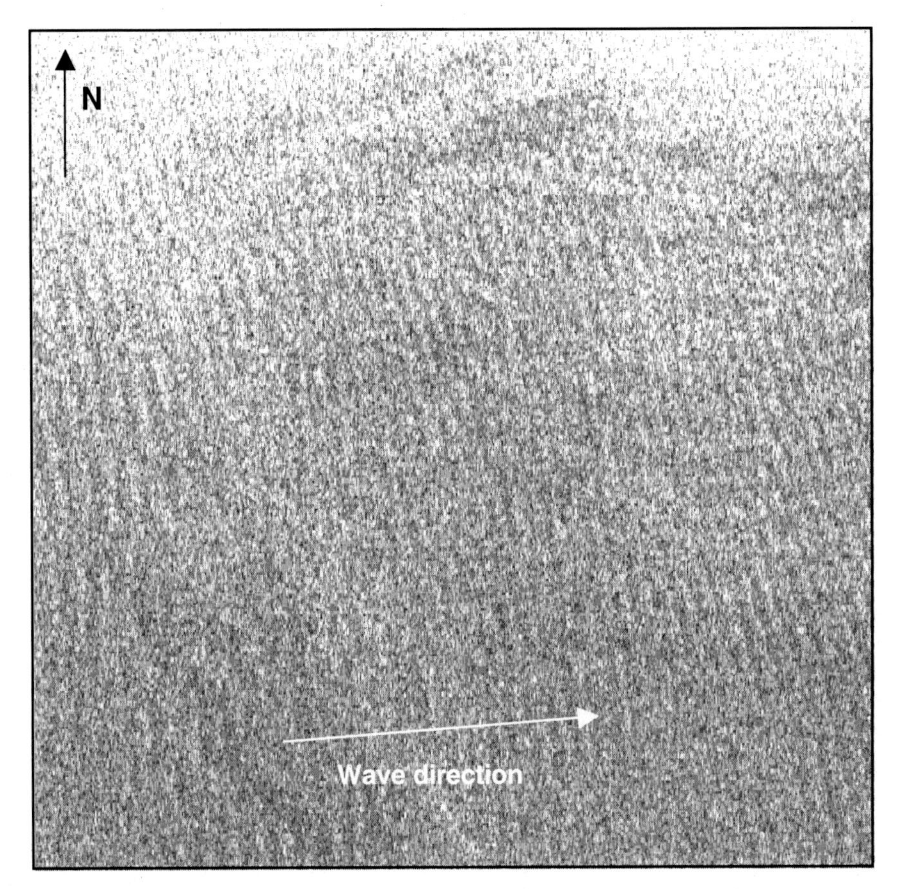

FIGURE 1.10
P-band span image of near-azimuth traveling (265°) swell in the Pacific Ocean off the coast of California near San Francisco.

data set because its location is near two NDBC buoys ("San Francisco" and "Half-Moon Bay"). Figure 1.10 gives a SPAN image of the ocean scene. The long-wavelength swell is clearly visible. The covariance matrix data was first Lee-filtered to reduce speckle noise [24] and was then corrected radiometrically.

A polarimetric signature was developed for a 512×512 segment of the image and some distortion was noted. Measuring the distribution of the phase between the HH-pol and VV-pol backscatter returns eliminated this distortion. For the ocean, this distribution should have a mean nearly equal to zero. The recalibration procedure set the mean to zero and the distortion in the polarimetric signature was corrected. Figure 1.11a gives a plot of the spectral intensity (cross-section modulation) versus the wave number in the direction of the dominant wave propagation. Figure 1.11b presents a spectrum of orientation angles versus the wave number. The major peak, caused by the visible swell, in both plots occurs at a wave number of $0.0175\ \mathrm{m}^{-1}$ or a wavelength of 359 m. Using Equation 1.22, the dominant wavelength was calculated at the San Francisco/Half Moon Bay buoy positions and depths. Estimates of the wave parameters developed from this data set using the orientation and alpha angle algorithms are presented in Table 1.2.

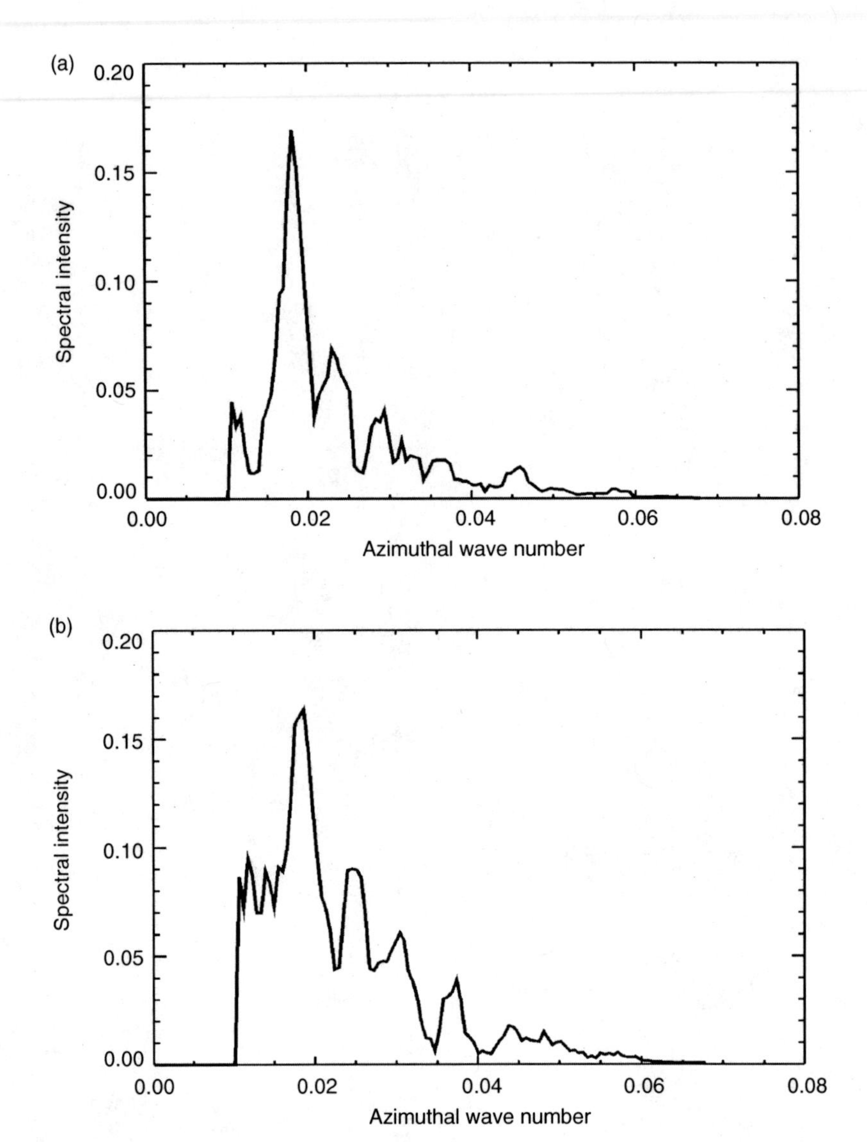

FIGURE 1.11
(a) Wave number spectrum of P-band intensity modulations and (b) a wave number spectrum of orientation angle modulations. The plots are taken in the propagation direction of the dominant wave (265°).

1.3 Polarimetric Measurement of Ocean Wave–Current Interactions

1.3.1 Introduction

Studies have been carried out on the use of polarization orientation angles to remotely sense ocean wave slope distribution changes caused by wave–current interactions. The wave–current features studied here involve the surface manifestations of internal waves [1,25,26–29,30–32] and wave modifications at oceanic current fronts.

Studies have shown that polarimetric SAR data may be used to measure bare surface roughness [33] and terrain topography [34,35]. Techniques have also been developed for measuring directional ocean wave spectra [36].

The polarimetric SAR image data used in all of the studies are NASA JPL/AIRSAR, P-, L-, and C-band, quad-pol microwave backscatter data. AIRSAR images of internal waves were obtained from the 1992 Joint US/Russia Internal Wave Remote Sensing Experiment (JUSREX'92) conducted in the New York Bight [25,26].

AIRSAR data on current fronts were obtained during the NRL Gulf Stream Experiment (NRL-GS'90). The NRL experiment is described in Ref. [20]. Extensive sea-truth is available for both of these experiments. These studies were motivated by the observation that strong perturbations occur in the polarization orientation angle θ in the vicinity of internal waves and current fronts. The remote sensing of orientation angle changes associated with internal waves and current fronts are applications that have only recently been investigated [27–29]. Orientation angle changes should also occur for the related SAR application involving surface expressions of shallow-water bathymetry [37].

In the studies outlined here, polarization orientation angle changes are shown to be associated with wave–current interaction features. Orientation angle changes are not, however, produced by all types of ocean surface features. For example, orientation angle changes have been successfully used here to discriminate internal wave signatures from other ocean features, such as surfactant slicks, which produce no mean orientation angle changes.

1.3.2 Orientation Angle Changes Caused by Wave–Current Interactions

A study was undertaken to determine the effect that several important types of wave–current interactions have on the polarization orientation angle. The study involved both actual SAR data and an NRL theoretical model described in Ref. [38].

An example of a JPL/AIRSAR VV-polarization, L-band image of several strong, intersecting, internal wave packets is given in Figure 1.12. Packets of internal waves are generated from parent solitons as the soliton propagates into shallower water at, in this case, the continental shelf break. The white arrow in Figure 1.12 indicates the propagation direction of a wedge of internal waves (bounded by the dashed lines). The packet members within the area of this wedge were investigated.

Radar cross-section (σ_0) intensity perturbations for the type of internal waves encountered in the New York Bight have been calculated in [30,31,39] and others. Related perturbations also occur in the ocean wave height and slope spectra. For the solitons often found in the New York Bight area, these perturbations become significantly larger for ocean wavelengths longer than about 0.25 m and shorter than 10–20 m. Thus, the study is essentially concerned with slope changes to meter-length wave scales. The AIRSAR slant range resolution cell size for these data is 6.6 m, and the azimuth resolution cell size is 12.1 m. These resolutions are fine enough for the SAR backscatter to be affected by perturbed wave slopes (meter-length scales).

The changes in the orientation angle caused by these wave perturbations are seen in Figure 1.13. The magnitude of these perturbations covers a range $\theta = [-1 \text{ to } +1]$. The orientation angle perturbations have a large spatial extent (>100 m for the internal wave soliton width).

The hypothesis assumed was that wave–current interactions make the meter wavelength slope distributions asymmetric. A profile of orientation angle perturbations caused by the internal wave study packet is given in Figure 1.14a. The values are obtained along the propagation vector line of Figure 1.12. Figure 1.14b gives a comparison of the orientation angle profile (solid line) and a normalized VV-pol backscatter intensity profile (dotted-dash line) along the same interval. Note that the orientation angle positive peaks (white stripe areas, Figure 1.13) align with the negative troughs

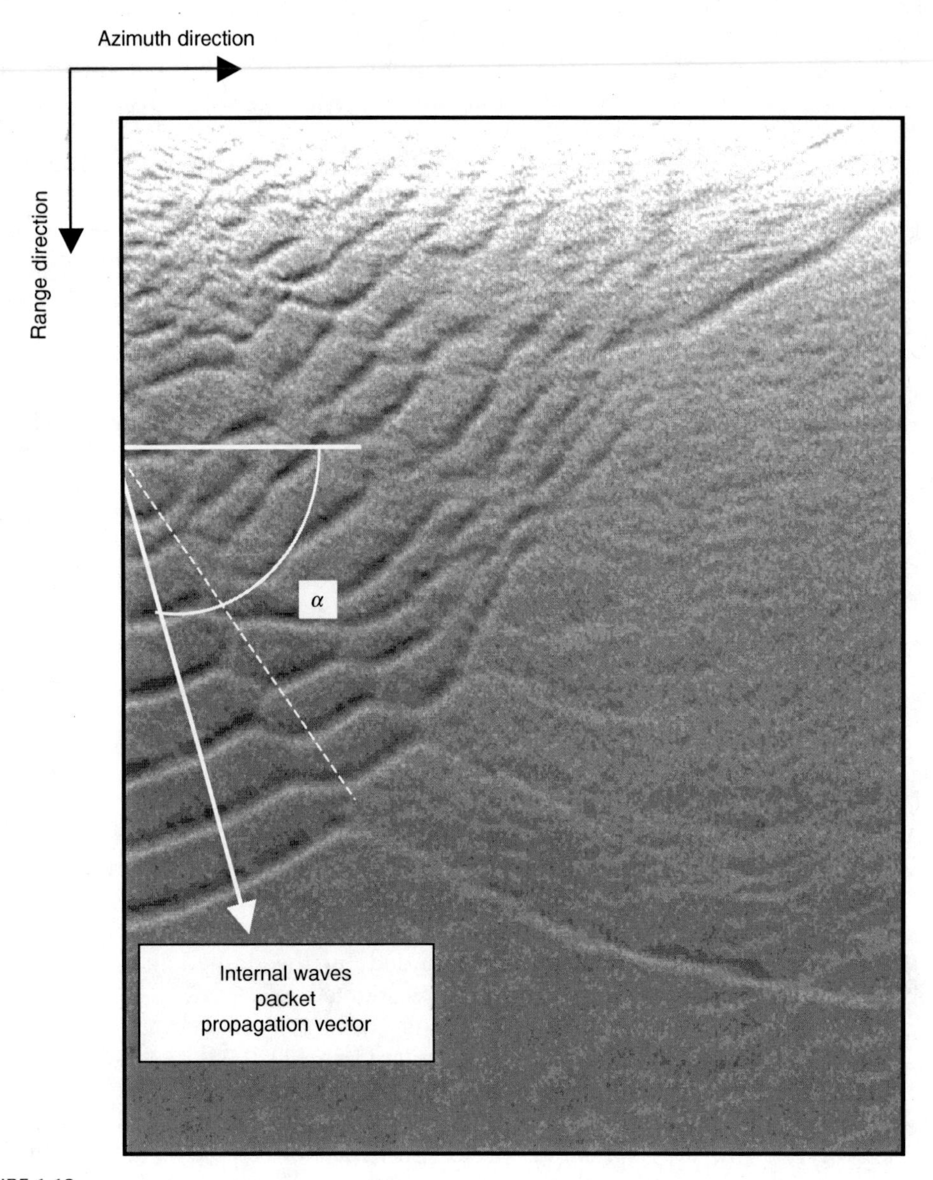

FIGURE 1.12
AIRSAR L-band, VV-pol image of internal wave intersecting packets in the New York Bight. The arrow indicates the propagation direction for the chosen study packet (within the dashed lines). The angle α relates to the SAR/packet coordinates. The image intensity has been normalized by the overall average.

(black areas, Figure 1.12). In the direction orthogonal to the propagation vector, every point is averaged 5×5 pixels along the profile. The ratio of the maximum of θ caused by the soliton to the average values of θ within the ambient ocean is quite large. The current-induced asymmetry creates a mean wave slope that is manifested as a mean orientation angle.

The relation between the tangent of the orientation angle θ, wave slopes in the radar azimuth and ground range directions ($\tan \omega$, $\tan \gamma$), and the radar look angle ϕ from [21] is given by Equation 1.1, and for a given look angle ϕ the average orientation angle tangent is

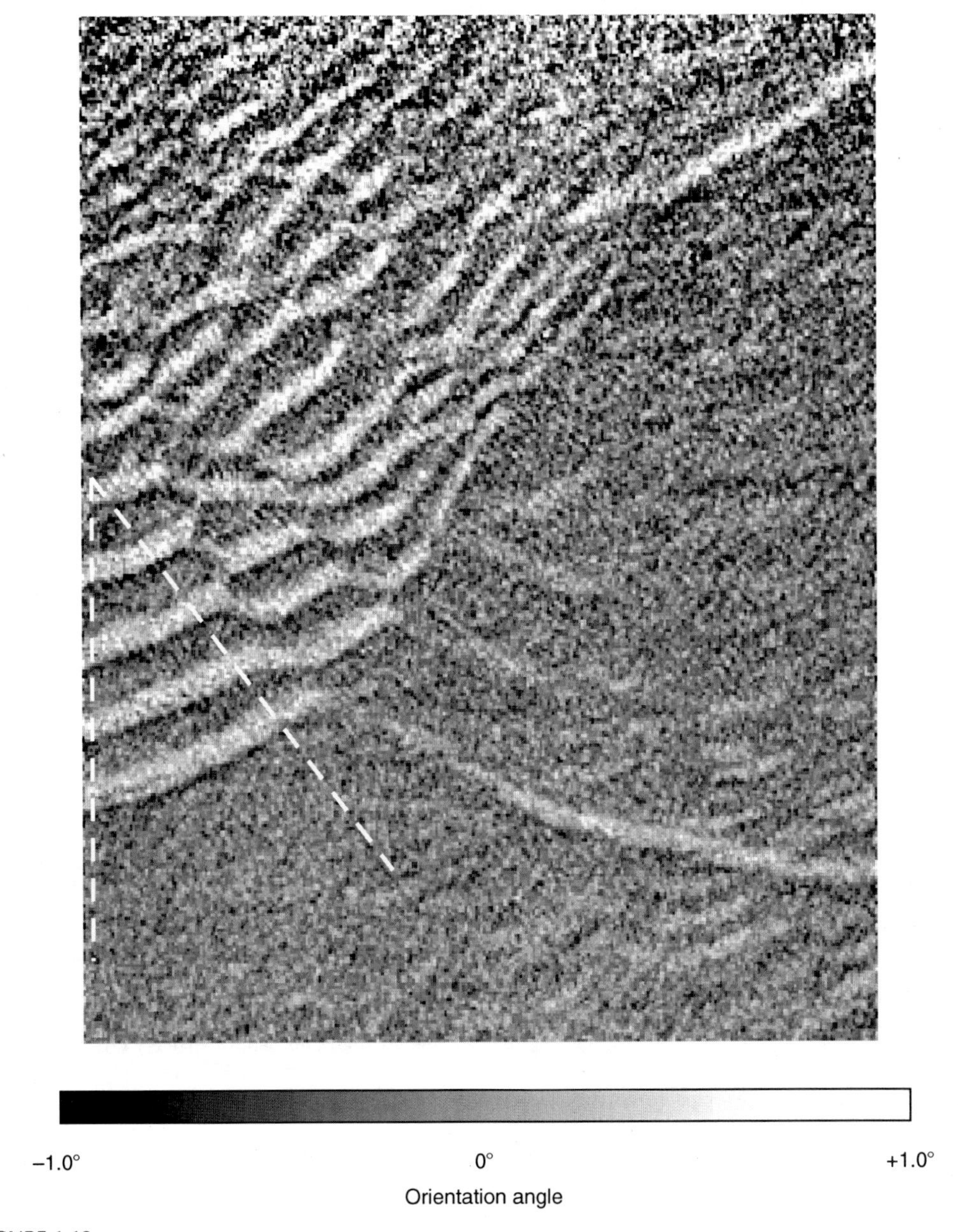

FIGURE 1.13
The orientation angle image of the internal wave packets in the New York Bight. The area within the wedge (dashed lines) was studied intensively.

$$\langle \tan\theta \rangle = \int_0^{\pi}\int_0^{\pi} \tan\theta(\omega,\gamma) \cdot P(\omega,\gamma)\, d\gamma\, d\omega \tag{1.23}$$

where $P(\omega,\gamma)$ is the joint probability distribution function for the surface slopes in the azimuth and range directions. If the slopes are zero-meaned, but $P(\omega,\gamma)$ is skewed, then the mean orientation angle may not be zero even though the mean azimuth and range slopes are zero. It is evident from the above equation that both the azimuth and the range

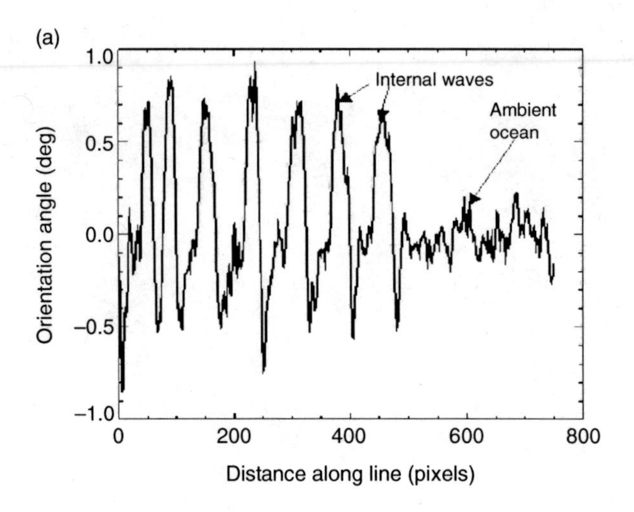

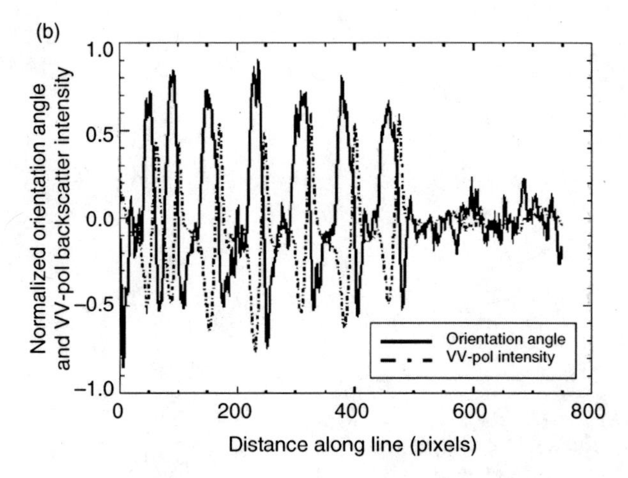

FIGURE 1.14
(a) The orientation angle value profile along the propagation vector for the internal wave study packet of Figure 1.12 and (b) a comparison of the orientation angle profile (solid line) and a normalized VV-pol backscatter intensity profile (dot–dash line). Note that the orientation angle positive peaks (white areas, Figure 1.13) align with the negative troughs (black areas, Figure 1.11).

slopes have an effect on the mean orientation angle. The azimuth slope effect is generally larger because it is not reduced by the cos ϕ term, which only affects the range slope. If, for instance, the meter-wavelength waves are produced by a broad wind-wave spectrum, then both ω and γ change locally. This yields a nonzero mean for the orientation angle. Figure 1.15 gives a histogram of orientation angle values (solid line) for a box inside the black area of the first packet member of the internal wave. A histogram for the ambient ocean orientation angle values for a similar-sized box near the internal wave is given by the dot–dash–dot line in Figure 1.15. Notice the significant difference in the mean value of these two distributions. The mean change in $\langle \tan(\theta) \rangle$ inferred from the bias for the perturbed area within the internal wave is 0.03 rad, corresponding to a θ value of 1.72°.

The mean water wave slope changes needed to cause such orientation angle changes are estimated from Equation 1.1. In the denominator of Equation 1.1, the value of $\tan(\gamma)\cos(\phi) \ll \sin(\phi)$ for the value ϕ $(=51°)$ at the packet member location. Using this approximation, the ensemble average of Equation 1.1 provides the mean azimuth slope value,

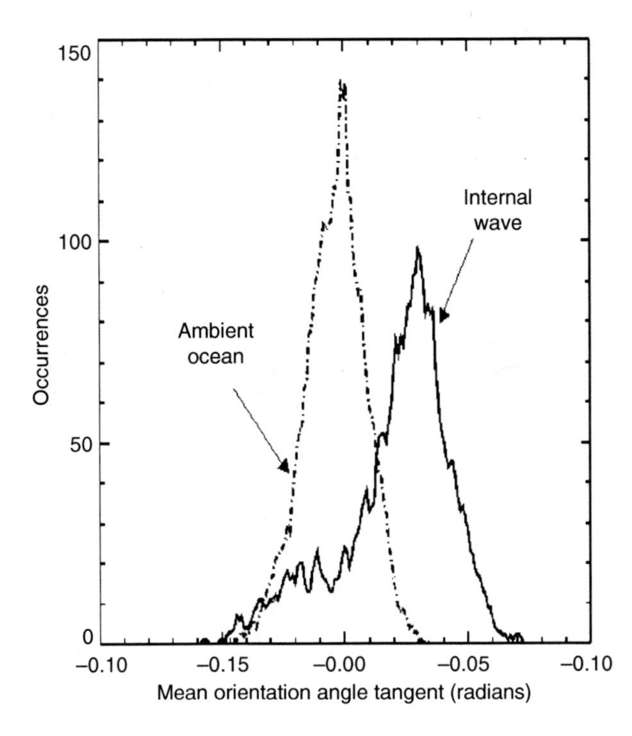

FIGURE 1.15
Distributions of orientation angles for the internal wave (solid line) and the ambient ocean (dot–dash–dot line).

$$\langle \tan(\omega) \rangle \cong \sin(\phi) \langle \tan(\theta) \rangle \tag{1.24}$$

From the data provided in Figure 1.15, $\langle \tan(\omega) \rangle = 0.0229$ rad or $\omega = 1.32°$. A slope value of this magnitude is in approximate agreement with slope changes predicted by Lyzenga et al. [32] for internal waves in the same area during an earlier experiment (SARSEX, 1988).

1.3.3 Orientation Angle Changes at Ocean Current Fronts

An example of orientation angle changes induced by a second type of wave–current interaction, the convergent current front, is given in Figure 1.16a and Figure 1.16b. This image was created using AIRSAR P-band polarimetric data.

The orientation angle response to this (NRL-GS'90) Gulf-Stream convergent-current front is the vertical white linear feature in Figure 1.16a and the sharp peak in Figure 1.16b. The perturbation of the orientation angle at, and near, the front location is quite strong relative to angle fluctuations in the ambient ocean. The change in the orientation angle maximum is $\cong 0.68°$. Other fronts in the same area of the Gulf Stream have similar changes in the orientation angle.

1.3.4 Modeling SAR Images of Wave–Current Interactions

To investigate wave–current-interaction features, a time-dependent ocean wave model has been developed that allows for general time-varying current, wind fields, and depth [20,38]. The model uses conservation of the wave action to compute the propagation of a statistical wind-wave system. The action density formalism that is used and an outline of the model are both described in Ref. [38]. The original model has been extended [1] to include calculations of polarization orientation angle changes due to wave–current interactions.

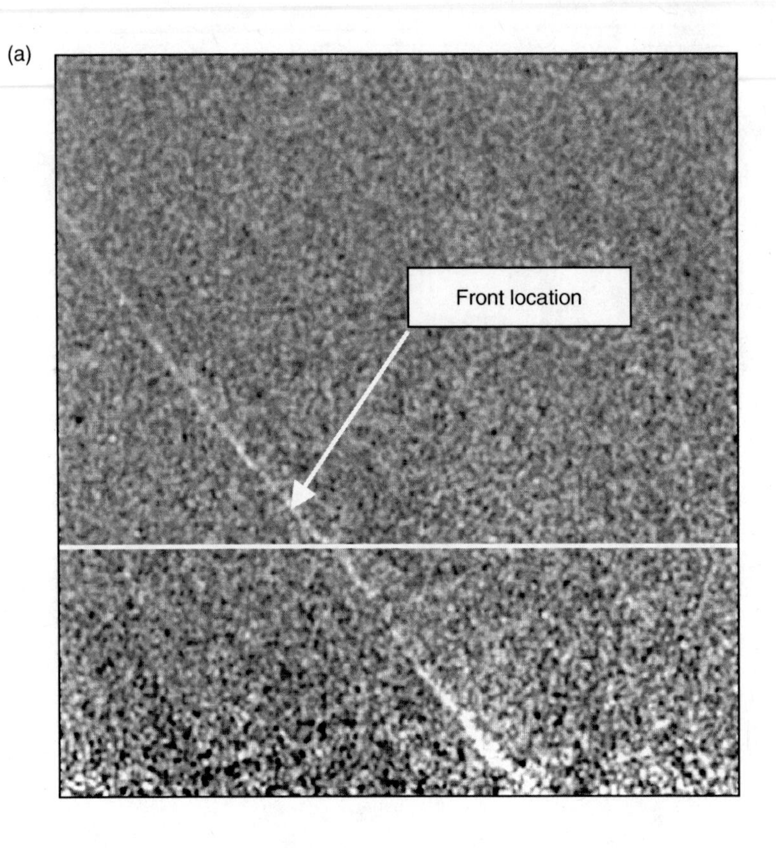

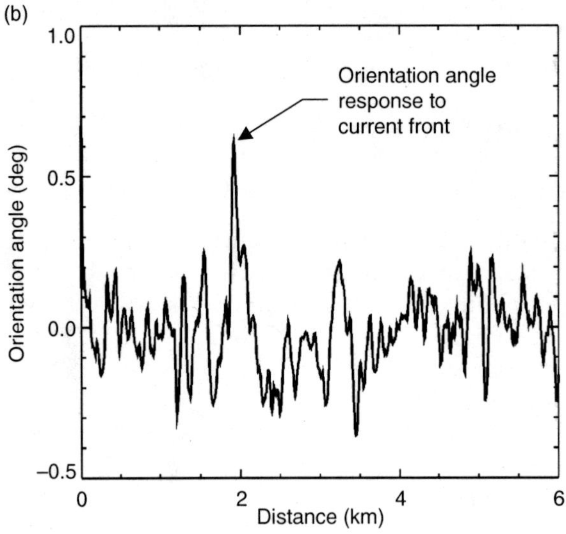

FIGURE 1.16
Current front within the Gulf Stream. An orientation angle image is given in (a) and orientation angle values are plotted in (b) (for values along the white line in (a)).

Model predictions have been made for the wind-wave field, radar return, and perturbation of the polarization orientation angle due to an internal wave. A model of the surface manifestation of an internal wave has also been developed. The algorithm used in the model has been modified from its original form to allow calculation of the polarization orientation angle and its variation throughout the extent of the soliton current field at the surface.

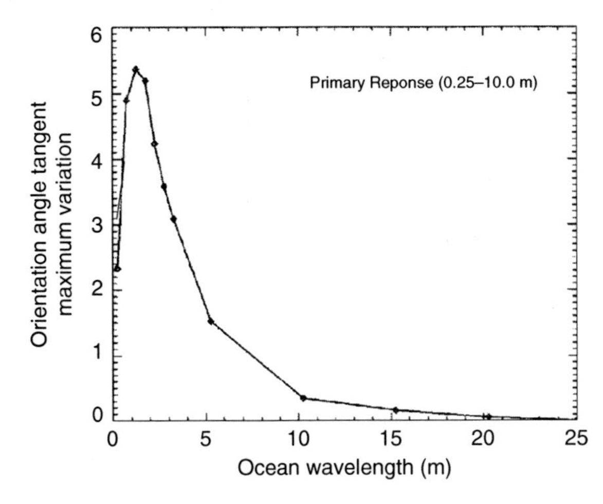

FIGURE 1.17
The internal wave orientation angle tangent maximum variation as a function of ocean wavelength as predicted by the model. The primary response is in the range of 0.25–10.0 m and is in good agreement with previous studies of sigma-0. (From Thompson, D.R., *J. Geophys. Res.*, 93, 12371, 1988.)

The values of both RCS ($\equiv\langle\sigma_0\rangle$) and $\langle\tan(\theta)\rangle$ are computed by the model. The dependence of $\langle\tan(\theta)\rangle$ on the perturbed ocean wavelength was calculated by the model. This wavelength dependence is shown in Figure 1.17. The waves resonantly perturb $\langle\tan(\theta)\rangle$ for wavelengths in the range of 0.25–10.0 m. This result is in good agreement with previous studies of sigma-0 resonant perturbations for the JUSREX'92 area [39].

Figure 1.18a and Figure 1.18b show the form of the soliton current speed dependence of $\langle\sigma_0\rangle$ and $\langle\tan(\theta)\rangle$. The potentially useful near-linear relation of $\langle\tan(\theta)\rangle_{\mathrm{v}}$ with current U (Figure 1.18b) is important in applications where determination of current gradients is the goal. The near-linear nature of this relationship provides the possibility that, from the value of $\langle\tan(\theta)\rangle_{\mathrm{v}}$, the current magnitude can be estimated. Examination of the model results has led to the following empirical model of the variation of $\langle\tan(\theta)\rangle$ as:

$$\langle\tan\theta\rangle = f(U, w, \theta_w) = (aU)\cdot(w^2 e^{-bw})\cdot\sin(\alpha|\psi_w| + \beta\psi_w^2) \quad (1.25)$$

where U, the surface current maximum speed (in m/s), w, the wind speed (in m/s) at (standard) 19.5 m height, and ψ_w, the wind direction (in radians) relative to the soliton propagation direction. The constants are $a = 0.00347$, $b = 0.365$, $\alpha = 0.65714$, and $\beta = 0.10913$. The range of ψ_w is over $[-\pi,\pi]$.

Using Equation 1.25, the dashed curve in Figure 1.18 can be generated to show good agreement relative to the complete model. The solid lines in Figure 1.18 represent results from the complete model and the dashed lines are results from the empirical relation of Equation 1.25. This relation is much simpler than conventional estimates based on perturbation of the backscatter intensity.

The scaling for the relationship is a relatively simple function of the wind speed and the direction of the locally wind-driven sea. If the orientation angle and wind measurements are available, then Equation 1.25 allows the internal wave current maximum U to be calculated.

1.4 Ocean Surface Feature Mapping Using Current-Driven Slick Patterns

1.4.1 Introduction

Biogenic and man-made slicks are widely dispersed throughout the oceans. Current driven surface features, such as spiral eddies, can be made visible by associated patterns of slicks [40]. A combined algorithm using the Cloude–Pottier decomposition and the Wishart classifier

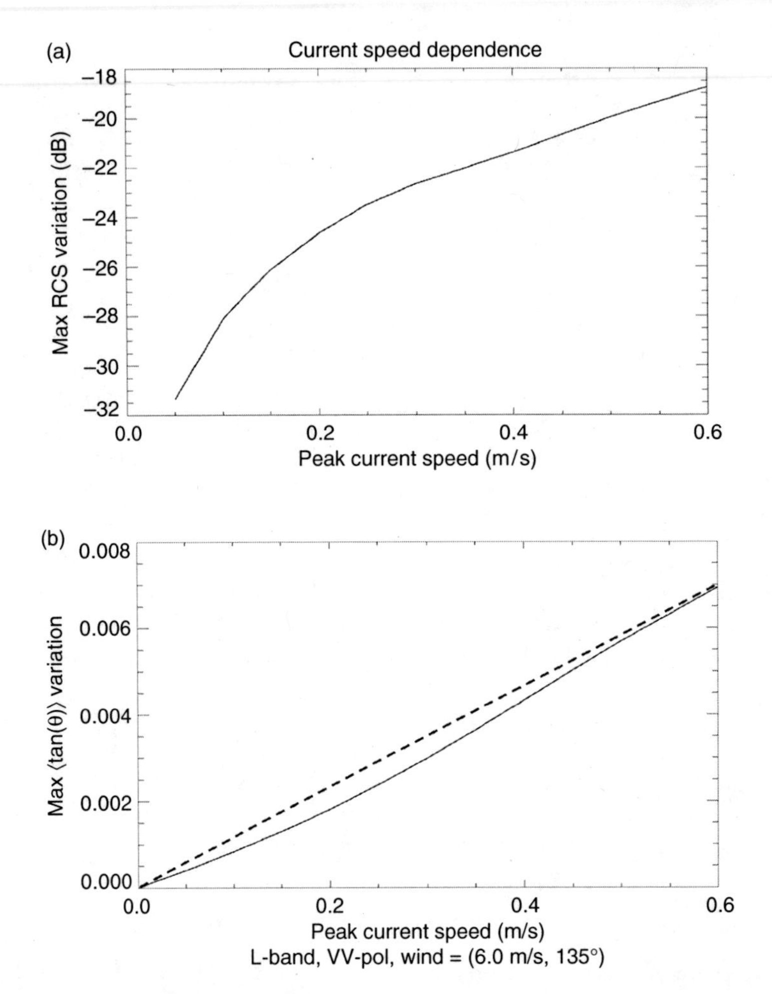

FIGURE 1.18
(a and b) Model development of the current speed dependence of the max RCS and $\langle \tan(\theta) \rangle$ variations. The dashed line in Figure 1.23b gives the values predicted by an empirical equation in Ref. [38].

[41] is utilized to produce accurate maps of slick patterns and to suppress the background wave field. This technique uses the classified slick patterns to detect spiral eddies. Satellite SAR instruments performing wave spectral measurements, or operating as wind scatterometers, regard the slicks as a measurement error term. The classification maps produced by the algorithm facilitate the flagging of slick-contaminated pixels within the image.

Aircraft L-band AIRSAR data (4/2003) taken in California coastal waters provided data on features that contained spiral eddies. The images also included biogenic slick patterns, internal wave packets, wind waves, and long wave swell.

The temporal and spatial development of spiral eddies is of considerable importance to oceanographers. Slick patterns are used as "markers" to detect the presence and extent of spiral eddies generated in coastal waters. In a SAR image, the slicks appear as black distributed patterns of lower return. The slick patterns are most prevalent during periods of low to moderate winds. The spatial distribution of the slicks is determined by local surface current gradients that are associated with the spiral eddies.

It has been determined that biogenic surfactant slicks may be identified and classified using SAR polarimetric decompositions. The purpose of the decomposition is to discriminate against other features such as background wave systems. The parameters entropy (H), anisotropy (A), and average alpha ($\bar{\alpha}$) of the Cloude–Pottier decomposition [23] were used in the classification. The results indicate that biogenic slick patterns, classified by the algorithm, can be used to detect the spiral eddies.

The decomposition parameters were also used to measure small-scale surface roughness as well as larger-scale rms slope distributions and wave spectra [4]. Examples of slope distributions are given in Figure 1.8b and that of wave spectra in Figure 1.9. Small-scale roughness variations that were detected by anisotropy changes are given in Figure 1.19. This figure shows variations in anisotropy at low wind speeds for a filament of colder, trapped water along the northern California coast. The air–sea stability has changed for the region containing the filament. The roughness changes are not seen in (a) the conventional VV-pol image but are clearly visible in (b) an anisotropy image. The data are from coastal waters near the Mendocino Co. town of Gualala.

Finally, the classification algorithm may also be used to create a flag for the presence of slicks. Polarimetric satellite SAR systems (e.g., RADARSAT-2, ALOS/PALSAR, SIR-C) attempting to measure wave spectra, or scatterometers measuring wind speed and direction can avoid using slick contaminated data.

In April 2003, the NRL and the NASA Jet Propulsion Laboratory (JPL) jointly carried out a series of AIRSAR flights over the Santa Monica Basin off the coast of California. Backscatter POLSAR image data at P-, L-, and C-bands were acquired. The purpose of the flights was to better understand the dynamical evolution of spiral eddies, which are

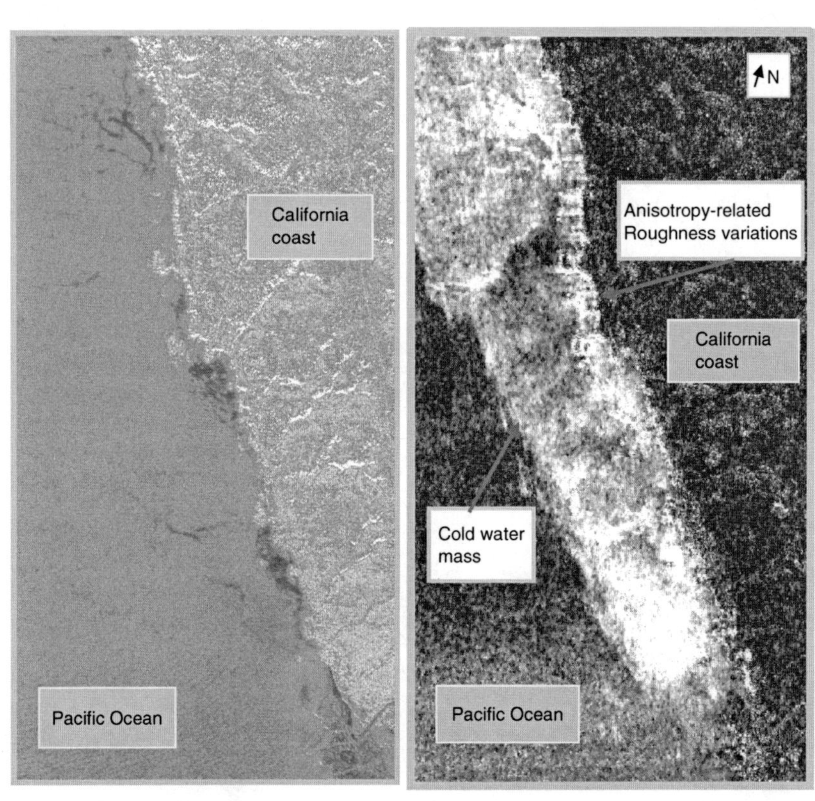

FIGURE 1.19 (See color insert following page 240.)
(a) Variations in anisotropy at low wind speeds for a filament of colder, trapped water along the northern California coast. The roughness changes are not seen in the conventional VV-pol image, but are clearly visible in (b) an anisotropy image. The data are from coastal waters near the Mendocino Co. town of Gualala.

generated in this area by interaction of currents with the Channel Islands. Sea-truth was gathered from a research vessel owned by the University of California at Los Angeles (UCLA). The flights yielded significant data not only on the time history of spiral eddies but also on surface waves, natural surfactants, and internal wave signatures. The data were analyzed using a polarimetric technique, the Cloude–Pottier $\langle H/A/\alpha \rangle$ decomposition given in Ref. [23]. In Figure 1.20a, the anisotropy is again mapped for a study site

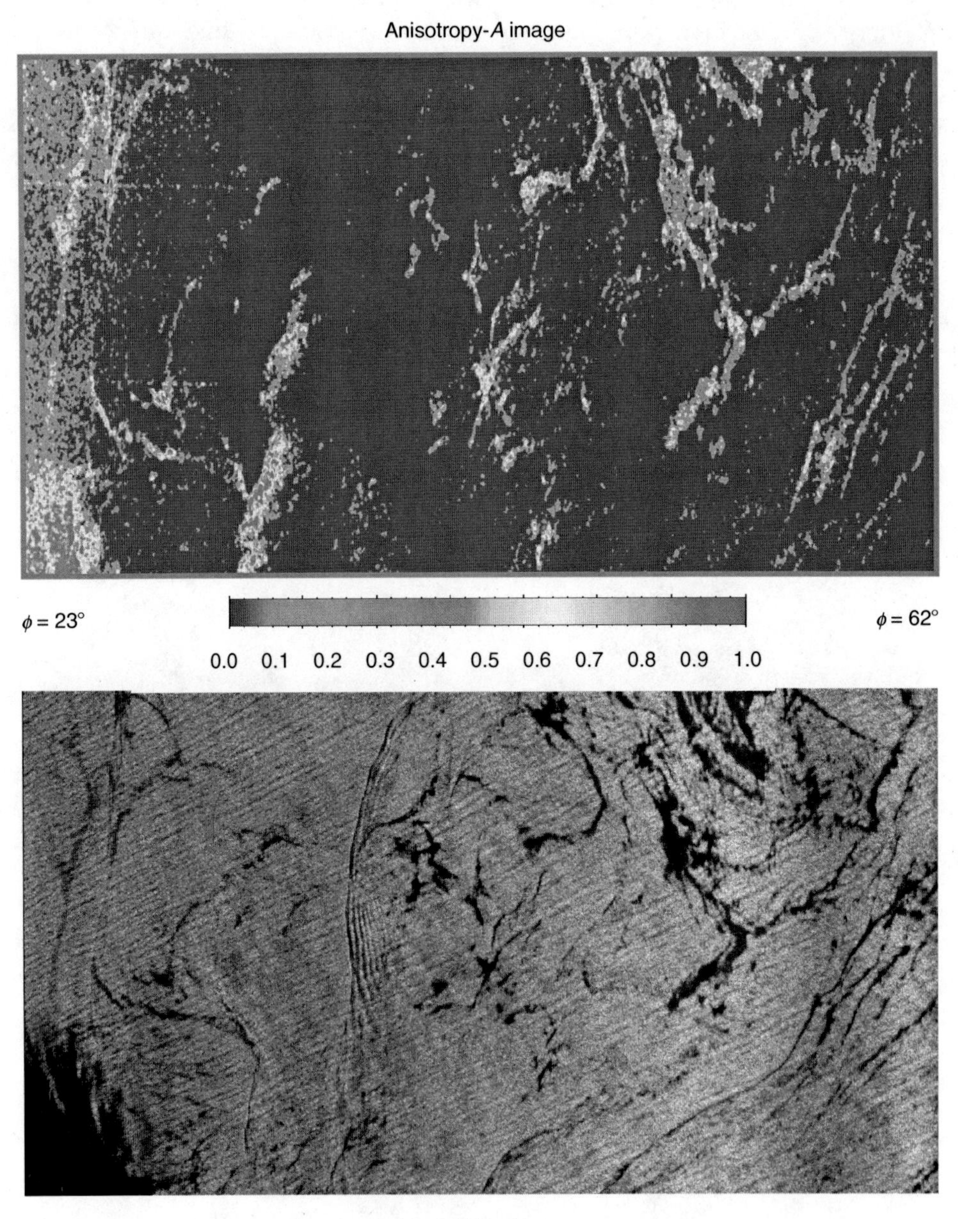

FIGURE 1.20 (See color insert following page 240.)
(a) Image of anisotropy values. The quantity, 1−A, is proportional to small-scale surface roughness and (b) a conventional L-band, VV-pol image of the study area.

east of Catalina Island, CA. For comparison, a VV-pol image is given in Figure 1.20b. The slick field is reasonably well mapped by anisotropy—but the image is noisy because of the difference in the two small second and third eigenvalues that are used to compute it.

1.4.2 Classification Algorithm

The overall purpose of the field research effort outlined in Section 1.4.1 was to create a means of detecting ocean features such as spiral eddies using biogenic slicks as markers, while suppressing other effects such as wave fields and wind-gradient effects. A polarimetric classification algorithm [41–43] was tested as a candidate means to create such a feature map.

1.4.2.1 Unsupervised Classification of Ocean Surface Features

Van Zyl [44] and Freeman–Durden [46] developed unsupervised classification algorithms that separate the image into four classes: odd-bounce, even bounce, diffuse (volume), and an in-determinate class. For an L-band image, the ocean surface typically is dominated by the characteristics of the Bragg-scattering odd (single) bounce. City buildings and structures have the characteristics of even (double) scattering, and heavy forest vegetation has the characteristics of diffuse (volume) scattering. Consequently, this classification algorithm provides information on the terrain scatterer type. For a refined separation into more classes, Pottier [6] proposed an unsupervised classification algorithm based on their target decomposition theory. The medium's scattering mechanisms, characterized by entropy H, $\bar{\alpha}$ average alpha angle, and later anisotropy A, were used for classification. The entropy H is a measure of randomness of the scattering mechanisms, and the alpha angle characterizes the scattering mechanism. The unsupervised classification is achieved by projecting the pixels of an image onto the H–$\bar{\alpha}$ plane, which is segmented into scattering zones. The zones for the Gualala study-site data are shown in Figure 1.21. Details of this segmentation are given in Ref. [6]. In the alpha–entropy scattering zone map of the decomposition, backscatter returns from the ocean surface normally occur in the lowest (dark blue color) zone of both alpha and entropy. Returns from slick covered areas have higher entropy H and average alpha $\bar{\alpha}$ values, and occur in both the lowest zone and higher zones.

1.4.2.2 Classification Using Alpha–Entropy Values and the Wishart Classifier

Classification of the image was initiated by creating an alpha–entropy zone scatterplot to determine the $\bar{\alpha}$ angle and level of entropy H for scatterers in the slick study area. Secondly, the image was classified into eight distinct classes using the Wishart classifier [41]. The alpha–entropy decomposition method provides good image segmentation based on the scattering characteristics.

The algorithm used is a combination of the unsupervised decomposition classifier and the supervised Wishart classifier [41]. One uses the segmented image of the decomposition method to form training sets as input for the Wishart classifier. It has been noted that multi-look data are required to obtain meaningful results in H and $\bar{\alpha}$, especially in the entropy H. In general, 4-look processed data are not sufficient. Normally, additional averaging (e.g., 5×5 boxcar filter), either of the covariance or of coherency matrices, has to be performed prior to the H and $\bar{\alpha}$ computation. This prefiltering is done on all the data. The filtered coherency matrix is then used to compute H and $\bar{\alpha}$. Initial classification is made using the eight zones. This initial classification map is then used to train the Wishart classification. The reclassified result shows improvement in retaining details.

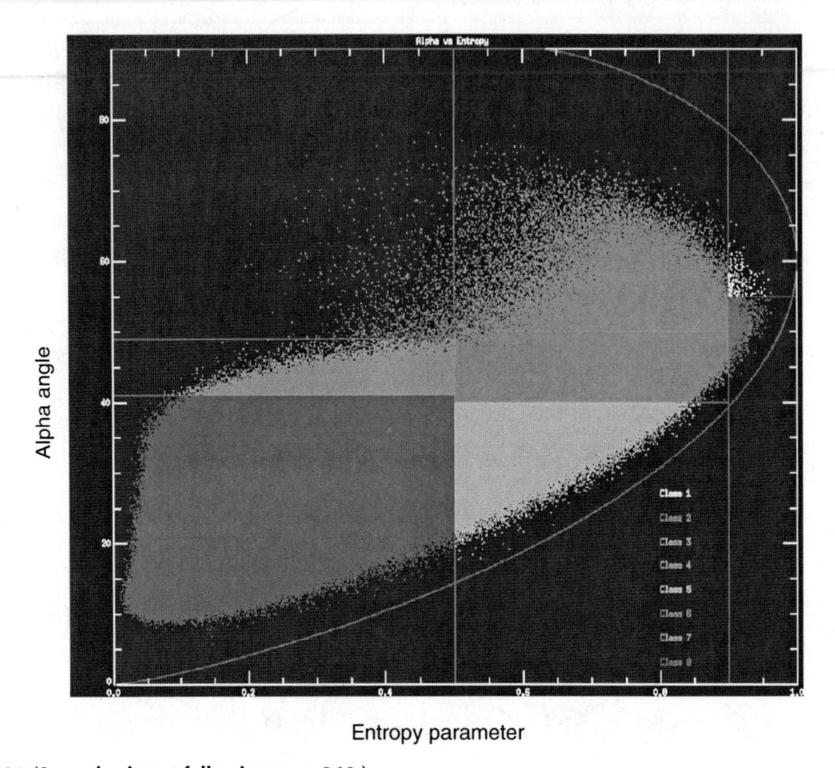

FIGURE 1.21 (See color insert following page 240.)
Alpha-entropy scatter plot for the image study area. The plot is divided into eight color-coded scattering classes for the Cloude–Pottier decomposition described in Ref. [6].

Further improvement is possible by using several iterations. The reclassified image is then used to update the cluster centers of the coherency matrices. For the present data, two iterations of this process were sufficient to produce good classifications of the complete biogenic fields. Figure 1.22 presents a completed classification map of the biogenic slick fields. Information is provided by the eight color-code classes in the image in Figure 1.22. The returns from within the largest slick (labeled as A) have classes that progressively increase in both average alpha and entropy as a path is made from clean water inward toward the center of the slick. Therefore, the scattering becomes less surfacelike ($\bar{\alpha}$ increase) and also becomes more depolarized (H increase) as one approaches the center of the slick (Figure 1.22, Label A).

The algorithm outlined above may be applied to an image containing large-scale ocean features. An image (JPL/CM6744) of classified slick patterns for two-linked spiral eddies near Catalina Island, CA, is given in Figure 1.23b. An L-band, HH-pol image is presented in Figure 1.23a for comparison. The Pacific swell is suppressed in areas where there are no slicks. The waves do, however, appear in areas where there are slicks because the currents associated with the orbital motion of the waves alternately compress or expand the slick-field density. Note the dark slick patch to the left of label A in Figure 1.23a and Figure 1.23b. This patch clearly has strongly suppressed the backscatter at HH-pol. The corresponding area of Figure 1.23b has been classified into three classes and colors (Class 7—salmon, Class 5—yellow, and Class 2—dark green), which indicate progressive increases in scattering complexity and depolarization as one moves from the perimeter of the slick toward its interior. A similar change in scattering occurs to the left of label B near the center of Figure 1.23a and Figure 1.23b. In this case, as one moves

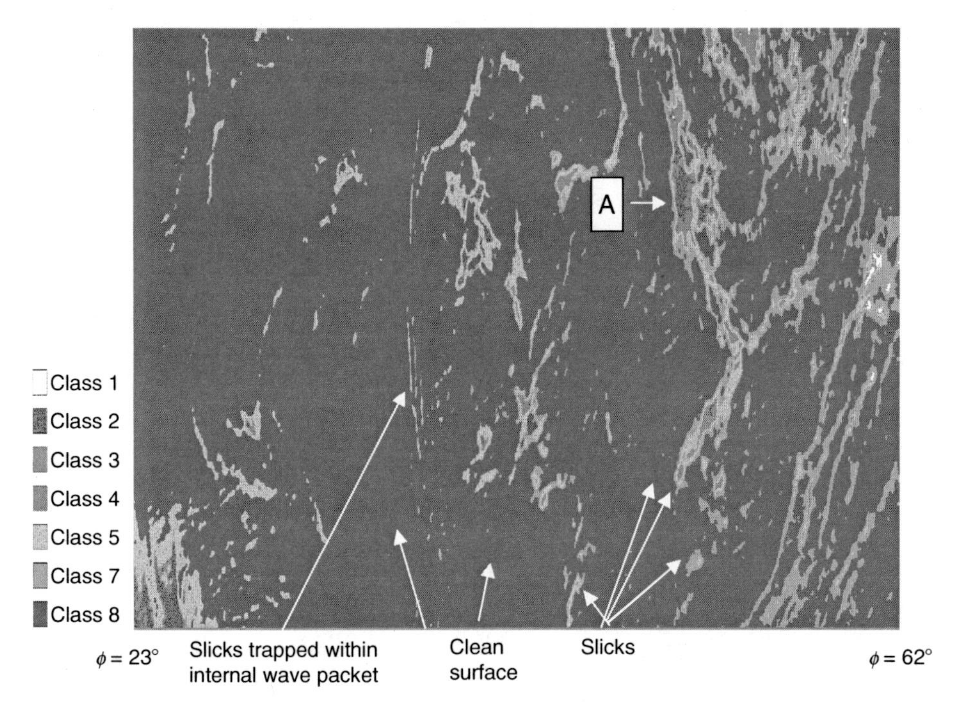

FIGURE 1.22 (See color insert following page 240.)
Classification of the slick-field image into H/$\bar{\alpha}$ scattering classes.

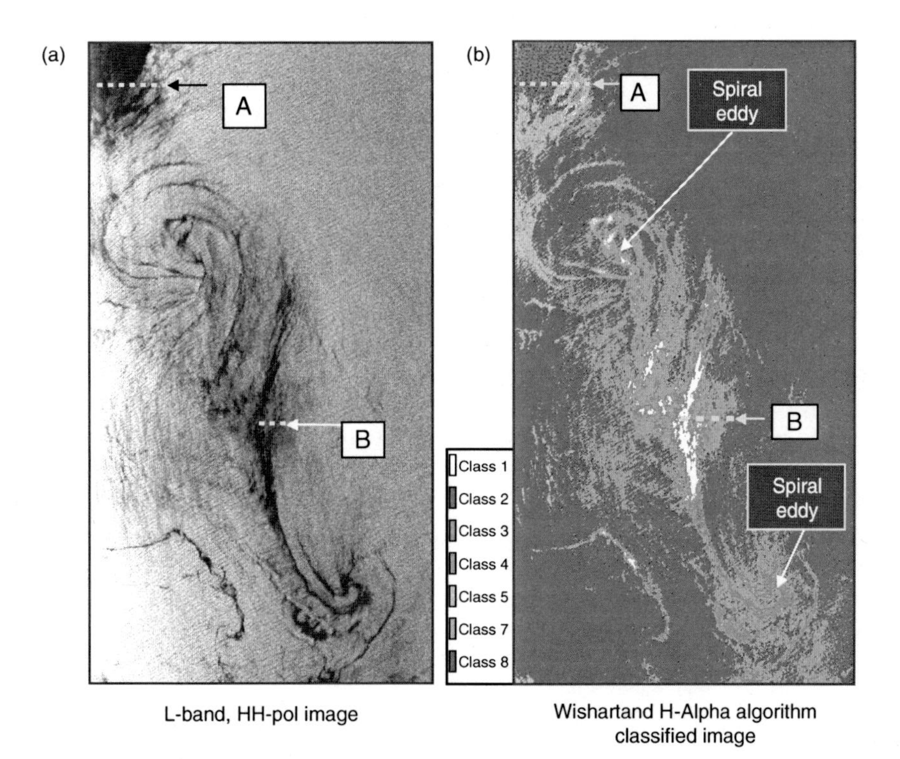

FIGURE 1.23 (See color insert following page 240.)
(a) L-band, HH-pol image of a second study image (CM6744) containing two strong spiral eddies marked by natural biogenic slicks and (b) classification of the slicks marking the spiral eddies. The image features were classified into eight classes using the H–$\bar{\alpha}$ values combined with the Wishart classifier.

from the perimeter into the slick toward the center, the classes and colors (Class 7—salmon, Class 4—light green, Class 1—white) also indicate progressive increases in scattering complexity and depolarization.

1.4.2.3 Comparative Mapping of Slicks Using Other Classification Algorithms

The question arises whether or not the algorithm using entropy-alpha values with the Wishart classifier is the best candidate for unsupervised detection and mapping of slick fields. Two candidate algorithms were suggested as possible competitive classification methods. These were (1) the Freeman–Durden decomposition [45] and (2) the $(H/A/\bar{\alpha})$–Wishart segmentation algorithm [42,43], which introduce anisotropy to the parameter mix because of its sensitivity to ocean surface roughness. Programs were developed to investigate the slick classification capabilities of these candidate algorithms. The same amount of averaging (5 × 5) and speckle reduction was done for all of the algorithms. The results with the Freeman–Durden classification were poor at both L- and C-bands. Nearly all of the returns were surface, single-bounce scatter. This is expected because the Freeman–Durden decomposition was developed on the basis of scattering models of land features. This method could not discriminate between waves and slicks and did not improve on the results using conventional VV or HH polarization.

The $(H/A/\bar{\alpha})$–Wishart segmentation algorithm was investigated to take advantage of the small-scale roughness sensitivity of the polarimetric anisotropy A. The anisotropy is shown (Figure 1.20b) to be very sensitive to slick patterns across the whole image. The $(H/A/\bar{\alpha})$–Wishart segmentation method expands the number of classes from 8 to 16 by including the anisotropy A. The best way to introduce information about A in the classification procedure is to carry out two successive Wishart classifier algorithms. The first classification only involves $H/\bar{\alpha}$. Each class in the $H/\bar{\alpha}$ plane is then further divided into two classes according to whether the pixel's anisotropy values are greater than 0.5 or less than 0.5. The Wishart classifier is then employed a second time. Details of this algorithm are given in Ref. [42,43]. The results of using the $(H/A/\bar{\alpha})$–Wishart method and iterating it twice are given in Figure 1.24.

Classification of the slick-field image using the $(H/A/\bar{\alpha})$–Wishart method resulted in 14 scattering classes. Two of the expected 16 classes were suppressed. The Classes 1–7 corresponded to anisotropy A values from 0.5 to 1.0 and the Classes 8–14 corresponded to anisotropy A values from 0.0 to 0.49. The new two lighter blue vertical features at the lower right of the image appeared in all images involving anisotropy and were thought to be a smooth slick of the lower surfactant material concentration. This algorithm was an improvement relative to the $H/\bar{\alpha}$–Wishart algorithm for slick mapping. All of the slick-covered areas were classified well and the unwanted wave field intensity modulations were suppressed.

1.5 Conclusions

Methods that are capable of measuring ocean wave spectra and slope distributions in both the range and azimuth directions were described. The new measurements are sensitive and provide nearly direct measurements of ocean wave spectra and slopes without the need for a complex MTF. The orientation modulation spectrum has a higher dominant wave peak and background ratio than the intensity-based spectrum. The results

FIGURE 1.24 (See color insert following page 240.)
Classification of the slick-field image into $H/A/\bar{\alpha}$ 14 scattering classes. The Classes 1–7 correspond to anisotropy A values 0.5 to 1.0 and the Classes 8–14 correspond to anisotropy A values 0.0 to 0.49. The two lighter blue vertical features at the lower right of the image appear in all images involving anisotropy and are thought to be smooth slicks of lower concentration.

determined for the dominant wave direction, wavelength, and wave height are comparable to the NDBC buoy measurements. The wave slope and wave spectra measurement methods that have been investigated may be developed further into fully operational algorithms. These algorithms may then be used by polarimetric SAR instruments, such as ALOS/PALSAR and RADARSAT II, to monitor sea-state conditions globally.

Secondly, this work has investigated the effect of internal waves and current fronts on the SAR polarization orientation angle. The results provide a potential (1) independent means for identifying these ocean features and (2) a method of estimating the mean value of the surface current and slope changes associated with an internal wave. Simulations of the NRL wave–current interaction model [38] have been used to identify and quantify the different variables such as current speed, wind speed, and wind direction, which determine changes in the SAR polarization orientation angle.

The polarimetric scattering properties of biogenic slicks have been found to be different from those of the clean surface wave field and the slicks may be separated from this background wave field. Damping of capillary waves, in the slick areas, lowers all of the eigenvalues of the decomposition and increases the average alpha angle, entropy, and the anisotropy.

The Cloude–Pottier polarimetric decomposition was also used as a new means of studying scattering properties of surfactant slicks perturbed by current-driven surface features. The features, for example, spiral eddies, were marked by filament patterns of slicks. These slick filaments were physically smoother. Backscatter from them was more complex (three eigenvalues nearly equal) and was more depolarized.

Anisotropy was found to be sensitive to small-scale ocean surface roughness, but was not a function of large-scale range or azimuth wave slopes. These unique properties provided an achievable separation of roughness scales on the ocean surface at low wind speeds. Changes in anisotropy due to surfactant slicks were found to be measurable across the entire radar swath.

Finally, polarimetric SAR decomposition parameters alpha, entropy, and anisotropy were used as an effective means for classifying biogenic slicks. Algorithms, using these parameters, were developed for the mapping of both slick fields and ocean surface features. Selective mapping of biogenic slick fields may be achieved using either the entropy or the alpha parameters with the Wishart classifier or, by the entropy, anisotropy, or the alpha parameters with the Wishart classifier. The latter algorithm gives the best results overall.

Slick maps made using this algorithm are of use for satellite scatterometers and wave spectrometers in efforts aimed at flagging ocean surface areas that are contaminated by slick fields.

References

1. Schuler, D.L., Jansen, R.W., Lee, J.S., and Kasilingam, D., Polarisation orientation angle measurements of ocean internal waves and current fronts using polarimetric SAR, *IEE Proc. Radar, Sonar Navigation*, 150(3), 135–143, 2003.
2. Alpers, W., Ross, D.B., and Rufenach, C.L., The detectability of ocean surface waves by real and synthetic aperture radar, *J. Geophys. Res.*, 86(C7), 6481, 1981.
3. Engen, G. and Johnsen, H., SAR-ocean wave inversion using image cross-spectra, *IEEE Trans. Geosci. Rem. Sens.*, 33, 1047, 1995.
4. Schuler, D.L., Kasilingam, D., Lee, J.S., and Pottier, E., Studies of ocean wave spectra and surface features using polarimetric SAR, *Proc. Int. Geosci. Rem. Sens. Symp.* (IGARSS'03), Toulouse, France, IEEE, 2003.
5. Schuler, D.L., Lee, J.S., and De Grandi, G., Measurement of topography using polarimetric SAR Images, *IEEE Trans. Geosci. Rem. Sens.*, 34, 1266, 1996.
6. Pottier, E., Unsupervised classification scheme and topography derivation of POLSAR data on the ≪H/A/α≫ polarimetric decomposition theorem, *Proc. 4th Int. Workshop Radar Polarimetry*, IRESTE, Nantes, France, 535–548, 1998.
7. Hasselmann, K. and Hasselmann, S., The nonlinear mapping of an ocean wave spectrum into a synthetic aperture radar image spectrum and its inversion, *J. Geophys. Res.*, 96(10), 713, 1991.
8. Vesecky, J.F. and Stewart, R.H., The observation of ocean surface phenomena using imagery from SEASAT synthetic aperture radar—an assessment, *J. Geophys. Res.*, 87, 3397, 1982.
9. Beal, R.C., Gerling, T.W., Irvine, D.E., Monaldo, F.M., and Tilley, D.G., Spatial variations of ocean wave directional spectra from the SEASAT synthetic aperture radar, *J. Geophys. Res.*, 91, 2433, 1986.
10. Valenzuela, G.R., Theories for the interaction of electromagnetic and oceanic waves—a review, *Boundary Layer Meteorol.*, 13, 61, 1978.
11. Keller, W.C. and Wright, J.W., Microwave scattering and straining of wind-generated waves, *Radio Sci.*, 10, 1091, 1975.
12. Alpers, W. and Rufenach, C.L., The effect of orbital velocity motions on synthetic aperture radar imagery of ocean waves, *IEEE Trans. Antennas Propagat.*, 27, 685, 1979.
13. Plant, W.J. and Zurk, L.M., Dominant wave directions and significant wave heights from SAR imagery of the ocean, *J. Geophys. Res.*, 102(C2), 3473, 1997.

14. Hasselmann, K., Raney, R.K., Plant, W.J., Alpers, W., Shuchman, R.A., Lyzenga, D.R., Rufenach, C.L., and Tucker, M.J., Theory of synthetic aperture radar ocean imaging: a MARSEN view, *J. Geophys. Res.*, 90, 4659, 1985.
15. Lyzenga, D.R., An analytic representation of the synthetic aperture radar image spectrum for ocean waves, *J. Geophys. Res.*, 93(13), 859, 1998.
16. Kasilingam, D. and Shi, J., Artificial neural network based-inversion technique for extracting ocean surface wave spectra from SAR images, *Proc. IGARSS'97*, Singapore, IEEE, 1193–1195, 1997.
17. Hasselmann, S., Bruning, C., Hasselmann, K., and Heimbach, P., An improved algorithm for the retrieval of ocean wave spectra from synthetic aperture radar image spectra, *J. Geophys. Res.*, 101, 16615, 1996.
18. Lehner, S., Schulz-Stellenfleth, Schattler, B., Breit, H., and Horstmann, J., Wind and wave measurements using complex ERS-2 SAR wave mode data, *IEEE Trans. Geosci. Rem. Sens.*, 38(5), 2246, 2000.
19. Dowd, M., Vachon, P.W., and Dobson, F.W., Ocean wave extraction from RADARSAT synthetic aperture radar inter-look image cross-spectra, *IEEE Trans. Geosci. Rem. Sens.*, 39, 21–37, 2001.
20. Lee, J.S., Jansen, R., Schuler, D., Ainsworth, T., Marmorino, G., and Chubb, S., Polarimetric analysis and modeling of multi-frequency SAR signatures from Gulf Stream fronts, *IEEE J. Oceanic Eng.*, 23, 322, 1998.
21. Lee, J.S., Schuler, D.L., and Ainsworth, T.L., Polarimetric SAR data compensation for terrain azimuth slope variation, *IEEE Trans. Geosci. Rem. Sens.*, 38, 2153–2163, 2000.
22. Lee, J.S., Schuler, D.L., Ainsworth, T.L., Krogager, E., Kasilingam, D., and Boerner, W.M., The estimation of radar polarization shifts induced by terrain slopes, *IEEE Trans. Geosci. Rem. Sens.*, 40, 30–41, 2001.
23. Cloude, S.R. and Pottier, E., A review of target decomposition theorems in radar polarimetry, *IEEE Trans. Geosci. Rem. Sens.*, 34(2), 498, 1996.
24. Lee, J.S., Grunes, M.R., and De Grandi, G., Polarimetric SAR speckle filtering and its implication for classification, *IEEE Trans. Geosci. Rem. Sens.*, 37, 2363, 1999.
25. Gasparovic, R.F., Apel, J.R., and Kasischke, E., An overview of the SAR internal wave signature experiment, *J. Geophys. Res.*, 93, 12304, 1998.
26. Gasparovic, R.F., Chapman, R., Monaldo, F.M., Porter, D.L., and Sterner, R.F., Joint U.S./Russia internal wave remote sensing experiment: interim results, Applied Physics Laboratory Report S1R-93U-011, Johns Hopkins University, 1993.
27. Schuler, D.L., Kasilingam, D., and Lee, J.S., Slope measurements of ocean internal waves and current fronts using polarimetric SAR, European Conference on Synthetic Aperture Radar (EUSAR'2002), Cologne, Germany, 2002.
28. Schuler, D.L., Kasilingam, D., Lee, J.S., Jansen, R.W., and De Grandi, G., Polarimetric SAR measurements of slope distribution and coherence changes due to internal waves and current fronts, *Proc. Int. Geosci. Rem. Sens.* (IGARSS'2002) *Symp.*, Toronto, Canada, 2002.
29. Schuler, D.L., Lee, J.S., Kasilingam, D., and De Grandi, G., Studies of ocean current fronts and internal waves using polarimetric SAR coherences, in *Proc. Prog. Electromagnetic Res. Symp.* (PIERS'2002), Cambridge, MA, 2002.
30. Alpers, W., Theory of radar imaging of internal waves, *Nature*, 314, 245, 1985.
31. Brant, P., Alpers, W., and Backhaus, J.O., Study of the generation and propagation of internal waves in the Strait of Gibraltar using a numerical model and synthetic aperture radar images of the European ERS-1 satellite, *J. Geophys. Res.*, 101(14), 14237, 1996.
32. Lyzenga, D.R. and Bennett, J.R., Full-spectrum modeling of synthetic aperture radar internal wave signatures, *J. Geophys. Res.*, 93(C10), 12345, 1988.
33. Schuler D.L., Lee, J.S., Kasilingam, D., and Nesti, G., Surface roughness and slope measurements using polarimetric SAR data, *IEEE Trans. Geosci. Rem. Sens.*, 40(3), 687, 2002.
34. Schuler, D.L., Ainsworth, T.L., Lee, J.S., and De Grandi, G., Topographic mapping using polarimetric SAR data, *Int. J. Rem. Sens.*, 35(5), 1266, 1998.
35. Schuler, D.L, Lee, J.S., Ainsworth, T.L., and Grunes, M.R., Terrain topography measurement using multi-pass polarimetric synthetic aperture radar data, *Radio Sci.*, 35(3), 813, 2002.
36. Schuler, D.L. and Lee, J.S., A microwave technique to improve the measurement of directional ocean wave spectra, *Int. J. Rem. Sens.*, 16, 199, 1995.

37. Alpers, W. and Hennings, I., A theory of the imaging mechanism of underwater bottom topography by real and synthetic aperture radar, *J. Geophys. Res.*, 89, 10529, 1984.
38. Jansen, R.W., Chubb, S.R., Fusina, R.A., and Valenzuela, G.R., Modeling of current features in Gulf Stream SAR imagery, Naval Research Laboratory Report NRL/MR/7234-93-7401, 1993
39. Thompson, D.R., Calculation of radar backscatter modulations from internal waves, *J. Geophys. Res.*, 93(C10), 12371, 1988.
40. Schuler, D.L., Lee, J.S., and De Grandi, G., Spiral eddy detection using surfactant slick patterns and polarimetric SAR image decomposition techniques, *Proc. Int. Geosci. Rem. Sens. Symp.* (IGARSS), Anchorage, Alaska, September, 2004.
41. Lee, J.S., Grunes, M.R., Ainsworth, T.L., Du, L.J., Schuler, D.L., and Cloude, S.R., Unsupervised classification using polarimetric decomposition and the complex Wishart classifier, *IEEE Trans. Geosci. Rem. Sens.*, 37(5), 2249, 1999.
42. Pottier, E. and Lee, J.S., Unsupervised classification scheme of POLSAR images based on the complex Wishart distribution and the polarimetric decomposition theorem, *Proc. 3rd Eur. Conf. Synth. Aperture Radar* (*EUSAR'2000*), Munich, Germany, 2000.
43. Ferro-Famil, L., Pottier, E., and Lee, J-S, Unsupervised classification of multifrequency and fully polarimetric SAR images based on the H/A/Alpha-Wishart classifier, *IEEE Trans. Geosci. Rem. Sens.*, 39(11), 2332, 2001.
44. Van Zyl, J.J., Unsupervised classification of scattering mechanisms using radar polarimetry data, *IEEE Trans. Geosci. Rem. Sens.*, 27, 36, 1989.
45. Freeman, A., and Durden, S.L., A three component scattering model for polarimetric SAR data, *IEEE Trans. Geosci. Rem. Sens.*, 36, 963, 1998.

2

MRF-Based Remote-Sensing Image Classification with Automatic Model Parameter Estimation

Sebastiano B. Serpico and Gabriele Moser

CONTENTS

2.1 Introduction

Within remote-sensing image analysis, Markov random field (MRF) models represent a powerful tool [1], due to their ability to integrate contextual information associated with the image data in the analysis process [2,3,4]. In particular, the use of a global model for the statistical dependence of all the image pixels in a given image-analysis scheme typically turns out to be an intractable task. The MRF approach offers a solution to this issue, as it allows expressing a global model of the contextual information by using only local relations among neighboring pixels [2]. Specifically, due to the Hammersley–Clifford theorem [3], a large class of global contextual models (i.e., the Gibbs random fields, GRFs [2]) can be proved to be equivalent to local MRFs, thus sharply reducing the related model complexity. In particular, MRFs have been used for remote-sensing image analysis, for single-date [5], multi-temporal [6–8], multi-source [9], and multi-resolution [10] classification, for denoising [1], segmentation [11–14], anomaly detection [15], texture extraction [2,13,16], and change detection [17–19].

Focusing on the specific problem of image classification, the MRF approach allows one to express a ''maximum-*a-posteriori*'' (MAP) decision task as the minimization of a suitable energy function. Several techniques have been proposed to deal with this minimization problem, such as the simulated annealing (SA), an iterative stochastic optimization algorithm converging to a global minimum of the energy [3] but typically involving long execution times [2,5], the iterative conditional modes (ICM), an iterative deterministic algorithm converging to a local (but usually good) minimum point [20] and requiring much shorter computation times than SA [2], and the maximization of posterior marginals (MPM), which approximates the MAP rule by maximizing the marginal posterior distribution of the class label of each pixel instead of the joint posterior distribution of all image labels [2,11,21].

However, an MRF model usually involves the use of one or more internal parameters, thus requiring a preliminary parameter-setting stage before the application of the model itself. In particular, especially in the context of supervised classification, interactive ''trial-and-error'' procedures are typically employed to choose suitable values for the model parameters [2,5,9,11,17,19], while the problem of fast automatic parameter-setting for MRF classifiers is still an open issue in the MRF literature. The lack of effective automatic parameter-setting techniques has represented a significant limitation on the operational use of MRF-based supervised classification architectures, although such methodologies are known for their ability to generate accurate classification maps [9]. On the other hand, the availability of the above-mentioned unsupervised parameter-setting procedures has contributed to an extensive use of MRFs for segmentation and unsupervised classification purposes [22,23].

In the present chapter, an automatic parameter optimization algorithm is proposed to overcome the above limitation in the context of supervised image classification. The method refers to a broad category of MRF models, characterized by energy functions expressed as linear combinations of different energy contributions (e.g., representing different typologies of contextual information) [2,7]. The algorithm exploits this linear dependence to formalize the parameter-setting problem as the solution of a set of linear inequalities, and addresses this problem by extending to the present context the Ho–Kashyap method for linear classifier training [24,25]. The well-known convergence properties of such a method and the absence of parameters to be tuned are among the good features of the proposed technique.

The chapter is organized as follows. Section 2.2 provides an overview of the previous work on the problem of MRF parameter estimation. Section 2.3 describes the methodological issues of the method, and Section 2.4 presents the results of the application of the technique to the classification of real (both single-date and multi-temporal) remote sensing images. Finally, conclusions are drawn in Section 2.5.

2.2 Previous Work on MRF Parameter Estimation

Several parameter-setting algorithms have been proposed in the context of MRF models for image segmentation (e.g., [12,22,26,27]) or unsupervised classification [23,28,29], although often resulting in a considerable computational burden. In particular, the usual maximum likelihood (ML) parameter estimation approach [30] exhibits good theoretical consistency properties when applied to GRFs and MRFs [31], but turns out to be computationally very expensive for most MRF models [22] or even intractable [20,32,33], due to the difficulty of analytical computation and numerical maximization of the normalization constants involved by the MRF-based distributions (the so-called "partition functions") [20,23,33]. Operatively, the use of ML estimates for MRF parameters turns out to be restricted just to specific typologies of MRFs, such as continuous Gaussian [1,13,15,34,35] or generalized Gaussian [36] fields, which allows an ML estimation task to be formulated in an analytically feasible way.

Beyond such case-specific techniques, essentially three approaches have been suggested in the literature to address the problem of unsupervised ML estimation of MRF parameters indirectly, namely the Monte Carlo methods, the stochastic gradient approaches, or the pseudo-likelihood approximations [23]. The combination of the ML criterion with Monte Carlo simulations has been proposed to overcome the difficulty of computation of the partition function [22,37] and it provides good estimation results, but it usually involves long execution times. Stochastic-gradient approaches aim at maximizing the log-likelihood function by integrating a Gibbs stochastic sampling strategy into the gradient ascent method [23,38]. Combinations of the stochastic-gradient approach with the iterated conditional expectation (ICE) and with an estimation–restoration scheme have also been proposed in Refs. [12,29], respectively. Approximate pseudo-likelihood functionals have been introduced [20,23], and they are numerically feasible, although the resulting estimates are not actual ML estimates (except in the trivial noncontextual case of pixel independence) [20]. Moreover the pseudo-likelihood approximation may underestimate the interactions between pixels and can provide unsatisfactory results unless the interactions are suitably weak [21,23]. Pseudo-likelihood approaches have also been developed in conjunction with mean-field approximations [26], with the expectation-maximization (EM) [21,39], the Metropolis–Hastings [40], and the ICE [21] algorithms, with Monte Carlo simulations [41], or with multi-resolution analysis [42]. In Ref. [20] a pseudo-likelihood approach is plugged into the ICM energy minimization strategy, by iteratively alternating the update of the contextual clustering map and the update of the parameter values. In Ref. [10] this method is integrated with EM in the context of multi-resolution MRFs. A related technique is the "coding method," which is based on a pseudo-likelihood functional computed over a subset of pixels, although these subsets depend on the choice of suitable coding strategies [34].

Several empirical or *ad hoc* estimators have also been developed. In Ref. [33] a family of MRF models with polynomial energy functions is proved to be dense in the space of all MRFs and is endowed with a case-specific estimation scheme based on the method of moments. In Ref. [43] a least-square approach is proposed for Ising-type and Potts-type MRF models [22], and it formulates an overdetermined system of linear equations relating the unknown parameters with a set of relative frequencies of pixel-label configurations. The combination of this method with EM is applied in Ref. [44] for sonar image segmentation purposes. A simpler but conceptually similar approach is adopted in Ref. [12] and is combined with ICE: a simple algebraic empirical estimator for a one-parameter spatial MRF model is developed and it directly relates the parameter value with the relative frequencies of several class-label configurations in the image.

On the other hand, the literature about MRF parameter-setting for supervised image classification is very limited. Any unsupervised MRF parameter estimator can also naturally be applied to a supervised problem, by simply neglecting the training data in the estimation process. However, only a few techniques have been proposed so far, effectively exploiting the available training information for MRF parameter-optimization purposes as well. In particular, a heuristic algorithm is developed in Ref. [7] aiming to optimize automatically the parameters of a multi-temporal MRF model for a joint supervised classification of two-date imagery, and a genetic approach is combined in Ref. [45] with simulated annealing [3] for the estimation of the parameters of a spatial MRF model for multi-source classification.

2.3 Supervised MRF-Based Classification

2.3.1 MRF Models for Image Classification

Let $I = \{x_1, x_2, \ldots, x_N\}$ be a given n-band remote-sensing image, modeled as a set of N identically distributed n-variate random vectors. We assume M thematic classes $\omega_1, \omega_2, \ldots, \omega_M$ to be present in the image and we denote the resulting set of classes by $\Omega = \{\omega_1, \omega_2, \ldots, \omega_M\}$ and the class label of the k-th image pixel ($k = 1, 2, \ldots, N$) by $s_k \in \Omega$. By operating in the context of supervised image classification, we assume a training set to be available, and we denote the index set of the training pixels by $T \subset \{1, 2, \ldots, N\}$ and the corresponding true class label of the k-th training pixel ($k \in T$) by s_k^*. When collecting all the feature vectors of the N image pixels in a single ($N \cdot n$)-dimensional column vector[1] $X = col[x_1, x_2, \ldots, x_N]$ and all the pixel labels in a discrete random vector $S = (s_1, s_2, \ldots, s_N) \in \Omega^N$, the MAP decision rule (i.e., the Bayes rule for minimum classification error [46]) assigns to the image data X the label vector $\tilde{S}$, which maximizes the joint posterior probability $P(S|X)$, that is,

$$\tilde{S} = \arg\max_{S \in \Omega^N} P(S|X) = \arg\max_{S \in \Omega^N} [p(X|S)P(S)] \tag{2.1}$$

where $p(X|S)$ and $P(S)$ are the joint probability density function (PDF) of the global feature vector X conditioned to the label vector S and the joint probability mass function (PMF) of the label vector itself, respectively. The MRF approach offers a computationally tractable solution to this maximization problem by passing from a global model for the statistical dependence of the class labels to a model of the local image properties, defined according to a given neighborhood system [2,3]. Specifically, for each k-th image pixel, a neighborhood $N_k \subset \{1, 2, \ldots, N\}$ is assumed to be defined, such that, for instance, N_k includes the four (first-order neighborhood) or the eight (second-order neighborhood) pixels surrounding the k-th pixel ($k = 1, 2, \ldots, N$). More formally, a neighborhood system is a collection $\{N_k\}_{k=1}^N$ of subsets of pixels such that each pixel is outside its neighborhood (i.e., $k \notin N_k$ for all $k = 1, 2, \ldots, N$) and neighboring pixels are always mutually neighbors (i.e., $k \in N_h$ if and only if $h \in N_k$ for all $k,h = 1, 2, \ldots, N$, $k \neq h$). This simple discrete topological structure attached to the image data is exploited in the MRF framework to model the statistical relationships between the class labels of spatially distinct pixels and to provide a computationally affordable solution to the global MAP classification problem of Equation 2.1.

Specifically, we assume the feature vectors $x_1, x_2, \ldots, x_N$ to be conditionally independent and identically distributed with PDF $p(x|s)$ ($x \in \mathbb{R}^n$, $s \in \Omega$), that is [2],

$$p(X|S) = \prod_{k=1}^{N} p(x_k|s_k) \tag{2.2}$$

and the joint prior PMF $P(S)$ to be a Markov random field with respect to the above-mentioned neighborhood system, that is [2,3],

- the probability distribution of each k-th image label, conditioned to all the other image labels, is equivalent to the distribution of the k-th label conditioned only to the labels of the neighboring pixels ($k = 1, 2, \ldots, N$):

[1] All the vectors in the chapter are implicitly assumed to be column vectors, and we denote by u_i the i-th component of an m-dimensional vector $u \in \mathbb{R}^m$ ($i = 1, 2, \ldots, m$), by "col" the operator of column vector juxtaposition (i.e., col $[u,v]$ is the vector obtained by stacking the two vectors $u \in \mathbb{R}^m$ and $v \in \mathbb{R}^n$ in a single $(m + n)$-dimensional column vector), and by the superscript "T" the matrix transpose operator.

$$P\{s_k = \omega_i | s_h : h \neq k\} = P\{s_k = \omega_i | s_h : h \in N_k\}, \quad i = 1, 2, \ldots, M \tag{2.3}$$

- the PMF of S is a strictly positive function on Ω^N, that is, $P(S) > 0$ for all $S \in \Omega^N$.

The Markov assumption expressed by Equation 2.3 allows restricting the statistical relationships among the image labels to the local relationships inside the predefined neighborhood, thus greatly simplifying the spatial-contextual model for the label distribution as compared to a generic global model for the joint PMF of all the image labels. However, as stated in the next subsection, due to the so-called Hammersley–Clifford theorem, despite this strong analytical simplification, a large class of contextual models can be accomplished under the MRF formalism.

2.3.2 Energy Functions

Given the neighborhood system $\{N_k\}_{k=1}^N$, we denote by "clique" a set Q of pixels ($Q \subset \{1, 2, \ldots, N\}$) such that, for each pair (k, h) of pixels in Q, k and h turn out to be mutually neighbors, that is,

$$k \in N_h \iff h \in N_k \quad \forall k, h \in Q, \quad k \neq h \tag{2.4}$$

By marking the collection of all the cliques in the adopted neighborhood system by $\mathcal{Q}$ and the vector of the pixel labels in the clique $Q \in \mathcal{Q}$ (i.e., $S_Q = \text{col}[s_k : k \in Q]$) by S_Q, the Hammersley–Clifford theorem states that the label configuration S is an MRF if and only if, for any clique $Q \in \mathcal{Q}$ there exists a real-valued function $V_Q(S_Q)$ (usually named "potential function") of the pixel labels in Q, so that the global PMF of S is given by the following Gibbs distribution [3]:

$$P(S) = \frac{1}{Z_{\text{prior}}} \exp\left[-\frac{U_{\text{prior}}(S)}{\Theta}\right], \quad \text{where: } U_{\text{prior}}(S) = \sum_{Q \in \mathcal{Q}} V_Q(S_Q) \tag{2.5}$$

Θ is a positive parameter, and Z_{prior} is a normalizing constant. Because of the formal similarity between the probability distribution in Equation 2.5 and the well-known Maxwell–Boltzmann distribution introduced in statistical mechanics for canonical ensembles [47], $U_{\text{prior}}(\cdot)$, Θ, and Z_{prior} are usually named "energy function," "temperature," and "partition function," respectively. A very large class of statistical interactions among spatially distinct pixels can be modeled in this framework, by simply choosing a suitable function $U_{\text{prior}}(\cdot)$, which makes the MRF approach highly flexible. In addition, when coupling the Hammersley–Clifford formulation of the prior PMF $P(S)$ with the conditional independence assumption stated for the conditional PDF $p(X|S)$ (see Equation 2.2), an energy representation also holds for the global posterior distribution $P(S|X)$, that is [2],

$$P(S|X) = \frac{1}{Z_{\text{post},X}} \exp\left[-\frac{U_{\text{post}}(S|X)}{\Theta}\right] \tag{2.6}$$

where $Z_{\text{post},X}$ is a normalizing constant and $U_{\text{post}}(\cdot)$ is a posterior energy function, given by:

$$U_{\text{post}}(S|X) = -\Theta \sum_{k=1}^{N} \ln p(x_k|s_k) + \sum_{Q \in \mathcal{Q}} V_Q(S_Q) \tag{2.7}$$

This formulation of the global posterior probability allows addressing the MAP classification task as the minimization of the energy function $U_{\text{post}}(\cdot|X)$, which is locally defined, according to the pixelwise PDF of the feature vectors conditioned to the class labels and to the collection of the potential functions. This makes the maximization problem of Equation 2.1 tractable and allows a contextual classification map of the image I to be feasibly generated.

However, the (prior and posterior) energy functions are generally parameterized by several real parameters $\lambda_1, \lambda_2, \ldots, \lambda_L$. Therefore, the solution of the minimum-energy problem requires the preliminary selection of a proper value for the parameter vector $\lambda = (\lambda_1, \lambda_2, \ldots, \lambda_L)^T$. As described in the next subsections, in the present chapter, we address this parameter-setting issue with regard to a broad family of MRF models, which have been employed in the remote-sensing literature, and to the ICM minimization strategy for the energy function.

2.3.3 Operational Setting of the Proposed Method

Among the techniques proposed in the literature to deal with the task of minimizing the posterior energy function (see Section 2.1), in the present chapter ICM is adopted as a trade-off between the effectiveness of the minimization process and the computation time [5,7], and it is endowed with an automatic parameter-setting stage. Specifically, ICM is initialized with a given label vector $S^0 = (s_1^0, s_2^0, \ldots, s_N^0)^T$ (e.g., generated by a previously applied noncontextual supervised classifier) and it iteratively modifies the class labels to decrease the energy function. In particular, by marking by C_k the set of the neighbor labels of the k-th pixel (i.e., the "context" of the k-th pixel, $C_k = \text{col}[s_h{:}h \in N_k]$), at the t-th ICM iteration, the label s_k is updated according to the feature vector x_k and to the current neighboring labels C_k^t, so that s_k^{t+1} is the class label ω_i that minimizes a local energy function $U_\lambda(\omega_i|x_k,C_k^t)$ (the subscript λ is introduced to stress the dependence of the energy on the parameter vector λ). More specifically, given the neighborhood system, an energy representation can also be proved for the local distribution of the class labels, that is, ($i = 1,2,\ldots,M; k = 1,2,\ldots,N$) [9]:

$$P\{s_k = \omega_i|x_k,\ C_k\} = \frac{1}{Z_{k\lambda}} \exp\left[-\frac{U_\lambda(\omega_i|x_k, C_k)}{\Theta}\right] \tag{2.8}$$

where $Z_{k\lambda}$ is a further normalizing constant and the local energy is defined by

$$U_\lambda(\omega_i|x_k,\ C_k) = -\Theta \ln p(x_k|\omega_i) + \sum_{Q \ni k} V_Q(S_Q) \tag{2.9}$$

Here we focus on the family of MRF models whose local energy functions $U_\lambda(\cdot)$ can be expressed as weighted sums of distinct energy contributions, that is,

$$U_\lambda(\omega_i|x_k,\ C_k) = \sum_{\ell=1}^{L} \lambda_\ell \mathcal{E}_\ell(\omega_i|x_k,\ C_k),\ k = 1,2,\ldots,N \tag{2.10}$$

where $\mathcal{E}_\ell(\cdot)$ is the ℓ-th contribution and the parameter λ_ℓ plays the role of the weight of $\mathcal{E}_\ell(\cdot)$ ($\ell = 1,2,\ldots,L$). A formal comparison between Equation 2.9 and Equation 2.10 suggests that one of the considered L contributions (say, $\mathcal{E}_1(\cdot)$) should be related to the pixelwise conditional PDF of the feature vector (i.e., $\mathcal{E}_1(\omega_i|x_k) \propto -\ln p(x_k|\omega_i)$), which formalizes the spectral information associated with each single pixel. From this viewpoint, postulating the presence of $(L-1)$ further contributions implicitly means that $(L-1)$ typologies of contextual information are modeled by the adopted MRF and are supposed to be separable (i.e., combined in a purely additive way) [7]. Formally speaking, Equation 2.10

represents a constraint on the energy function, but most MRF models employed in remote sensing for classification, change detection, or segmentation purposes belong to this category. For instance, the pairwise interaction model described in Ref. [2], the well-known spatial Potts MRF model employed in Ref. [17] for change detection purposes and in Ref. [5] for hyperspectral data classification, the spatio-temporal MRF model introduced in Ref. [7] for multi-date image classification, and the multi-source classification models defined in Ref. [9] belong to this family of MRFs.

2.3.4 The Proposed MRF Parameter Optimization Method

With regard to the class of MRFs identified by Equation 2.10, the proposed method employs the training data to identify a set of parameters for the purpose of maximizing the classification accuracy by exploiting the linear relation between the energy function and the parameter vector. When focusing on the first ICM iteration, the k-th training sample ($k \in T$) is correctly classified by the minimum-energy rule if:

$$U_\lambda(s_k^*|x_k, C_k^0) \leq U_\lambda(\omega_i|x_k, C_k^0) \quad \forall \omega_i \neq s_k^* \tag{2.11}$$

or equivalently:

$$\sum_{\ell=1}^{L} \lambda_\ell \left[\mathcal{E}_\ell(\omega_i|x_k, C_k^0) - \mathcal{E}_\ell(s_k^*|x_k, C_k^0)\right] \geq 0 \quad \forall \omega_i \neq s_k^* \tag{2.12}$$

We note that the inequalities in Equation 2.12 are linear with respect to λ, that is, the correct classification of a training pixel is expressed as a set of $(M - 1)$ linear inequalities with respect to the model parameters. More formally, when collecting the energy differences contained in Equation 2.12 in a single L-dimensional column vector:

$$\varepsilon_{ki} = \text{col}\left[\mathcal{E}_\ell(\omega_i|x_k, C_k^0) - \mathcal{E}_\ell(s_k^*|x_k, C_k^0) \colon \ell = 1, 2, \ldots, L\right] \tag{2.13}$$

the k-th training pixel is correctly classified by ICM if $\varepsilon_{ki}^T \lambda \geq 0$ for all the class labels $\omega_i \neq s_k^*$ ($k \in T$). By denoting by E the matrix obtained by juxtaposing all the row vectors ε_{ki}^T ($k \in T$, $\omega_i \neq s_k^*$), we conclude that ICM correctly classifies the whole training set T if and only if[2]:

$$E\lambda \geq 0 \tag{2.14}$$

Operatively, the matrix E includes all the coefficients (i.e., the energy differences) of the linear inequalities in Equation 2.12; hence, E has L columns and has a row for each inequality, that is, it has $(M - 1)$ rows for each training pixel (corresponding to the $(M - 1)$ class labels, which are different from the true label). Therefore, denoting by $|T|$ the number of training samples (i.e., the cardinality of T), E is an $(R \times L)$-sized matrix, with $R = |T|\ (M - 1)$. Since $R \gg L$, the system in Equation 2.14 presents a number of inequalities, which is much larger than the number of unknown variables. In particular, a feasible approach would lie in formulating the matrix inequality in Equation 2.14 as an equality $E\ \lambda = b$, where $bin\mathbb{R}^R$ is a "margin" vector with positive components, to solve such equality by a minimum square error (MSE) approach [24]. However, MSE would require a preliminary manual choice of margin vector b. Therefore, we avoid using this approach, and note that this problem is formally identical to the problem of computing

[2] Given two m-dimensional vectors, u and v ($u, v \in \mathbb{R}^m$), we write $u \geq v$ to mean $u_i \geq v_i$ for all $i = 1, 2, \ldots, m$.

the weight parameters of a linear binary discriminant function [24]. Hence, we propose to address the solution of Equation 2.14 by extending to the present context the methods described in the literature to compute such a linear discriminant function. In particular, we adopt the Ho–Kashyap method [24]. This approach jointly optimizes both λ and b according to the following quadratic programing problem [24]:

$$\begin{cases} \min\limits_{\lambda,b} \|E\lambda - b\|^2 \\ \lambda \in \mathbb{R}^L, b \in \mathbb{R}^R \\ b_r > 0 \text{ for } r = 1,2,\ldots,R \end{cases} \tag{2.15}$$

which is solved by alternating iteratively an MSE step to update λ and a gradient-like descent step to update b [24]. Specifically, the following operations are performed at the t-th step ($t = 0,1,2,\ldots$) of the Ho–Kashyap procedure [24]:

- given the current margin vector b^t, compute a corresponding parameter vector λ^t by minimizing $\|E\,\lambda - b^t\|^2$ with respect to λ, that is, compute:

$$\lambda^t = E^{\#}b^t, \text{ where } E^{\#} = (E^TE)^{-1}E^T \tag{2.16}$$

is the so-called pseudo-inverse of E (provided that E^TE is nonsingular) [48]

- compute the error vector $e^t = E\lambda^t - b^t \in R^R$
- update each component of the margin vector by minimizing $\|E\lambda^t - b\|^2$ with respect to b by a gradient-like descent step that allows the margin components only to be increased [24], that is, compute

$$b_r^{t+1} = \begin{cases} b_r^t + \rho e_r^t & \text{if } e_r^t > 0 \\ b_r^t & \text{if } e_r^t \le 0 \end{cases} \quad r = 1,2,\ldots,R \tag{2.17}$$

where ρ is a convergence parameter.

In particular, the proposed approach to the MRF parameter-setting problem allows exploiting the known theoretical results about the Ho–Kashyap method in the context of linear discriminant functions. Specifically, one can prove that, if the matrix inequality in Equation 2.14 has solutions and if $0 < \rho < 2$, then the Ho–Kashyap algorithm converges to a solution of Equation 2.14 in a finite number τ of iterations (i.e., we obtain $e^\tau = 0$ and consequently $b^{\tau+1} = b^\tau$ and $E\lambda^\tau = b^\tau \geq 0$) [24]. On the other hand, if the inequality in Equation 2.14 has no solution, it is possible to prove that each component e_r^t of the error vector ($r = 1,2,\ldots,R$) either vanishes for $t \to +\infty$ or takes on nonpositive values, while the error magnitude $\|e^t\|$ converges to a positive limit, which is bounded away from zero; in the latter situation, according to the Ho–Kashyap iterative procedure, the algorithm stops, and one can conclude that the matrix inequality in Equation 2.14 has no solution [46]. Therefore, the convergence (either finite-time or asymptotic) of the proposed parameter-setting method is guaranteed in any case. However, it is worth noting that the number τ of iterations required to reach convergence in the first case or the number τ' of iterations needed to detect the nonexistence of a solution of Equation 2.14 in the second case are not known in advance. Hence, specific stop conditions are usually adopted for the Ho–Kashyap procedure, for instance, by stopping the iterative process when $|\lambda_\ell^{t+1} - \lambda_\ell^t| < \varepsilon_{\text{stop}}$ for all $\ell = 1,2,\ldots,L$ and $|b_r^{t+1} - b_r^t| < \varepsilon_{\text{stop}}$ for all $r = 1,2,\ldots,R$, where $\varepsilon_{\text{stop}}$ is a given threshold (in the present chapter, $\varepsilon_{stop} = 0.0001$ is used). The algorithm has been initialized by setting unitary values for all the components of λ and b and by choosing $\rho = 1$.

Coupling the proposed Ho–Kashyap-based parameter-setting method with the ICM classification approach results in an automatic contextual supervised classification approach (hereafter marked by HK–ICM), which performs the following processing steps:

- *Noncontextual step*: generate an initial noncontextual classification map (i.e., an initial label vector S^0) by applying a given supervised noncontextual (e.g., Bayesian or neural) classifier;
- *Energy-difference step*: compute the energy-difference matrix E according to the adopted MRF model and to the label vector S^0;
- *Ho–Kashyap step*: compute an optimal parameter vector λ^* by running the Ho–Kashyap procedure (applied to the matrix E) up to convergence; and
- *ICM step*: generate a contextual classification map by running ICM (fed with the parameter vector λ^*) up to convergence.

A block diagram of the proposed contextual classification scheme is shown in Figure 2.1. It is worth noting that, according to the role of the parameters as weights of energy contributions, negative values for $\lambda_1^*, \lambda_2^*, \ldots, \lambda_L^*$ would be undesirable. To prevent this issue, we note that, if all the entries in a given row of the matrix E are negative,

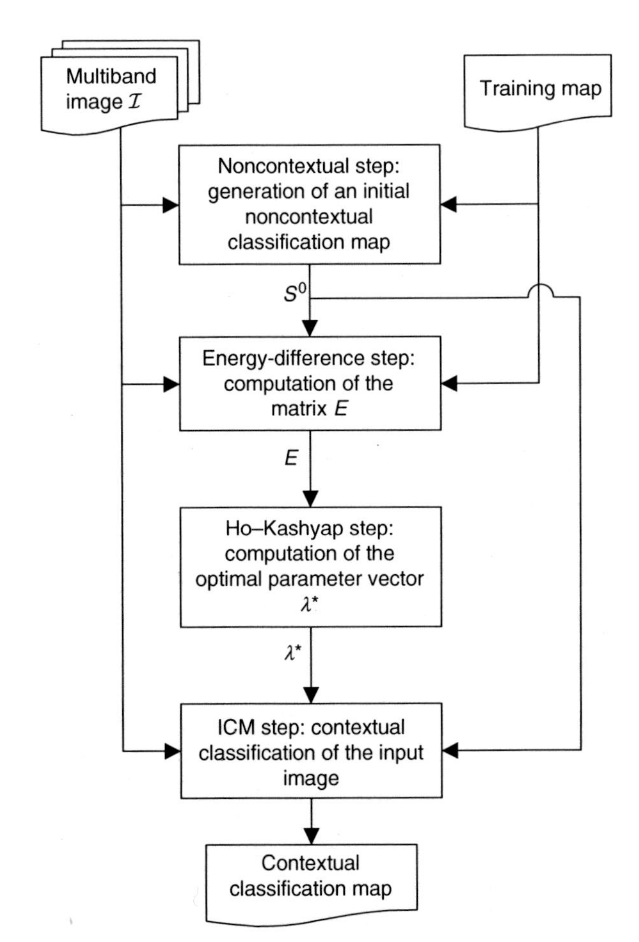

FIGURE 2.1
Block diagram of the proposed contextual classification scheme.

the corresponding inequality cannot be satisfied by a vector λ with positive components. Therefore, before the application of the Ho–Kashyap step, the rows of E containing only negative entries are cancelled, as such rows would represent linear inequalities that could not be solved by positive parameter vector solutions. In other words, each deleted row corresponds to a training pixel $k \in T$, so that a class $\omega_i \neq s_k^*$ has a lower energy than the true class label s_k^* for all possible positive values of the parameters $\lambda_1, \lambda_2, \ldots, \lambda_L$ (therefore, its classification is wrong).

2.4 Experimental Results

The proposed HK–ICM method was tested on three different MRF models, each applied to a distinct real data set. In all the cases, the results provided by ICM, endowed with the proposed automatic parameter-optimization method, are compared with the ones that can be achieved by performing an exhaustive grid search for the parameter values yielding the highest accuracies on the training set; operationally, such parameter values can be obtained by a "trial-and-error" procedure (TE–ICM in the following).

Specifically, in all the experiments, spatially distinct training and test fields were available for each class. We note that the highest training-set (and not test-set) accuracy is searched for by TE–ICM to perform a consistent comparison with the proposed method, which deals only with the available training samples. A comparison between the classification accuracies of HK–ICM and TE–ICM on such a set of samples aims at assessing the performances of the proposed method from the viewpoint of the MRF parameter-optimization problem. On the other hand, an analysis of the test-set accuracies allows one to assess the quality of the resulting classification maps. Correspondingly, Table 2.1 through Table 2.3 show, for the three experiments, the classification accuracies provided by HK–ICM and those given by TE–ICM on both the training and the test sets.

In all the experiments, 10 ICM iterations were sufficient to reach convergence, and the Ho–Kashyap algorithm was initialized by setting a unitary value for all the MRF parameters (i.e., by initially giving the same weight to all the energy contributions).

TABLE 2.1

Experiment I: Training and Test-Set Cardinalities and Classification Accuracies Provided by the Noncontextual GMAP Classifier, by HK–ICM, and by TE–ICM

Class	Training Samples	Test Samples	Test-Set Accuracy			Training-Set Accuracy	
			GMAP (%)	HK–ICM (%)	TE–ICM (%)	HK–ICM (%)	TE–ICM (%)
Wet soil	1866	750	93.73	96.93	96.80	98.66	98.61
Urban	4195	2168	94.10	99.31	99.45	97.00	97.02
Wood	3685	2048	98.19	98.97	98.97	96.85	96.85
Water	1518	875	91.66	92.69	92.69	98.16	98.22
Bare soil	3377	2586	97.53	99.54	99.54	99.20	99.20
Overall accuracy			95.86	98.40	98.42	97.80	97.81
Average accuracy			95.04	97.49	97.49	97.97	97.98

TABLE 2.2

Experiment II: Training and Test-Set Cardinalities and Classification Accuracies Provided by the Noncontextual DTC Classifier, by HK–ICM, and by TE–ICM

Image	Class	Training Samples	Test Samples	Test-Set Accuracy			Training-Set Accuracy	
				DTC (%)	HK–ICM (%)	TE–ICM (%)	HK–ICM (%)	TE–ICM (%)
October 2000	Wet soil	1194	895	79.29	92.63	91.84	96.15	94.30
	Urban	3089	3033	88.30	96.41	99.04	94.59	96.89
	Wood	4859	3284	99.15	99.76	99.88	99.96	99.98
	Water	1708	1156	98.70	98.70	98.79	99.82	99.82
	Bare soil	4509	3781	97.70	99.00	99.34	98.54	98.91
	Overall accuracy			93.26	98.06	98.81	98.15	98.59
	Average accuracy			92.63	97.30	97.78	97.81	97.98
June 2001	Wet soil	1921	1692	96.75	98.46	98.94	99.64	99.74
	Urban	3261	2967	89.25	92.35	94.00	98.25	99.39
	Wood	1719	1413	94.55	98.51	98.51	96.74	98.08
	Water	1461	1444	99.72	99.72	99.79	99.93	100
	Bare soil	3168	3052	97.67	99.34	99.48	98.77	98.90
	Agricultural	2773	2431	82.48	83.42	83.67	99.89	99.96
	Overall accuracy			92.68	94.61	95.13	98.86	99.34
	Average accuracy			93.40	95.30	95.73	98.87	99.34

TABLE 2.3

Experiment III: Training and Test-Set Cardinalities and Classification Accuracies Provided by the Noncontextual DTC Classifier, by HK–ICM, and by TE–ICM

				Test-Set Accuracy			Training-Set Accuracy	
Image	**Class**	**Training Samples**	**Test Samples**	**DTC (%)**	**HK–ICM (%)**	**TE–ICM (%)**	**HK–ICM (%)**	**TE–ICM (%)**
April 16	Wet soil	6205	5082	90.38	99.19	96.79	100	100
	Bare soil	1346	1927	91.70	92.42	95.28	99.93	100
	Overall accuracy			90.74	97.33	96.38	99.99	100
	Average accuracy			91.04	95.81	96.04	99.96	100
April 17	Wet soil	6205	5082	93.33	99.17	97.93	100	100
	Bare soil	1469	1163	99.66	100	100	99.52	100
	Overall accuracy			94.51	99.33	98.32	99.91	100
	Average accuracy			96.49	99.59	98.97	99.76	100
April 18	Wet soil	6205	5082	86.80	97.28	92.84	100	100
	Bare soil	1920	2122	97.41	96.80	98.26	98.70	99.95
	Overall accuracy			89.92	97.14	94.43	99.69	99.99
	Average accuracy			92.10	97.04	95.55	99.35	99.97

2.4.1 Experiment I: Spatial MRF Model for Single-Date Image Classification

The spatial MRF model described in Refs. [5,17] is adopted: it employs a second-order neighborhood (i.e., C_k includes the labels of the 8 pixels surrounding the k-th pixel; $k=1,2,\ldots,N$) and a parametric Gaussian $N(m_i, \Sigma_i)$ model for the PDF $p(x_k|\omega_i)$ of a feature vector x_k conditioned to the class ω_i ($i = 1,2, \ldots, M$; $k = 1,2, \ldots, N$). Specifically, two energy contributions are defined, that is, a "spectral" energy term $E_1(\cdot)$, related to the information conveyed by the conditional statistics of the feature vector of each single pixel and a "spatial" term $E_2(\cdot)$, related to the information conveyed by the correlation among the class labels of neighboring pixels, that is, ($k=1,2,\ldots,N$):

$$\mathcal{E}_1(\omega_i \mid x_k) = (x_k - \hat{m}_i)^T \hat{\Sigma}_i^{-1}(x_k - \hat{m}_i) + \ln\det\hat{\Sigma}_i, \qquad \mathcal{E}_2(\omega_i \mid C_k) = -\sum_{\omega_j \in C_k} \delta(\omega_i, \omega_j) \tag{2.18}$$

where $\hat{m}_i$ and $\hat{\Sigma}_i$ are a sample-mean estimate and a sample-covariance estimate of the conditional mean m_i and of the conditional covariance matrix Σ_i, respectively, computed using the training samples of ω_i ($i = 1,2, \ldots, M$); $\delta(\cdot, \cdot)$ is the Kronecker delta function (i.e., $\delta(a,b)=1$ for $a=b$ and $\delta(a,b)=0$ for $a \neq b$). Hence, in this case, two parameters, λ_1 and λ_2, have to be set.

The model was applied to a 6 -band 870 × 498 pixel-sized Landsat-5 TM image acquired in April, 1994, over an urban and agricultural area around the town of Pavia (Italy) (Figure 2.2a), and presenting five thematic classes (i.e., "wet soil," "urban," "wood," "water," and "bare soil;" see Table 2.1). The above-mentioned normality assumption for the conditional PDFs is usually accepted as a model for the class statistics in optical data [49,50]. A standard noncontextual MAP classifier with Gaussian classes (hereafter denoted simply by GMAP) [49,50] was adopted in the noncontextual step. The Ho–Kashyap step required 31716 iterations to reach convergence and selected $\lambda_1^*=0.99$ and $\lambda_2^*=10.30$. The classification accuracies provided by GMAP and HK–ICM on the test set are given in Table 2.1. GMAP already provides good classification performances, with accuracies higher than 90% for all the classes. However, as expected, the contextual HK–ICM classifier further improves the classification result, in particular, yielding, a 3.20% accuracy increase for "wet soil" and a 5.21% increase for "urban," which result in a 2.54% increase in the overall accuracy (OA, i.e., the percentage of correctly classified test samples) and in a 2.45% increase in the average accuracy (AA, i.e., the average of the accuracies obtained on the five classes).

Furthermore, for the sake of comparison, Table 2.1 also shows the results obtained by the contextual MRF–ICM classifier, applied by using a standard "trial-and-error" parameter-setting procedure (TE–ICM). Specifically, fixing[3] $\lambda_1=1$, the value of λ_2 yielding the highest value of OA on the training set has been searched exhaustively in the range [0,20] with discretization step 0.2. The value selected exhaustively by TE–ICM was $\lambda_2=11.20$, which is quite close to the value computed automatically by the proposed procedure. Accordingly, the classification accuracies obtained by HK–ICM and TE–ICM on both the training and test sets were very similar (Table 2.1).

2.4.2 Experiment II: Spatio-Temporal MRF Model for Two-Date Multi-Temporal Image Classification

As a second experiment, we addressed the problem of supervised multi-temporal classification of two co-registered images, I_0 and I_1, acquired over the same ground area at

[3] According to the minimum-energy decision rule, this choice causes no loss of generality, as the classification result is affected only by the relative weight λ_2/λ_1 between the two parameters and not by their absolute values.

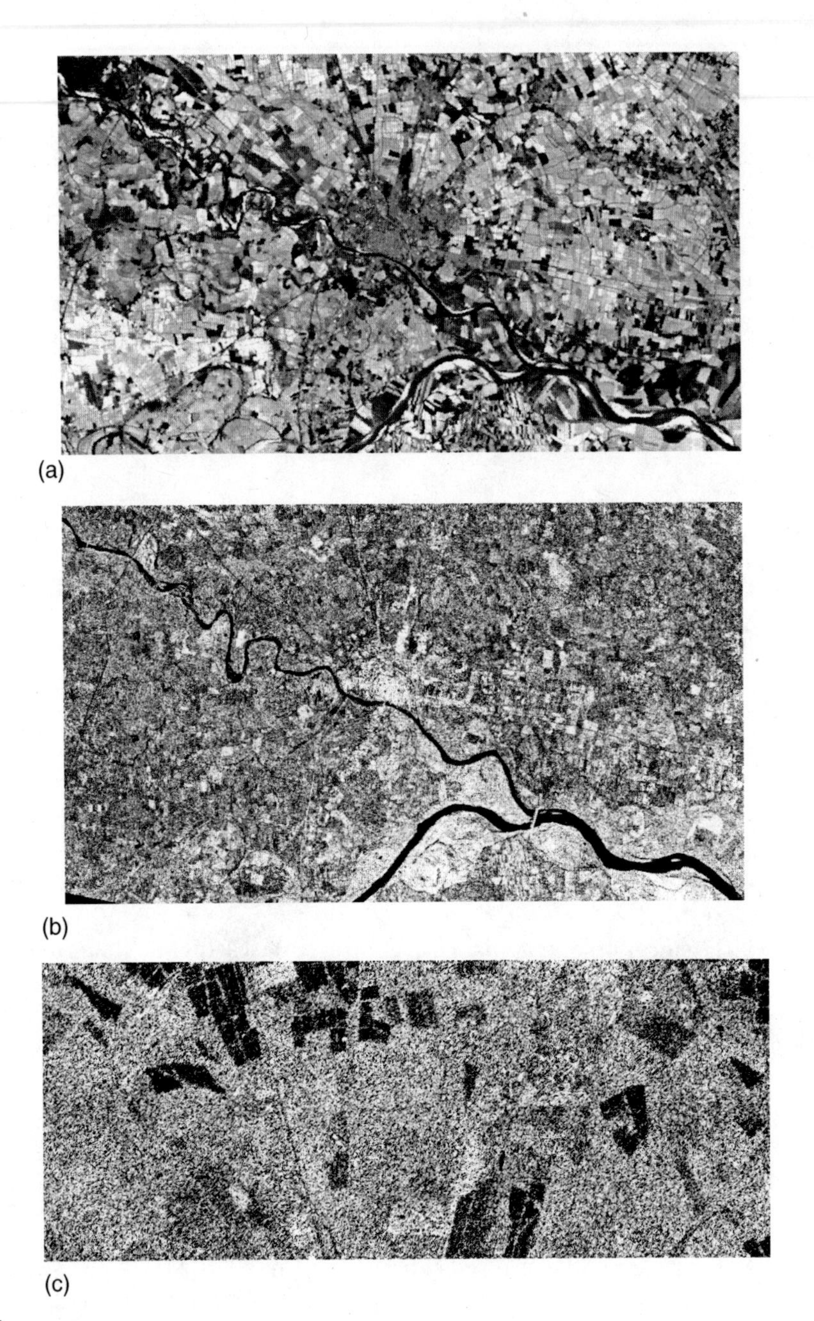

FIGURE 2.2
Data sets employed for the experiments: (a) band TM-3 of the multi-spectral image acquired in April, 1994, and employed in Experiment I; (b) ERS-2 channel (after histogram equalization) of the multi-sensor image acquired in October, 2000, and employed in Experiment II; (c) XSAR channel (after histogram equalization) of the SAR multi-frequency image acquired in April 16, 1994, and employed in Experiment III.

different times, t_0 and t_1 ($t_1 > t_0$). Specifically, the spatio-temporal mutual MRF model proposed in Ref. [7] is adopted that introduces an energy function taking into account both the spatial context (related to the correlation among neighboring pixels in the same image) and the temporal context (related to the correlation between distinct images of the same area). Let us denote by M_j, x_{jk}, and ω_{ji} the number of classes in I_j, the feature vector of the k-th pixel in I_j ($k = 1, 2, \ldots, N$), and the i-th class in I_j ($i = 1, 2, \ldots, M_j$), respectively

($j=0,1$). Focusing, for instance, on the classification of the first-date image I_0, the context of the k-th pixel includes two distinct sets of labels: (1) the set S_{0k} of labels of the spatial context, that is, the labels of the 8 pixels surrounding the k-th pixel in I_0, and (2) the set T_{0k} of labels of the temporal context, that is, the labels of the pixels lying in I_1 inside a 3×3 window centered on the k-th pixel ($k=1,2,\ldots,N$) [7]. Accordingly, three energy contributions are introduced, that is, a spectral and a spatial term (as in Experiment I), and an additional temporal term, that is, ($i=1,2,\ldots,M_0$; $k=1,2,\ldots,N$):

$$\mathcal{E}_1(\omega_{0i}|x_{0k}) = -\ln \hat{p}(x_{0k}|\omega_{0i}), \qquad \mathcal{E}_2(\omega_{0i}|S_{0k}) = -\sum_{\omega_{0j}\in S_{0k}} \delta(\omega_{0i},\omega_{0j})$$

$$\mathcal{E}_3(\omega_{0i}|T_{0k}) = -\sum_{\omega_{1j}\in T_{0k}} \hat{P}(\omega_{0i}|\omega_{1j}) \tag{2.19}$$

where $\hat{p}(x_{0k}|\omega_{0i})$ is an estimate of the PDF $p(x_{0k}|\omega_{0i})$ of a feature vector x_{0k} ($k=1,2,\ldots,N$) conditioned to the class ω_{0i}, and $\hat{P}(\omega_{0i}|\omega_{1j})$ is an estimate of the transition probability between class ω_{1j} at time t_1 and class ω_{0i} at time t_0 ($i=1,2,\ldots,M_0$; $j=1,2,\ldots,M_1$) [7]. A similar 3-component energy function is also introduced at the second date; hence, three parameters have to be set at date t_0 (namely, $\lambda_{01},\lambda_{02}$, and λ_{03}) and three other parameters at date t_1 (namely, λ_{11}, λ_{12}, and λ_{13}).

In the present experiment, HK–ICM was applied to a two-date multi-temporal data set, consisting of two 870×498 pixel-sized images acquired over the same ground area as in Experiment I in October, 2000 and July, 2001, respectively. At both acquisition dates, eight Landsat-7 ETM+ bands were available, and a further ERS-2 channel (C-band, VV polarization) was also available in October, 2000 (Figure 2.2b). The same five thematic classes, as considered in Experiment I were considered in the October, 2000 scene, while the July, 2001 image also presented a further "agricultural" class. For all the classes, training and test data were available (Table 2.2).

Due to the multi-sensor nature of the October, 2000 image, a nonparametric technique was employed in the noncontextual step. Specifically, the dependence tree classifier (DTC) approach was adopted that approximates each multi-variate class-conditional PDF as a product of automatically selected bivariate PDFs [51]. As proposed in Ref. [7], the Parzen window method [30], applied with Gaussian kernels, was used to model such bivariate PDFs. In particular, to avoid the usual "trial-and-error" selection of the kernel-width parameters involved by the Parzen method, we adopted the simple "reference density" automatization procedure, which computes the kernel width by asymptotically minimizing the "mean integrated square error" functional, according to a given reference model for the unknown PDF (for further details on the automatic kernel-width selection for Parzen density estimation, we refer the reader to Refs. [52,53]). The "reference density" approach is chosen for its simplicity and short computation time; in particular, it is applied according to a Gaussian reference density for the kernel widths of the ETM+ features [53] and to a Rayleigh density for the kernel widths of the ERS-2 feature (for further details on this SAR-specific kernel-width selection approach, we refer the reader to Ref. [54]). The resulting PDF estimates were fed to an MAP classifier together with prior-probability estimates (computed as relative frequencies on the training set) to generate an initial noncontextual classification map for each date.

During the noncontextual step, transition-probability estimates were computed as relative frequencies on the two initial maps (i.e., $\hat{P}(\omega_{0i}|\omega_{1j})$ was computed as the ratio n_{ij}/m_j between the number n_{ij} of pixels assigned both to ω_{0i} in I_0 and to ω_{1j} in I_1 and the total number m_j of pixels assigned to ω_{1j} in I_1; $i=1,2,\ldots,M_0$; $j=1,2,\ldots,M_1$). The energy-difference and the Ho–Kashyap steps were applied first to the October,

2000 image, converging to (1.06,0.95,1.97) in 2632 iterations, and then to the July, 2001 image, converging to (1.00,1.00,1.01) in only 52 iterations. As an example, the convergent behavior of the parameter values during the Ho–Kashyap iterations for the October, 2000 image is shown in Figure 2.3.

The classification accuracies obtained by ICM, applied to the spatio-temporal model of Equation 2.19 with these parameter values, are shown in Table 2.2, together with the results of the noncontextual DTC algorithm. Also in this experiment, HK–ICM provided good classification accuracies at both dates, specifically yielding large accuracy increases for several classes as compared with the noncontextual DTC initialization (in particular, "urban" at both dates, "wet soil" in October, 2000, and "wood" in June, 2001).

Also for this experiment, we compared the results obtained by HK–ICM with the ones provided by the MRF–ICM classifier with a "trial-and-error" parameter setting (TE–ICM, Table 2.2). Fixing $\lambda_{01} = \lambda_{11} = 1$ (as in Experiment I) at both dates, a grid search was performed to identify the values of the other parameters that yielded the highest overall accuracies on the training set at the two dates. In general, this would require an interactive optimization of four distinct parameters (namely, λ_{02}, λ_{03}, λ_{12}, and λ_{13}). To reduce the time taken by this interactive procedure, the restrictions $\lambda_{02} = \lambda_{12}$ and $\lambda_{03} = \lambda_{13}$ were adopted, thus assigning the same weight λ_2 to the two spatial energy contributions and the same weight λ_3 to the two temporal contributions, and performing a grid search in a two-dimensional (and not four-dimensional) space. The adopted search range was [0,10] for both parameters with discretization step 0.2: for each combination of (λ_2, λ_3), ICM was run until convergence and the resulting classification accuracies on the training set were computed.

A global maximum of the average of the two training-set OAs obtained at the two dates was reached for $\lambda_2 = 9.8$ and $\lambda_3 = 2.6$: the corresponding accuracies (on both the training and test sets) are shown in Table 2.2. This optimal parameter vector turns out to be quite different from the solution automatically computed by the proposed method. The training-set accuracies of TE–ICM are better than the ones of HK–ICM; however, the difference between the OAs provided on the training set by the two approaches is only 0.44% in October, 2000 and 0.48% in June, 2001, and the differences between the AAs are only 0.17% and 0.47%, respectively. This suggests that, although identifying different parameter solutions, the proposed automatic approach and the standard interactive one allow achieving similar classification performances. A similar small difference of performance can also be noted between the corresponding test-set accuracies.

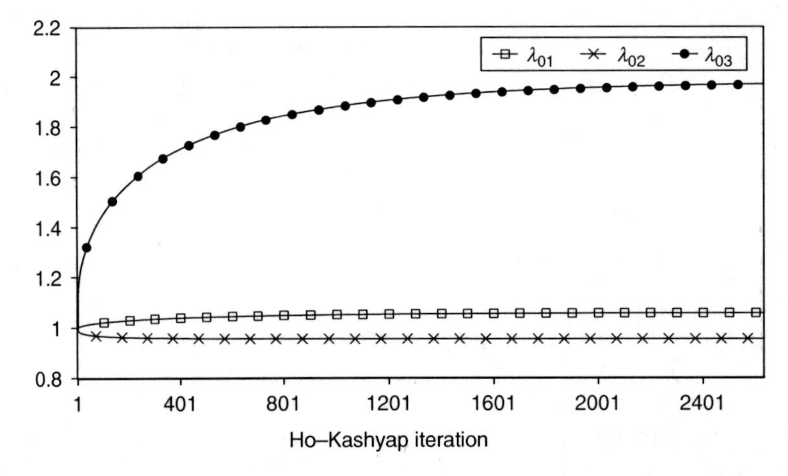

FIGURE 2.3
Experiment II: Plots of behaviors of the "spectral" parameter λ_{01}, the "spatial" parameter λ_{02}, and the "temporal" parameter λ_{03} versus the number of Ho–Kashyap iterations for the October, 2000, image.

2.4.3 Experiment III: Spatio-Temporal MRF Model for Multi-Temporal Classification of Image Sequences

As a third experiment, HK–ICM was applied to the multi-temporal mutual MRF model (see Section 2.2) extended to the problem of supervised classification of image sequences [7]. In particular, considering, a sequence $\{I_0, I_1, I_2\}$ of three co-registered images acquired over the same ground area at times t_0, t_1, t_2, respectively ($t_0 < t_1 < t_2$), the mutual model is generalized by introducing an energy function with four contributions at the intermediate date t_1, that is, a spectral and a spatial energy term (namely, $E_1(\cdot)$ and $E_2(\cdot)$) expressed as in Equation 2.19 and two distinct temporal energy terms, $E_3(\cdot)$ and $E_4(\cdot)$, related to the backward temporal correlation of I_1 with the previous image I_0 and to the forward temporal correlation of I_1 with I_2, respectively [7]. For the first and last images in the sequence (i.e., I_0 and I_2), only one typology of temporal energy is well defined (in particular, only the forward temporal energy $E_3(\cdot)$ for I_0 and only the backward energy $E_4(\cdot)$ for I_2). All the temporal energy terms are computed in terms of transition probabilities, as in Equation 2.19. Therefore, four parameters ($\lambda_{11}, \lambda_{12}, \lambda_{13}$, and λ_{14}) have to be set at date t_1, while only three parameters are needed at date t_0 (namely, λ_{01}, λ_{02}, and λ_{03}) or at date t_2 (namely, λ_{21}, λ_{22}, and λ_{24}).

Specifically, a sequence of three 700 × 280 pixel-sized co-registered multi-frequency SAR images, acquired over an agricultural area near the city of Pavia in April 16, 17, and 18, 1994, was used in the experiment (the ground area is not the same as the one considered in Experiments I and II, although the two regions are quite close to each other). At each date, a 4-look XSAR band (VV polarization) and three 4-look SIR-C-bands (HH, HV, and TP (total power)) were available. The observed scene at the three dates presented two main classes, that is, "wet soil" and "bare soil," and the temporal evolution of these classes between April 16 and April 18 was due to the artificial flooding processes caused by rice cultivation. The PDF estimates and the initial classification maps were computed using the DTC-Parzen method (as in Experiment II) applied here with a kernel-width optimized according again to the SAR-specific method developed in Ref. [54] and based on a Rayleigh reference density. As in Experiment II, the transition probabilities were estimated as relative frequencies on the initial noncontextual maps. Specifically, the Ho–Kashyap step was applied separately to set the parameters at each date, converging to (0.97,1.23,1.56) in 11510 iterations for April 16, to (1.12,1.03,0.99,1.28) in 15092 iterations for April 17, and to (1.00,1.04,0.97) in 2112 iterations for April 18.

As shown in Table 2.3, the noncontextual classification stage already provided good accuracies, but the use of the contextual information allowed a sharp increase in accuracy to be obtained at all the three dates (i.e., +6.59%, +4.82%, and +7.22% in OA for April 16, 17, and 18, respectively, and +4.77%, +3.09%, and +4.94% in AA, respectively), thus suggesting the effectiveness of the parameter values selected by the developed algorithm. In particular, as shown in Figure 2.4 with regard to the April 16 image, the noncontextual DTC maps (e.g., Figure 2.4a) were very noisy, due to the impact of the presence of speckle in the SAR data on the classification results, whereas the contextual HK–ICM maps (e.g., Figure 2.4b) were far less affected by speckle, because of the integration of contextual information in the classification process. It is worth noting that no preliminary despeckling procedure was applied to the input SIR-C/XSAR images before classifying them.

Also for this experiment, in Table 2.3, we present the results obtained by searching exhaustively for the MRF parameters yielding the best training-set accuracies. As in Experiment II, for the "trial-and-error" parameter optimization the following restrictions were adopted: $\lambda_{01} = \lambda_{11} = \lambda_{21} = 1$, $\lambda_{02} = \lambda_{12} = \lambda_{22} = \lambda_2$, and $\lambda_{03} = \lambda_{13} = \lambda_{14} = \lambda_{24} = \lambda_3$ (i.e., a unitary value was used for the "spectral" parameters, the same weight λ_2 was assigned to all the spatial energy contributions employed at the three dates, and the same weight λ_3 was associated with all the temporal energy terms). Therefore, the number of

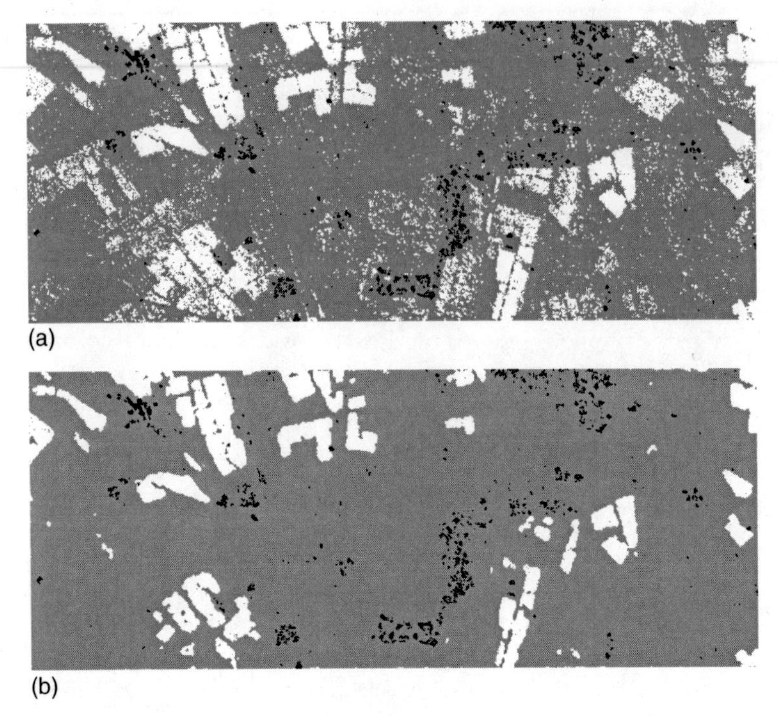

FIGURE 2.4
Experiment III: Classification maps for the April 16, image provided by: (a) the noncontextual DTC classifier; (b) the proposed contextual HK–ICM classifier. Color legend: white = "wet soil," grey = "dry soil," black = not classified (not classified pixels are present where the DTC-Parzen PDF estimates exhibit very low values for both classes).

independent parameters to be set interactively was reduced from 7 to 2. Again the search range [0,10] (discretized with step 0.2) was adopted for both parameters and, for each couple (λ_2, λ_3), ICM was run up to convergence. Then, the parameter vector yielding the best classification result was chosen. In particular, the average of the three OAs on the training set at the three dates exhibited a global maximum for $\lambda_2 = 1$ and $\lambda_3 = 0.4$. The corresponding accuracies are given in Table 2.3 and denoted by TE–ICM. As noted with regard to Experiment II, although the parameter vectors selected by the proposed automatic procedure and by the interactive "trial-and-error" one were quite different, the classification performances achieved by the two methods on the training set were very similar (and very close or equal to 100%). On the other hand, in most cases the test-set accuracies obtained by HK–ICM are even better than the ones given by TE–ICM. These results can be interpreted as a consequence of the fact that, to make the computation time for the exhaustive parameter search tractable, the above-mentioned restrictions on the parameter values were used when applying TE–ICM, whereas HK–ICM optimized all the MRF parameters, without any need for additional restrictions. The resulting TE–ICM parameters were effective for the classification of the training set (as they were specifically optimized toward this end), but they yielded less accurate results on the test set.

2.5 Conclusions

In the present chapter, an innovative algorithm has been proposed that addresses the problem of the automatization of the parameter-setting operations involved by MRF models for ICM-based supervised image classification. The method is applicable to a wide class of MRF models (i.e., the models whose energy functions can be expressed as weighted sums of distinct energy contributions), and expresses the parameter optimization as a linear discrimination problem, solved by using the Ho–Kashyap method. In particular, the theoretical properties known for this algorithm in the context of linear classifier design still holds also for the proposed method, thus ensuring a good convergent behavior.

The numerical results proved the capability of HK–ICM to select automatically parameter values that can yield accurate contextual classification maps. These results have been obtained for several different MRF models, dealing both with single-date supervised classification and with multi-temporal classification of two-date or multi-date imagery, which suggests a high flexibility of the method. This is further confirmed by the fact that good classification results were achieved on different typologies of remote sensing data (i.e., multi-spectral, optical-SAR multi-source, and multi-frequency SAR) and with different PDF estimation strategies (both parametric and nonparametric).

In particular, an interactive choice of the MRF parameters that selects through an exhaustive grid search the parameter values that provide the best performances on the training set yielded results very similar to the ones obtained by the proposed HK–ICM methodology in terms of classification accuracy. Specifically, in the first experiment, HK–ICM automatically identified parameter values quite close to the ones selected by this "trial-and-error" procedure (thus providing a very similar classification map), while in the other experiments, HK–ICM chose parameter values different from the ones obtained by the "trial-and-error" strategy, although generating classification results with similar (Experiment II) or even better (Experiment III) accuracies. The application of the "trial-and-error" approach in acceptable execution times required the definition of suitable restrictions on the parameter values to reduce the dimension of the parameter space to be explored. Such restrictions are not needed by the proposed method, which turns out to be a good compromise between accuracy and level of automatization, as it allows obtaining very good classification results, although avoiding completely the time-consuming phase of "trial-and-error" parameter setting.

In addition, in all the experiments, the adopted MRF models yielded a significant increase in accuracy, as compared with the initial noncontextual results. This further highlights the advantages of MRF models, which effectively exploit the contextual information in remote sensing image classification, and further confirms the usefulness of the proposed technique in automatically setting the parameters of such models. The proposed algorithm allows one to overcome the usual limitation on the use of MRF techniques in supervised image classification, consisting in the lack of automatization of such techniques, and supports a more extensive use of this powerful classification approach in remote sensing.

It is worth noting that HK–ICM optimizes the MRF parameter vector according to the initial classification maps provided as an input to ICM: the resulting optimal parameters are then employed to run ICM up to convergence. A further generalization of the method would adapt automatically the MRF parameters also during the ICM iterative process by applying the proposed Ho–Kashyap-based method at each ICM iteration or according to a different predefined iterative scheme. This could allow a further improvement in accuracy, although requiring one to run the Ho–Kashyap algorithm not once but several times, thus increasing the total computation time of the contextual classification process. The effect of this combined optimization strategy on the convergence of ICM is an issue worth being investigated.

Acknowledgments

This research was carried out within the framework of the PRIN-2002 project entitled "Processing and analysis of multitemporal and hypertemporal remote-sensing images for environmental monitoring," which was funded by the Italian Ministry of Education, University, and Research (MIUR). The support is gratefully acknowledged. The authors would also like to thank Dr. Paolo Gamba from the University of Pavia, Italy, for providing the SAR images employed in Experiments II and III.

References

1. M. Datcu, K. Seidel, and M. Walessa, Spatial information retrieval from remote sensing images: Part I. Information theoretical perspective, *IEEE Trans. Geosci. Rem. Sens.*, 36(5), 1431–1445, 1998.
2. R.C. Dubes and A.K. Jain, Random field models in image analysis, *J. Appl. Stat.*, 16, 131–163, 1989.
3. S. Geman and D. Geman, Stochastic relaxation, Gibbs distributions, and the Bayesian restoration, *IEEE Trans. Pattern Anal. Machine Intell.*, 6, 721–741, 1984.
4. A.H.S. Solberg, Flexible nonlinear contextual classification, *Pattern Recogn. Lett.*, 25(13), 1501–1508, 2004.
5. Q. Jackson and D. Landgrebe, Adaptive Bayesian contextual classification based on Markov random fields, *IEEE Trans. Geosci. Rem. Sens.*, 40(11), 2454–2463, 2002.
6. M.De Martino, G. Macchiavello, G. Moser., and S.B. Serpico, Partially supervised contextual classification of multitemporal remotely sensed images, *In Proc. of IEEE-IGARSS 2003, Toulouse*, 2, 1377–1379, 2003.
7. F. Melgani and S.B. Serpico, A Markov random field approach to spatio-temporal contextual image classification, *IEEE Trans. Geosci. Rem. Sens.*, 41(11), 2478–2487, 2003.
8. P.H. Swain, Bayesian classification in a time-varying environment, *IEEE Trans. Syst., Man, Cybern.*, 8(12), 880–883, 1978.
9. A.H.S. Solberg, T. Taxt, and A.K. Jain, A Markov Random Field model for classification of multisource satellite imagery, *IEEE Trans. Geosci. Rem. Sens.*, 34(1), 100–113, 1996.
10. G. Storvik, R. Fjortoft, and A.H.S. Solberg, A bayesian approach to classification of multiresolution remote sensing data, *IEEE Trans. Geosci. Rem. Sens.*, 43(3), 539–547, 2005.
11. M.L. Comer and E.J. Delp, The EM/MPM algorithm for segmentation of textured images: Analysis and further experimental results, *IEEE Trans. Image Process.*, 9(10), 1731–1744, 2000.
12. Y. Delignon, A. Marzouki, and W. Pieczynski, Estimation of generalized mixtures and its application to image segmentation, *IEEE Trans. Image Process.*, 6(10), 1364–1375, 2001.
13. X. Descombes, M. Sigelle, and F. Preteux, Estimating Gaussian Markov random field parameters in a nonstationary framework: Application to remote sensing imaging, *IEEE Trans. Image Process.*, 8(4), 490–503, 1999.
14. P. Smits and S. Dellepiane, Synthetic aperture radar image segmentation by a detail preserving Markov random field approach, *IEEE Trans. Geosci. Rem. Sens.*, 35(4), 844–857, 1997.
15. G.G. Hazel, Multivariate Gaussian MRF for multispectral scene segmentation and anomaly detection, *IEEE Trans. Geosci. Rem. Sens.*, 38(3), 1199–1211, 2000.
16. G. Rellier, X. Descombes, F. Falzon, and J. Zerubia, Texture feature analysis using a Gauss–Markov model in hyperspectral image classification, *IEEE Trans. Geosci. Rem. Sens.*, 42(7), 1543–1551, 2004.
17. L. Bruzzone and D.F. Prieto, Automatic analysis of the difference image for unsupervised change detection, *IEEE Trans. Geosci. Rem. Sens.*, 38(3), 1171–1182, 2000.
18. L. Bruzzone and D.F. Prieto, An adaptive semiparametric and context-based approach to unsupervised change detection in multitemporal remote-sensing images, *IEEE Trans. Image Process.*, 40(4), 452–466, 2002.

19. T. Kasetkasem and P.K. Varshney, An image change detection algorithm based on markov random field models, *IEEE Trans. Geosci. Rem. Sens.*, 40(8), 1815–1823, 2002.
20. J. Besag, On the statistical analysis of dirty pictures, *J. R. Statist. Soc.*, 68, 259–302, 1986.
21. Y. Cao, H. Sun, and X. Xu, An unsupervised segmentation method based on MPM for SAR images, *IEEE Geosci. Rem. Sens. Lett.*, 2(1), 55–58, 2005.
22. X. Descombes, R.D. Morris, J. Zerubia, and M. Berthod, Estimation of Markov Random Field prior parameters using Markov Chain Monte Carlo maximum likelihood, *IEEE Trans. Image Process.*, 8(7), 954–963, 1999.
23. M.V. Ibanez and A. Simo′, Parameter estimation in Markov random field image modeling with imperfect observations. A comparative study, *Pattern Rec. Lett.*, 24(14), 2377–2389, 2003.
24. J.T. Tou and R.C. Gonzalez, *Pattern Recognition Principles*, Addison-Wesley, MA, 1974.
25. Y.-C. Ho and R.L. Kashyap, An algorithm for linear inequalities and its applications, *IEEE Trans. Elec. Comp.*, 14, 683–688, 1965.
26. G. Celeux, F. Forbes, and N. Peyrand, EM procedures using mean field-like approximations for Markov model-based image segmentation, *Pattern Recogn.*, 36, 131–144, 2003.
27. Y. Zhang, M. Brady, and S. Smith, Segmentation of brain MR images through a hidden Markov random field model and the expectation-maximization algorithm, *IEEE Trans. Med. Imag.*, 20(1), 45–57, 2001.
28. R. Fiortoft, Y. Delignon, W. Pieczynski, M. Sigelle, and F. Tupin, Unsupervised classification of radar images using hidden Markov models and hidden Markov random fields, *IEEE Trans. Geosci. Rem. Sens.*, 41(3), 675–686, 2003.
29. Z. Kato, J. Zerubia, and M. Berthod, Unsupervised parallel image classification using Markovian models, *Pattern Rec.*, 32, 591–604, 1999.
30. K. Fukunaga, *Introduction to Statistical Pattern Recognition*, 2nd edition, Academic Press, New York, 1990.
31. F. Comets and B. Gidas, Parameter estimation for Gibbs distributions from partially observed data, *Ann. Appl. Prob.*, 2(1), 142–170, 1992.
32. W. Qian and D.N. Titterington, Estimation of parameters of hidden Markov models, *Phil. Trans.: Phys. Sci. and Eng.*, 337(1647), 407–428, 1991.
33. X. Descombes, A dense class of Markov random fields and associated parameter estimation, *J. Vis. Commun. Image R.*, 8(3), 299–316, 1997.
34. J. Besag, Spatial interaction and the statistical analysis of lattice systems, *J. R. Statist. Soc.*, 36, 192–236, 1974.
35. L. Bedini, A. Tonazzini, and S. Minutoli, Unsupervised edge-preserving image restoration via a saddle point approximation, *Image and Vision Computing*, 17, 779–793, 1999.
36. S.S. Saquib, C.A. Bouman, and K. Sauer, ML parameter estimation for Markov random fields with applications to Bayesian tomography, *IEEE Trans. Image Process.*, 7(7), 1029–1044, 1998.
37. C. Geyer and E. Thompson, Constrained Monte Carlo maximum likelihood for dependent data, *J. Roy. Statist. Soc. Ser. B*, 54, 657–699, 1992.
38. L. Younes, Estimation and annealing for Gibbsian fields, *Annales de l'Institut Henri Poincaré. Probabilités et Statistiques*, 24, 269–294, 1988.
39. B. Chalmond, An iterative Gibbsian technique for reconstruction of mary images, *Pattern Rec.*, 22(6), 747–761, 1989.
40. Y. Yu and Q. Cheng, MRF parameter estimation by an accelerated method, *Patt. Rec. Lett.*, 24, 1251–1259, 2003.
41. L. Wang, J. Liu, and S.Z. Li, MRF parameter estimation by MCMC method, *Pattern Rec.*, 33, 1919–1925, 2000.
42. L. Wang and J. Liu, Texture segmentation based on MRMRF modeling, *Patt. Rec. Lett.*, 21, 189–200, 2000.
43. H. Derin and H. Elliott, Modeling and segmentation of noisy and textured images using Gibbs random fields, *IEEE Trans. Pattern Anal. Machine Intell.*, 9(1), 39–55, 1987.
44. M. Mignotte, C. Collet, P. Perez, and P. Bouthemy, Sonar image segmentation using an unsupervised hierarchical MRF model, *IEEE Trans. Image Process.*, 9(7), 1216–1231, 2000.
45. B.C.K. Tso and P.M. Mather, Classification of multisource remote sensing imagery using a genetic algorithm and Markov Random Fields, *IEEE Trans. Geosci. Rem. Sens.*, 37(3), 1255–1260, 1999.

46. R.O. Duda, P.E. Hart, and D.G. Stork, *Pattern Classification*, 2nd edition, Wiley, New York, 2001.
47. M.W. Zemansky, *Heat and Thermodynamics*, 5th edition, McGraw-Hill, New York, 1968.
48. J. Stoer and R. Bulirsch, *Introduction to Numerical Analysis*, 2nd edition, Springer-Verlag, New York, 1992.
49. J. Richards and X. Jia, Remote sensing digital image analysis, 3rd edition, Springer-Verlag, Berlin, 1999.
50. D.A. Landgrebe, *Signal Theory Methods in Multispectral Remote Sensing*. Wiley-InterScience, New York, 2003.
51. M. Datcu, F. Melgani, A. Piardi, and S.B. Serpico, Multisource data classification with dependence trees, *IEEE Trans. Geosci. Rem. Sens.*, 40(3), 609–617, 2002.
52. A. Berlinet and L. Devroye, A comparison of kernel density estimates, *Publications de l'Institut de Statistique de l'Université de Paris*, 38(3), 3–59, 1994.
53. K.-D. Kim and J.-H. Heo, Comparative study of flood quantiles estimation by nonparametric models, *J. Hydrology*, 260, 176–193, 2002.
54. G. Moser, J. Zerubia, and S.B. Serpico, Dictionary-based stochastic expectation-maximization for SAR amplitude probability density function estimation, Research Report 5154, INRIA, Mar. 2004.

3

Random Forest Classification of Remote Sensing Data

Sveinn R. Joelsson, Jon Atli Benediktsson, and Johannes R. Sveinsson

CONTENTS

3.1 Introduction

Ensemble classification methods train several classifiers and combine their results through a voting process. Many ensemble classifiers [1,2] have been proposed. These classifiers include consensus theoretic classifiers [3] and committee machines [4]. Boosting and bagging are widely used ensemble methods. Bagging (or bootstrap aggregating) [5] is based on training many classifiers on bootstrapped samples from the training set and has been shown to reduce the variance of the classification. In contrast, boosting uses iterative re-training, where the incorrectly classified samples are given more weight in successive training iterations. This makes the algorithm slow (much slower than bagging) while in most cases it is considerably more accurate than bagging. Boosting generally reduces both the variance and the bias of the classification and has been shown to be a very accurate classification method. However, it has various drawbacks: it is computationally demanding, it can overtrain, and is also sensitive to noise [6]. Therefore, there is much interest in investigating methods such as random forests.

In this chapter, random forests are investigated in the classification of hyperspectral and multi-source remote sensing data. A random forest is a collection of classification trees or treelike classifiers. Each tree is trained on a bootstrapped sample of the training data, and at each node in each tree the algorithm only searches across a random subset of the features to determine a split. To classify an input vector in a random forest, the vector is submitted as an input to each of the trees in the forest. Each tree gives a classification, and it is said that the tree *votes* for that class. In the classification, the forest chooses the class having the most votes (over all the trees in the forest). Random forests have been shown to be comparable to boosting in terms of accuracies, but without the drawbacks of boosting. In addition, the random forests are computationally much less intensive than boosting.

Random forests have recently been investigated for classification of remote sensing data. Ham et al. [7] applied them in the classification of hyperspectral remote sensing data. Joelsson et al. [8] used random forests in the classification of hyperspectral data from urban areas and Gislason et al. [9] investigated random forests in the classification of multi-source remote sensing and geographic data. All studies report good accuracies, especially when computational demand is taken into account.

The chapter is organized as follows. Firstly random forest classifiers are discussed. Then, two different building blocks for random forests, that is, the classification and regression tree (CART) and the binary hierarchical classifier (BHC) approaches are reviewed. In Section 3.4, random forests with the two different building blocks are discussed. Experimental results for hyperspectral and multi-source data are given in Section 3.5. Finally, conclusions are given in Section 3.6.

3.2 The Random Forest Classifier

A random forest classifier is a classifier comprising a collection of treelike classifiers. Ideally, a random forest classifier is an *i.i.d.* randomization of weak learners [10]. The classifier uses a large number of individual decision trees, all of which are trained (grown) to tackle the same problem. A sample is decided to belong to the most frequently occurring of the classes as determined by the individual trees.

The individuality of the trees is maintained by three factors:

1. Each tree is trained using a random subset of the training samples.
2. During the growing process of a tree the best split on each node in the tree is found by searching through m randomly selected features. For a data set with M features, m is selected by the user and kept much smaller than M.
3. Every tree is grown to its fullest to diversify the trees so there is no pruning.

As described above, a random forest is an ensemble of treelike classifiers, each trained on a randomly chosen subset of the input data where final classification is based on a majority vote by the trees in the forest.

Each node of a tree in a random forest looks to a random subset of features of fixed size m when deciding a split during training. The trees can thus be viewed as random vectors of integers (features used to determine a split at each node). There are two points to note about the parameter m:

1. Increasing the correlation between the trees in the forest by increasing m, *increases* the error rate of the forest.
2. Increasing the classification accuracy of every individual tree by increasing m, *decreases* the error rate of the forest.

An optimal interval for m is between the somewhat fuzzy extremes discussed above. The parameter m is often said to be the only adjustable parameter to which the forest is sensitive and the "optimal" range for m is usually quite wide [10].

3.2.1 Derived Parameters for Random Forests

There are three parameters that are derived from the random forests. These parameters are the out-of-bag (OOB) error, the variable importance, and the proximity analysis.

3.2.1.1 Out-of-Bag Error

To estimate the test set accuracy, the out-of-bag samples (the remaining training set samples that are not in the bootstrap for a particular tree) of each tree can be run down through the tree (cross-validation). The OOB error estimate is derived by the classification error for the samples left out for each tree, averaged over the total number of trees. In other words, for all the trees where case n was OOB, run case n down the trees and note if it is correctly classified. The proportion of times the classification is in error, averaged over all the cases, is the OOB error estimate. Let us consider an example. Each tree is trained on a random 2/3 of the sample population (training set) while the remaining 1/3 is used to derive the OOB error rate for that tree. The OOB error rate is then averaged over all the OOB cases yielding the final or total OOB error. This error estimate has been shown to be unbiased in many tests [10,11].

3.2.1.2 Variable Importance

For a single tree, run it on its OOB cases and count the votes for the correct class. Then, repeat this again after randomly permuting the values of a single variable in the OOB cases. Now subtract the correctly cast votes for the randomly permuted data from the number of correctly cast votes for the original OOB data. The average of this value over all the forest is the raw importance score for the variable [5,6,11].

If the values of this score from tree to tree are independent, then the standard error can be computed by a standard computation [12]. The correlations of these scores between trees have been computed for a number of data sets and proved to be quite low [5,6,11]. Therefore, we compute standard errors in the classical way: divide the raw score by its standard error to get a z-score, and assign a significance level to the z-score assuming normality [5,6,11].

3.2.1.3 Proximities

After a tree is grown all the data are passed through it. If cases k and n are in the same terminal node, their proximity is increased by one. The proximity measure can be used (directly or indirectly) to visualize high dimensional data [5,6,11]. As the proximities are indicators on the "distance" to other samples this measure can be used to detect outliers in the sense that an outlier is "far" from all other samples.

3.3 The Building Blocks of Random Forests

Random forests are made up of several trees or building blocks. The building blocks considered here are CART, which partition the input data, and the BHC trees, which partition the labels (the output).

3.3.1 Classification and Regression Tree

CART is a decision tree where splits are made on a variable/feature/dimension resulting in the greatest change in impurity or minimum impurity given a split on a variable in the data set at a node in the tree [12]. The growing of a tree is maintained until either the change in impurity has stopped or is below some bound or the number of samples left to split is too small according to the user.

CART trees are easily overtrained, so a single tree is usually pruned to increase its generality. However, a collection of unpruned trees, where each tree is trained to its fullest on a subset of the training data to diversify individual trees can be very useful. When collected in a multi-classifier ensemble and trained using the random forest algorithm, these are called RF-CART.

3.3.2 Binary Hierarchy Classifier Trees

A binary hierarchy of classifiers, where each node is based on a split regarding labels and output instead of input as in the CART case, are naturally organized in trees and can as such be combined, under similar rules as the CART trees, to form RF-BHC. In a BHC, the best split on each node is based on (meta-) class separability starting with a single meta-class, which is split into two meta-classes and so on; the *true* classes are realized in the leaves. Simultaneously to the splitting process, the Fisher discriminant and the corresponding projection are computed, and the data are projected along the Fisher direction [12]. In "Fisher space," the projected data are used to estimate the likelihood of a sample belonging to a meta-class and from there the probabilities of a *true* class belonging to a meta-class are estimated and used to update the Fisher projection. Then, the data are projected using this updated projection and so forth until a user-supplied level of separation is acquired. This approach utilizes *natural* class affinities in the data, that is, the most *natural* splits occur early in the growth of the tree [13]. A drawback is the possible instability of the split algorithm. The Fisher projection involves an inverse of an estimate of the *within-class* covariance matrix, which can be unstable at some nodes of the tree, depending on the data being considered and so if this matrix estimate is singular (to numerical precision), the algorithm fails.

As mentioned above, the BHC trees can be combined to an RF-BHC where the best splits on classes are performed on a subset of the features in the data to diversify individual trees and stabilize the aforementioned inverse. Since the number of leaves in a BHC tree is the same as the number of classes in the data set the trees themselves can be very informative when compared to CART-like trees.

3.4 Different Implementations of Random Forests

3.4.1 Random Forest: Classification and Regression Tree

The RF-CART approach is based on CART-like trees where trees are grown to minimize an impurity measure. When trees are grown using a minimum Gini impurity criterion [12], the impurity of two descendent nodes in a tree is less than the parents. Adding up the decrease in the Gini value for each variable over all the forest gives a variable importance that is often very consistent with the permutation importance measure.

3.4.2 Random Forest: Binary Hierarchical Classifier

RF-BHC is a random forest based on an ensemble of BHC trees. In the RF-BHC, a split in the tree is based on the best separation between meta-classes. At each node the best separation is found by examining m features selected at random. The value of m can be selected by trials to yield optimal results. In the case where the number of samples is small enough to induce the "curse" of dimensionality, m is calculated by looking to a user-supplied ratio R between the number of samples and the number of features; then m is either used unchanged as the supplied value or a new value is calculated to preserve the ratio R, whichever is smaller at the node in question [7]. An RF-BHC is uniform regarding tree size (depth) because the number of nodes is a function of the number of classes in the dataset.

3.5 Experimental Results

Random forests have many important qualities of which many apply directly to multi- or hyperspectral data. It has been shown that the volume of a hypercube concentrates in the corners and the volume of a hyper ellipsoid concentrates in an outer shell, implying that with limited data points, much of the hyperspectral data space is empty [17]. Making a collection of trees is attractive, when each of the trees looks to minimize or maximize some information content related criteria given a *subset* of the features. This means that the random forest can arrive at a good decision boundary without deleting or extracting features explicitly while making the most out of the training set. This ability to handle thousands of input features is especially attractive when dealing with multi- or hyperspectral data, because more often than not it is composed of tens to hundreds of features and a limited number of samples. The unbiased nature of the OOB error rate can in some cases (if not all) eliminate the need for a validation dataset, which is another plus when working with a limited number of samples.

In experiments, the RF-CART approach was tested using a FORTRAN implementation of random forests supplied on a web page maintained by Leo Breiman and Adele Cutler [18].

3.5.1 Classification of a Multi-Source Data Set

In this experiment we use the Anderson River data set, which is a multi-source remote sensing and geographic data set made available by the Canada Centre for Remote Sensing

(CCRS) [16]. This data set is very difficult to classify due to a number of mixed forest type classes [15].

Classification was performed on a data set consisting of the following six data sources:

1. Airborne multispectral scanner (AMSS) with 11 spectral data channels (ten channels from 380 nm to 1100 nm and one channel from 8 μm to 14 μm)
2. Steep mode synthetic aperture radar (SAR) with four data channels (X-HH, X-HV, L-HH, and L-HV)
3. Shallow mode SAR with four data channels (X-HH, X-HV, L-HH, and L-HV)
4. Elevation data (one data channel, where elevation in meters pixel value)
5. Slope data (one data channel, where slope in degrees pixel value)
6. Aspect data (one data channel, where aspect in degrees pixel value)

There are 19 information classes in the ground reference map provided by CCRS. In the experiments, only the six largest ones were used, as listed in Table 3.1. Here, training samples were selected uniformly, giving 10% of the total sample size. All other known samples were then used as test samples [15].

The experimental results for random forest classification are given in Table 3.2 through Table 3.4. Table 3.2 shows line by line, how the parameters (number of split variables m and number of trees) are selected. First, a forest of 50 trees is grown for various number of split variables, then the number yielding the highest train accuracy (OOB) is selected, and then growing more trees until the overall accuracy stops increasing is tried. The overall accuracy (see Table 3.2) was seen to be insensitive to variable settings on the interval 10–22 split variables. Growing the forest larger than 200 trees improves the overall accuracy insignificantly, so a forest of 200 trees, each of which considers all the input variables at every node, yields the highest accuracy. The OOB accuracy in Table 3.2 seems to support the claim that overfitting is next to impossible using random forests in this manner. However the "best" results were obtained using 22 variables so there is no random selection of input variables at each node of every tree here because all variables are being considered on every split. This might suggest that a boosting algorithm using decision trees might yield higher overall accuracies.

The highest overall accuracies achieved with the Anderson River data set, known to the authors at the time of this writing, have been reached by boosting using j4.8 trees [17]. These accuracies were 100% training accuracy (vs. 77.5% here) and 80.6% accuracy for test data, which are not dramatically higher than the overall accuracies observed here (around 79.0%) with a random forest (about 1.6 percentage points difference). Therefore, even though m is not much less than the total number of variables (in fact equal), the

TABLE 3.1

Anderson River Data: Information Classes and Samples

Class No.	Class Description	Training Samples	Test Samples
1	Douglas fir (31–40 m)	971	1250
2	Douglas fir (21–40 m)	551	817
3	Douglas fir + Other species (31–40 m)	548	701
4	Douglas fir + Lodgepole pine (21–30 m)	542	705
5	Hemlock + Cedar (31–40 m)	317	405
6	Forest clearings	1260	1625
	Total	4189	5503

TABLE 3.2

Anderson River Data: Selecting *m* and the Number of Trees

Trees	Split Variables	Runtime (min:sec)	OOB acc. (%)	Test Set acc. (%)
50	1	00:19	68.42	71.58
50	5	00:20	74.00	75.74
50	10	00:22	75.89	77.63
50	15	00:22	76.30	78.50
50	20	00:24	76.01	78.14
50	22	00:24	76.63	78.10
100	22	00:38	77.18	78.56
200	22	01:06	77.51	79.01
400	22	02:06	77.56	78.81
1000	22	05:09	77.68	78.87
100	10	00:32	76.65	78.39
200	10	00:52	77.04	78.34
400	10	01:41	77.54	78.41
1000	10	04:02	77.66	78.25

22 split variables selected as the "best" choice

random forest ensemble performs rather well, especially when running times are taken into consideration. Here, in the random forest, each tree is an expert on a subset of the data but all the experts look to the same number of variables and do not, in the strictest sense, utilize the strength of random forests. However, the fact remains that the results are among the best ones for this data set.

The training and test accuracies for the individual classes using random forests with 200 trees and 22 variables at each node are given in Table 3.3 and Table 3.4, respectively. From these tables, it can be seen that the random forest yields the highest accuracies for classes 5 and 6 but the lowest for class 2, which is in accordance with the outlier analysis below.

A variable importance estimate for the training data can be seen in Figure 3.1, where each data channel is represented by one variable. The first 11 variables are multi-spectral data, followed by four steep-mode SAR data channels, four shallow-mode synthetic aperture radar, and then elevation, slope, and aspect measurements, one channel each. It is interesting to note that variable 20 (elevation) is the most important variable, followed by variable 22 (aspect), and spectral channel 6 when looking at the raw importance (Figure 3.1a), but slope when looking at the *z*-score (Figure 3.1b). The variable importance for each individual class can be seen in Figure 3.2. Some interesting conclusions can be drawn from Figure 3.2. For example, with the exception of class 6, topographic data

TABLE 3.3

Anderson River Data: Confusion Matrix for Training Data in Random Forest Classification (Using 200 Trees and Testing 22 Variables at Each Node)

Class No.	1	2	3	4	5	6	%
1	764	126	20	35	1	57	78.68
2	75	289	38	8	1	43	52.45
3	32	62	430	21	0	51	78.47
4	11	3	11	423	42	25	78.04
5	8	2	9	39	271	14	85.49
6	81	69	40	16	2	1070	84.92

TABLE 3.4

Anderson River Data: Confusion Matrix for Test Data in Random Forest Classification (Using 200 Trees and Testing 22 Variables at Each Node)

Class No.	1	2	3	4	5	6	%
1	1006	146	29	44	2	60	80.48
2	87	439	40	23	0	55	53.73
3	26	67	564	18	3	51	80.46
4	19	65	12	565	44	22	80.14
5	7	6	7	45	351	14	86.67
6	105	94	49	10	5	1423	87.57

(channels 20–22) are of high importance and then come the spectral channels (channels 1–11). In Figure 3.2, we can see that the SAR channels (channels 12–19) seem to be almost irrelevant to class 5, but seem to play a more important role for the other classes. They always come third after the topographic and multi-spectral variables, with the exception of class 6, which seems to be the only class where this is not true; that is, the topographic variables score lower than an SAR channel (Shallow-mode SAR channel number 17 or X-HV).

These findings can then be verified by classifying the data set according to only the most important variables and compared to the accuracy when all the variables are

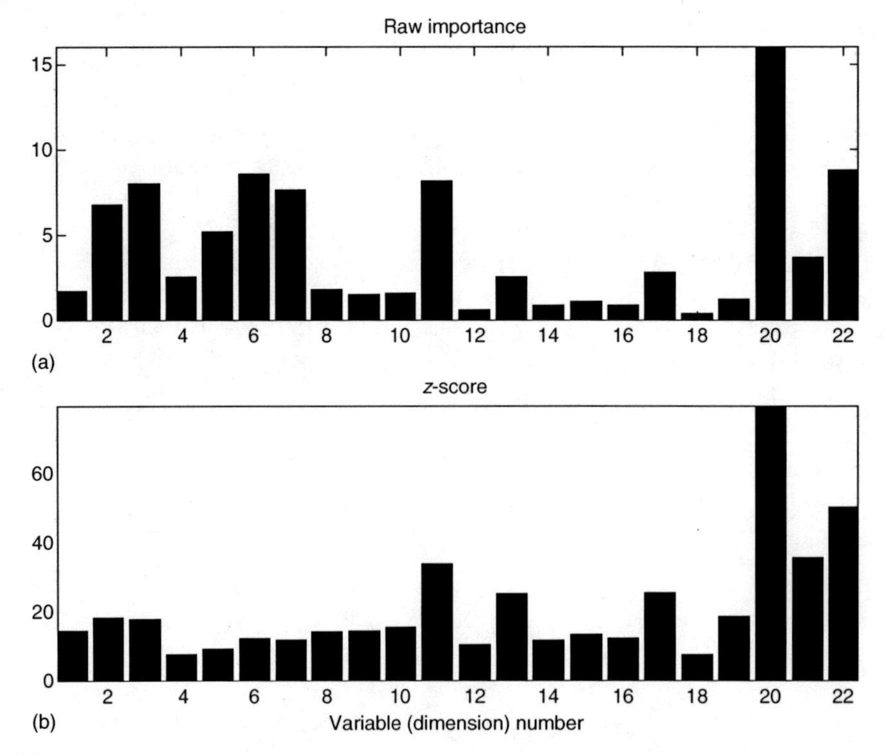

FIGURE 3.1
Anderson river training data: (a) variable importance and (b) z-score on raw importance.

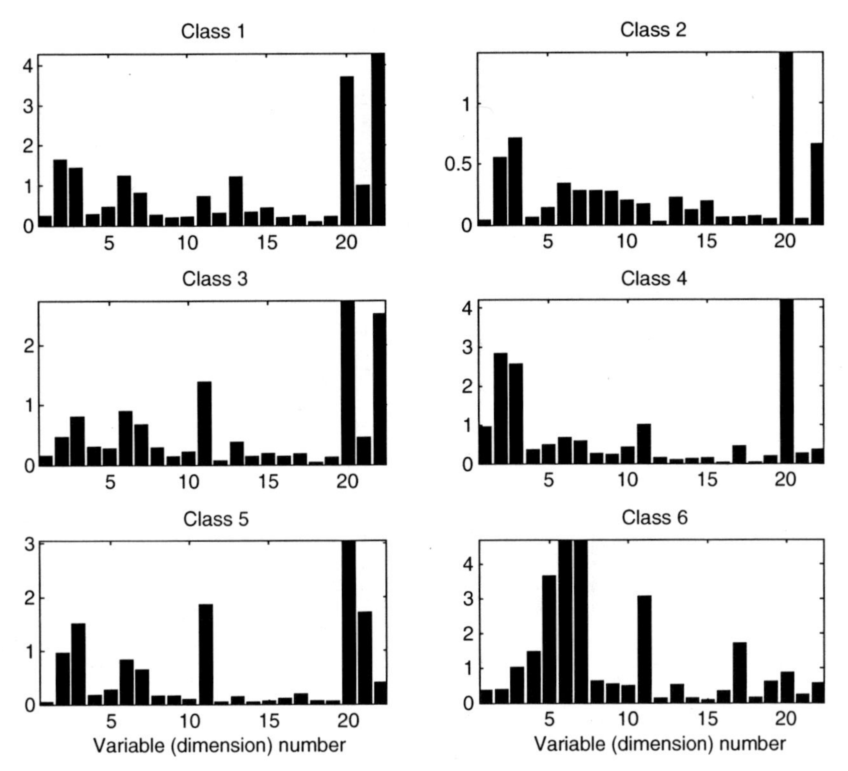

FIGURE 3.2
Anderson river training data: variable importance for each of the six classes.

included. For example leaving out variable 20 should have less effect on classification accuracy in class 6 than on all the other classes.

A proximity matrix was computed for the training data to detect outliers. The results of this outlier analysis are shown in Figure 3.3, where it can be seen that the data set is difficult for classification as there are several outliers. From Figure 3.3, the outliers are spread over all classes—with a varying degree. The classes with the least amount of outliers (classes 5 and 6) are indeed those with the highest classification accuracy (Table 3.3 and Table 3.4). On the other hand, class 2 has the lowest accuracy and the highest number of outliers.

In the experiments, the random forest classifier proved to be fast. Using an Intel® Celeron® CPU 2.20-GHz desktop, it took about a minute to read the data set into memory, train, and classify the data set, with the settings of 200 trees and 22 split variables when the FORTRAN code supplied on the random forest web site was used [18]. The running times seem to indicate a linear time increase when considering the number of trees. They are seen along with a least squares fit to a line in Figure 3.4.

3.5.1.1 The Anderson River Data Set Examined with a Single CART Tree

We look to all of the 22 features when deciding a split in the RF-CART approach above, so it is of interest here to examine if the RF-CART performs any better than a single CART tree. Unlike the RF-CART, a single CART is easily overtrained. Here we prune the CART tree to reduce or eliminate any overtraining features of the tree and hence use three data sets, a training set, testing set (used to decide the level of pruning), and a validation set to estimate the performance of the tree as a classifier (Table 3.5 and Table 3.6).

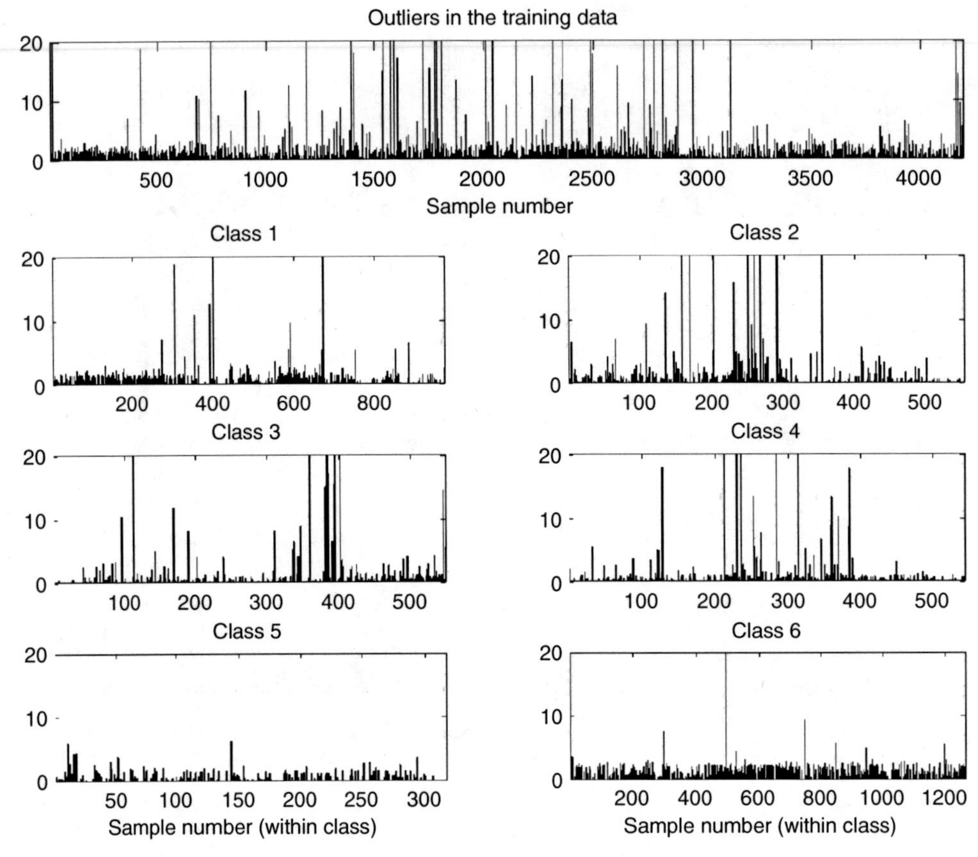

FIGURE 3.3
Anderson River training data: outlier analysis for individual classes. In each case, the x-axis (index) gives the number of a training sample and the y-axis the outlier measure.

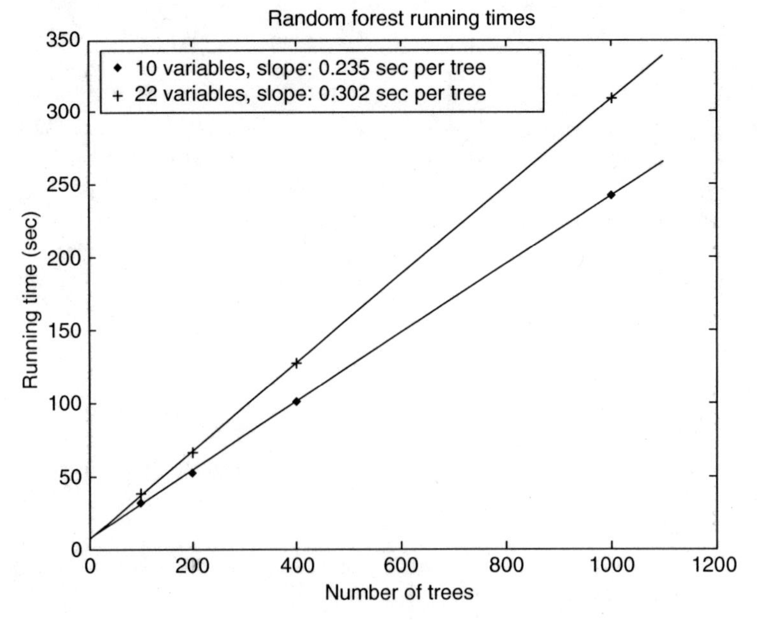

FIGURE 3.4
Anderson river data set: random forest running times for 10 and 22 split variables.

TABLE 3.5

Anderson River Data Set: Training, Test, and Validation Sets

Class No.	Class Description	Training Samples	Test Samples	Validation Samples
1	Douglas fir (31–40 m)	971	250	1000
2	Douglas fir (21–40 m)	551	163	654
3	Douglas fir + Other species (31–40 m)	548	140	561
4	Douglas fir + Lodgepole pine (21–30 m)	542	141	564
5	Hemlock + Cedar (31–40 m)	317	81	324
6	Forest clearings	1260	325	1300
	Total samples	4189	1100	4403

As can be seen in Table 3.6 and from the results of the RF-CART runs above (Table 3.2), the overall accuracy is about 8 percentage points higher $((78.8/70.8-1)*100 = 11.3\%)$ than the overall accuracy for the validation set in Table 3.6. Therefore, a boosting effect is present even though we need all the variables to determine a split in every tree in the RF-CART.

3.5.1.2 The Anderson River Data Set Examined with the BHC Approach

The same procedure was used to select the variable m when using the RF-BHC as in the RF-CART case. However, for the RF-BHC, the separability of the data set is an issue. When the number of randomly selected features was less than 11, it was seen that a singular matrix was likely for the Anderson River data set. The best overall performance regarding the realized classification accuracy turned out to be the same as for the RF-CART approach or for $m = 22$. The R parameter was set to 5, but given the number of samples per (meta-)class in this data set, the parameter is not necessary. This means 22 is always at least 5 times smaller than the number of samples in a (meta-)class during the growing of the trees in the RF-BHC. Since all the trees were trained using all the available features, the trees are more or less the same, the only difference is that the trees are trained on different subsets of the samples and thus the RF-BHC gives a very similar result as a single BHC. It can be argued that the RF-BHC is a more general classifier due to the nature of the error or accuracy estimates used during training, but as can be seen in Table 3.7 and Table 3.8 the differences are small, at least for this data set and no boosting effect seems to be present when using the RF-BHC approach when compared to a single BHC.

TABLE 3.6

Anderson River Data Set: Classification Accuracy (%) for Training, Test, and Validation Sets

Class Description	Training	Test	Validation
Douglas fir (31–40 m)	87.54	73.20	71.30
Douglas fir (21–40 m)	77.50	47.24	46.79
Douglas fir + Other species (31–40 m)	87.96	70.00	72.01
Douglas fir + Lodgepole pine (21–30 m)	84.69	68.79	69.15
Hemlock + Cedar (31–40 m)	90.54	79.01	77.78
Forest clearings	90.08	81.23	81.00
Overall accuracy	86.89	71.18	70.82

TABLE 3.7

Anderson River Data Set: Classification Accuracies in Percentage for a Single BHC Tree Classifier

Class Description	Training	Test
Douglas fir (31–40 m)	50.57	50.40
Douglas fir (21–40 m)	47.91	43.57
Douglas fir + Other species (31–40 m)	58.94	59.49
Douglas fir + Lodgepole pine (21–30 m)	72.32	67.23
Hemlock + Cedar (31–40 m)	77.60	73.58
Forest clearings	71.75	72.80
Overall accuracy	62.54	61.02

3.5.2 Experiments with Hyperspectral Data

The data used in this experiment were collected in the framework of the HySens project, managed by Deutschen Zentrum fur Luft-und Raumfahrt (DLR) (the German Aerospace Center) and sponsored by the European Union. The optical sensor reflective optics system imaging spectrometer (ROSIS 03) was used to record four flight lines over the urban area of Pavia, northern Italy. The number of bands of the ROSIS 03 sensor used in the experiments is 103, with spectral coverage from 0.43 μm through 0.86 μm. The flight altitude was chosen as the lowest available for the airplane, which resulted in a spatial resolution of 1.3 m per pixel.

The ROSIS data consist of nine classes (Table 3.9): The data were composed of 43923 samples, split up into 3921 training samples and 40002 for testing. Pseudo color image of the area along with the ground truth mask (training and testing samples) are shown in Figure 3.5.

This data set was classified using a BHC tree, an RF-BHC, a single CART, and an RF-CART.

The BCH, RF-BHC, CART, and RF-CART were applied on the ROSIS data. The forest parameters, m and R (for RF-BHC), were chosen by trials to maximize accuracies. The growing of trees was stopped when the overall accuracy did not improve using additional trees. This is the same procedure as for the Anderson River data set (see Table 3.2). For the RF-BHC, R was chosen to be 5, m chosen to be 25 and the forest was grown to only 10 trees. For the RF-CART, m was set to 25 and the forest was grown to 200 trees. No feature extraction was done at individual nodes in the tree when using the single BHC approach.

TABLE 3.8

Anderson River Data Set: Classification Accuracies in Percentage for an RF-BHC, $R = 5$, $m = 22$, and 10 Trees

Class Description	Training	Test
Douglas fir (31–40 m)	51.29	51.12
Douglas fir (21–40 m)	45.37	41.13
Douglas fir + Other species (31–40 m)	59.31	57.20
Douglas fir + Lodgepole pine (21–30 m)	72.14	67.80
Hemlock + Cedar (31–40 m)	77.92	71.85
Forest clearings	71.75	72.43
Overall accuracy	62.43	60.37

TABLE 3.9

ROSIS University Data Set: Classes and Number of Samples

Class No.	Class Description	Training Samples	Test Samples
1	Asphalt	548	6,304
2	Meadows	540	18,146
3	Gravel	392	1,815
4	Trees	524	2,912
5	(Painted) metal sheets	265	1,113
6	Bare soil	532	4,572
7	Bitumen	375	981
8	Self-blocking bricks	514	3,364
9	Shadow	231	795
	Total samples	3,921	40,002

Classification accuracies are presented in Table 3.11 through Table 3.14. As in the single CART case for the Anderson River data set, approximately 20% of the samples in the original test set were randomly sampled into a new test set to select a pruning level for the tree, leaving 80% of the original test samples for validation as seen in Table 3.10. All the other classification methods used the training and test sets as described in Table 3.9.

From Table 3.11 through Table 3.14 we can see that the RF-BHC give the highest overall accuracies of the tree methods where the single BHC, single CART, and the RF-CART methods yielded lower and comparable overall accuracies. These results show that using many weak learners as opposed to a few stronger ones is not always the best choice in classification and is dependent on the data set. In our experience the RF-BHC approach is as accurate or more accurate than the RF-CART when the data set consists of moderately

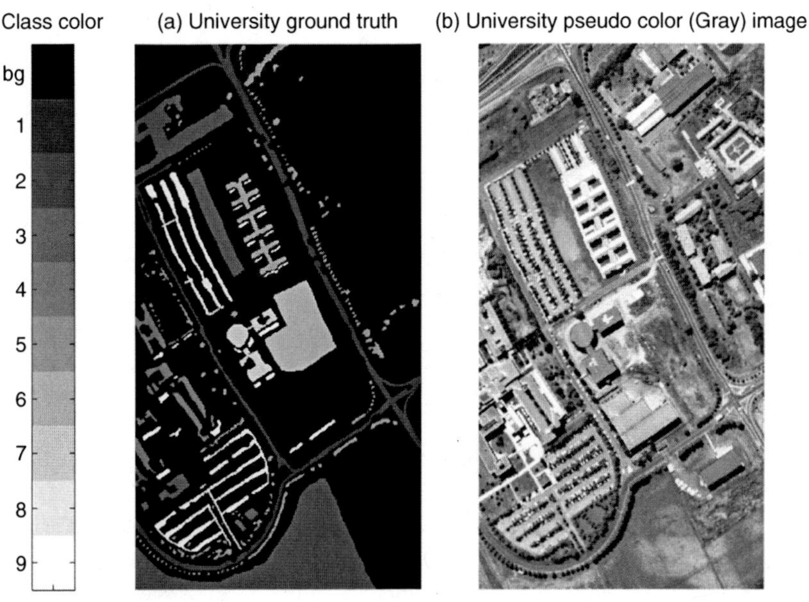

FIGURE 3.5
ROSIS University: (a) reference data and (b) gray scale image.

TABLE 3.10

ROSIS University Data Set: Train, Test, and Validation Sets

Class No.	Class Description	Training Samples	Test Samples	Validation Samples
1	Asphalt	548	1,261	5,043
2	Meadows	540	3,629	14,517
3	Gravel	392	363	1,452
4	Trees	524	582	2,330
5	(Painted) metal sheets	265	223	890
6	Bare soil	532	914	3,658
7	Bitumen	375	196	785
8	Self-blocking bricks	514	673	2,691
9	Shadow	231	159	636
	Total samples	3,921	8,000	32,002

to highly separable (meta-)classes, but for difficult data sets the partitioning algorithm used for the BHC trees can fail to converge (inverse of the *within-class* covariance matrix becomes singular to numerical precision) and thus no BHC classifier can be realized. This is not a problem when using the CART trees as the building blocks partition the input and simply minimize an impurity measure given a split on a node. The classification results for the single CART tree (Table 3.11), especially for the two classes *gravel* and *bare-soil*, may be considered unacceptable when compared to the other methods that seem to yield more balanced accuracies for all classes. The classified images for the results given in Table 3.12 through Table 3.14 are shown in Figure 3.6a through Figure 3.6d.

Since BHC trees are of fixed size regarding the number of leafs it is worth examining the tree in the single case (Figure 3.7).

Notice the siblings on the tree (nodes sharing a parent): *gravel* (3)/*shadow* (9), *asphalt* (1)/*bitumen* (7), and finally *meadows* (2)/*bare soil* (6). Without too much stretch of the imagination, one can intuitively decide that these classes are related, at least *asphalt/bitumen* and *meadows/bare soil*. When comparing the *gravel* area in the ground truth image (Figure 3.5a) and the same area in the gray scale image (Figure 3.5b), one can see it has gray levels ranging from bright to relatively dark, which might be interpreted as an intuitive relation or overlap between the *gravel* (3) and the *shadow* (9) classes. The *self-blocking bricks* (8) are the class closest to the *asphalt-bitumen* meta-class, which again looks very similar in the pseudo color image. So the tree more or less seems to place ''naturally''

TABLE 3.11

Single CART: Training, Test, and Validation Accuracies in Percentage for ROSIS University Data Set

Class Description	Training	Test	Validation
Asphalt	80.11	70.74	72.24
Meadows	83.52	75.48	75.80
Gravel	0.00	0.00	0.00
Trees	88.36	97.08	97.00
(Painted) metal sheets	97.36	91.03	86.07
Bare soil	46.99	24.73	26.60
Bitumen	85.07	82.14	80.38
Self-blocking bricks	84.63	92.42	92.27
Shadow	96.10	100.00	99.84
Overall accuracy	72.35	69.59	69.98

TABLE 3.12

BHC: Training and Test Accuracies in Percentage for ROSIS University Data Set

Class	Training	Test
Asphalt	78.83	69.86
Meadows	93.33	55.11
Gravel	72.45	62.92
Trees	91.60	92.20
(Painted) metal sheets	97.74	94.79
Bare soil	94.92	89.63
Bitumen	93.07	81.55
Self-blocking bricks	85.60	88.64
Shadow	94.37	96.35
Overall accuracy	88.52	69.83

TABLE 3.13

RF-BHC: Training and Test Accuracies in Percentage for ROSIS University Data Set

Class	Training	Test
Asphalt	76.82	71.41
Meadows	84.26	68.17
Gravel	59.95	51.35
Trees	88.36	95.91
(Painted) metal sheets	100.00	99.28
Bare soil	75.38	78.85
Bitumen	92.53	87.36
Self-blocking bricks	83.07	92.45
Shadow	96.10	99.50
Overall accuracy	82.53	75.16

TABLE 3.14

RF-CART: Train and Test Accuracies in Percentage for ROSIS University Data Set

Class	Training	Test
Asphalt	86.86	80.36
Meadows	90.93	54.32
Gravel	76.79	46.61
Trees	92.37	98.73
(Painted) metal sheets	99.25	99.01
Bare soil	91.17	77.60
Bitumen	88.80	78.29
Self-blocking bricks	83.46	90.64
Shadow	94.37	97.23
Overall accuracy	88.75	69.70

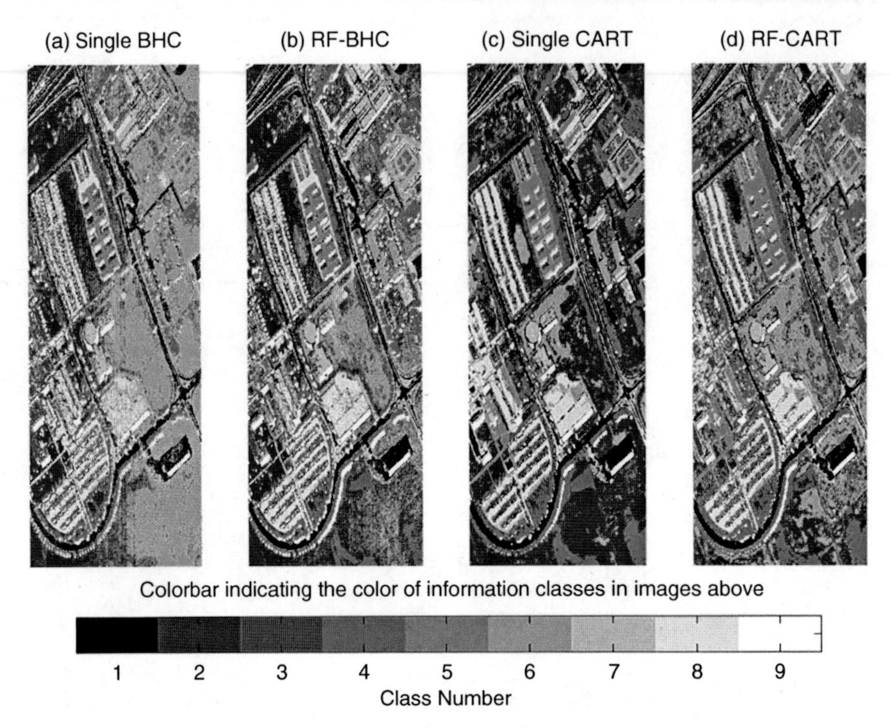

FIGURE 3.6
ROSIS University: image classified by (a) single BHC, (b) RF-BHC, (c) single CART, and (d) RF-CART.

related classes close to one another in the tree. That would mean that classes 2, 6, 4, and 5 are more related to each other than to classes 3, 9, 1, 7, or 8. On the other hand, it is not clear if *(painted) metal sheets* (5) are ''naturally'' more related to *trees* (4) than to *bare soil* (6) or *asphalt* (1). However, the point is that the partition algorithm finds the ''clearest'' separation between meta-classes. Therefore, it may be better to view the tree as a separation hierarchy rather than a relation hierarchy. The single BHC classifier finds that class 5 is the most separable class within the first right meta-class of the tree, so it might not be related to meta-class 2–6–4 in any ''natural'' way, but it is more separable along with these classes when the whole data set is split up to two meta-classes.

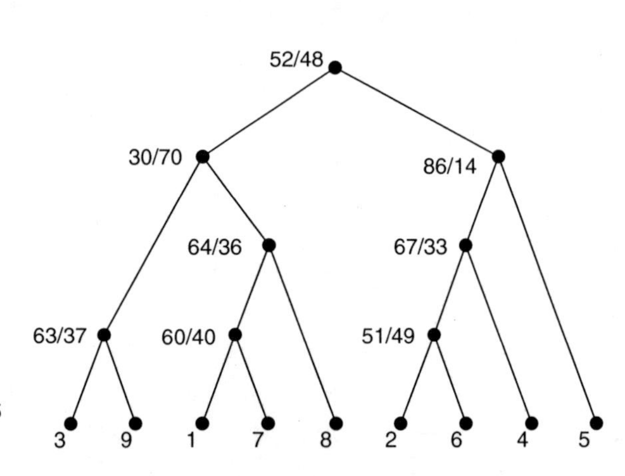

FIGURE 3.7
The BHC tree used for classification of Figure 3.5 with *left/right* probabilities (%).

3.6 Conclusions

The use of random forests for classification of multi-source remote sensing data and hyperspectral remote sensing data has been discussed. Random forests should be considered attractive for classification of both data types. They are both fast in training and classification, and are distribution-free classifiers. Furthermore, the problem with the curse of dimensionality is naturally addressed by the selection of a low m, without having to discard variables and dimensions completely. The only parameter random forests are truly sensitive to is the number of variables m, the nodes in every tree draw at random during training. This parameter should generally be much smaller than the total number of available variables, although selecting a high m can yield good classification accuracies, as can be seen above for the Anderson River data (Table 3.2).

In experiments, two types of random forests were used, that is, random forests based on the CART approach and random forests that use BHC trees. Both approaches performed well in experiments. They gave excellent accuracies for both data types and were shown to be very fast.

Acknowledgment

This research was supported in part by the Research Fund of the University of Iceland and the Assistantship Fund of the University of Iceland. The Anderson River SAR/MSS data set was acquired, preprocessed, and loaned by the Canada Centre for Remote Sensing, Department of Energy Mines and Resources, Government of Canada.

References

1. L.K. Hansen and P. Salamon, Neural network ensembles, *IEEE Transactions on Pattern Analysis and Machine Intelligence*, 12, 993–1001, 1990.
2. L.I. Kuncheva, Fuzzy versus nonfuzzy in combining classifiers designed by Boosting, *IEEE Transactions on Fuzzy Systems*, 11, 1214–1219, 2003.
3. J.A. Benediktsson and P.H. Swain, Consensus Theoretic Classification Methods, *IEEE Transactions on Systems, Man and Cybernetics*, 22(4), 688–704, 1992.
4. S. Haykin, *Neural Networks, A Comprehensive Foundation*, 2nd ed., Prentice-Hall, Upper Saddle River, NJ, 1999.
5. L. Breiman, Bagging predictors, *Machine Learning*, 24I(2), 123–140, 1996.
6. Y. Freund and R.E. Schapire: Experiments with a new boosting algorithm, *Machine Learning: Proceedings of the Thirteenth International Conference*, 148–156, 1996.
7. J. Ham, Y. Chen, M.M. Crawford, and J. Ghosh, Investigation of the random forest framework for classification of hyperspectral data, *IEEE Transactions on Geoscience and Remote Sensing*, 43(3), 492–501, 2005.
8. S.R. Joelsson, J.A. Benediktsson, and J.R. Sveinsson, Random forest classifiers for hyperspectral data, *IEEE International Geoscience and Remote Sensing Symposium (IGARSS '05)*, Seoul, Korea, 25–29 July 2005, pp. 160–163.
9. P.O. Gislason, J.A. Benediktsson, and J.R. Sveinsson, Random forests for land cover classification, *Pattern Recognition Letters*, 294–300, 2006.

10. L. Breiman, Random forests, *Machine Learning*, 45(1), 5–32, 2001.
11. L. Breiman, Random forest, Readme file. Available at: http://www.stat.berkeley.edu/~ briman/RandomForests/cc.home.htm Last accessed, 29 May, 2006.
12. R.O. Duda, P.E. Hart, and D.G. Stork, *Pattern Classification*, 2nd ed., John Wiley & Sons, New York, 2001.
13. S. Kumar, J. Ghosh, and M.M. Crawford, Hierarchical fusion of multiple classifiers for hyperspectral data analysis, *Pattern Analysis & Applications*, 5, 210–220, 2002.
14. http://oz.berkeley.edu/users/breiman/RandomForests/cc_home.htm (Last accessed, 29 May, 2006.)
15. G.J. Briem, J.A. Benediktsson, and J.R. Sveinsson, Multiple classifiers applied to multisource remote sensing data, *IEEE Transactions on Geoscience and Remote Sensing*, 40(10), 2291–2299, 2002.
16. D.G. Goodenough, M. Goldberg, G. Plunkett, and J. Zelek, The CCRS SAR/MSS Anderson River data set, *IEEE Transactions on Geoscience and Remote Sensing*, GE-25(3), 360–367, 1987.
17. L. Jimenez and D. Landgrebe, Supervised classification in high-dimensional space: Geometrical, statistical, and asymptotical properties of multivariate data, *IEEE Transactions on Systems, Man, and Cybernetics, Part. C*, 28, 39–54, 1998.
18. http://www.stat.berkeley.edu/users/breiman/RandomForests/cc_software.htm (Last accessed, 29 May, 2006.)

4

Supervised Image Classification of Multi-Spectral Images Based on Statistical Machine Learning

Ryuei Nishii and Shinto Eguchi

CONTENTS

4.1 Introduction

Image classification for geostatistical data is one of the most important issues in the remote-sensing community. Statistical approaches have been discussed extensively in the literature. In particular, Markov random fields (MRFs) are used for modeling distributions of land-cover classes, and contextual classifiers based on MRFs exhibit efficient performances. In addition, various classification methods were proposed. See Ref. [3] for an excellent review paper on classification. See also Refs. [1,4–7] for a general discussion on classification methods, and Refs. [8,9] for backgrounds on spatial statistics.

In a paradigm of supervised learning, AdaBoost was proposed as a machine learning technique in Ref. [10] and has been widely and rapidly improved for use in pattern recognition. AdaBoost linearly combines several weak classifiers into a strong classifier. The coefficients of the classifiers are tuned by minimizing an empirical exponential risk. The classification method exhibits high performance in various fields [11,12]. In addition, fusion techniques have been discussed [13–15].

In the present chapter, we consider contextual classification methods based on statistics and machine learning. We review AdaBoost with binary class labels as well as multi-class labels. The procedures for deriving coefficients for classifiers are discussed, and robustness for loss functions is emphasized here. Next, contextual image classification methods including Switzer's smoothing method [1], MRF-based methods [16], and spatial boosting [2,17] are introduced. Relationships among them are also pointed out. Spatial parallel boost by meta-learning for multi-source and multi-temporal data classification is proposed.

The remainder of the chapter is organized as follows. In Section 4.2, AdaBoost is briefly reviewed. A simple example with binary class labels is provided to illustrate AdaBoost. Then, we proceed to the case with multi-class labels. Section 4.3 gives general boosting methods to obtain the robustness property of the classifier. Then, contextual classifiers including Switzer's method, an MRF-based method, and spatial boosting are discussed. Relationships among them are shown in Section 4.5. The exact error rate and the properties of the MRF-based classifier are given. Section 4.6 proposes spatial parallel boost applicable to classification of multi-source and multi-temporal data sets. The methods treated here are applied to a synthetic data set and two benchmark data sets, and the performances are examined in Section 4.7. Section 4.8 concludes the chapter and mentions future problems.

4.2 AdaBoost

We begin this section with a simple example to illustrate AdaBoost [10]. Later, AdaBoost with multi-class labels is mentioned.

4.2.1 Toy Example in Binary Classification

Suppose that a q-dimensional feature vector $\mathbf{x} \in R^q$ observed by a supervised example labeled by $+1$ or -1 is available. Furthermore, let $g_k(\mathbf{x})$ be functions (*classifiers*) of the feature vector $\mathbf{x}$ into label set $\{+1, -1\}$ for $k = 1, 2, 3$. If these three classifiers are equally efficient, a new function, $\text{sign}(f_1(\mathbf{x}) + f_2(\mathbf{x}) + f_3(\mathbf{x}))$, is a combined classifier based on a majority vote, where sign(z) is the sign of the argument z. Suppose that classifier f_1 is the most reliable, f_2 has the next greatest reliability, and f_3 is the least reliable. Then, a new function sign $(\beta_1 f_1(\mathbf{x}) + \beta_2 f_2(\mathbf{x}) + \beta_3 f_3(\mathbf{x}))$ is a boosted classifier based on a weighted vote, where $\beta_1 > \beta_2 > \beta_3$ are positive constants to be determined according to efficiencies of the classifiers. Constants β_k are tuned by minimizing the empirical risk, which will be defined shortly.

In general, let y be the true label of feature vector $\mathbf{x}$. Then, label y is estimated by a signature, sign$(F(\mathbf{x}))$, of a classification function $F(\mathbf{x})$. Actually, if $F(\mathbf{x}) > 0$, then $\mathbf{x}$ is classified into the class with label 1, otherwise into -1. Hence, if $yF(\mathbf{x}) < 0$, vector $\mathbf{x}$ is misclassified. For evaluating classifier F, AdaBoost in Ref. [10] takes the exponential loss function defined by

$$L_{\exp}(F \mid \mathbf{x}, y) = \exp\{-yF(\mathbf{x})\} \tag{4.1}$$

The loss function $L_{\exp}(t) = \exp(-t)$ vs. $t = yF(\mathbf{x})$ is given in Figure 4.1. Note that the exponential function assigns a heavy loss to an outlying example that is misclassified. AdaBoost is apt to overlearn misclassified examples.

Let $\{(x_i, y_i) \in R^q \times \{+1, -1\} \mid i = 1, 2, \ldots, n\}$ be a set of training data. The classification function, F, is determined to minimize the empirical risk:

$$R_{\exp}(F) = \frac{1}{n}\sum_{i=1}^{n} L_{\exp}(F \mid x_i, y_i) = \frac{1}{n}\sum_{i=1}^{n} \exp\{-y_i F(\mathbf{x}_i)\} \tag{4.2}$$

In the toy example above, $F(x)$ is $\beta_1 f_1(x) + \beta_2 f_2(x) + \beta_3 f_3(\mathbf{x})$ and coefficients β_1, β_2, β_3 are tuned by minimizing the empirical risk in Equation 4.2. A fast sequential procedure for minimizing the empirical risk is well known [11]. We will provide a new understanding of the procedure in the binary class case as well as in the multi-class case in Section 4.2.3.

A typical classifier is a decision stump defined by a function δ sign$(x^j - t)$, where $\delta = \pm 1$, $t \in R$ and x^j denotes the j-th coordinate of the feature vector $\mathbf{x}$. Nevertheless, each decision stump is poor. Finally, a linearly combined function of many stumps is expected to be a strong classification function.

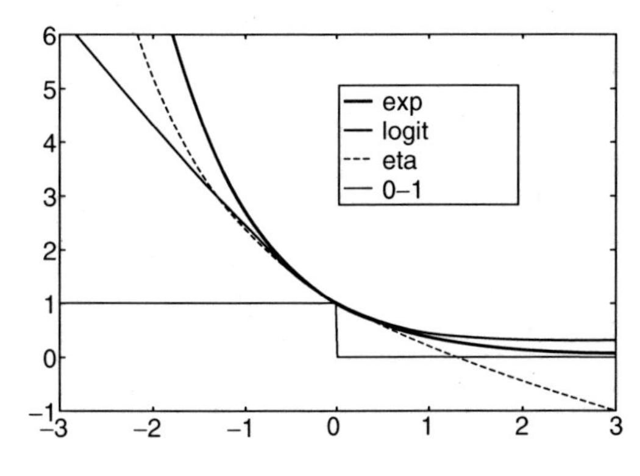

FIGURE 4.1
Loss functions (loss vs. $yF(\mathbf{x})$).

4.2.2 AdaBoost for Multi-Class Problems

We will give an extension of loss and risk functions to cases with multi-class labels. Suppose that there are g possible land-cover classes $C_1, \ldots, C_g$, for example, coniferous forest, broad leaf forest, and water area. Let $D = \{1, \ldots, n\}$ be a training region with n pixels over some scene. Each pixel i in region D is supposed to belong to one of the g classes. We denote a set of all class labels by $G = \{1, \ldots, g\}$. Let $x_i \in R^q$ be a q-dimensional feature vector observed at pixel i, and y_i be its true label in label set G. Note that pixel i in region D is a numbered small area corresponding to the observed unit on the earth.

Let $F(\mathbf{x}, k)$ be a classification function of feature vector $\mathbf{x} \in R^q$ and label k in set G. We allocate vector $\mathbf{x}$ into the class with label $\hat{y}_F \in G$ by the following maximizer:

$$\hat{y}_F = \underset{k \in G}{\arg\ \max}\ F(\mathbf{x}, k) \tag{4.3}$$

Typical examples of the strong classification function would be given by posterior probability functions. Let $p(\mathbf{x} \mid k)$ be a class-specific probability density function of the k-th class, C_k. Thus, the posterior probability of the label, $Y = k$, given feature vector $\mathbf{x}$, is defined by

$$p(k \mid \mathbf{x}) = p(\mathbf{x} \mid k) / \sum_{\ell \in G} p(\mathbf{x} \mid \ell) \quad \text{with } k \in G \tag{4.4}$$

which gives a strong classification function, where the prior distribution of class C_k is assumed to be uniform. Note that the label estimated by posteriors $p(k \mid \mathbf{x})$, or equivalently by log posteriors $\log p(k \mid \mathbf{x})$, is just the Bayes rule of classification. Note also that $p(k \mid \mathbf{x})$ is a measure of the confidence of the current classification and is closely related to logistic discriminant functions [18].

Let $y \in G$ be the true label of feature vector $\mathbf{x}$ and $F(\mathbf{x})$ a classification function. Then, the loss by misclassification into class label k is assessed by the following exponential loss function:

$$L_{\exp}(F, k \mid \mathbf{x}, y) = \exp\{F(\mathbf{x}, k) - F(\mathbf{x}, y)\} \quad \text{for } k \neq y \text{ with } k \in G \tag{4.5}$$

This is an extension of the exponential loss (Equation 4.1) with binary classification. The empirical risk is defined by averaging the loss functions over the training data set $\{(x_i, y_i) \in R^q \times G \mid i \in D\}$ as

$$R_{\exp}(F) = \frac{1}{n} \sum_{i \in D} \sum_{k \neq y_i} L_{\exp}(F, k \mid \mathbf{x}_i, y_i) = \frac{1}{n} \sum_{i \in D} \sum_{k \neq y_i} \exp\{F(\mathbf{x}_i, k) - F(\mathbf{x}_i, y_i)\} \tag{4.6}$$

AdaBoost determines the classification function F to minimize exponential risk $R_{\exp}(F)$, in which F is a linear combination of base functions.

4.2.3 Sequential Minimization of Exponential Risk with Multi-Class

Let f and F be fixed classification functions. Then, we obtain the optimal coefficient, β_*, which gives the minimum value of empirical risk $R_{\text{emp}}(F + \beta f)$:

$$\beta_* = \underset{\beta \in R}{\arg\ \min} \{R_{\exp}(F + \beta f)\}; \tag{4.7}$$

Applying procedure in Equation 4.7 sequentially, we combine classifiers $f_1, f_2, \ldots, f_T$ as

$$F^{(0)} \equiv 0,\ F^{(1)} = \beta_1 f_1,\ F^{(2)} = \beta_1 f_1 + \beta_2 f_2, \ldots,\ F^{(T)} = \beta_1 f_1 + \beta_2 f_2 + \cdots + \beta_T f_T$$

where β_T is defined by the formula given in Equation 4.7 with $F = F^{(t-1)}$ and $f = f_t$.

4.2.3.1 Case 1

Suppose that function $f(\cdot, k)$ takes values 0 or 1, and it takes 1 only once. In this case, coefficient β_* in Equation 4.7 is given by the closed form as follows. Let $\hat{y}_{i,f}$ be the label of pixel i selected by classifier f, and D_f be a subset of D classified correctly by f. Define

$$V_i(k) = F(\mathbf{x}_i, k) - F(\mathbf{x}_i, y_i) \quad \text{and} \quad v_i(k) = f(\mathbf{x}_i, k) - f(\mathbf{x}_i, y_i) \tag{4.8}$$

Then, we obtain

$$\begin{aligned} R_{\exp}(F + \beta f) &= \sum_{i=1}^{n} \sum_{k \neq y_i} \exp\left[V_i(k) + \beta v_i(k)\right] \\ &= e^{-\beta} \sum_{i \in D_f} \sum_{k \neq y_i} \exp\{V_i(k)\} + e^{\beta} \sum_{j \notin D_f} \exp\left\{V_j(\hat{y}_{jf})\right\} + \sum_{j \notin D_f} \sum_{k \neq y_i,\, \hat{y}_{jf}} \exp\{V_i(k)\}] \\ &\geqq 2\sqrt{\sum_{i \in D_f} \sum_{k \in y_i} \exp\{V_i(k)\} \sum_{j \notin D_f} \exp\{V_j(\hat{y}_{jf})\}} + \sum_{j \notin D_f} \sum_{k \neq y_i,\, \hat{y}_{jf}} \exp\{V_j(k)\} \end{aligned} \tag{4.9}$$

The last inequality is due to the relationship between the arithmetic and the geometric means. The equality holds if and only if $\beta = \beta_*$, where

$$\beta_* = \frac{1}{2} \log\left[\sum_{i \in D_f} \sum_{k \neq y_i} \exp\{V_i(k)\} / \sum_{j \notin D_f} \exp\{V_j(\hat{y}_{jf})\}\right] \tag{4.10}$$

The optimal coefficient, β_*, can be expressed as

$$\beta_* = \frac{1}{2} \log\left\{\frac{1 - \varepsilon_F(f)}{\varepsilon_F(f)}\right\} \quad \text{with } \varepsilon_F(f) = \frac{\sum_{j \notin D_f} \exp\left\{V_j(\hat{y}_{jf})\right\}}{\sum_{j \notin D_f} \exp\left\{V_j(\hat{y}_{jf})\right\} + \sum_{i \in D_f} \sum_{k \neq y_i} \exp\left\{V_i(k)\right\}} \tag{4.11}$$

In the binary class case, $\varepsilon_F(f)$ coincides with the error rate of classifier f.

4.2.3.2 Case 2

If $f(\cdot, k)$ takes real values, there is no closed form of coefficient β_*. We must perform an iterative procedure for the optimization of risk R_{emp}. Using the Newton-like method, we update estimate $\beta^{(t)}$ at the t-th step as follows:

$$\beta^{(t+1)} = \beta^{(t)} - \sum_{i=1}^{n} \sum_{k \neq y_i} v_i(k) \exp\left[V_i(k) + \beta^{(t)} v_i(k)\right] / \sum_{i=1}^{n} \sum_{k \neq y_i} v_i^2(k) \exp\left[V_i(k) + \beta^{(t)} v_i(k)\right] \tag{4.12}$$

where $v_i(k)$ and $V_i(k)$ are defined in the formulas in Equation 4.8. We observe that the convergence of the iterative procedure starting from $\beta^{(0)} = 0$ is very fast. In numerical examples in Section 4.7, the procedure converges within five steps in most cases.

4.2.4 AdaBoost Algorithm

Now, we summarize an iterative procedure of AdaBoost for minimizing the empirical exponential risk. Let $\{F\} = \{f : R^q \longrightarrow G\}$ be a set of classification functions, where $G = \{1,\ldots,g\}$ is the label set. AdaBoost combines classification functions as follows:

- Find classification function f in F and coefficient β that jointly minimize empirical risk $R_{\exp}(\beta f)$ defined in Equation 4.6, for example, f_1 and β_1.
- Consider empirical risk $R_{\exp}(\beta_1 f_1 + \beta f)$ with $\beta_1 f_1$ given from the previous step. Then, find classification function $f \in \{F\}$ and coefficient β that minimize the empirical risk, for example, f_2 and β_2.
- This procedure is repeated T-times and the final classification function $F_T = \beta_1 f_1 + \cdots + \beta_T f_T$ is obtained.
- Test vector $x \in R^q$ is classified into the label maximizing the final function $F_T(x, k)$ with respect to $k \in G$.

Substituting exponential risk $R_{\exp}$ for other risk functions, we have different classification methods. Risk functions R_{logit} and R_{eta} will be defined in the next section.

4.3 LogitBoost and EtaBoost

AdaBoost was originally designed to combine *weak* classifiers for deriving a *strong* classifier. However, if we combine strong classifiers with AdaBoost, the exponential loss assigns an extreme penalty for misclassified data. It is well known that AdaBoost is not robust. In the multi-class case, this seems more serious than the binary class case. Actually, this is confirmed by our numerical example in Section 4.7.3. In this section, we consider robust classifiers derived by a loss function that is more robust than the exponential loss function.

4.3.1 Binary Class Case

Consider binary class problems such that feature vector $\mathbf{x}$ with true label $y \in \{1,-1\}$ is classified into class label sign $(F(\mathbf{x}))$. Then, we take the logit and the eta loss functions defined by

$$L_{\text{logit}}(F|\mathbf{x}, y) = \log[1 + \exp\{-yF(x)\}] \tag{4.13}$$

$$L_{\text{eta}}(F \mid \mathbf{x}, y) = (1-\eta)\log[1 + \exp\{-yF(x)\}] + \eta\{-yF(\mathbf{x})\} \quad \text{for } 0 < \eta < 1 \tag{4.14}$$

The logit loss function is derived by the log posterior probability of a binomial distribution. The eta loss function, an extension of the logit loss, was proposed by Takenouchi and Eguchi [19].

Three loss functions given in Figure 4.1 are defined as follows:

$$L_{\text{exp}}(t) = \exp(-t), \quad L_{\text{logit}}(t) = \log\{1 + \exp(-2t)\} - \log 2 + 1$$

and

$$L_{\text{eta}}(t) = (1/2)L_{\text{logit}}(t) + (1/2)(-t) \tag{4.15}$$

We see that the logit and the eta loss functions assign less penalty for misclassified data than the exponential loss function does. In addition, the three loss functions are convex and differentiable with respect to t. The convexity assures the uniqueness of the coefficient minimizing $R_{\text{emp}}(F + \beta f)$ with respect to β, where R_{emp} denotes an empirical risk function under consideration. The convexity makes the sequential minimization of the empirical risk feasible.

For corresponding empirical risk functions, we define the empirical risks as follows:

$$R_{\text{logit}}(F) = \frac{1}{n}\sum_{i=1}^{n} \log[1 + \exp\{-y_i F(x_i)\}], \text{ and} \tag{4.16}$$

$$R_{\text{eta}}(F) = \frac{1-\eta}{n}\sum_{i=1}^{n} \log[1 + \exp\{-y_i F(x_i)\}] + \frac{\eta}{n}\sum_{i=1}^{n}\{-y_i F(x_i)\} \tag{4.17}$$

4.3.2 Multi-Class Case

Let y be the true label of feature vector $\mathbf{x}$, and $F(\mathbf{x}, k)$ a classification function. Then, we define the following function in a similar manner to that of posterior probabilities:

$$p_{\text{logit}}(y \mid \mathbf{x}) = \frac{\exp\{F(\mathbf{x}, y)\}}{\sum_{k \in G} \exp\{F(\mathbf{x}, k)\}}$$

Using the function, we define the loss functions in the multi-class case as follows:

$$L_{\text{logit}}(F \mid x, y) = -\log\ p_{\text{logit}}(y \mid x)$$

and

$$L_{\text{eta}}(F \mid \mathbf{x}, y) = \{1 - (g-1)\eta\}\{-\log\ p_{\text{logit}}(y \mid \mathbf{x})\} + \eta \sum_{k \neq y} \log p_{\text{logit}}(k \mid \mathbf{x})$$

where η is a constant with $0 < \eta < 1/(g-1)$. Then empirical risks are defined by the average of the loss functions evaluated by training data set $\{(x_i, y_i) \in R^q \times G \mid i \in D\}$ as

$$R_{\text{logit}}(F) = \frac{1}{n}\sum_{i=1}^{n} L_{\text{logit}}(F \mid x_i, y_i) \quad \text{and} \quad R_{\text{eta}}(F) = \frac{1}{n}\sum_{i=1}^{n} L_{\text{eta}}(F \mid x_i, y_i) \tag{4.18}$$

LogitBoost and EtaBoost aim to minimize logit risk function $R_{\text{logit}}(F)$ and eta risk function $R_{\text{eta}}(F)$, respectively. These risk functions are expected to be more robust than the exponential risk function. Actually, EtaBoost is more robust than LogitBoost in the presence of mislabeled training examples.

4.4 Contextual Image Classification

Ordinary classifiers proposed for independent samples are of course utilized for image classification. However, it is known that contextual classifiers show better performance than noncontextual classifiers. In this section, contextual classifiers: the smoothing method by Switzer [1], the MRF-based classifiers, and spatial boosting [17], will be discussed.

4.4.1 Neighborhoods of Pixels

In this subsection, we define notations related to observations and two sorts of neighborhoods. Let $D = \{1, \ldots, n\}$ be an *observed area* consisting of n pixels. A q-dimensional feature vector and its observation at pixel i are denoted as $\mathbf{X}_i$ and $\mathbf{x}_i$, respectively, for i in area D. The class label covering pixel i is denoted by random variable Y_i, where Y_i takes an element in the label set $G = \{1, \ldots, g\}$. All feature vectors are expressed in vector form as

$$\mathbf{X} = (\mathbf{X}_1^T, \ldots, \mathbf{X}_n^T)^T : nq \times 1 \tag{4.19}$$

In addition, we define random label vectors as

$$\mathbf{Y} = (\mathrm{Y}_1, \ldots, \mathrm{Y}_n)^T : n \times 1 \quad \text{and} \quad \mathbf{Y}_{-i} = \mathbf{Y} \text{ with deleted } Y_i : (n-1) \times 1 \tag{4.20}$$

Recall that class-specific density functions are defined by $p(\mathbf{x} \mid k)$ with $\mathbf{x} \in R^q$ for deriving the posterior distribution in Equation 4.4. In the numerical study in Section 4.7, the densities are fitted by homoscedastic q-dimensional Gaussian distributions, $N_q(\boldsymbol{\mu}(k), \Sigma)$, with common variance–covariance matrix Σ, or heteroscedastic Gaussian distributions, $N_q(\boldsymbol{\mu}(k), \Sigma_k)$, with class-specific variance–covariance matrices Σ_k.

Here, we define neighborhoods to provide contextual information. Let $d(i, j)$ denote the distance between centers of pixels i and j. Then, we define two kinds of neighborhoods of pixel i as follows:

$$U_r(i) = \{j \in D \mid d(i, j) = r\} \quad \text{and} \quad N_r(i) = \{j \in D \mid \leqq d(i, j) \leqq r\} \tag{4.21}$$

where $r = 1, \sqrt{2}, 2, \ldots$, which denotes the radius of the neighborhood. Note that subset $U_r(i)$ constitutes an isotropic ring region. Subsets $U_r(i)$ with $r = 0, 1, \sqrt{2}, 2$ are shown in Figure 4.2. Here, we find that $U_0(i) = \{i\}$, $N_1(i) = U_1(i)$ is the first-order neighborhood, and $N_{\sqrt{2}}(i) = U_1(i) \cup U_{\sqrt{2}}(i)$ forms the second-order neighborhood of pixel i. In general, we have $N_r(i) = \cup_{1 \leqq r' \leqq r} U_{r'}(i)$ for $r \geqq 1$.

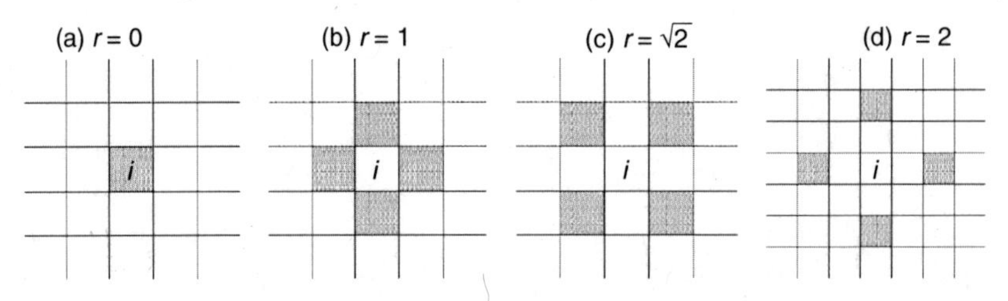

FIGURE 4.2
Isotropic neighborhoods $U_r(i)$ with center pixel i and radius r.

4.4.2 MRFs Based on Divergence

Here, we will discuss the spatial distribution of the classes. A pairwise dependent MRF is an important model for specifying the field. Let $D(k, \ell) > 0$ be a divergence between two classes, C_k and C_ℓ ($k \neq \ell$), and put $D(k, k) = 0$. The divergence is employed for modeling the MRF. In Potts model, $D(k, \ell)$ is defined by $D_0\ (k, \ell) := 1$ if $k \neq \ell := 0$ otherwise. Nishii [18] proposed to take the squared Mahalanobis distance between homoscedastic Gaussian distributions $N_q(\boldsymbol{\mu}(k), \Sigma)$ defined by

$$D_1(\boldsymbol{\mu}(k), \boldsymbol{\mu}(\ell)) = \{\boldsymbol{\mu}(k) - \boldsymbol{\mu}(\ell)\}^T \Sigma^{-1} \{\boldsymbol{\mu}(k) - \boldsymbol{\mu}(\ell)\} \tag{4.22}$$

Nishii and Eguchi (2004) proposed to take Jeffreys divergence $\int\{p(\mathbf{x} \mid k) - p(\mathbf{x}|\ell)\} \log\{p(\mathbf{x} \mid k)/p(\mathbf{x} \mid \ell)\}d\mathbf{x}$ between densities $p(\mathbf{x} \mid k)$. The models are called *divergence models.*

Let $\Delta_i(g)$ be the average of divergences in the neighborhood $N_r(i)$ defined by Equation 4.21 as follows:

$$\Delta_i(k) = \begin{cases} \frac{1}{|N_r(i)|} \sum_{j \in N_r(i)} D(k, y_j), & \text{if } |N_r(i)| \geq 1 \\ 0 & \text{otherwise} \end{cases} \quad \text{for } (i, k) \in D \times G \tag{4.23}$$

where $|S|$ denotes the cardinality of set S. Then, random variable $\mathbf{Y}_i$ conditional on all the other labels $\mathbf{Y}_{-i} = \mathbf{y}_{-i}$ is assumed to follow a multinomial distribution with the following probabilities:

$$\Pr\{Y_i = k \mid \mathbf{Y}_{-i} = \mathbf{y}_{-i}\} = \frac{\exp\{-\beta\Delta_i(k)\}}{\sum_{\ell \in G} \exp\{-\beta\Delta_i(\ell)\}} \quad \text{for } k \in G \tag{4.24}$$

Here, β is a non-negative constant called the clustering parameter, or the granularity of the classes, and $\Delta_i(k)$ is defined by the formula given in Equation 4.23.

Parameter β characterizes the degree of the spatial dependency of the MRF. If $\beta = 0$, then the classes are spatially independent. Here, radius r of neighborhood $U_r(i)$ denotes the extent of spatial dependency. Of course, β, as well as r, are parameters that need to be estimated. Due to the Hammersley–Clifford theorem, conditional distribution in Equation 4.24 is known to specify the distribution of test label vector, $\mathbf{Y}$, under the mild condition. The joint distribution of test labels, however, cannot be obtained in a closed form. This causes a difficulty in estimating the parameters specifying the MRF.

Geman and Geman [6] developed a method for the estimation of test labels by simulated annealing. However, the procedure is time consuming. Besag [4] proposed an iterative conditional mode (ICM) method, which is reviewed in Section 4.4.5.

4.4.3 Assumptions

Now, we make the following assumptions for deriving classifiers.

4.4.3.1 Assumption 1 (Local Continuity of the Classes)

If a class label of a pixel is $k \in G$, then pixels in the neighborhood have the same class label k. Furthermore, this is true for any pixel.

4.4.3.2 Assumption 2 (Class-Specific Distribution)

A feature vector of a sample from class C_k follows a class-specific probability density function $p(x \mid k)$ for label k in G.

4.4.3.3 Assumption 3 (Conditional Independence)

The conditional distribution of vector $\mathbf{X}$ in Equation 4.19 given label vector $\mathbf{Y} = \mathbf{y}$ in Equation 4.20 is given by $\Pi_{i \in D}\, p(\mathbf{x}_i \mid y_i)$.

4.4.3.4 Assumption 4 (MRFs)

Label vector $\mathbf{Y}$ defined by Equation 4.20 follows an MRF specified by divergence (*quasi-distance*) between the classes.

4.4.4 Switzer's Smoothing Method

Switzer [1] derived the contextual classification method (*the smoothing method*) under Assumptions 1–3 with homoscedastic Gaussian distributions $N_q(\boldsymbol{\mu}(k), \Sigma)$. Let $\psi(\mathbf{x} \mid k)$ be its probability density function. Assume that Assumption 1 holds for neighborhoods $N_r(\cdot)$. Then, he proposed to estimate label y_i of pixel i by maximizing the following joint probability densities:

$$\psi(\mathbf{x}_i \mid k) \times \Pi_{j \in N_r(i)} \psi(\mathbf{x}_j \mid k) \quad \left(\psi(\mathbf{x} \mid k) \equiv (2\pi)^{-q/2} \, |\Sigma|^{-1/2} \exp\{-D_1(\mathbf{x}, \boldsymbol{\mu}(k))/2\}\right)$$

with respect to label $k \in G$, where $D_1(\cdot,\cdot)$ stands for the squared Mahalanobis distance in Equation 4.22. The maximization problem is equivalent to minimizing the following quantity:

$$D_1(\mathbf{x}_i, \boldsymbol{\mu}(k)) + \sum_{j \in N_r(i)} D_1(\mathbf{x}_j, \boldsymbol{\mu}(k)) \tag{4.25}$$

Obviously, Assumption 1 does not hold for the whole image. However, the method still exhibits good performance, and the classification is performed very quickly. Thus, the method is a pioneering work of contextual image classification.

4.4.5 ICM Method

Under Assumptions 2–4 with conditional distribution in Equation 4.24, the posterior probability of $Y_i = k$ given feature vector $\mathbf{X} = \mathbf{x}$ and label vector $\mathbf{Y}_{-i} = \mathbf{y}_{-i}$ is expressed by

$$\Pr\{Y_i = k \mid \mathbf{X} = \mathbf{x}, \mathbf{Y}_{-i} = \mathbf{y}_{-i}\} = \frac{\exp\{-\beta \Delta_i(k)\} p(x_i \mid k)}{\sum_{\ell \in G} \exp\{-\beta \Delta_i(\ell)\} p(x_i \mid \ell)} \equiv p_i(k \mid r, \beta) \tag{4.26}$$

Then, the posterior probability $\Pr\{\mathbf{Y} = \mathbf{y} \mid \mathbf{X} = \mathbf{x}\}$ of label vector $\mathbf{y}$ is approximated by the pseudo-likelihood

$$\mathrm{PL}(\mathbf{y} \mid r, \beta) = \prod_{i=1}^{n} p_i(y_i \mid r, \beta) \tag{4.27}$$

where posterior probability $p_i\,(y_i \mid r, \beta)$ is defined by Equation 4.26. Pseudo-likelihood in Equation 4.27 is used for accuracy assessment of the classification as well as for parameter estimation. Here, class-specific densities $p(\mathbf{x} \mid k)$ are estimated using the training data.

When radius r and clustering parameter β are given, the optimal label vector $\mathbf{y}$, which maximizes the pseudo-likelihood $\mathrm{PL}(\mathbf{y} \mid r, \beta)$ defined by Equation 4.27, is usually estimated by the ICM procedure [4]. At the $(t+1)$-st step, ICM finds the optimal label, $y_i^{(t+1)}$, given $\mathbf{y}_{-i}^{(t)}$ for all test pixels $i \in D$. This procedure is repeated until the convergence of the label vector, for example $\mathbf{y} = \mathbf{y}(r, \beta) : n \times 1$. Furthermore, we must optimize a pair of parameters (r, β) by maximizing pseudo-likelihood $\mathrm{PL}(\mathbf{y}(r, \beta) \mid r, \beta)$.

4.4.6 Spatial Boosting

As shown in Section 4.2, AdaBoost combines classification functions defined over the feature space. Of course, the classifiers give noncontextual classification. We extend AdaBoost to build contextual classification functions, which we call spatial AdaBoost.

Define an averaged logarithm of the posterior probabilities (Equation 4.4) in neighborhood $U_r(i)$ (Equation 4.21) by

$$f_r(x, k \mid i) = \begin{cases} \dfrac{1}{\mid U_r(i) \mid} \sum_{j \in U_r(i)} \log p(k \mid \mathbf{x}_j) & \text{if } \mid U_r(i) \mid \geq 1 \\ 0 & \text{otherwise} \end{cases} \quad \text{for } r = 0, 1, \sqrt{2}, \ldots \tag{4.28}$$

where $\mathbf{x} = (\mathbf{x}_1{}^T, \ldots, \mathbf{x}_n^T)^T : qn \times 1$. Therefore, the averaged log posterior $f_0\,(\mathbf{x}, k \mid i)$ with radius $r = 0$ is equal to the log posterior $\log p(k \mid \mathbf{x}_i)$ itself. Hence, the classification due to function $f_0(\mathbf{x}, k \mid i)$ is equivalent to a noncontextual classification based on the maximum-*a-posteriori* (MAP) criterion. If the spatial dependency among the classes is not negligible, then the averaged log posteriors $f_1(\mathbf{x}, k \mid i)$ in the first-order neighborhood may have information for classification. If the spatial dependency becomes stronger, then $f_r(\mathbf{x}, k \mid i)$ with a larger r is also useful. Thus, we adopt the average of the log posteriors $f_r(\mathbf{x}, k \mid i)$ as a classification function of center pixel i.

The efficiency of the averaged log posteriors as classification functions would be intuitively arranged in the following order:

$$f_0(\mathbf{x}, k \mid i), f_1(\mathbf{x}, k \mid i), f_{\sqrt{2}}(\mathbf{x}, k \mid i), f_2(\mathbf{x}, k \mid i), \cdots, \text{where } \mathbf{x} = (\mathbf{x}_1^T, \cdots, \mathbf{x}_n^T)^T \tag{4.29}$$

The coefficients for the above classification functions can be tuned by minimizing the empirical risk given by Equation 4.6 or Equation 4.18. See Ref. [2] for possible candidates for contextual classification functions.

The following is the contextual classification procedure based on the spatial boosting method.

- Fix an empirical risk function, $R_{\mathrm{emp}}(F)$, of classification function F evaluated over training data set $\{(\mathbf{x}_i, y_i) \in R^q \times G \mid i \in D\}$.
- Let $f_0\,(\mathbf{x}, k \mid i), f_1(\mathbf{x}, k \mid i), f_{\sqrt{2}}\,(\mathbf{x}, k \mid i), \ldots, f_r\,(\mathbf{x}, k \mid i)$ be the classification functions defined by Equation 4.28.
- Find coefficient β that minimizes empirical risk $R_{\mathrm{emp}}\,(\beta f_0)$. Put the optimal value to β_0.
- If coefficient β_0 is negative, quit the procedure. Otherwise, consider empirical risk $R_{\mathrm{emp}}(\beta_0 f_0 + \beta f_1)$ with $\beta_0 f_0$ obtained by the previous step. Then, find coefficient β, which minimizes the empirical risk. Put the optimal value to β_1.
- If β_1 is negative, quit the procedure. Otherwise, consider empirical risk R_{emp} $(\beta_0 f_0 + \beta_1 f_1 + \beta f_{\sqrt{2}})$. This procedure is repeated, and we obtain a sequence of positive coefficients $\beta_0, \beta_1, \ldots, \beta_r$ for the classification functions.

Finally, the classification function is derived by

$$F_r(\mathbf{x}, k \mid i) = \beta_0 f_0(\mathbf{x}, k \mid i) + \beta_1 f_1(\mathbf{x}, k \mid i) + \cdots + \beta_r f_r(\mathbf{x}, k \mid i), \ \mathbf{x} = (\mathbf{x}_1^T, \ \ldots, \mathbf{x}_n^T)^T \tag{4.30}$$

Test label y_* of test vector $\mathbf{x}_* \in R^q$ is estimated by maximizing classification function in Equation 4.30 with respect to label $k \in G$. Note that the pixel is classified by the feature vector at the pixel as well as feature vectors in neighborhood $N_r(i)$ in the test area only. There is no need to estimate labels of neighbors, whereas the ICM method requires estimated labels of neighbors and needs an iterative procedure for the classification. Hence, we claim that spatial boosting provides a very fast classifier.

4.5 Relationships between Contextual Classification Methods

Contextual classifiers discussed in the chapter can be regarded as an extension of Switzer's method from a unified viewpoint, cf. [16] and [2].

4.5.1 Divergence Model and Switzer's Model

Let us consider the divergence model in Gaussian MRFs (GMRFs), where feature vectors follow homoscedastic Gaussian distributions $N_q(\boldsymbol{\mu}(k), \Sigma)$. The divergence model can be viewed as a natural extension of Switzer's model.

The image with center pixel 1 and its neighbors is shown in Figure 4.3. First-order and second-order neighborhoods of the center pixel are given by sets of pixel numbers $N_1(1) = \{2, 4, 6, 8\}$ and $N_{\sqrt{2}}(1) = \{2, 3, \ldots, 9\}$, respectively. We focus our attention on center pixel 1 and its neighborhood $N_r(1)$ of size $2K$ in general and discuss the classification problem of center pixel 1 when labels y_j of $2K$ neighbors are observed.

Let $\hat{\beta}$ be a non-negative estimated value of the clustering parameter β. Then, label y_1 of center pixel 1 is estimated by the ICM algorithm. In this case, the estimate is derived by maximizing conditional probability (Equation 4.26) with $p(\mathbf{x} \mid k) = \psi(\mathbf{x} \mid k)$. This is equivalent to finding label $\hat{Y}_{\text{Div}}$ defined by

$$\hat{Y}_{\text{Div}} = \underset{k \in G}{\arg\min} \ \{D_1(\mathbf{x}_1, \boldsymbol{\mu}(k)) + \frac{\hat{\beta}}{K} \sum_{j \in N_r(1)} D_1(\boldsymbol{\mu}(y_j), \boldsymbol{\mu}(k))\}, \quad |N_r(1)| = 2K \tag{4.31}$$

where $D_1(s, t)$ is the squared Mahalanobis distance (Equation 4.22).

Switzer's method [1] classifies the center pixel by minimizing formula given in Equation 4.25 with respect to label k. Here, the method can be slightly extended by changing the coefficient for $\sum_{j \in N_{r(1)}} D_1(\mathbf{x}_j, \boldsymbol{\mu}(k))$ from 1 to $\hat{\beta}/K$. Thus, we define the estimate due to Switizer's method as follows:

5	4	3
6	1	2
7	8	9

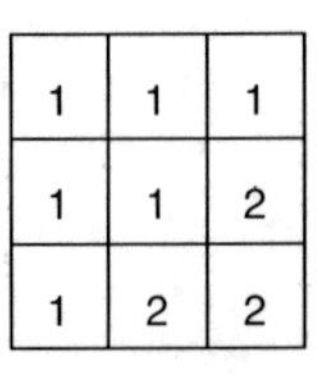

FIGURE 4.3
Pixel numbers (left) and pixel labels (right).

$$\hat{Y}_{\text{Switzer}} = \underset{k \in G}{\arg\min} \left\{ D_1(\mathbf{x}_1, \boldsymbol{\mu}(k)) + \frac{\hat{\beta}}{K} \sum_{j \in N_r(1)} D_1(\mathbf{x}_j, \boldsymbol{\mu}(k)) \right\} \tag{4.32}$$

Estimates in Equation 4.31 and Equation 4.32 of label y_1 are given by the same formula except for $\boldsymbol{\mu}(y_j)$ and $\mathbf{x}_j$ in the respective last terms. Note that $\mathbf{x}_j$ itself is a primitive estimate of $\boldsymbol{\mu}(y_j)$. Hence, the classification method based on the divergence model is an extension of Switzer's method.

4.5.2 Error Rates

We derive the exact error rate due to the divergence model and Switzer's model in the previous local region with two classes $G = \{1, 2\}$. In the two-class case, the only positive quasi-distance is $D(1, 2)$. Substituting $\beta D(1, 2)$ for β, we note that the MRF based on Jeffreys divergence is reduced to the Potts model.

Let δ be the Mahalanobis distance between distributions $N_q(\boldsymbol{\mu}(k), \Sigma)$ for $k = 1, 2$, and N_r (1) be a neighborhood consisting of $2K$ neighbors of center pixel 1, where K is a fixed natural number. Furthermore, suppose that the number of neighbors with label 1 or 2 is randomly changing. Our aim is to derive the error rate of pixel 1 given features x_1, x_j, and labels y_j of neighbors j in $N_r(1)$. Recall that $\hat{Y}_{\text{Div}}$ is the estimated label of y_1 obtained by formula in Equation 4.31. Then, the exact error rate, $\Pr\{\hat{Y}_{\text{Div}} \neq Y_1\}$, is given by

$$e\,(\hat{\beta};\, \beta, \delta) = \pi_0 \Phi(-\delta/2) + \sum_{k=1}^{K} \pi_k \left\{ \frac{\Phi(-\delta/2 - k\hat{\beta}\delta/g)}{1 + e^{-k\beta\delta^2/g}} + \frac{\Phi(-\delta/2 + k\hat{\beta}\delta/g)}{1 + e^{k\beta\delta^2/g}} \right\} \tag{4.33}$$

where $\Phi(x)$ is the cumulative standard Gaussian distribution function, and $\hat{\beta}$ is an estimate of clustering parameter β. Here, π_k gives a prior probability such that the number of neighbors with label 1, W_1, is equal to $K+k$ or $K - k$ in the neighborhood $N_r(1)$ for $k = 0, 1, \ldots, K$. In Figure 4.3, first-order neighborhood $N_1(1)$ is given by $\{2, 4, 6, 8\}$ with $(W_1, K) = (2, 2)$, and second-order neighborhood $N_{\sqrt{2}}(1)$ is given by $\{2, 3, \ldots, 9\}$ with $(W_1, K) = (3, 4)$; see Ref. [16].

If prior probability π_0 is equal to one, K pixels in neighborhood $N_r(1)$ are labeled 1 and the remaining K pixels are labeled 2 with probability one. In this case, the majority vote of the neighbors does not work. Hence, we assume that π_0 is less than one. Then, we have the following properties of the error rate, $e(\hat{\beta}; \beta, \delta)$; see Ref. [16].

- P1. $e(0; \beta, \delta) = \Phi(-\delta/2)$, $\lim_{\hat{\beta} \to \infty} e(\hat{\beta}; \beta, \delta) = \pi_0 \Phi(-\delta/2) + \sum_{k=1}^{K} \dfrac{\pi_k}{1 + e^{k\beta\delta^2/g}}$.
- P2. The function, $e(\hat{\beta}; \beta, \delta)$, of $\hat{\beta}$ takes a minimum at $\hat{\beta} = \beta$ (Bayes rule), and the minimum $e(\beta; \beta, \delta)$ is a monotonically decreasing function of the Mahalanobis distance, δ, for any fixed positive clustering parameter β.
- P3. The function, $e(\hat{\beta}; \beta, \delta)$, is a monotonically decreasing function of β for any fixed positive constants $\hat{\beta}$ and δ.
- P4. We have the inequality: $e(\hat{\beta}; \beta, \delta) < \Phi(-\delta/2)$ for any positive $\hat{\beta}$ if the inequality $\beta \geq \frac{1}{\delta^2} \log\left\{ \frac{1 - \Phi(-\delta/2)}{\Phi(-\delta/2)} \right\}$ holds.

Note that the value $e(0; \beta, \delta) = \Phi(-\delta/2)$ is simply the error rate due to Fisher's linear discriminant function (LDF) with uniform prior probabilities on the labels. The asymptotic value $\sum_{k=1}^{K} \pi_k / \left(1 + e^{k\beta\delta^2/g}\right)$ given in P1 is the error rate due to the vote-for-majority rule when the number of neighbors, W_1, is not equal to K. The property, P2, recommends that we use the true parameter, β, if it is known, and this is quite natural. P3 means that the classification becomes more efficient when δ or β becomes large. Note that δ is a distance in the feature space and β is a distance in the image. P4 implies that the use of spatial information *always improves the performance of noncontextual discrimination* even if estimate $\hat{\beta}$ is far from the true value β.

The error rate due to Switzer's method is obtained in the same form as that of Equation 4.33 by replacing δ with $\delta * \equiv \delta/\sqrt{1+4\hat{\beta}^2/g} < \delta$ appearing in $\Phi(\cdot)$.

The comparison of these two error rates is illustrated in Figure 4.4 with $4 = 2K$ neighbors. The error rates due to rules in Equation 4.5 and Equation 4.6 vs. the estimated clustering parameter $\hat{\beta}$ are shown for cases (a) $\delta = 1.5$ and $\beta = 0.5$ and (b) $\delta = 1.5$ and $\beta = 1$. The prior probability of random variable W_1 is defined by a binomial distribution with $\Pr\{W_1 = 2 \pm k\} = \binom{2}{k}/4$ for $k = 0, 1, 2$.

We see that the divergence model is superior to Switzer's method, and both y-intercepts are the same value $\Phi(-1.5/2) = 0.2266$. Furthermore, the error rate due to the MRF takes the minimum value at the true value $\beta = 0.5$ or 1 (recall property P2). The parameters yielding Figure 4.4a do not meet the sufficient condition of P4 because $\beta = 0.5 < 0.5455 = \log\left[\frac{1-\Phi(-\delta/2)}{\Phi(-\delta/2)}\right]/\delta^2$ as in P3. Actually, the error rate exceeds the 1y-intercept $\Phi(-\delta/2)$ for large $\hat{\beta}$ (Figure 4.4a). On the other hand, the parameters of (b) meet the condition because $\beta = 1$ and δ is the same. Hence, the error rate is always less than $\Phi(-\delta/2)$ for any $\hat{\beta}$, and this is shown in Figure 4.4b.

4.5.3 Spatial Boosting and the Smoothing Method

We consider the proposed classification function in Equation 4.30. When the radius is equal to zero ($r = 0$), classification function in Equation 4.28 is given by

$$F_0(\mathbf{x}, k \mid i) = \beta_0 \log\, p(k \mid \mathbf{x}_i) = \beta_0 \log\, p(\mathbf{x}_i \mid k) + C, \quad \mathbf{x} = (\mathbf{x}_1^T, \ldots, \mathbf{x}_n^T)^T : nq \times 1 \tag{4.34}$$

where $p(k \mid \mathbf{x}_i)$ is the posterior probability given in Equation 4.4, and C denotes a constant independent of label k. Constant β_0 is positive, so the maximization of function $F_0(\mathbf{x}, k \mid i)$ with respect to label k is equivalent to that of class-specific density $p(\mathbf{x}_i \mid k)$. Hence, classification function F_0 yields the usual noncontextual Bayes classifier.

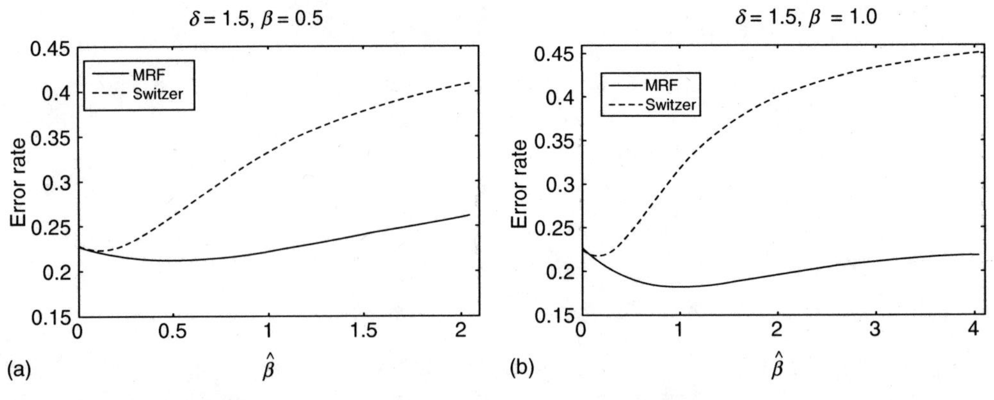

FIGURE 4.4
Error rates due to MRF and Switzer's method (x-axis: $\hat{\beta}$).

When the radius is equal to one ($r = 1$), the classification function in Equation 4.18 is given by

$$F_1(\mathbf{x}, k \mid i) = \beta_0 \log\, p(k \mid \mathbf{x}_i) + \frac{\beta_1}{\mid U_1(i) \mid} \sum_{j \in U_1(i)} \log\, p(k \mid \mathbf{x}_j)$$

$$= \beta_0 \log\, p(\mathbf{x}_i \mid k) + \frac{\beta_1}{|U_1(i)|} \sum_{j \in U_1(i)} \log p(\mathbf{x}_j \mid k) + C' \qquad (4.35)$$

where C' denotes a constant independent of k, and set $U_1(i)$ is defined by Equation 4.21, which is the first-order neighborhood of pixel i. Hence, the label of pixel i is estimated by a convex combination of the log-likelihood of the feature vector $\mathbf{x}_i$ and that of $\mathbf{x}_j$ with pixel j in neighborhood $U_1(i)$. In other words, the label is estimated by a majority vote based on the weighted sum of the log-likelihoods.

Thus, the proposed method is an extension of the smoothing method proposed by Switzer [1] for multivariate Gaussian distributions with an isotropic correlation function. However, the derivation is completely different.

4.5.4 Spatial Boosting and MRF-Based Methods

Usually, the MRF-based method uses training data for the estimation of parameters that specify the class-specific densities $p(\mathbf{x} \mid k)$. Clustering parameter β appearing in conditional probability in Equation 4.24 and the test labels are jointly estimated. The ICM method maximizes conditional probability in Equation 4.26 sequentially with a given clustering parameter β. Furthermore, β is chosen to maximize pseudo-likelihood in Equation 4.27 calculated over the test data. This procedure requires a great deal of CPU time.

On the other hand, spatial boosting uses the training data for the estimation of coefficients β_r and parameters of densities $p(\mathbf{x} \mid k)$. The estimation cost of the parameters of the densities is the same as that of the MRF-based classifier. Although estimation of the coefficients requires iterative procedure (Equation 4.12), the number of iterations is very low. After estimation of the coefficients and the parameters, the test labels can be estimated without any iterative procedure.

Thus, the required computational resources, such as CPU times, should be significantly different between the two methods, which needs further scrutiny. Although the training data and the test data are numerous, spatial AdaBoost classifies the test data very quickly.

4.6 Spatial Parallel Boost by Meta-Learning

Spatial boosting is extended for classification of spatio-temporal data. Suppose we have S sets of training data over regions $D^{(s)} \subset D$ for $s = 1, 2, \ldots, S$. Let $\{(x_i^{(s)}, y_i^{(s)} \in R^q \times G \mid i \in D^{(s)}\}$ be a set of training data with $n^{(s)} = \mid D^{(s)} \mid$ samples. Parameters specifying the class-specific densities are estimated by each training data set. Then, we estimate the posterior distribution in Equation 4.4 according to each training data set. Thus, we derive the averaged log posteriors (Equation 4.28). We assume that an appropriate calibration for feature vectors $\mathbf{x}_i^{(s)}$ throughout the training data sets is preprocessed.

Fix loss function L, and let $F^{(s)}$ be a set of classification functions of the s-th training data set. Now, we propose spatial boosting based on S training data sets (*Parallel Boost*). Let F be a classification function applicable to all training data sets. Then, we propose an empirical risk function of F:

$$R_{\text{emp}}(F) = \sum_{s=1}^{S} R_{\text{emp}}^{(s)}(F) \tag{4.36}$$

where $R_{\text{emp}}^{(s)}(F)$ denotes empirical risk given in Equation 4.6 or Equation 4.18 due to loss function L evaluated over the s-th training data set. The following steps describe parallel boost:

- Let $\{f_0{}^{(s)}, f_1{}^{(s)}, \ldots, f_r{}^{(s)}\}$ be a set of averages of the log posteriors (Equation 4.28) evaluated through the s-th training data set for $s = 1, \ldots, S$.
- Stage for radius 0

 (a) Obtain the optimal coefficient and the training data set defined by

$$(a_{01}, \beta_{01}) = \arg \min_{(s, \beta) \in \{1, \ldots, S\} \times R} R_{\text{emp}}(\beta f_0^{(s)})$$

 (b) If coefficient β_{01} is negative, quit the procedure. Otherwise, find the minimizer of empirical risk $R_{\text{emp}}\,(\beta_{01} f_0{}^{(a_{01})} + \beta f_0{}^{(s)})$. Let $s = a_{02}$ and $\beta = \beta_{02} \in R$ be the solution.
 (c) If coefficient β_{02} is negative, quit the procedure. Repeat this at most S times and define the convex combination, $F_0 \equiv \beta_{01}\, f_0{}^{(a_{01})} + \cdots + \beta_{0m_0}\, f_0{}^{(a_{0m0})}$, of the classification functions, where m_0 denotes the number of replications of the substeps.

- Stage for radius 1 (first-order neighborhood)

 (a) Let F_0 be a function obtained by the previous stage. Then, find classification function $f \in F_1$ and coefficient $\beta \in R$ that minimize the empirical risk $R_{\text{emp}}(F_0 + \beta\, f_0{}^{(s)})$, for example, $s = a_{11}$ and $\beta = \beta_{11}$.
 (b) If coefficient β_{11} is negative, quit the procedure. Otherwise, minimize the empirical risk, $R_{\text{emp}}(F_0 + \beta_{11}\, f_1{}^{(a_{11})} + \beta f_1{}^{(s)})$, given F_0 and $\beta_{11}\, f_1{}^{(a_{11})}$. This is repeated and we define a convex combination $F_1 \equiv \beta_{11}\, f_1{}^{(a_{11})} + \cdots + \beta_{1a_1} f_1{}^{(a_{1m_1})}$. The function $F_0 + F_1$ constitutes a contextual classifier.
- The stage is repeated up to radius r, and we obtain the classification function $\bar{F}_r \equiv F_0 + F_1 + \cdots + F_r$.

Note that spatial parallel boost is based on the assumption that training data sets and test data are not significantly different, especially in the spatial distribution of the classes.

4.7 Numerical Experiments

Various classification methods discussed here will be examined by applying them to three data sets. Data set 1 is a synthetic data set, and Data sets 2 and 3 are real data sets offered in Ref. [20] as benchmarks. Legends of the data sets are given in the following section; then, the classification methods are applied to the data sets for evaluating the efficiency of the methods.

4.7.1 Legends of Three Data Sets

4.7.1.1 Data Set 1: Synthetic Data Set

Data set 1 is constituted by multi-spectral images generated over the image shown in Figure 4.5a. There are three classes ($g = 3$), and the labels correspond to black, white, and gray. Sample sizes from these classes are 3330, 1371, and 3580 ($n = 8281$), respectively. We simulate four-dimensional spectral images ($q = 4$) at each pixel of the true image of Figure 4.4a following independent multi-variate Gaussian distributions with mean vectors $\mu_1 = (\mathbf{0\ 0\ 0\ 0})^T$, $\mu_2 = (\mathbf{1\ 1\ 0\ 0})^T/\sqrt{2}$, and $\mu_3 = (1.0498{-}0.6379\ 0\ 0)^T$ and with common variance–covariance matrix $\sigma^2 \mathrm{I}$, where I denotes the identity matrix. The mean vectors were chosen to maximize pseudo-likelihood (Equation 4.27) of the training data based on the squared Mahalanobis distance used as the divergence in Equation 4.22. Test data of size 8281 are similarly generated over the same image. Gaussian density functions with the common variance–covariance matrix are used to derive the posterior probabilities.

4.7.1.2 Data Set 2: Benchmark Data Set grss_dfc_0006

Data set grss_dfc_0006 is a benchmark data set for supervised image classification. The data set can be accessed from the IEEE GRS-S Data Fusion reference database [20], which consists of samples acquired by ATM and SAR (six and nine bands, respectively; $q = 15$) observed at Feltwell, UK. There are five agricultural classes of sugar beets, stubble, bare soil, potatoes, and carrots ($g = 5$). Sample sizes of the training data and test data are $5072 = 1436 + 1070 + 341 + 1411 + 814$ and $5760 = 2017 + 1355 + 548 + 877 + 963$, respectively, in a rectangular region of size 350×250. The data set was analyzed in Ref. [21] through various methods based on statistics and neural networks. In this case, the labels of the test and training data are observed sparsely.

4.7.1.3 Data Set 3: Benchmark Data Set grss_dfc_0009

Data set grss_dfc_0009 is also taken from the IEEE GRS-S Data Fusion reference database, which consists of examples acquired by Landsat 7 TM+ ($q = 7$) with 11 agricultural classes ($g = 11$) in Flevoland, Netherlands. Training data and test data sizes are 2891 and 2890, respectively, in a square region of size 512×512. We note that the labels of the test and training data of Data sets 2 and 3 are observed sparsely in the region.

4.7.2 Potts Models and the Divergence Models

We compare the performance of the Potts and the divergence models applied to Data sets 1 and 2 (see Table 4.1 and Table 4.2). Gaussian distributions with a common variance–covariance matrix (*homoscedastic*), and Gaussian distributions with class-specific matrices (*heteroscedastic*) are employed for class-specific distributions. The former corresponds to the LDF and the latter to the quadratic discriminant function (QDF). Test errors due to GMRF-based classifiers with several radii r are listed in the tables. Numerals in boldface denote the optimal values in the respective columns. We see that the divergence models give the minimum values in Table 4.1 and Table 4.2.

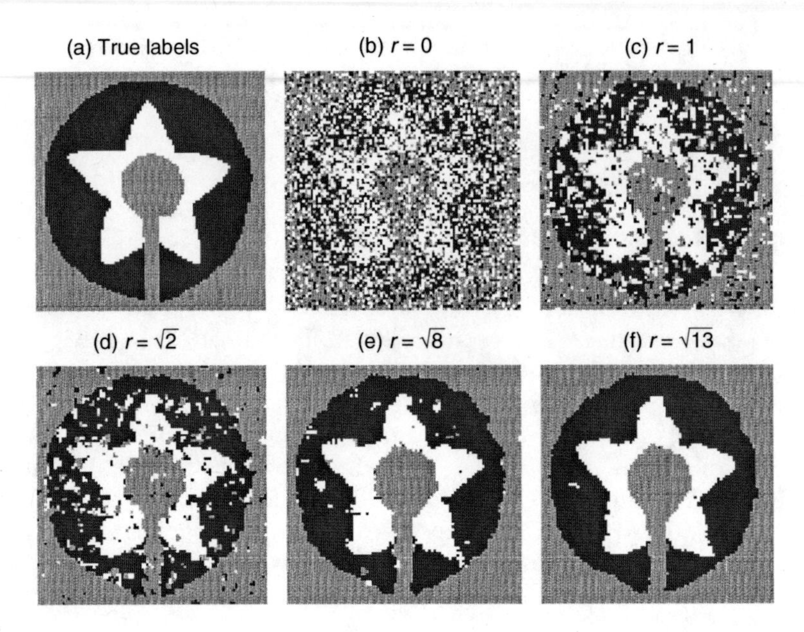

FIGURE 4.5
(a) True labels of data set 1, (b) labels estimated by LDF, (c)–(f) labels estimated by spatial AdaBoost with $\beta_0 f_0 + \beta_1 f_1 + \cdots + \beta_r f_r$.

In the homoscedastic cases of Table 4.1 and Table 4.2, the divergence model gives a performance similar to that of the Potts model. In the heteroscedastic case, the divergence model is superior to the Potts model to a small degree. Classified images with $\sigma^2 = 1$ are given in Figure 4.5b through Figure 4.5f. Figure 4.5b is an image classified by the LDF, which exhibits very poor performance. As radius r becomes larger, images classified by

TABLE 4.1

Test Error Rates (%) Due to Gaussian MRFs with Variance–Covariance Matrices σ^2I with $\sigma^2 = 1,2$ for Data Set 1

	Homoscedastic Gaussian MRFs							
	$\sigma^2 = 1$		$\sigma^2 = 2$		$\sigma^2 = 3$		$\sigma^2 = 4$	
Radius r^2	**Potts**	**Jeffreys**	**Potts**	**Jeffreys**	**Potts**	**Jeffreys**	**Potts**	**Jeffreys**
0	41.66	41.66	41.57	41.57	52.63	52.63	54.45	54.45
1	9.35	9.25	9.97	9.68	25.60	27.41	30.76	31.07
2	3.90	3.61	3.90	4.46	13.26	12.23	17.49	15.58
4	3.47	**2.48**	**3.02**	3.12	9.82	9.20	11.83	14.70
5	**3.19**	2.72	3.32	**2.96**	8.03	**7.75**	10.72	**9.12**
8	3.39	3.06	3.35	3.37	**7.76**	8.57	**10.14**	10.61
9	3.65	3.30	3.51	4.06	8.28	8.20	10.58	11.00
10	3.89	3.79	3.86	4.09	8.88	10.08	10.57	12.66
13	4.46	4.13	4.36	4.66	10.12	10.20	13.25	14.06
16	4.49	4.17	5.00	4.99	11.16	11.68	14.25	14.95
17	4.96	4.67	5.30	5.54	12.04	11.65	15.34	15.76
18	5.00	4.94	5.92	5.83	12.64	12.67	16.07	15.89
20	5.68	6.11	6.50	6.56	14.89	13.55	17.18	17.00

TABLE 4.2

Test Error Rates (%) Due to Homoscedastic and Heteroscedastic Gaussian MRFs with Two Sorts of Divergences for Data Set 2

	Gaussian MRFs			
	Homoscedastic		Heteroscedastic	
Radius r^2	Potts	Jeffreys	Potts	Jeffreys
0	8.61	8.61	14.67	14.67
4	6.44	6.09	12.93	11.44
9	6.16	5.69	13.09	10.19
16	5.99	**5.35**	13.28	9.67
25	6.16	5.69	**12.80**	**9.39**
36	6.49	5.64	13.16	9.55
49	6.37	6.18	12.90	9.90
64	**6.30**	6.65	12.92	9.76

spatial AdaBoost becomes clearer. Therefore, we see that spatial AdaBoost improves the noncontextual classification result ($r = 0$) significantly.

4.7.3 Spatial AdaBoost and Its Robustness

Spatial AdaBoost is applied to the three data sets. The error rates due to spatial AdaBoost and the MRF-based classifier for Data set 1 are listed in Table 4.3. Here, homoscedastic Gaussian distributions are employed for class-specific distributions in both classifiers. Hence, classification with radius $r = 0$ indicates noncontextual classification by the LDF. We used the GMRF with the divergence model discussed in Section 4.4.3.

From Table 4.3, we see that the minimum error rate due to AdaBoost is higher than that due to the GMRF-based method, but the value is still comparable to that due to the

TABLE 4.3

Error Rates (%) Due to Homoscedastic GMRF and Spatial AdaBoost for Data Set 1 with Variance–Covariance Matrix σ^2I

	$\sigma^2 = 1$		$\sigma^2 = 4$	
Radius r^2	GMRF	AdaBoost	GMRF	AdaBoost
0	41.66	41.65	54.45	54.45
1	9.25	20.40	31.07	40.88
2	3.61	12.15	15.58	33.79
4	**2.48**	8.77	14.70	28.78
5	2.72	5.72	**9.12**	22.71
8	3.06	5.39	10.61	20.50
9	3.30	4.82	11.00	18.96
10	3.79	4.38	12.66	16.58
13	4.13	4.34	14.06	14.99
16	4.17	4.31	14.95	14.14
17	4.67	**4.19**	15.76	13.02
18	4.94	4.34	15.89	12.57
20	6.11	4.43	17.00	**12.16**

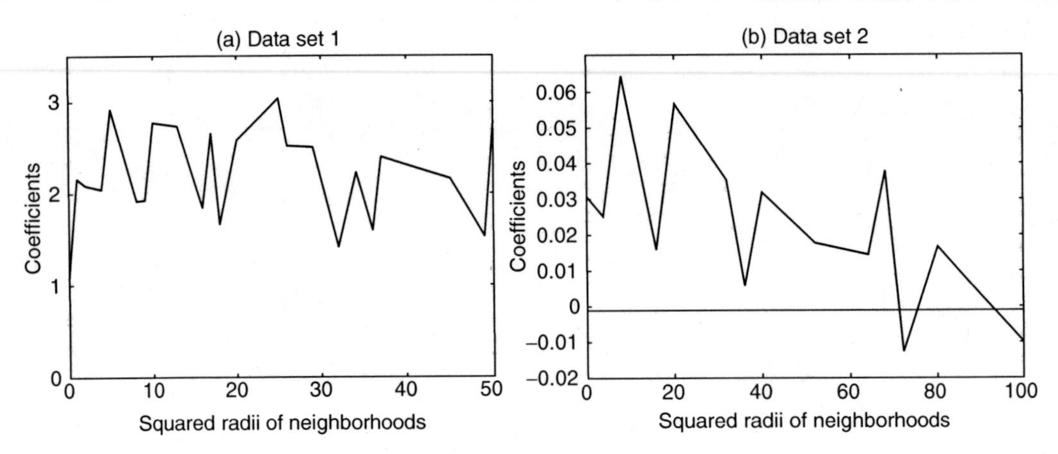

FIGURE 4.6
Estimated coefficients for the log posteriors in the ring regions of data set 1 (left) and data set 2 (right).

GMRF-based method. The left figure in Figure 4.6 is a plot of the estimated coefficients for the averaged log posterior f_r vs. r^2. All the coefficients are positive. Therefore, the problem of stopping to combine classification function arises. Training and test errors as radius r becomes large are given in Figure 4.7. The test error takes the minimum value at $r^2 = 17$, whereas the training error is stable for large r.

The same study as the above is applied to Data set 2 (Table 4.4). We obtain a similar result as that observed in Table 4.3. The left figure in Figure 4.6 gives the estimated coefficient. We see that coefficient β_r with $r = \sqrt{72}$ is $-$ 0.012975, a negative value. This implies that averaged log posterior f_r with $r = \sqrt{72}$ is not appropriate for the classification. Therefore, we must exclude functions of radius $\sqrt{72}$ or larger.

The application to Data set 3 leads to a quite different result (Table 4.5). We fit homoscedastic and heteroscedastic Gaussian distributions for feature vectors. Hence, classification with radius $r = 0$ is performed by the LDF and the QDF. The error rates due to noncontextual classification function $\log p(k \mid \mathbf{x})$ and the coefficients are tabulated in Table 4.5. The test error due to the QDF is 1.66%, and this is excellent. Unfortunately, coefficients β_0 of functions $\log p(k \mid \mathbf{x})$ in both cases are estimated by the negative values.

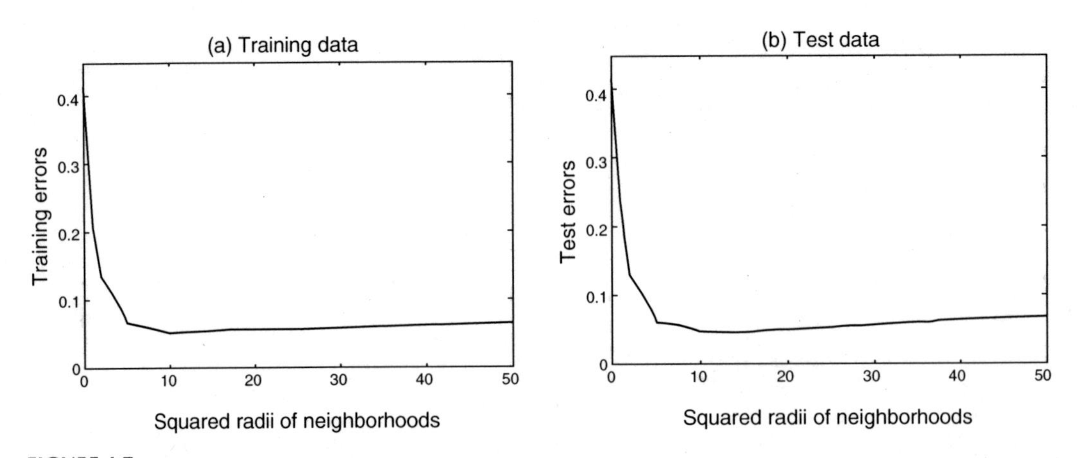

FIGURE 4.7
Error rates with respect to squared radius, r^2, for training data (left) and test data (right) due to classifiers $\beta_0 f_0 + \beta_1 f_1 + \cdots + \beta_r f_r$ for data set 1.

TABLE 4.4

Error Rates (%) Due to GMRF and Spatial AdaBoost for Data Set 2

	Homoscedastic	
Radius r^2	**GMRF**	**AdaBoost**
0	8.61	8.61
4	6.09	7.08
9	5.69	6.11
16	5.35	5.95
25	5.69	5.83
36	5.69	5.85
49	6.13	5.99
64	6.65	6.01

Therefore, the classification functions $\beta_0 \log p(k \mid \mathbf{x})$ are not applicable. This implies that the exponential loss defined by Equation 4.5 is too sensitive to misclassified data.

Error rates for spatial AdaBoost $\beta_0 f_0 + \beta_1 f_1$ are listed in Table 4.6 when coefficient β_0 is predetermined and β_1 is tuned by the empirical risk $R_{\text{emp}}(\beta_0 f_0 + \beta_1 f_1)$. The LDF and the QDF columns correspond to the posterior probabilities based on the homoscedastic Gaussian distributions and the heteroscedastic Gaussian distributions, respectively. For both of these cases, information in the first-order neighborhood ($r = 1$) improves the classification slightly. We note that homoscedastic and heteroscedastic GMRFs have minimum error rates of 4.01% and 0.52%, respectively.

4.7.4 Spatial AdaBoost and Spatial LogitBoost

We apply classification functions $F_0, F_1, \ldots, F_r$ with $r = \sqrt{20}$ to the test data, where function F_r is defined by Equation 4.30, and coefficients $\beta_0, \beta_1, \ldots, \beta_r$ with $r = \sqrt{20}$ are sequentially tuned by minimizing exponential risk (Equation 4.6) or logit risk (Equation 4.18). Error rates due to GMRF-based classifiers and spatial boosting methods for error variances $\sigma^2 = 1$ and 4 for Data set 1 are compared in Table 4.7.

We see that GMRF is superior to the spatial boosting methods to a small degree. Furthermore, spatial AdaBoost and spatial LogitBoost exhibit similar performances for Data set 1. Note that spatial boosting is very fast compared with GMRF-based classifiers. This observation is also true for Data set 2 (Table 4.8). Then, GMRF and spatial LogitBoost were applied to Data set 3 (Table 4.9). Spatial LogitBoost still works well in spite of the failure of spatial AdaBoost. Note that spatial AdaBoost and spatial LogitBoost use the same classification functions: $f_0, f_1, \ldots, f_r$. However, the loss function used by LogitBoost

TABLE 4.5

Optimal Coefficients Tuned by Minimizing Exponential Risks Based on LDF and QDF for Data Set 3

	LDF	**QDF**
Error rate by f_0	10.69 %	1.66 %
Tuned β_0 to f_0	−0.006805	−0.000953
Error rate by $\beta_0 f_0$	Nearly 100%	Nearly 100%

TABLE 4.6

Error Rates (%) Due to Contextual Classification Function $\beta_0 f_0 + \beta_1 f_1$ Given β_0 for Data Set 3. β_1 Is Tuned by Minimizing the Empirical Exponential Risk

β_0	Homoscedastic	Heteroscedastic
10^{-7}	9.97	1.35
10^{-6}	10.00	1.35
10^{-5}	10.00	1.35
10^{-4}	10.00	1.25
10^{-3}	**9.10**	1.00
10^{-2}	9.45	1.38
10^{-1}	10.24	1.66
1	10.69	1.66
∞	10.69	1.66

TABLE 4.7

Error Rates (%) Due to GMRF and Spatial Boosting for Data Set 1 with Variance–Covariance Matrix σ^2I

	$\sigma^2 = 1$			$\sigma^2 = 4$		
		Spatial Boosting			Spatial Boosting	
Radius r^2	Homoscedastic GMRF	AdaBoost	LogitBoost	Homoscedastic GMRF	AdaBoost	LogitBoost
0	41.66	41.66	41.66	54.45	54.45	54.45
1	9.25	20.40	19.45	31.07	40.88	40.77
2	3.61	12.15	11.63	15.58	33.79	33.95
4	**2.48**	8.77	8.36	14.70	28.78	28.74
5	2.72	5.72	5.93	**9.12**	22.71	22.44
8	3.06	5.39	5.43	10.61	20.50	20.55
9	3.30	4.82	4.96	11.00	18.96	19.01
10	3.79	4.38	4.65	12.66	16.58	16.56
13	4.13	4.34	4.54	14.06	14.99	15.08
16	4.17	4.31	4.43	14.95	14.14	14.12
17	4.67	**4.19**	4.31	15.76	13.02	13.22
18	4.94	4.34	4.58	15.89	12.57	12.69
20	6.11	4.43	4.66	17.00	**12.16**	12.11

TABLE 4.8

Error Rates (%) Due to GMRF and Spatial Boosting for Data Set 2

		Homoscedastic	
Radius r^2	Homoscedastic GMRF	AdaBoost	LogitBoost
0	8.61	8.61	8.61
4	6.09	7.08	7.93
9	5.69	6.11	7.33
16	**5.35**	5.95	6.74
25	5.69	**5.83**	6.35
36	5.69	5.85	**6.13**
49	6.13	5.99	6.32
64	6.65	6.01	6.48

TABLE 4.9

Error Rates (%) Due to GMRF and Spatial LogitBoost for Data Set 3

	Homoscedastic		Heteroscedastic	
Radius r^2	**GMRF**	**LogitBoost**	**GMRF**	**LogitBoost**
0	10.69	10.69	1.66	1.66
1	8.58	9.48	1.97	1.31
4	7.06	6.51	1.69	0.93
8	**3.81**	5.43	0.55	**0.76**
9	4.01	5.40	**0.52**	0.76
10	4.08	5.29	0.52	0.83
13	4.01	**5.16**	0.52	0.87
16	4.01	5.26	0.52	0.90

assigns less penalty to misclassified data. Hence, spatial LogitBoost seems to have a robust property.

4.7.5 Spatial Parallel Boost

We examine parallel boost by applying it to Data set 1. Three rectangular regions found in Figure 4.8a give training areas. Now, we consider the following four subsets of the training data set:

- Near the center of Figure 4.8a with sample size 75 ($D^{(1)}$)
- North-west of Figure 4.8a with sample size 160 ($D^{(2)}$)

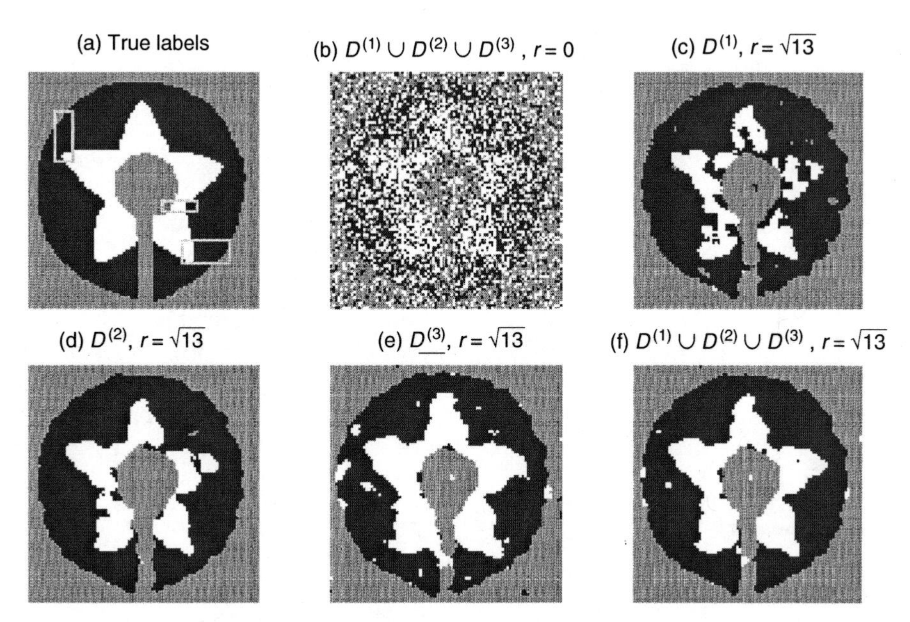

FIGURE 4.8

True labels, training regions, and classified images. The n denotes the size of training samples. (a) True labels with three training regions. (b) Parallel boost with $r = 0$ (error rate $= 42.29\%$, $n = 435$) noncontextual classification. (c) Spatial boost by $D^{(1)}$ with $r = \sqrt{13}$ (error rate $= 10.26\%$, $n = 75$). (d) Spatial boost by $D^{(2)}$ with $r = \sqrt{13}$ (error rate $= 6.29\%$, $n = 160$). (e) Spatial boost by $D^{(3)}$ with $r = \sqrt{13}$ (error rate $= 7.46\%$, $n = 200$). (f) Parallel boost by $D^{(1)} \cup D^{(2)} \cup D^{(3)}$ with $r = \sqrt{13}$ (error rate $= 4.96\%$, $n = 435$).

TABLE 4.10

Error Rates (%) Due to Spatial AdaBoost Based on Four Training Data Sets and Spatial Parallel Boost Based on Union of Three Subsets in Data Set 1

	Spatial AdaBoost				Parallel Boost
Radius r^2	$D^{(1)}$	$D^{(2)}$	$D^{(3)}$	Whole	$D^{(1)} \cup D^{(2)} \cup D^{(3)}$
0	43.13	41.67	45.88	41.75	42.29
1	23.84	20.23	26.51	20.40	21.18
2	17.16	12.84	18.92	12.15	13.03
4	13.88	9.44	15.25	8.77	9.46
5	10.88	7.34	11.21	5.72	6.39
8	10.49	6.97	10.26	5.39	5.78
9	10.48	6.64	9.26	4.82	5.41
10	10.41	6.35	8.08	4.38	5.19
13	**10.26**	**6.29**	**7.46**	**4.34**	**4.96**

- South-east of Figure 4.8a with sample size 200 ($D^{(3)}$)
- All training data with sample size 8281

The test data set is also of sample size 8281. The training region $D^{(1)}$ is relatively small, so we only consider radius r up to $\sqrt{13}$. Gaussian density functions with a common variance–covariance matrix are fitted to each training data set, and we obtain four types of posterior probabilities for each pixel in training data sets and the test data set. Spatial AdaBoost derived by the four training data sets and spatial parallel boost derived by the region $D^{(1)} \cup D^{(2)} \cup D^{(3)}$ are applied to the test region shown in Figure 4.2a.

Error rates due to five classifiers examined with the test data using 8281 samples are compared in Table 4.10. Each row corresponds to spatial boost based on radius r in $\{0, 1, \ldots, \sqrt{13}\}$. The second to fourth columns correspond to Spatial AdaBoost based on the four training data sets. The last column is the error rate due to spatial parallel boost. We see that spatial AdaBoost exhibits a better performance as the training data size increases. The classifier based on $D^{(3)}$ seems exceptional, but it gives better results for a large r. Spatial parallel boost exhibits a performance similar to that of spatial AdaBoost based on all the training data. Here, we note again that our method uses only 435 samples, whereas all the training data have 8281 samples.

A noncontextual classification result based on spatial parallel boost is given in Figure 4.8b, and it is very poor. An excellent classification result is obtained as radius r becomes bigger. The result with $r = \sqrt{13}$, which is better than the results (b), (d), and (f) due to spatial AdaBoost using three training data sets, is given in Figure 4.8e. The error rates corresponding to figures (b), (d), (e), and (f) are given in Table 4.10 in bold.

4.8 Conclusion

We have considered contextual classifiers, Switzer's smoothing method, MRF-based methods, and spatial boosting methods. We showed that the last two methods can be regarded as extensions of the Switzer's method. The MRF-based method is known to be efficient, and it exhibited the best performance in our numerical studies given here. However, the method requires extra computation time for (1) the estimation of parameters specifying the MRF and (2) the estimation of test labels, even if the ICM procedure is utilized for (2). Both estimations are accomplished by iterative procedures. The method has already been fully discussed in the literature. The following are our remarks on spatial boosting and its possible future development.

Spatial boosting is derived as long as posterior probabilities are available. Hence, posteriors need not be defined by statistical models. Posteriors due to statistical support vector machines (SVM) [22,23] could be utilized. Spatial boosting is derived by fewer assumptions than those of an MRF-based method. Features of spatial boosting are as follows:

(a) Various types of posteriors can be implemented.

   Pairwise coupling [24] is a popular strategy for multi-class classification that involves estimating posteriors for each pair of classes, and then coupling the estimates together.

(b) Directed neighborhoods can be implemented.

   See Figure 4.9 for the directed neighborhoods. They would be suitable for classification of textured images.

(c) The method is applicable to a situation with multi-source and multi-temporal data sets.

(d) Classification is performed non-iteratively, and the performance is similar to that of MRF-based classifiers.

The following new questions arise:

- How do we define the posteriors? This is closely related to feature (a).
- How do we choose the loss function?
- How do we determine the maximum radius for regarding log posteriors into the classification function.

This is an old question in boosting: How many classification functions are sufficient for classification?

Spatial boosting should be used in cases where estimation of coefficients is performed through test data only. This is a first step for possible development into unsupervised classification. In addition, the method will be useful for unmixing. Originally, classification function F defined by Equation 4.30 is given by a convex combination of averaged

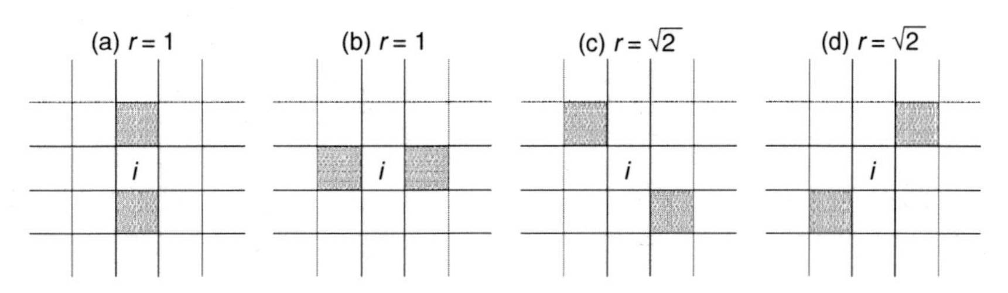

FIGURE 4.9
Directed subsets with center pixel i.

log posteriors. Therefore, F would be useful for estimation of posterior probabilities, which is directly applicable to unmixing.

Acknowledgment

Data sets grss_dfc_0006 and grss_dfc_0009 were provided by the IEEE GRS-S Data Fusion Committee.

References

1. Switzer, P., (1980) Extensions of linear discriminant analysis for statistical classification of remotely sensed satellite imagery, *Mathematical Geology*, 12(4), 367–376.
2. Nishii, R. and Eguchi, S. (2005), Supervised image classification by contextual AdaBoost based on posteriors in neighborhoods, *IEEE Transactions on Geoscience and Remote Sensing*, 43(11), 2547–2554.
3. Jain, A.K., Duin, R.P.W., and Mao, J. (2000), Statistical pattern recognition: a review, *IEEE Transaction on Pattern Analysis and Machine Intelligence*, PAMI-22, 4–37.
4. Besag, J. (1986), On the statistical analysis of dirty pictures, *Journal of the Royal Statistical Society, Series B*, 48, 259–302.
5. Duda, R.O., Hart, P.E., and Stork, D.G. (2000), *Pattern Classification* (2nd ed.), John Wiley & Sons, New York.
6. Geman, S. and Geman, D. (1984), Stochastic relaxation, Gibbs distribution, and Bayesian restoration of images, *IEEE Transactions on Pattern Analysis and Machine Intelligence*, PAMI-6, 721–741.
7. McLachlan, G.J. (2004), *Discriminant Analysis and Statistical Pattern Recognition* (2nd ed.), John Wiley & Sons, New York.
8. Chilès, J.P. and Delfiner, P. (1999), *Geostatistics*, John Wiley & Sons, New York.
9. Cressie, N. (1993), *Statistics for Spatial Data*, John Wiley & Sons, New York.
10. Freund, Y. and Schapire, R.E. (1997), A decision-theoretic generalization of on-line learning and an application to boosting, *Journal of Computer and System Sciences*, 55(1), 119–139.
11. Friedman, J., Hastie, T., and Tibshirani, R. (2000), Additive logistic regression: a statistical view of boosting (with discussion), *Annals of Statistics*, 28, 337–407.
12. Hastie, T., Tibshirani, R., and Friedman, J., (2001), *The Elements of Statistical Learning: Data Mining, Inference, and Prediction*, Springer, New York.
13. Benediktsson, J.A. and Kanellopoulos, I. (1999), Classification of multisource and hyperspectral data based on decision fusion, *IEEE Transactions on Geoscience and Remote Sensing*, 37, 1367–1377.
14. Melgani, F. and Bruzzone, L. (2004), Classification of hyperspectral remote sensing images with support vector machines, *IEEE Transactions on Geoscience and Remote Sensing*, 42(8), 1778–1790.
15. Nishii, R. (2003), A Markov random field-based approach to decision level fusion for remote sensing image classification, *IEEE Transactions on Geoscience and Remote Sensing*, 41(10), 2316–2319.
16. Nishii, R. and Eguchi, S. (2002), Image segmentation by structural Markov random fields based on Jeffreys divergence, *Res. Memo., Institute of Statistical Mathematics*, No. 849.
17. Nishii, R. and Eguchi, S. (2005), Robust supervised image classifiers by spatial AdaBoost based on robust loss functions, *Proc. SPIE*, Vol. 5982.
18. Eguchi, S. and Copas, J. (2002), A class of logistic-type discriminant functions, *Biometrika*, 89, 1–22.
19. Takenouchi, T. and Eguchi, S. (2004), Robustifying AdaBoost by adding the naive error rate, *Neural Computation*, 16, 767–787.

20. IEEE GRSS Data Fusion reference database, Data sets GRSS_DFC_0006, GRSS_DFC_0009 Online. http://www.dfc-grss.org/, 2001.
21. Giacinto, G., Roli, F., and Bruzzone, L. (2000), Combination of neural and statistical algorithms for supervised classification of remote-sensing images, *Pattern Recognition Letters*, 21, 385–397.
22. Kwok, J.T. (2000), The evidence framework applied to support vector machine, *IEEE Transactions on Neural Networks*, 11(5), 1162–1173.
23. Wahba, G. (1999), Support vector machines, reproducing kernel Hilbert spaces and the randomized GACV, In *Advances in Kernel Methods—Support Vector Learning*, Schölkopf, Burges and Smola, eds., MIT press, Cambridge, MA, 69–88.
24. Hastie, T. and Tibshirani, R. (1998), Classification by pairwise coupling, *Annals of Statistics*, 26(2), 451–471.

5

Unsupervised Change Detection in Multi-Temporal SAR Images

Lorenzo Bruzzone and Francesca Bovolo

CONTENTS

5.1 Introduction

The recent natural disasters (e.g., tsunami, hurricanes, eruptions, earthquakes, etc.) and the increasing amount of anthropogenic changes (e.g., due to wars, pollution, etc.) gave prominence to the topics related to environment monitoring and damage assessment. The study of environmental variations due to the time evolution of the above phenomena is of fundamental interest from a political point of view. In this context, the development of effective change-detection techniques capable of automatically identifying land-cover variations occurring on the ground by analyzing multi-temporal remote-sensing images assumes an important relevance for both the scientific community and the end-users. The change-detection process considers images acquired at different times over the same geographical area of interest. These images acquired from repeat-pass satellite sensors are an effective input for addressing change-detection problems. Several different Earth-observation satellite missions are currently operative, with different kinds of sensors mounted on board (e.g., MODIS and ASTER on board NASA's TERRA satellite, MERIS and ASAR on board ESA's ENVISAT satellite, Hyperion on board EO-1 NASA's satellite, SAR sensors on board RADARSAT-1 and RADARSAT-2 CSA's satellites, Ikonos and Quickbird satellites that acquire very high resolution pancromatic and multi-spectral (MS) images, etc.). Each sensor has specific properties with respect to the image acquisition mode (e.g., passive or active), geometrical, spectral, and radiometric resolutions, etc. In the development of automatic change-detection techniques, it is mandatory to take into account the properties of the sensors to properly extract information from the considered data.

Let us discuss the main characteristics of different kinds of sensors in detail (Table 5.1 summarizes some advantages and disadvantages of different sensors for change-detection applications according to their characteristics).

Images acquired from passive sensors are obtained by measuring the land-cover reflectance on the basis of the energy emitted from the sun and reflected from the ground[1]. Usually, the measured signal can be modeled as the desired reflectance (measured as a radiance) altered from an additive Gaussian noise. This noise model enables relatively easy processing of the signal when designing data analysis techniques. Passive sensors can acquire two different kinds of images [panchromatic (PAN) images and MS images] by defining different trade-offs between geometrical and spectral resolutions according to the radiometric resolution of the adopted detectors. PAN images are characterized by poor spectral resolution but very high geometrical resolution, whereas MS images have medium geometrical resolution but high spectral resolution. From the perspective of change detection, PAN images should be used when the expected size of the changed area is too small for adopting MS data. For example, in the case of the analysis of changes in urban areas, where detailed urban studies should be carried out, change detection in PAN images requires the definition of techniques capable of capturing the richness of information present both in the spatial-context relations between neighboring pixels and in the geometrical shapes of objects. MS data should be used

TABLE 5.1

Advantages and Disadvantages of Different Kinds of Sensors for Change-Detection Applications

Sensor	Advantages	Disadvantages
Multispectral (passive)	✓ Characterization of the spectral signature of land-covers	✓ Atmospheric conditions strongly affect the acquisition phase
	✓ The noise has an additive model	
Panchromatic (passive)	✓ High geometrical resolution	✓ Atmospheric conditions strongly affect the acquisition phase
	✓ High content of spatial-context information	✓ Poor characterization of the spectral signature of land-covers
SAR (active)	✓ Not affected by sunlight and atmospheric conditions	✓ Complexity of data preprocessing
		✓ Presence of multiplicative speckle noise

[1]Also, the emission of Earth affects the measurements in the infrared portion of the spectrum.

when a medium geometrical resolution (i.e., 10–30 m) is sufficient for characterizing the size of the changed areas and a detailed modeling of the spectral signature of the land-covers is necessary for identifying the change investigated. Change-detection methods in MS images should be able to properly exploit the available MS information in the change detection process. A critical problem related to the use of passive sensors in change detection consists in the sensitivity of the image-acquisition phase to atmospheric conditions. This problem has two possible effects: (1) atmospheric conditions may not be conducive to measure land-cover spectral signatures, which depends on the presence of clouds; and (2) variations in illumination and atmospheric conditions at different acquisition times may be a potential source of errors, which should be taken into account to avoid the identification of false changes (or the missed detection of true changes).

The working principle of active synthetic aperture radar (SAR) sensors is completely different from that of the passive ones and allows overcoming some of the drawbacks that affect optical images. The signal measured by active sensors is the Earth backscattering of an electromagnetic pulse emitted from the sensor itself. SAR instruments acquire different kinds of signals that result in different images: medium or high-resolution images, single-frequency or multi-frequency, and single-polarimetric or fully polarimetric images. As for optical data, the proper geometrical resolution should be chosen according to the size of the expected investigated changes. The SAR signal has different geometrical resolutions and a different penetration capability depending on the signal wavelength, which is usually included between band X and band P (i.e., between 2 and 100 cm). In other words, shorter wavelengths should be used for measuring vegetation changes and longer and more penetrating wavelengths for studying changes that have occurred on or under the terrain. All the wavelengths adopted for SAR sensors neither suffer from atmospheric and sunlight conditions nor from the presence of clouds; thus multi-temporal radar backscattering does not change with atmospheric conditions. The main problem related to the use of active sensors is the coherent nature of the SAR signal, which results in a multiplicative speckle noise that makes acquired data intrinsically complex to be analyzed. A proper handling of speckle requires both an intensive preprocessing phase and the development of effective data analysis techniques.

The different properties and statistical behaviors of signals acquired by active and passive sensors require the definition of different change-detection techniques capable of properly exploiting the specific data peculiarities.

In the literature, many different techniques for change detection in images acquired by passive sensors have been presented [1–8], and many applications of these techniques have been reported. This is because of both the amount of information present in MS images and the relative simplicity of data analysis, which results from the additive noise model adopted for MS data (the radiance of natural classes can be approximated with a Gaussian distribution). Less attention has been devoted to change detection in SAR images. This is explained by the intrinsic complexity of SAR data, which require both an intensive preprocessing phase and the development of effective data analysis techniques capable of dealing with multiplicative speckle noise. Nonetheless, in the past few years the remote-sensing community has shown more interest in the use of SAR images in change-detection problems, due to their independence from atmospheric conditions that results in excellent operational properties. The recent technological developments in sensors and satellites have resulted in the design of more sophisticated systems with increased geometrical resolution. Apart from the active or passive nature of the sensor, the very high geometrical resolution images acquired by these systems (e.g., PAN images) require the development of specific techniques capable of taking advantage of the richness of the geometrical information they contain. In particular, both the high correlation between neighboring pixels and the object shapes should be considered in the design of data analysis procedures.

In the above-mentioned context, two main challenging issues of particular interest in the development of automatic change-detection techniques are: (1) the definition of advanced and effective techniques for change detection in SAR images, and (2) the development of proper methods for the detection of changes in very high geometrical resolution images. A solution for these issues lies in the definition of multi-scale and multi-resolution change-detection techniques, which can properly analyze the different components of the change signal at their optimal scale[2]. On the one hand, the multi-scale analysis allows one to better handle the noise present in medium-resolution SAR images, resulting in the possibility of obtaining accurate change-detection maps characterized by a high spatial fidelity. On the other hand, multi-scale approaches are intrinsically suitable to exploit the information present in very high geometrical resolution images according to effective modeling (at different resolution levels) of the different objects present at the scene.

According to the analysis mentioned above, after a brief survey on change detection and on unsupervised change detection in SAR images, we present, in this chapter, a novel adaptive multi-scale change detection technique for multi-temporal SAR images. This technique exploits a proper scale-driven analysis to obtain a high sensitivity to geometrical features (i.e., details and borders of changed areas are well preserved) and a high robustness to noisy speckle components in homogeneous areas. Although explicitly developed and tested for change detection in medium-resolution SAR images, this technique can be easily extended to the analysis of very high geometrical resolution images.

The chapter is organized into five sections. Section 5.2 defines the change-detection problem in multi-temporal remote-sensing images and focuses attention on unsupervised techniques for multi-temporal SAR images. Section 5.3 presents a multi-scale approach to change detection in multi-temporal SAR images recently developed by the authors. Section 5.4 gives an example of the application of the proposed multi-scale technique to a real multi-temporal SAR data set and compares the effectiveness of the presented method with those of standard single-scale change-detection techniques. Finally, in Section 5.5, results are discussed and conclusions are drawn.

5.2 Change Detection in Multi-Temporal Remote-Sensing Images: Literature Survey

5.2.1 General Overview

A very important preliminary step in the development of a change-detection system, based on automatic or semi-automatic procedures, consists in the design of a proper phase of data collection. The phase of data collection aims at defining: (1) the kind of satellite to be used (on the basis of the repetition time and on the characteristics of the sensors mounted on-board), (2) the kind of sensor to be considered (on the basis of the desired properties of the images and of the system), (3) the end-user requirements (which are of basic importance for the development of a proper change-detection

[2]It is worth noting that these kinds of approaches have been successfully exploited in image classification problems [9–12].

technique), and (4) the kinds of available ancillary data (all the available information that can be used for constraining the change-detection procedure).

The outputs of the data-collection phase should be used for defining the automatic change-detection technique. In the literature, many different techniques have been proposed. We can distinguish between two main categories: supervised and unsupervised methods [9,13].

When performing supervised change detection, in addition to the multi-temporal images, multi-temporal ground-truth information is also needed. This information is used for identifying, for each possible land-cover class, spectral signature samples for performing supervised data classification and also for explicitly identifying what kinds of land-cover transitions have taken place. Three main general approaches to supervised change detection can be found in the literature: *postclassification comparison, supervised direct multi-data classification* [13], and *compound classification* [14–16]. Postclassification comparison computes the change-detection map by comparing the classification maps obtained by classifying independently two multi-temporal remote-sensing images. On the one hand, this procedure avoids data normalization aimed at reducing atmospheric conditions, sensor differences, etc. between the two acquisitions; on the other hand, it critically depends on the accuracies of the classification maps computed at the two acquisition dates. As postclassification comparison does not take into account the dependence existing between two images of the same area acquired at two different times, the global accuracy is close to the product of the accuracies yielded at the two times [13]. Supervised direct multi-data classification [13] performs change detection by considering each possible transition (according to the available *a priori* information) as a class and by training a classifier to recognize the transitions. Although this method exploits the temporal correlation between images in the classification process, its major drawback is that training pixels should be related to the same points on the ground at the two times and should accurately represent the proportions of all the transitions in the whole images. Compound classification overcomes the drawbacks of supervised multi-date classification technique by removing the constraint that training pixels should be related to the same area on the ground [14–16]. In general, the approach based on supervised classification is more accurate and detailed than the unsupervised one; nevertheless, the latter approach is often preferred in real-data applications. This is due to the difficulties in collecting proper ground-truth information (necessary for supervised techniques), which is a complex, time consuming, and expensive process (in many cases this process is not consistent with the application constraints).

Unsupervised change-detection techniques are based on the comparison of the spectral reflectances of multi-temporal raw images and a subsequent analysis of the comparison output. In the literature, the most widely used unsupervised change-detection techniques are based on a three-step procedure [13,17]: (1) preprocessing, (2) pixel-by-pixel comparison of two raw images, and (3) image analysis and thresholding (Figure 5.1).

The aim of the preprocessing step is to make the two considered images as comparable as possible. In general, preprocessing operations include: co-registration, radiometric and geometric corrections, and noise reduction. From the practical point of view, co-registration is a fundamental step as it allows obtaining a pair of images where corresponding pixels are associated to the same position on the ground[3]. Radiometric corrections reduce differences between the two acquisitions due to sunlight and atmospheric conditions. These procedures are applied to optical images, but they are not necessary for SAR

[3]It is worth noting that usually it is not possible to obtain a perfect alignment between temporal images. This may considerably affect the change-detection process [18]. Consequently, if the amount of residual misregistration noise is significant, proper techniques aimed at reducing its effects should be used for change detection [1,4].

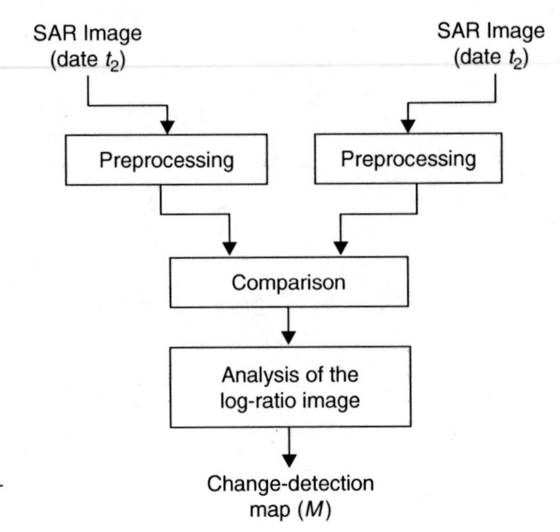

FIGURE 5.1
Block scheme of a standard unsupervised change-detection approach.

images (as SAR data are not affected by atmospheric conditions). Also noise reduction is performed differently according to the kind of remote-sensing images considered. In optical images common low-pass filters can be used, whereas in SAR images proper despeckling filters should be applied.

The comparison step aims at producing a further image where differences between the two acquisitions considered are highlighted. Different mathematical operators (see Table 5.2 for a summary) can be adopted for performing image comparison; this choice gives rise to different kinds of techniques [13,19–23]. One of the most widely used operators is the difference one. The difference can be applied to: (1) a single spectral band (*univariate image differencing*) [13,21–23], (2) multiple spectral bands (*change vector analysis*) [13,24], and (3) vegetation indices (*vegetation index differencing*) [13,19] or other linear (e.g., *tasselled cap transformation* [22]) or nonlinear combinations of spectral bands. Another widely used operator is the ratio operator (*image ratioing*) [13], which can be successfully used in SAR image processing [17,25,26]. A different approach is based on the use of the principal component analysis (PCA) [13,20,23]. PCA can be applied separately to the feature space at single times or jointly to both images. In the first case, comparison should be performed in the transformed feature space before performing change detection; in the second case, the minor components of the transformed feature space contains change information.

TABLE 5.2

Summary of the Most Widely Used Comparison Operators. (f_k is the considered feature at time t_k that can be: (1) a single spectral band X_k^b, (2) a vector of m spectral bands $[X_k^1,\ldots,X_k^m]$, (3) a vegetation index V_k, or (4) a vector of features $[P_k^1,\ldots,P_k^m]$ obtained after PCA. $\mathbf{X}_\mathrm{D}$ and $\mathbf{X}_\mathrm{R}$ are the images after comparison with the difference or ratio operators, respectively)

Technique	Feature Vector f_k at the Time t_k	Comparison Operator
Univariate image differencing	$f_k = X_k^b$	$X_\mathrm{D} = f_2 - f_1$
Vegetation index differencing	$f_k = V_k$	$X_\mathrm{D} = f_2 - f_1$
Image rationing	$f_k = X_k^b$	$X_\mathrm{R} = f_2/f_1$
Change vector analysis	$f_k = [X_k^1,\ldots,X_k^m]$	$X_\mathrm{D} = \lVert f_2 - f_1 \rVert$
Principal component analysis	$f_k = [P_k^1,\ldots,P_k^n]$	$X_\mathrm{D} = \lVert f_2 - f_1 \rVert$

Performances of the above-mentioned techniques could be degraded by several factors (like differences in illumination at two dates, differences in atmospheric conditions, and in sensor calibration) that make a direct comparison between raw images acquired at different times difficult. These problems related to unsupervised change detection disappear when dealing with SAR images instead of optical data.

Once image comparison is performed, the decision threshold can be selected either with a manual trial-and-error procedure (according to the desired trade-off between false and missed alarms) or with automatic techniques (e.g., by analyzing the statistical distribution of the image obtained after comparison, by fixing the desired false alarm probability [27,28], or following a Bayesian minimum-error decision rule [17]).

Since the remote-sensing community has devoted more attention to passive sensors [6–8] rather than active SAR sensors, in the following section we focus our attention on change-detection techniques for SAR data. Although change-detection techniques based on different architectures have been proposed for SAR images [29–37], we focus on the most widely used techniques, which are based on the three-step procedure described above (see Figure 5.1).

5.2.2 Change Detection in SAR Images

Let us consider two co-registered intensity SAR images, $X_1 = \{X_1(i,j), 1 \leq i \leq I, 1 \leq j \leq J\}$ and $X_2 = \{X_2(i,j), 1 \leq i \leq I, 1 \leq j \leq J\}$, of size $I \cdot J$, acquired over the same area at different times t_1 and t_2. Let $\Omega = \{\omega_c, \omega_u\}$ be the set of classes associated with changed and unchanged pixels. Let us assume that no ground-truth information is available for the design of the change-detection algorithm, i.e., the statistical analysis of change and no-change classes should be performed only on the basis of the raw data. The change-detection process aims at generating a change-detection map representing changes on the ground between the two considered acquisition dates. In other words, one of the possible labels in Ω should be assigned to each pixel (i,j) in the scene.

5.2.2.1 Preprocessing

The first step for properly performing change detection based on direct image comparison is image preprocessing. This procedure aims at generating two images that are as similar as possible unless in changed areas. As SAR data are not corrupted by differences in atmospheric and sunlight conditions, preprocessing usually comprises three steps: (1) geometric correction, (2) co-registration, and (3) noise reduction. The first procedure aims at reducing distortions that are strictly related to the active nature of the SAR signal, as layover, foreshortening, and shadowing due to ground topography. The second step is very important, as it allows aligning temporal images to ensure that corresponding pixels in the spatial domain are associated to the same geographical position on the ground. Co-registration in SAR images is usually carried out by maximizing cross-correlation between the multi-temporal images [38,39]. The major drawback of this process is the need for performing interpolation of backscattering values, which is a time-consuming process. Finally, the last step is aimed at reducing the speckle noise. Many different techniques have been developed in the literature for reducing the speckle. One of the most attractive techniques for speckle reduction is multi-looking [25]. This procedure, which is used for generating images with the same resolution along the azimuth and range directions, allows reduction of the effect of the coherent speckle components. However, a further filtering step is usually applied to the images for making them suitable to the desired analysis. Usually, adaptive despeckling procedures are applied. Among these procedures we mention the following filtering techniques:

Frost [40], Lee [41], Kuan [42], Gamma Map [43,44], and Gamma WMAP [45] (i.e., the Gamma MAP filter applied in the wavelet domain). As the description of despeckling filters is outside the scope of this chapter, we refer the reader to the literature for more details.

5.2.2.2 Multi-Temporal Image Comparison

As described in Section 5.1, image pixel-by-pixel comparison can be performed by means of different mathematical operators. In general, the most widely used operators are the difference and the ratio (or log-ratio). Depending on the selected operator, the image resulting from the comparison presents different behaviors with respect to the change-detection problem and to the signal statistics. To analyze this issue, let us consider two multi-look intensity images. It is possible to show that the measured backscattering of each image follows a Gamma distribution [25,26], that is,

$$p(X_k) = \frac{L^L X_k^{L-1}}{m_k^L (L-1)!} \exp\left(-\frac{LX_k}{m_k}\right), \quad k = 1, 2 \tag{5.1}$$

where X_k is a random variable that represents the value of the pixels in image X_k ($k = 1, 2$), m_k is the average intensity of a homogeneous region at time t_k, and L is the equivalent number of looks (ENL) of the considered image. Let us also assume that the intensity images X_1 and X_2 are statistically independent. This assumption, even if not entirely realistic, simplifies the analytical derivation of the pixel statistical distribution in the image after comparison. In the following, we analyze the effects of the use of the difference and ratio (log-ratio) operators on the statistical distributions of the signal.

5.2.2.2.1 *Difference Operator*

The difference image $\mathbf{X}_D$ is computed subtracting the image acquired before the change from the image acquired after, i.e.,

$$\mathbf{X}_D = \mathbf{X}_2 - \mathbf{X}_1 \tag{5.2}$$

Under the stated conditions, the distribution of the difference image $\mathbf{X}_D$ is given by [25,26]:

$$p(X_D) = \frac{L^L}{(L-1)!} \frac{\exp\left\{-L\frac{X_D}{m_2}\right\}}{(m_1 + m_2)^L} \times \sum_{j=0}^{L-1} \frac{(L-1+j)!}{j!(L-1-j)!} \cdot X_D^{L-1-j} \left[\frac{m_1 m_2}{L(m_1 + m_2)}\right]^j \tag{5.3}$$

where X_D is a random variable that represents the values of the pixels in $\mathbf{X}_D$. As can be seen, the difference-image distribution depends on both the relative change between the intensity values in the two images and also a reference intensity value (i.e., the intensity at t_1 or t_2). It is possible to show that the distribution variance of X_D increases with the reference intensity level. From a practical point of view, this leads to a higher change-detection error for changes that have occurred in high intensity regions of the image than in low intensity regions. Although in some applications the difference operator was used with SAR data [46], this behavior is an undesired effect that renders the difference operator intrinsically not suited to the statistics of SAR images.

5.2.2.2.2 Ratio Operator

The ratio image $\mathbf{X}_\mathrm{R}$ is computed by dividing the image acquired after the change by the image acquired before (or *vice-versa*), i.e.,

$$\mathbf{X}_\mathrm{R} = \mathbf{X}_2/\mathbf{X}_1 \tag{5.4}$$

It is possible to prove that the distribution of the ratio image $\mathbf{X}_\mathrm{R}$ can be written as follows [25,26]:

$$p(X_\mathrm{R}) = \frac{(2L-1)!}{(L-1)!^2} \frac{\bar{X}_\mathrm{R}^L X_\mathrm{R}^{L-1}}{(\bar{X}_\mathrm{R} + X_\mathrm{R})^{2L}} \tag{5.5}$$

where X_R is a random variable that represents the values of the pixels in $\mathbf{X}_\mathrm{R}$ and $\bar{X}_\mathrm{R}$ is the true change in the radar cross section. The ratio operator shows two main advantages over the difference operator. The first one is that the ratio-image distribution depends only on the relative change $\bar{X}_\mathrm{R} = m_2/m_1$ in the average intensity between the two dates and not on a reference intensity level. Thus changes are detected in the same manner both in high- and low-intensity regions. The second advantage is that the ratioing allows reduction in common multiplicative error components (which are due to both multiplicative sensor calibration errors and the multiplicative effects of the interaction of the coherent signal with the terrain geometry [25,47]), as far as these components are the same for images acquired with the same geometry. It is worth noting that, in the literature, the ratio image is usually expressed in a logarithmic scale. With this operation the distribution of the two classes of interest (ω_c and ω_u) in the ratio image can be made more symmetrical and the residual multiplicative speckle noise can be transformed to an additive noise component [17]. Thus the log-ratio operator is typically preferred when dealing with SAR images and change detection is performed analyzing the log-ratio image $\mathbf{X}_\mathrm{LR}$ defined as:

$$\mathbf{X}_\mathrm{LR} = \log \mathbf{X}_\mathrm{R} = \log \frac{\mathbf{X}_2}{\mathbf{X}_1} = \log \mathbf{X}_2 - \log \mathbf{X}_1 \tag{5.6}$$

Based on the above considerations, the ratio and log-ratio operators are more used than the difference one in SAR change-detection applications [17,26,29,47–49]. It is worth noting that for keeping the changed class on one side of the histogram of the ratio (or log-ratio) image, a normalized ratio can be computed pixel-by-pixel, i.e.,

$$\mathbf{X}_\mathrm{NR} = \min\left\{\frac{\mathbf{X}_1}{\mathbf{X}_2}, \frac{\mathbf{X}_2}{\mathbf{X}_1}\right\} \tag{5.7}$$

This operator allows all changed areas (independently of the increasing or decreasing value of the backscattering coefficient) to play a similar role in the change-detection problem.

5.2.2.3 Analysis of the Ratio and Log-Ratio Image

The most widely used approach to extract change information from the ratio and log-ratio image is based on histogram thresholding[4]. In this context, the most difficult task is to

[4]For simplicity, in the following, we will refer to the log-ratio image.

properly define the threshold value. Typically, changed pixels are identified as those pixels that modified their backscattering more than $\pm x$ dB, where x is a real number depending on the considered scene. The value of x is fixed according to the kind of change and the expected magnitude variation to obtain a desired probability of correct detection P_d (which is the probability to be over the threshold if a change occurred) or false alarm P_{fa} (which is the probability to be over the threshold if no change occurred). It has been shown that the value of x can be analytically defined as a function of the true change in the radar backscattering $\bar{X}_R$ and of the ENL L [25,26], once P_d and P_{fa} are fixed. This means that there exists a value of L such that the given constraints on P_d and P_{fa} are satisfied. The major drawback of this approach is that, as the desired change intensity decreases and the detection probability increases, a ratio image with an even higher ENL is required for constraint satisfaction. This is due to the sensitivity of the ratio to the presence of speckle; thus a complex preprocessing procedure is required for increasing the ENL. A similar approach is presented in Ref. [46]; it identifies the decision threshold on the basis of predefined values on the cumulative histogram of the difference image. It is worth noting that these approaches are not fully automatic and objective from an application point of view, as they depend on the user's sensibility in constraint definition with respect to the considered kind of change.

Recently, extending the work previously carried out for MS passive images [3,5,23,50], a novel Bayesian framework has been developed for performing automatic unsupervised change detection in the log-ratio image derived from SAR data. The aim of this framework is to use the well-known Bayes decision theory in unsupervised problems for deriving decision thresholds that optimize the separation between changed and unchanged pixels. The main problems to be solved for the application of the Bayes decision theory consist in the estimation of both the probability density functions $p(X_{LR}/\omega_c)$ and $p(X_{LR}/\omega_u)$ and the *a-priori* probabilities $P(\omega_c)$ and $P(\omega_u)$ of the classes ω_c and ω_u, respectively [51], without any ground-truth information (i.e., without any training set). The starting point of such kinds of methodologies is the hypothesis that the statistical distributions of pixels in the log-ratio image can be modeled as a mixture of two densities associated with the classes of changed and unchanged pixels, i.e.,

$$p(X_{LR}) = p(X_{LR}/\omega_u)P(\omega_u) + p(X_{LR}/\omega_c)P(\omega_c) \tag{5.8}$$

Under this hypothesis, two different approaches to estimate class statistical parameters have been proposed in the literature: (1) an implicit approach [17] and (2) an explicit approach [49]. The first approach derives the decision threshold according to an implicit and biased parametric estimation of the statistical model parameters, carried out on the basis of simple cost functions. In this case, the change-detection map is computed in a one-step procedure. The second approach separates the image analysis in two steps: (1) estimation of the class statistical parameters and (2) definition of the decision threshold based on the estimated statistical parameters. Both techniques require the selection of a proper statistical model for the distributions of the change and no-change classes. In Ref. [17], it has been shown that the generalized Gaussian distribution is a flexible statistical model that allows handling the complexity of the log-ratio images better than the more commonly used Gaussian distribution. Based on this consideration, in Refs. [17] the well-known Kittler and Illingworth (KI) thresholding technique (which is an implicit estimation approach) [52–55] was reformulated under the generalized Gaussian assumption for the statistical distributions of classes. Despite its simplicity, the KI technique produces satisfactory change-detection results. The alternative approach, proposed in Refs. [56–58], which is based on a theoretically more precise explicit procedure for the estimation of statistical parameters of classes, exploits the combined use of the expectation–maximization (EM) algorithm and of the Bayesian decision theory for producing the change-detection map. In Ref. [49], the iterative EM algorithm (reformulated under the hypothesis of generalized Gaussian data distribution) was successfully applied to the analysis of the log-ratio image. Once the statistical parameters are computed, pixel-based or context-based decision rules can be applied. In the former group, we find the Bayes rule for minimum error, the Bayes rule for minimum cost, the Neyman–Pearson criterion, etc. [3,5,24,59]. In the latter group, we find the contextual Bayes rule for minimum error formulated in the Markov random field (MRF) framework [24,60]. In both cases (implicit and explicit parameter estimation), change-detection accuracy increases as the ENL increases. This means that, depending on the data and the application, an intensive despeckling phase may be required to achieve good change-detection accuracies [17]. It is worth noting that Equation 5.8 assumes that only one kind of change[5] has occurred in the area under investigation between the two acquisition dates. However, techniques that can automatically identify the number of changes and the related threshold values have been recently proposed in the literature (both in the context of implicit [61] and explicit [60] approaches). We refer the reader to the literature for greater details on these approaches.

Depending on the kind of preprocessing applied to the multi-temporal images, the standard thresholding techniques can achieve different trade-offs between the preservation of detail and accuracy in the representation of homogeneous areas in change-detection maps. In most of the applications, both properties need to be satisfied; however, they are contrasting to each other. On the one hand, high accuracy in homogeneous areas usually requires an intensive despeckling phase; on the other hand, intensive despeckling degrades the geometrical details in the SAR images. This is due to both the smoothing effects of the filter and the removal of the informative components of the speckle (which are related to the coherent properties of the SAR signal).

To address the above limitations of standard methods, in the next section we present an adaptive scale-driven approach to the analysis of the log-ratio image recently developed by the authors.

5.3 Advanced Approaches to Change Detection in SAR Images: A Detail-Preserving Scale-Driven Technique

In this chapter we present a scale-driven approach to unsupervised change detection in SAR images, which is based on a multi-scale analysis of the log-ratio image. This approach can be suitably applied to medium- and high-resolution SAR images to produce change-detection maps characterized by high accuracy both in modeling details present in the scene (e.g., border of changed areas) and in homogeneous regions. Multi-temporal SAR images intrinsically contain different areas of changes in the spatial domain that can be modeled at different spatial resolutions. The identification of these areas with high accuracy requires the development of proper change-detection techniques capable of handling information at different scales. The rationale of the presented scale-driven unsupervised change-detection method is to exploit only high-resolution levels in the analysis of the expected edge (or detail) pixels and to use low-resolution levels also in the processing of pixels in homogeneous areas, to improve both preservation of geometrical detail and accuracy in homogeneous areas in the final change-detection map. In

[5]Or different kinds of changes that can be represented with a single generalized Gaussian distribution.

the following, we present the proposed multi-scale change-detection technique in the context of the analysis of multi-temporal SAR images. However, we expect that the methodology has general validity and can also be applied successfully to change detection in very high resolution optical and SAR images with small modifications in the implementation procedures.

The presented scale-driven technique is based on three main steps: (1) multi-scale decomposition of the log-ratio image, (2) adaptive selection of the reliable scales for each pixel (i.e., the scales at which the considered pixel can be represented without border details problems) according to an adaptive analysis of its local statistics, and (3) scale-driven combination of the selected scales (Figure 5.2). Scale-driven combination can be performed by following three different strategies: (1) fusion at the decision level by an "optimal" scale selection, (2) fusion at the decision level of all reliable scales, and (3) fusion at the feature level of all reliable scales.

The first step of the proposed method aims at building a multi-scale representation of the change information in the considered test site. The desired scale-dependent representation can be obtained by applying different decomposition techniques to the data, such as Laplacian–Gaussian pyramid decomposition [62], wavelet transform [63,64], recursively upsampled bicubic filter [65], etc. Given the computational cost and the assumption of the additive noise model required by the above techniques, we chose to apply the multiresolution decomposition process to the log-ratio image $\mathbf{X}_{LR}$, instead of decomposing the two original images $\mathbf{X}_1$ and $\mathbf{X}_2$ separately. At the same time this allows a reduction in the computational cost and satisfies the additive noise model hypothesis. The selection of the most appropriate multi-resolution technique is related to the statistical behaviors of $\mathbf{X}_{LR}$ and will be discussed in the next section. The multi-resolution decomposition step produces a set of images $\mathbf{X}_{MS} = \{\mathbf{X}_{LR}^0, \ldots, \mathbf{X}_{LR}^n, \ldots, \mathbf{X}_{LR}^{N-1}\}$, where the superscript n ($n = 0, 1, \ldots, N-1$) indicates the resolution level. As we consider a dyadic decomposition process, the scale corresponding to each resolution level is given by 2^n. In our notation, the output at resolution level 0 corresponds to the original image, i.e., $\mathbf{X}_{LR}^0 \equiv \mathbf{X}_{LR}$. For n ranging from 0 to $N-1$, the obtained images are distinguished by different trade-offs between preservation of spatial detail and speckle reduction. In particular, images with a low value of n are strongly affected by speckle, but they are characterized by a large amount of geometrical details, whereas images identified by a high value of n show significant speckle reduction and contain degraded geometrical details (high frequencies are smoothed out).

In the second step, local and global statistics are evaluated for each pixel at different resolution levels. At each level and for each spatial position, by comparing the local and

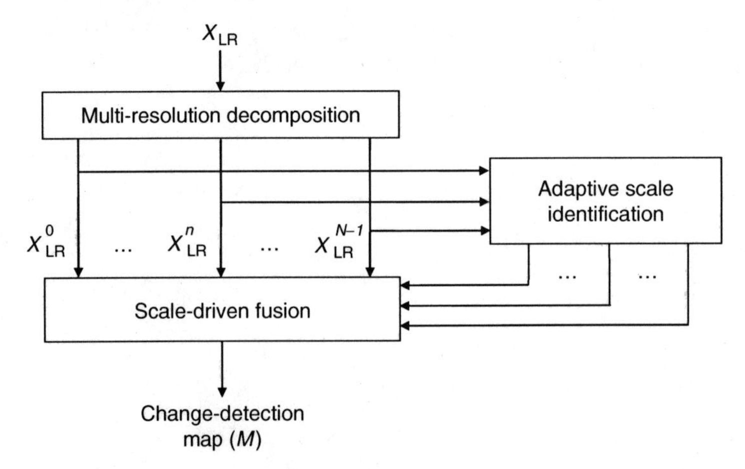

FIGURE 5.2
General scheme of the proposed approach.

global statistical behaviors it is possible to identify adaptively whether the considered scale is reliable for the analyzed pixel.

The selected scales are used to drive the last step, which consists of the generation of the change-detection map according to a scale-driven fusion. In this paper, three different scale-driven combination strategies are proposed and investigated. Two perform fusion at the decision level, while the third performs it at the feature level. Fusion at the decision level can either be based on "optimal" scale selection or on the use of all reliable scales; fusion at the feature level is carried out by analyzing all reliable scales.

5.3.1 Multi-Resolution Decomposition of the Log-Ratio Image

As mentioned in the previous section, our aim is to handle the information at different scales (resolution levels) to improve both preservation of geometrical detail and accuracy in homogeneous areas in the final change-detection map. Images included in the set $\mathbf{X}_{\mathrm{MS}}$ are computed by adopting a multi-resolution decomposition process of the log-ratio image $\mathbf{X}_{\mathrm{LR}}$. In the SAR literature [45,66–70], image multi-resolution representation has been applied extensively to image de-noising. Here, a decomposition based on the two-dimensional discrete stationary wavelet transform (2D-SWT) has been adopted, as in our image analysis framework it has a few advantages (as described in the following) over the standard discrete wavelet transform (DWT) [71]. As the log-ratio operation transforms the SAR signal multiplicative model into an additive noise model, SWT can be applied to $\mathbf{X}_{\mathrm{LR}}$ without any additional processing. 2D-SWT applies appropriate level-dependent high- and low-pass filters with impulse response $h^n(\cdot)$ and $l^n(\cdot)$, $(n = 0, 1, \ldots, N-1)$, respectively, to the considered signal at each resolution level. A one-step wavelet decomposition is based on both level-dependent high- and low-pass filtering, first along rows and then along columns to produce four different images at the next scale. After each convolution step, unlike DWT, SWT avoids down-sampling the filtered signals. Thus, according to the scheme in Figure 5.3, the image $\mathbf{X}_{\mathrm{LR}}$ is decomposed into four images of the same size as the original. In particular, decomposition produces: (1) a lower resolution version $\mathbf{X}_{\mathrm{LR}}^{\mathrm{LL}_1}$ of image $\mathbf{X}_{\mathrm{LR}}$, which is called the approximation sub-band, and contains low spatial frequencies both in the horizontal and the vertical direction at resolution level 1; and (2) three high-frequency images $\mathbf{X}_{\mathrm{LR}}^{\mathrm{LH}_1}$, $\mathbf{X}_{\mathrm{LR}}^{\mathrm{HL}_1}$, and $\mathbf{X}_{\mathrm{LR}}^{\mathrm{HH}_1}$, which correspond to the horizontal, vertical, and diagonal detail sub-bands at resolution level 1, respectively. Note that, superscripts LL, LH, HL, and HH specify the order in which high-(H) and low-(L) pass filters have been applied to obtain the considered sub-band.

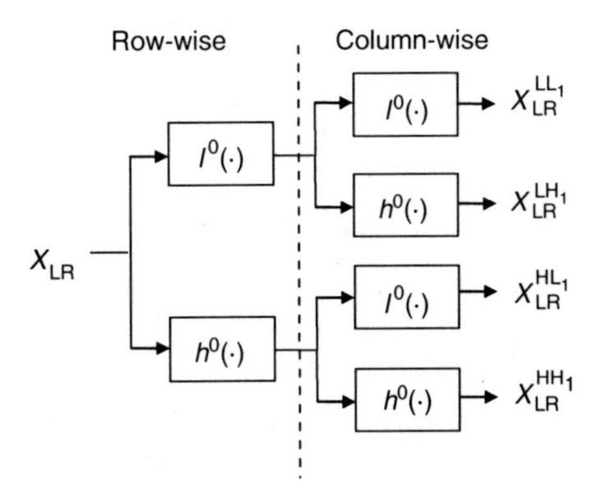

FIGURE 5.3
Block scheme of the stationary wavelet decomposition of the log-ratio image $\mathbf{X}_{\mathrm{LR}}$.

Multi-resolution decomposition is obtained by recursively applying the described procedure to the approximation sub-band $\mathbf{X}_{\mathrm{LR}}^{\mathrm{LL}_n}$ obtained at each scale 2^n. Thus, the outputs at a generic resolution level n can be expressed analytically as follows:

$$
\begin{aligned}
\mathbf{X}_{\mathrm{LR}}^{\mathrm{LL}_{(n+1)}}(i,j) &= \sum_{p=0}^{D^n-1}\sum_{q=0}^{D^n-1} l^n[p]l^n[q]\mathbf{X}_{\mathrm{LR}}^{\mathrm{LL}_n}(i+p,j+q)\\
\mathbf{X}_{\mathrm{LR}}^{\mathrm{LH}_{(n+1)}}(i,j) &= \sum_{p=0}^{D^n-1}\sum_{q=0}^{D^n-1} l^n[p]h^n[q]\mathbf{X}_{\mathrm{LR}}^{\mathrm{LL}_n}(i+p,j+q)\\
\mathbf{X}_{\mathrm{LR}}^{\mathrm{HL}_{(n+1)}}(i,j) &= \sum_{p=0}^{D^n-1}\sum_{q=0}^{D^n-1} h^n[p]l^n[q]\mathbf{X}_{\mathrm{LR}}^{\mathrm{LL}_n}(i+p,j+q)\\
\mathbf{X}_{\mathrm{LR}}^{\mathrm{HH}_{(n+1)}}(i,j) &= \sum_{p=0}^{D^n-1}\sum_{q=0}^{D^n-1} h^n[p]h^n[q]\mathbf{X}_{\mathrm{LR}}^{\mathrm{LL}_n}(i+p,j+q)
\end{aligned}
\tag{5.9}
$$

where D^n is the length of the wavelet filters at resolution level n. At each decomposition step, the length of the impulse response of both high- and low-pass filters is up-sampled by a factor 2. Thus, filter coefficients for computing sub-bands at resolution level $n+1$ can be obtained by applying a dilation operation to the filter coefficients used to compute level n. In particular, 2^{n-1} zeros are inserted between the filter coefficients used to compute sub-bands at the lower resolution levels [71]. This allows a reduction in the bandwidth of the filters by a factor two between subsequent resolution levels.

Filter coefficients of the first decomposition step for $n = 0$ depend on the selected wavelet family and on the length of the chosen wavelet filter. According to an analysis of the literature [68,72], we selected the *Daubechies* wavelet family and set the filter length to 8. The impulse response of *Daubechies* of order 4 low-pass filter prototype is given by the following coefficients set: {0.230378, 0.714847, 0.630881, −0.0279838, −0.187035, 0.0308414, 0.0328830, −0.0105974}.

The finite impulse response of the high-pass filter for the decomposition step is obtained by satisfying the properties of the quadrature mirror filters. This is done by reversing the order of the low-pass decomposition filter coefficients and by changing the sign of the even indexed coefficients [73].

To adopt the proposed multi-resolution fusion strategies, one should return to the original image domain. This is done by applying the two-dimensional inverse stationary wavelet transform (2D-ISWT) at each computed resolution level independently. For further detail about the stationary wavelet transform, the reader is referred to Ref. [71].

To obtain the desired image set $\mathbf{X}_{\mathrm{MS}}$ (where each image contains information at a different scale), for each resolution level a one-step inverse stationary wavelet transform is applied in the reconstruction phase as many times as in the decomposition phase. The reconstruction process can be performed by applying the 2D-ISWT either to the approximation and thresholded detail sub-bands at the considered level (this is usually done in wavelet-based speckle filters [69]) or only to the approximation sub-bands at each resolution level[6]. Since the change-detection phase considers all the different levels, all the geometrical detail is in $\mathbf{X}_{\mathrm{MS}}$ even when detail coefficients at a particular scale are

[6] It is worth noting that the approximation sub-band contains low frequencies in both horizontal and vertical directions. It represents the input image at a coarser scale and contains most informative components, whereas detail sub-bands contain information related to high frequencies (i.e., both geometrical detail information and noise components) each in a preferred direction. According to this observation, it is easy to understand how proper thresholding of detail coefficients allows noise reduction [69].

neglected (in other words, the details removed at a certain resolution level are recovered at a higher level without removing them from the decision process). Once all resolution levels have been brought back to the image domain, the desired multi-scale sequence of images X_{LR}^{n} $(n = 0, 1, \ldots, N-1)$ is complete and each element in X_{MS} has the same size as the original image.

It is important to point out that unlike DWT, SWT avoids decimating. Thus, this multi-resolution decomposition strategy "fills in the gaps" caused by the decimation step in the standard wavelet transform [71]. In particular, the SWT decomposition preserves translation invariance and allows avoiding aliasing effects during synthesis without providing high-frequency components.

5.3.2 Adaptive Scale Identification

Based on the set of multi-scale images X_{LR}^{n} $(n = 0, 1, \ldots, N-1)$ obtained, we must identify reliable scales for each considered spatial position to drive the next fusion stage with this information. By using this information we can obtain change-detection maps characterized by high accuracy in homogeneous and border areas.

Reliable scales are selected according to whether the considered pixel belongs to a border or a homogeneous area at different scales. It is worth noting that the information at low-resolution levels is not reliable for pixels belonging to the border area, because at those scales details and edge information has been removed from the decomposition process. Thus, a generic scale is reliable for a given pixel, if the pixel at this scale is not in a border region or if it does not represent a geometrical detail.

To define whether a pixel belongs to a border or a homogeneous area at a given scale n, we use a multi-scale local coefficient of variation (LCV^n), as typically done in adaptive speckle de-noising algorithms [70,74]. This allows better handling of any residual multiplicative noise that may still be present in the scale selection process after ratioing[7]. As the coefficient of variation cannot be computed on the multi-scale log-ratio image sequence, the analysis is applied to the multi-resolution ratio image sequence, which can be easily obtained from the former by inverting the logarithm operation. Furthermore, it should be mentioned that by working on the multi-resolution ratio sequence we can design a homogeneity test capable of identifying border regions (or details) and no-border regions related to the presence of changes on the ground. This is different from applying the same test to the original images (which would result in identifying border and no-border regions with respect to the original scene but not with respect to the change signal).

The LCV^n is defined as:

$$\text{LCV}^n(i, j) = \frac{\sigma^n(i, j)}{\mu^n(i, j)} \tag{5.10}$$

where $\sigma^n(i,j)$ and $\mu^n(i,j)$ are the local standard deviation and the local mean, respectively, computed for the spatial position (i,j) at resolution level n $(n = 0, 1, \ldots, N-1)$, on a moving window of a user-defined size. Windows that are too small reduce the reliability of the local statistical parameters, while those that are too large decrease in sensitivity to identify geometrical details. Thus the selected size should be a trade-off between the above properties. The normalization operation defined in Equation 5.10 helps in adapting the standard deviation to the multiplicative speckle model. This coefficient is a measure of

[7] An alternative choice could be to use the standard deviation computed on the log-ratio image. However, in this way we would neglect possible residual effects of the multiplicative noise component.

the scene heterogeneity [74]: low values correspond to homogeneous areas, while high values refer to heterogeneous areas (e.g., border areas and point targets). To separate the homogeneous from the heterogeneous regions, a threshold value must be defined. In a homogeneous region the degree of homogeneity can be expressed in relation to the global coefficient of variation (CV^n) of the considered image at resolution level n, which is defined as:

$$\mathrm{CV}^n = \frac{\sigma^n}{\mu^n} \tag{5.11}$$

where σ^n and μ^n are the mean and the standard deviation computed over a homogeneous region at resolution level n, ($n = 0, 1, \ldots, N-1$). Homogeneous regions at each scale can be defined as those regions that satisfy the following condition:

$$\mathrm{LCV}^n(i,j) \leq \mathrm{CV}^n \tag{5.12}$$

In detail, a resolution level r ($r = 0, 1, \ldots, N-1$) is said to be reliable for a given pixel if Equation 5.12 is satisfied for all resolution levels t ($t = 0, 1, \ldots, r$). Thus, for the pixel (i,j), the set $\mathbf{X}_{\mathrm{MS}}^{Rij}$ of images with a reliable scale is defined as:

$$\mathbf{X}_{\mathrm{MS}}^{Rij} = \left\{\mathbf{X}_{\mathrm{LR}}^{0}, \ldots, \mathbf{X}_{\mathrm{LR}}^{n}, \ldots, \mathbf{X}_{\mathrm{LR}}^{Sij}\right\}, \quad \text{with } S_{ij} \leq N-1 \tag{5.13}$$

where S_{ij} is the level with the lowest resolution (identified by the highest value of n), such that the pixel can be represented without any border problems and therefore it satisfies the definition of a reliable scale shown in Equation 5.12 (note that the value of S_{ij} is pixel-dependent).

It is worth noting that, if the scene contains different kinds of changes with different radiometry (e.g., with increasing and decreasing radiometry), the above analysis should be applied to the normalized ratio image $\mathbf{X}_{\mathrm{NR}}$ (Equation 5.7) (rather than to the standard ratio image $\mathbf{X}_{\mathrm{R}}$). This makes the identification of border areas independent of the order with which the images are considered in the ratio, thus allowing all changed areas (independently of the related radiometry) to play a similar role in the definition of border pixels.

5.3.3 Scale-Driven Fusion

Once the set $\mathbf{X}_{\mathrm{MS}}^{Rij}$ has been defined for each spatial position, selected reliable scales are used to drive the fourth step, which consists of the generation of the change-detection map according to a scale-driven fusion. In this chapter three different fusion strategies are reported: two of them perform fusion at the decision level, while the third performs it at the feature level. Fusion at the decision level can either be based on "optimal" scale selection (FDL-OSS) or on the use of all reliable scales (i.e., scales included in $\mathbf{X}_{\mathrm{MS}}^{Rij}$) (FDL-ARS); fusion at the feature level is carried out by analyzing all reliable scales (FFL-ARS).

For each pixel, the FDL-OSS strategy only considers the reliable level with the lowest resolution, that is, the "optimal" resolution level S_{ij}. The rationale of this strategy is that the reliable level with the lowest resolution presents an "optimal" trade-off between speckle reduction and detail preservation for the considered pixel. In detail, each scale-dependent image in the set $\mathbf{X}_{\mathrm{MS}}$ is analyzed independently to discriminate between the two classes ω_{c} and ω_{u} associated with change and no-change classes, respectively. The desired partitioning for the generic scale n can be obtained by thresholding $\mathbf{X}_{\mathrm{LR}}^{n}$. It is worth noting that since the threshold value is scale-dependent, given the set of images $\boldsymbol{X}_{\text{MS}} = \{\boldsymbol{X}_{\text{LR}}^{0}, \ldots, \boldsymbol{X}_{\text{LR}}^{n}, \ldots, \boldsymbol{X}_{\text{LR}}^{N-1}\}$, we should determine (either automatically [17,52,59] or manually) a set of threshold values $T = \{T^0, \ldots, T^n, \ldots, T^{N-1}\}$. Regardless of the threshold-selection method adopted, a sequence of change-detection maps $\boldsymbol{M}_{\text{MS}} = \{\boldsymbol{M}^0, \ldots, \boldsymbol{M}^n, \ldots, \boldsymbol{M}^{N-1}\}$ is obtained from the images in $\boldsymbol{X}_{\text{MS}} = \{\boldsymbol{X}_{\text{LR}}^{0}, \ldots, \boldsymbol{X}_{\text{LR}}^{n}, \ldots, \boldsymbol{X}_{\text{LR}}^{N-1}\}$. A generic pixel $M(i,j)$ in the final change-detection map M is assigned to the class it belongs to in the map $\boldsymbol{M}^{S_{ij}}$ ($\in \boldsymbol{M}_{\text{MS}}$) computed at its optimal selected scale, i.e.,

$$M(i,j) \in \omega_k \Leftrightarrow M^{S_{ij}}(i,j) \in \omega_k, \quad \text{with } k = \{\text{c}, \text{u}\} \quad \text{and} \quad S_{ij} \leq N-1 \tag{5.14}$$

The accuracy of the resulting change-detection map depends both on the accuracy of the maps in the multi-resolution sequence and on the effectiveness of the procedure adopted to select the optimal resolution level. Both aspects are affected by the amount of residual noise in $\boldsymbol{X}_{\text{LR}}^{S_{ij}}$.

The second approach that considers FDL-ARS, makes the decision process more robust to noise. For each pixel, the set $M_{\text{MS}}^{R}(i,j) = \{M^0(i,j), \ldots, M^n(i,j), \ldots, M^{S_{ij}}(i,j)\}$ of the related reliable multi-resolution labels is considered. Each label in M_{MS}^{R} can be seen as a decision of a member of a pool of experts. Thus the pixel is assigned to the class that obtains the highest number of votes. In fact, the final change-detection map $\boldsymbol{M}$ is computed by applying a majority voting rule to the set $M_{\text{MS}}^{R}(i,j)$, at each spatial position. The class that receives the largest number of votes $V_{\omega_k}(i,j)$, $k = \{\text{c}, \text{u}\}$, represents the final decision for the considered input pattern, i.e.,

$$M(i,j) \in \omega_k \Leftrightarrow \omega_k = \arg\max_{\omega_h \in 0} \left\{V_{\omega_h}(i,j)\right\}, \quad k = \{\text{c}, \text{u}\} \tag{5.15}$$

The main disadvantage of the FDL-ARS strategy is that it only considers the final classification of each pixel at different reliable scales. A better utilization of the information in the multi-resolution sequence $\boldsymbol{X}_{\text{MS}}$ can be obtained by considering a fusion at feature level strategy (FFL-ARS). To accomplish the fusion process at different scales, a new set of images $\overline{\boldsymbol{X}}_{\text{MS}}^{R} = \{\overline{\boldsymbol{X}}_{\text{MS}}^{0}, \ldots, \overline{\boldsymbol{X}}_{\text{MS}}^{n}, \ldots, \overline{\boldsymbol{X}}_{\text{MS}}^{N-1}\}$ is computed by averaging all possible sequential combinations of images in $\boldsymbol{X}_{\text{MS}}$, i.e.,

$$\overline{\boldsymbol{X}}_{\text{MS}}^{n} = \frac{1}{n+1} \sum_{h=0}^{n} \boldsymbol{X}_{\text{LR}}^{h}, \quad \text{with } n = 0, 1, \ldots, N-1 \tag{5.16}$$

where the superscript n identifies the highest scale included in the average operation. When low values of n are considered, the image $\overline{\boldsymbol{X}}_{\text{MS}}^{n}$ contains a large amount of both geometrical details and speckle components, whereas when n increases, the image $\overline{\boldsymbol{X}}_{\text{MS}}^{n}$ contains a smaller amount of both geometrical details and speckle components. A pixel in position (i,j) is assigned to the class obtained by applying a standard thresholding procedure to the image $\overline{\boldsymbol{X}}_{\text{MS}}^{S_{ij}}$, $S_{ij} \leq N-1$, computed by averaging on the reliable scales selected for that spatial position, i.e.,

$$M(i,j) \in \begin{cases} \omega_{\text{u}} & \text{if } \overline{\boldsymbol{X}}_{\text{MS}}^{S_{ij}}(i,j) \leq T^{S_{ij}} \\ \omega_{\text{c}} & \text{if } \overline{\boldsymbol{X}}_{\text{MS}}^{S_{ij}}(i,j) > T^{S_{ij}} \end{cases} \tag{5.17}$$

where $T^{S_{ij}}$ is the decision threshold optimized (either automatically [17,52,59] or manually) for the considered image $\overline{\boldsymbol{X}}_{\text{MS}}^{S_{ij}}$. The latter strategy is capable of exploiting also the information component in the speckle, as it considers all the high frequencies in the decision process. It is worth noting that in the FFL-ARS strategy, as the information present at a given scale r is also contained in all images X_{LR}^{n} with $n < r$, the components characterizing the optimal scale S_{ij} (and the scales closer to the optimal one) in the fusion process are implicitly associated with greater weights than those associated with other considered levels. This seems reasonable, given the importance of these components for the analyzed spatial position.

5.4 Experimental Results and Comparisons

In this section experimental results obtained by applying the proposed multi-scale change-detection technique to a data set of multi-temporal remote-sensing SAR images are reported. These results are compared with those obtained by applying standard single-scale change-detection techniques to the same data set. In all trials involving image thresholding, the optimal threshold value was obtained according to a manual trial-and-error procedure. We selected for each image the threshold value (among all possible) that showed the minimum overall error in the change-detection map compared to the available reference map. This ensured it is possible to properly compare method performances without any bias due to human operator subjectivity or to the kind of automatic thresholding algorithm adopted. However, any type of automatic threshold-selection technique can be used with this technique (see Ref. [17] for more details about automatic thresholding of the log-ratio image). Performance assessment was accomplished both quantitatively (in terms of overall errors, false, and missed alarms) and qualitatively (according to a visual comparison of the produced change-detection maps with reference data).

5.4.1 Data Set Description

The data set used in the experiments is made up of two SAR images acquired by the ERS-1 SAR sensor (C-band and VV-polarization) in the province of Saskatchewan (Canada) before (1st July) and after (14th October) the 1995 fire season. The two images considered are characterized by a geometrical resolution of 25 m in both directions and by a nominal number of looks equal to 5. The selected test site (see in Figure 5.4a and Figure 5.4b) is a section (350×350 pixels) of the entire available scene. A fire caused by a lightning event destroyed a large portion of the vegetation in the considered area between the two dates mentioned above.

The two multi-look intensity images were geocoded using the digital elevation model (DEM) GTOPO30; no speckle reduction algorithms were applied to the images. The log-ratio image was computed from the above data according to Equation 5.4.

To enable a quantitative evaluation of the effectiveness of the proposed approach, a reference map was defined manually (see Figure 5.5b). To this end, we used the available ground-truth information provided by the Canadian Forest Service (CFS) and by the fire agencies of the individual Canadian provinces. Ground-truth information is coded in a vector format and includes information about fires (e.g., geographical coordinates, final size, cause, etc.) that occurred from 1981 to 1995 and in areas greater than 200 ha in final size. CFS ground truth was used for a rough localization of the burned areas as it shows a medium geometrical resolution. An accurate identification of the boundaries of the burned areas was obtained from a detailed visual analysis of the two original 5-look

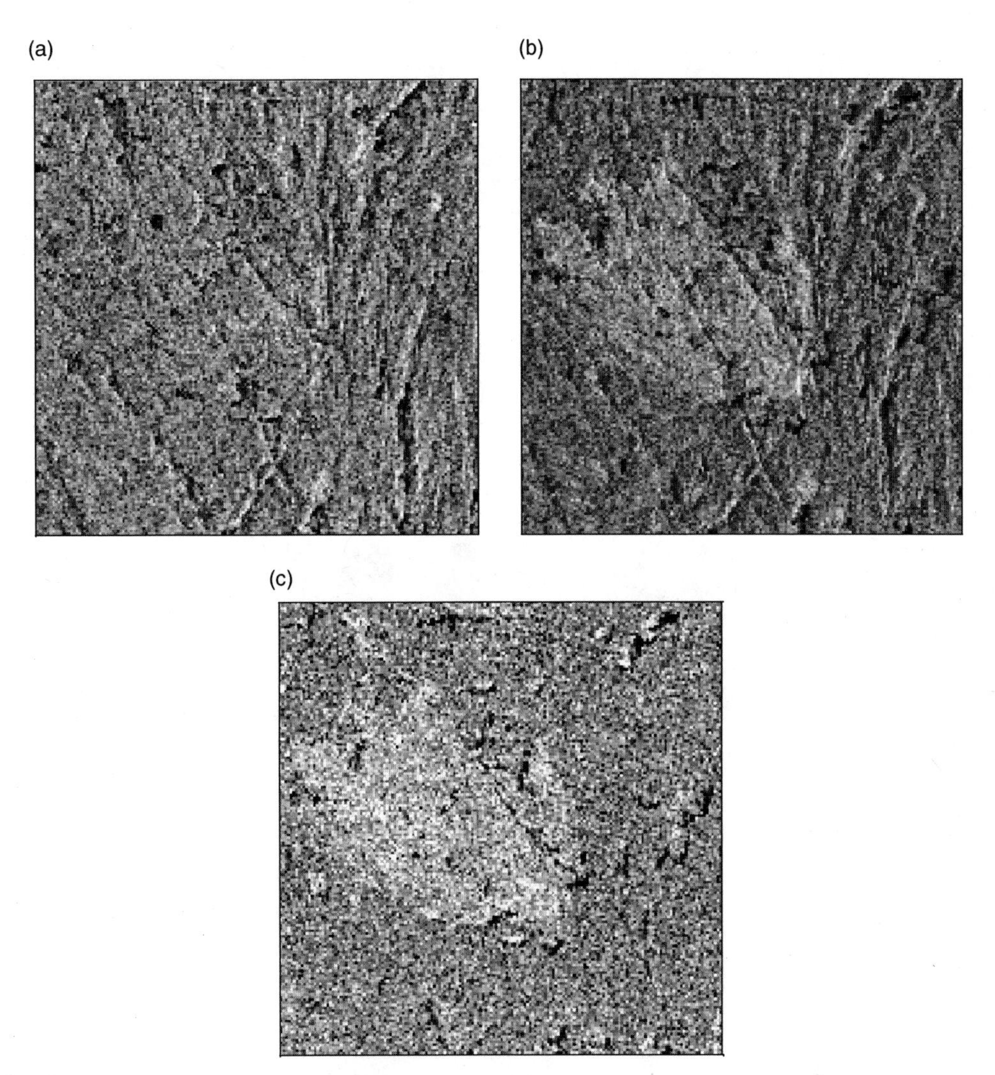

FIGURE 5.4
Images of the Saskatchewan province, Canada, used in the experiments. (a) Image acquired from the ERS-1 SAR sensor in July 1995, (b) image acquired from the ERS-1 SAR sensor in October 1995, and (c) analyzed log-ratio image.

intensity images (Figure 5.4a and Figure 5.4b), the ratio image, and the log-ratio image (Figure 5.4c), carried out accurately in cooperation with experts in SAR-image interpretation. In particular, different color composites of the above-mentioned images were used to highlight all the portions of the changed areas in the best possible way. It is worth noting that no despeckling or wavelet-based analysis was applied to the images exploited to generate the reference map for this process to be as independent as possible of the methods adopted in the proposed change-detection technique. In generating the reference map, the irregularities of the edges of the burned areas were faithfully reproduced to enable accurate assessment of the effectiveness of the proposed change-detection approach. At the end of the process, the obtained reference map contained 101219 unchanged pixels and 21281 changed pixels. Our goal was to obtain, with the proposed automatic technique, a change-detection map as similar as possible to the reference map obtained according to the above-mentioned time-consuming manual process driven by ground-truth information and by experts in SAR image interpretation.

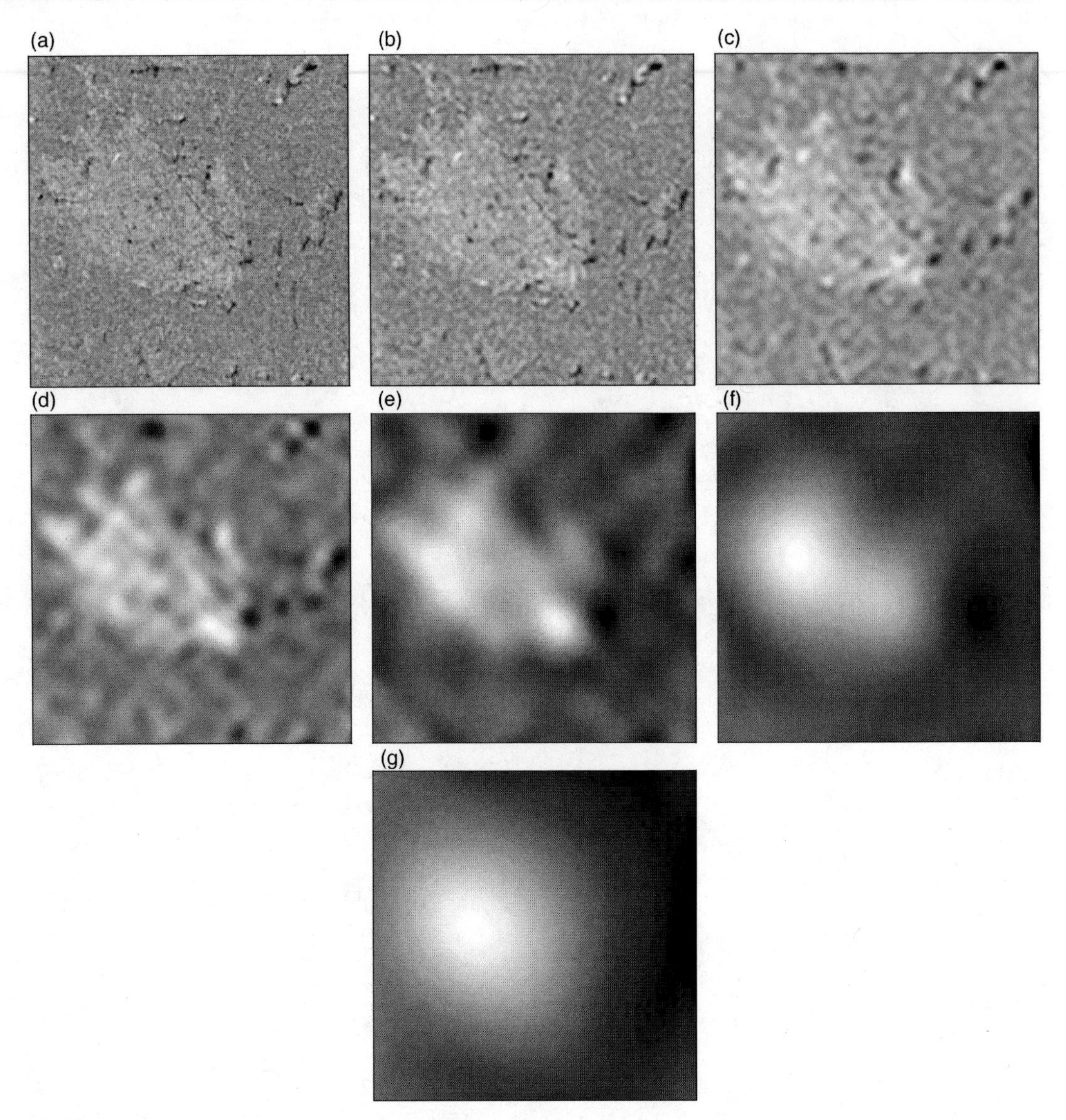

FIGURE 5.5
Multi-scale image sequence obtained by applying the wavelet decomposition procedure to the log-ratio image $X_{MS} = \{X^1_{LR}, \ldots, X^7_{LR}\}$.

5.4.2 Results

Several experiments were carried out to assess the effectiveness of the proposed change-detection technique (which is based on scale-driven fusion strategies) with respect to classical methods (which are based on thresholding of the log-ratio image).

In all trials involving image thresholding, the optimal threshold value was obtained according to a manual trial-and-error procedure. Among all possible values, we selected for each image the threshold value that showed the minimum overall error in the change-detection map compared to the reference map. Subsequently, it was possible to evaluate the optimal performance of the proposed methodology without any bias due to human operator subjectivity or to the fact that the selection was made by an automatic thresholding algorithm. However, any type of automatic threshold-selection technique can be used with this technique (see Ref. [17] for more details about automatic thresholding of the log-ratio image). As the described procedure is independently optimized for each considered image, it leads to different threshold values in each case. Performance assessment was accomplished both quantitatively (in terms of overall errors, false, and missed alarms) and qualitatively (according to a visual comparison of the produced change-detection maps with reference data).

To apply the three scale-driven fusion strategies (see Section 5.3.3), the log-ratio image was first decomposed into seven resolution levels by applying the *Daubechies*-4 SWT. Each computed approximation sub-band was used to construct different scales, that is, $X_{MS} = \{X_{LR}^1, \ldots, X_{LR}^7\}$ (see Figure 5.5). For simplicity, only approximation sub-bands have been involved in the reconstruction phase (it is worth noting that empirical experiments on real data have confirmed that details sub-band elimination does not affect the change-detection accuracy). Observe that the full-resolution original image X_{LR} ($\equiv X_{LR}^0$) was discarded from the analyzed set, as it was affected by a strong speckle noise. In particular, empirical experiments pointed out that when X_{LR}^0 is used on this data set the accuracy of the proposed change-detection technique gets degraded. Nevertheless, in the general case, resolution level 0 can also be considered and should not be discarded *a priori*. A number of trials were carried out to identify the optimal window size to compute the local coefficient of variation (LCV^n) used to detect detail pixels (e.g., border) at different resolution levels. The optimal size (i.e., the one that gives the minimum overall error) was selected for all analyzed strategies (see Table 5.3).

Table 5.3 summarizes the quantitative results obtained with the different fusion strategies. As can be seen from the analysis of the overall error, the FFL-ARS strategy gave the lowest error, i.e., 5557 pixels, while the FDL-ARS strategy gave 6223, and the FDL-OSS strategy 7603 (the highest overall error). As expected, by including all the reliable scales in the fusion phase it was possible to improve the change-detection accuracy compared to a single "optimal" scale. The FFL-ARS strategy gave the lowest false and missed alarms, decreasing their values by 1610 and 436 pixels, respectively, compared to the FDL-OSS strategy. This is because on the one hand the FDL-OSS procedure is penalized both by the change-detection accuracy at a single resolution level (which is significantly affected by noise when fine scales are considered) and by residual errors in identifying the optimal scale of a given pixel; on the other hand, the use of the entire subset of reliable scales allows a better exploitation of the information at the highest resolution levels of the multi-resolution sequence in the change-detection process. It is worth noting that FFL-ARS outperformed FDL-ARS also in terms of false (2181 vs. 2695) and missed (3376 vs. 3528) alarms. This is mainly due to its ability to handle better all the information in the scale-dependent images before the decision process. This leads to a more accurate recognition of critical pixels (i.e., pixels that are very close to the boundary between the changed and unchanged classes on the log-ratio image) that

TABLE 5.3

Overall Error, False Alarms, and Missed Alarms (in Number of Pixels and Percentage) Resulting from the Proposed Adaptive Scale-Driven Fusion Approaches

	False Alarms		Missed Alarms		Overall Errors		
Fusion Strategy	Pixels	%	Pixels	%	Pixels	%	LCV Window Size
FDL-OSS	3791	3.75%	3812	17.91%	7603	6.21%	23 × 23
FDL-ARS	2695	2.66%	3528	16.58%	6223	5.08%	7 × 7
FFL-ARS	2181	2.15%	3376	15.86%	5557	4.54%	5 × 5

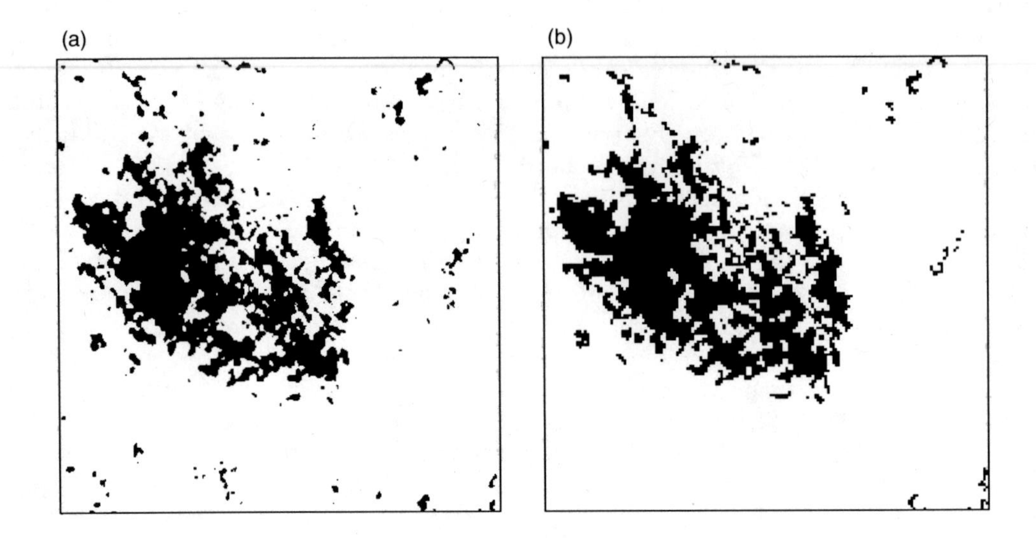

FIGURE 5.6
(a) Change-detection map obtained for the considered data set using the FFL-ARS strategy on all reliable scales and (b) reference map of the changed area used in the experiment.

exploit the joint consideration of all the information present at the different scales in the decision process. For a better understanding of the results achieved, we made a visual analysis of the change-detection maps obtained. Figure 5.6a shows the change-detection map obtained with the FFL-ARS strategy (which proved to be the most accurate), while Figure 5.6b is the reference map. As can be seen, the considered strategy produced a change-detection map that was very similar to the reference map. In particular, the change-detection map obtained with the proposed approach shows good properties both in terms of detail preservation and in terms of high accuracy in homogeneous areas.

To assess the effectiveness of the scale-driven change-detection approach, the results obtained with the FFL-ARS strategy were compared with those obtained with a classical change-detection algorithm. In particular, we computed a change-detection map by an optimal (in the sense of minimum error) thresholding of the log-ratio image obtained after despeckling with the adaptive enhanced Lee filter [74]. The enhanced Lee filter was applied to the two original images (since a multiplicative speckle model is required). Several trials were carried out while varying the window size, to find the value that leads to the minimum overall error. The best result for the considered test site (see Table 5.4) was obtained with a 7×7 window size. The thresholding operation gave an overall error of 8053 pixels. This value is significantly higher than the overall error obtained with the

TABLE 5.4

Overall Error, False Alarms, and Missed Alarms (in Number of Pixels and Percentage) Resulting from Classical Change-Detection Approaches

	False Alarms		Missed Alarms		Total Errors		
Applied Filtering Technique	Pixels	%	Pixels	%	Pixels	%	Filter Window Size
Enhanced Lee filter	3725	3.68%	4328	20.34%	8053	6.57%	7 × 7
Gamma MAP filter	3511	3.47%	4539	21.33%	8050	6.57%	7 × 7
Wavelet de-noising	2769	2.74%	4243	19.94%	7012	5.72%	—

FFL-ARS strategy (i.e., 5557). In addition the proposed scale-driven fusion technique also decreased both the false (2181 vs. 3725) and the missed alarms (3376 vs. 4328) compared to the considered classical procedure. From a visual analysis of Figure 5.7a and Figure 5.6a and Figure 5.6b, it is clear that the change-detection map obtained after the Lee-based despeckling procedure significantly reduces the geometrical detail content in the final change-detection map compared to that obtained with the FFL-ARS approach. This is mainly due to the use of the filter, which not only results in a significant smoothing of the images but also strongly reduces the information component present in the speckle. Similar results and considerations both from a quantitative and qualitative point of view were obtained by filtering the image with the Gamma MAP filter (compare Table 5.3 and Table 5.4).

Furthermore, we also analyzed the effectiveness of classical thresholding of the log-ratio image after de-noising with a recently proposed more advanced despeckling procedure. In particular, we investigated a DWT-based de-noising [69,75] technique (not used previously in change-detection problems). This technique achieves noise reduction in three steps: (1) image decomposition (DWT), (2) thresholding of wavelet coefficients, and (3) image reconstruction by inverse wavelet transformation (IDWT) [69,75]. It is worth noting also that this procedure is based on the multi-scale decomposition of the images. We can therefore better evaluate the effectiveness of the scale-driven procedure in exploiting the multi-scale information obtained by the DWT decomposition. Wavelet-based de-noising was applied to the log-ratio image because an additive speckle model was required. Several trials were carried out varying the wavelet-coefficient de-noising algorithm while keeping the type of wavelet fixed, that is, *Daubechies*-4 (the same used for multi-level decomposition). The best change-detection result (see Table 5.4) was obtained by soft thresholding detail coefficients according to the universal threshold $T = \sqrt{2\sigma^2 \log(I \cdot J)}$, where $I \cdot J$ is the image size and σ^2 is the estimated noise variance [75]. The soft thresholding procedure sets detail coefficients that fall between T and $-T$ to zero, and shrinks the module of coefficients that fall out of this interval by a factor T. The noise variance estimation was performed by computing the variance of the diagonal-detail sub-band at the first decomposition level, given the above thresholding approach. However, in this case also the error (i.e., 7012 pixels) obtained was significantly higher than the overall error obtained with the multi-scale change-detection technique based on the FFL-ARS strategy (i.e., 5557 pixels). Moreover, the multi-scale method performs better also in terms of false and missed alarms, which were reduced from 4243 to 3376, and from 2769 to 2181, respectively (see Table 5.3 and Table 5.4). By analyzing Figure 5.7b, Figure 5.6a, and Figure 5.6b, it can be seen that the change-detection map obtained by thresholding the log-ratio image after applying the DWT-based de-noising algorithm preserves geometrical information well.

Nevertheless, on observing the map in greater detail, it can be concluded qualitatively that the spatial fidelity obtained with this procedure is lower than that obtained with the proposed approach. This is confirmed, for example, when we look at the right part of the burned area (circles in Figure 5.7b), where some highly irregular areas saved from the fire are properly modeled by the proposed technique, but smoothed out by the procedure based on DWT de-noising. This confirms the quantitative results and thus the effectiveness of the proposed approach in exploiting information from multi-level image decomposition.

It is worth noting that the improvement in performance shown by the proposed approach was obtained without any additional computational burden compared to the thresholding procedure after wavelet de-noising. In particular, both methods require analysis and synthesis steps (though for different purposes). The main difference between the two considered techniques is the scale-driven combination step, which

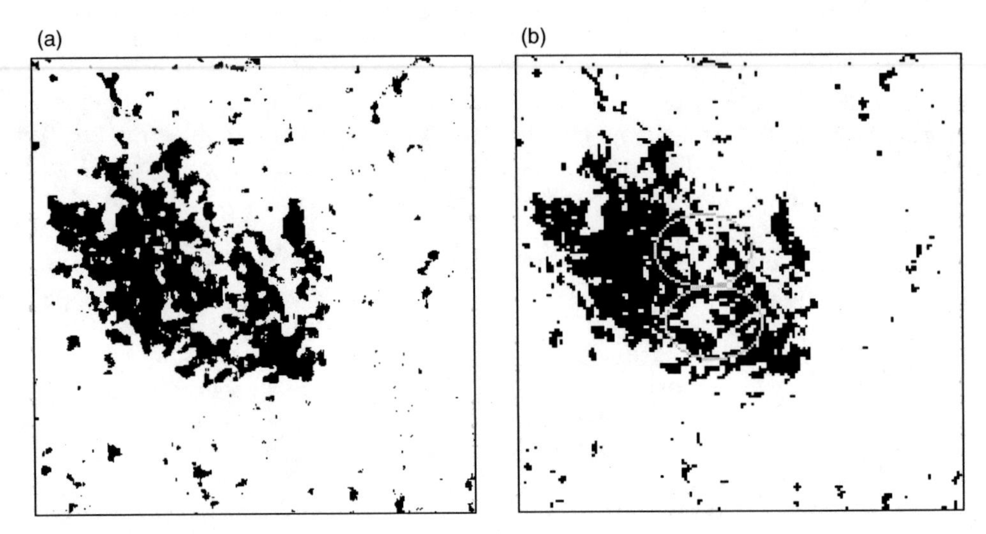

FIGURE 5.7
Change-detection maps obtained for the considered data set by optimal manual thresholding of the log-ratio image after the despeckling with (a) the Lee-enhanced filter and (b) the DWT-based technique.

does not increase the computational time required by the thresholding of detail coefficients according to the standard wavelet-based de-noising procedure.

5.5 Conclusions

In this chapter the problem of unsupervised change detection in multi-temporal remote-sensing images has been addressed. In particular, after a general overview on the unsupervised change-detection problem, attention has been focused on multi-temporal SAR images. A brief analysis of the literature on unsupervised techniques in SAR images has been reported, and a novel adaptive scale-driven approach to change detection in multi-temporal SAR data (recently developed by the authors) has been proposed. Unlike classical methods, this approach exploits information at different scales (obtained by a wavelet-based decomposition of the log-ratio image) to improve the accuracy and geometric fidelity of the change-detection map.

Three different fusion strategies that exploit the subset of reliable scales for each pixel have been proposed and tested: (1) fusion at the decision level by an optimal scale selection (FDL-OSS), (2) fusion at the decision level of all reliable scales (FDL-ARS), and (3) fusion at the feature level of all reliable scales (FFL-ARS). As expected, a comparison among these strategies showed that fusion at the feature level led to better results than the other two procedures, in terms both of geometrical detail preservation and accuracy in homogeneous areas. This is due to a better intrinsic capability of this technique to exploit the information present in all the reliable scales for the analyzed spatial position, including the amount of information present in the speckle.

Experimental results confirmed the effectiveness of the proposed scale-driven approach with the FFL-ARS strategy on the considered data set. This approach outperformed a classical change-detection technique based on the thresholding of the log-ratio image after a proper despeckling based on the application of the enhanced Lee filter and also of the Gamma filter. In particular, change detection after despeckling resulted in a higher overall error, more false alarms and missed alarms, and significantly lower geometrical fidelity. To further assess the validity of the proposed approach, the standard technique based on the thresholding of the log-ratio image was applied after a despeckling phase applied according to an advanced DWT-based de-noising procedure (which has not been used previously in change-detection problems). The obtained results suggest that the proposed approach performs slightly better in terms of spatial fidelity and significantly increases the overall accuracy of the change-detection map. This confirms that on the considered data set and for solving change-detection problems, the scale-driven fusion strategy exploits the multi-scale decomposition better than standard de-noising methods.

It is worth noting that all experimental results were carried out applying an optimal manual trial-and-error threshold selection procedure, to avoid any bias related to the selected automatic procedure in assessing the effectiveness of both the proposed and standard techniques. Nevertheless, this step can be performed adopting automatic thresholding procedures [17,50].

As a final remark, it is important to point out that the proposed scale-driven approach is intrinsically suitable to be used with very high geometrical resolution active (SAR) and passive (PAN) images, as it properly handles and processes the information present at different scales. Furthermore, we think that the use of multi-scale and multi-resolution approaches represents a very promising avenue for investigation to develop advanced automatic change-detection techniques for the analysis of very high spatial resolution images acquired by the previous generation of remote-sensing sensors.

Acknowledgments

This work was partially supported by the Italian Ministry of Education, University and Research. The authors are grateful to Dr. Francesco Holecz and Dr. Paolo Pasquali (SARMAP s.a.®, Cascine di Barico, CH-6989 Purasca, Switzerland) for providing images of Canada and advice about ground truth.

References

1. Bruzzone, L. and Serpico, S.B., Detection of changes in remotely-sensed images by the selective use of multi-spectral information, *Int. J. Rem. Sens.*, 18, 3883–3888, 1997.
2. Bruzzone, L. and Fernández Prieto, D., An adaptive parcel-based technique for unsupervised change detection, *Int. J. Rem. Sens.*, 21, 817–822, 2000.
3. Bruzzone, L. and Fernández Prieto, D., A minimum-cost thresholding technique for unsupervised change detection, *Int. J. Rem. Sens.*, 21, 3539–3544, 2000.
4. Bruzzone, L. and Cossu, R., An adaptive approach for reducing registration noise effects in unsupervised change detection, *IEEE Trans. Geosci. Rem. Sens*, 41, 2455–2465, 2003.
5. Bruzzone, L. and Cossu, R., Analysis of multitemporal remote-sensing images for change detection: bayesian thresholding approaches, in *Geospatial Pattern Recognition*, Eds. Binaghi, E., Brivio, P.A., and Serpico, S.B., Research Signpost/Transworld Research, Kerala, India, 2002, Chap. 15.

6. *Proc. First Int. Workshop Anal. Multitemporal Rem. Sens. Images*, Eds. Bruzzone, L. and Smits, P.C., World Scientific Publishing, Singapore, 2002, ISBN: 981-02-4955-1, Book Code: 4997 hc.
7. *Proc. Second Int. Workshop Anal. Multitemporal Rem. Sens. Images*, Eds. Smits, P.C. and Bruzzone, L., World Scientific Publishing, Singapore, 2004, ISBN: 981-238-915-6, Book Code: 5582 hc.
8. Special Issue on Analysis of Multitemporal Remote Sensing Images, *IEEE Trans. Geosci. Rem. Sens.*, Guest Eds. Bruzzone, L. and Smits, P.C., Tilton, J.C., 41, 2003.
9. Carlin, L. and Bruzzone, L., A scale-driven classification technique for very high geometrical resolution images, *Proc. SPIE Conf. Image Signal Proc. Rem. Sens. XI*, 2005.
10. Baraldi, A. and Bruzzone, L., Classification of high-spatial resolution images by using Gabor wavelet decomposition, *Proc. SPIE Conf. Image Signal Proc. Rem. Sens. X*, 5573, 19–29, 2004.
11. Binaghi, E., Gallo, I., and Pepe, M., A cognitive pyramid for contextual classification of remote sensing images, *IEEE Trans. Geosci. Rem. Sens.*, 41, 2906–2922, 2003.
12. Benz, U.C. et al., Multi-resolution, object-oriented fuzzy analysis of remote sensing data for GIS-ready information, *ISPRS J. Photogramm. Rem. Sens.*, 58, 239–258, 2004.
13. Singh, A., Digital change detection techniques using remotely-sensed data, *Int. J. Rem. Sens.*, 10, 989–1003, 1989.
14. Bruzzone, L. and Serpico, S.B., An iterative technique for the detection of land-cover transitions in multitemporal remote-sensing images, *IEEE Trans. Geosci. Rem. Sens.*, 35, 858–867, 1997.
15. Bruzzone, L., Fernández Prieto, D., and Serpico, S.B., A neural-statistical approach to multitemporal and multisource remote-sensing image classification, *IEEE Trans. Geosci. Rem. Sens.*, 37, 1350–1359, 1999.
16. Serpico, S.B. and Bruzzone, L., Change detection, in *Information Processing for Remote Sensing*, Ed. Chen, C.H., World Scientific Publishing, Singapore, Chap. 15.
17. Bazi, Y., Bruzzone, L., and Melgani, F., An unsupervised approach based on the generalized Gaussian model to automatic change detection in multitemporal SAR images, *IEEE Trans. Geosci. Rem. Sens.*, 43, 874–887, 2005.
18. Dai, X. and Khorram, S., The effects of image misregistration on the accuracy of remotely sensed change detection, *IEEE Trans. Geosci. Rem. Sens.*, 36, 1566–1577, 1998.
19. Townshend, J.R.G. and Justice, C.O., Spatial variability of images and the monitoring of changes in the normalized difference vegetation index, *Int. J. Rem. Sens.*, 16, 2187–2195, 1995.
20. Fung, T. and LeDrew, E., Application of principal component analysis to change detection, *Photogramm. Eng. Rem. Sens.*, 53, 1649–1658, 1987.
21. Chavez, P.S. Jr., and MacKinnon, D.J., Automatic detection of vegetation changes in the southwestern united states using remotely sensed images, *Photogramm. Eng. Rem. Sens.*, 60, 571–583, 1994.
22. Fung, T., An assessment of TM imagery for land-cover change detection, *IEEE Trans. Geosci. Rem. Sens.*, 28, 681–684, 1990.
23. Muchoney, D.M. and Haack, B.N., Change detection for monitoring forest defoliation, *Photogramm. Eng. Rem. Sens.*, 60, 1243–1251, 1994.
24. Bruzzone, L. and Fernández Prieto, D., Automatic analysis of the difference image for unsupervised change detection, *IEEE Trans. Geosci. Rem. Sens.*, 38, 1170–1182, 2000.
25. Oliver, C.J. and Quegan, S., *Understanding Synthetic Aperture Radar Images*, Artech House, Norwood, MA, 1998.
26. Rignot, E.J.M. and van Zyl, J.J., Change detection techniques for ERS-1 SAR data, *IEEE Trans. Geosci. Rem. Sens.*, 31, 896–906, 1993.
27. Quegan, S. et al., Multitemporal ERS SAR analysis applied to forest mapping, *IEEE Trans. Geosci. Rem. Sens.*, 38, 741–753, 2000.
28. Dierking, W. and Skriver, H., Change detection for thematic mapping by means of airborne multitemporal polarimetric SAR imagery, *IEEE Trans. Geosci. Rem. Sens.*, 40, 618–636, 2002.
29. Grover, K., Quegan, S., and da Costa Freitas, C., Quantitative estimation of tropical forest cover by SAR, *IEEE Trans. Geosci. Rem. Sens.*, 37, 479–490, 1999.
30. Liu, J.G. et al., Land surface change detection in a desert area in Algeria using multi-temporal ERS SAR coherence images, *Int. J. Rem. Sens.*, 2, 2463–2477, 2001.
31. Lombardo, P. and Macrì Pellizzeri, T., Maximum likelihood signal processing techniques to detect a step pattern of change in multitemporal SAR images, *IEEE Trans. Geosci. Rem. Sens.*, 40, 853–870, 2002.

32. Lombardo, P. and Oliver, C.J., Maximum likelihood approach to the detection of changes between multitemporal SAR images, *IEEE Proc. Radar, Sonar, Navig.*, 148, 200–210, 2001.
33. Engeset, R.V. and Weydahl, D.J., Analysis of glaciers and geomorphology on svalbard using multitemporal ERS-1 SAR images, *IEEE Trans. Geosci. Rem. Sens.*, 36, 1879–1887, 1998.
34. White, R.G. and Oliver, C.J., Change detection in SAR imagery, *Rec. IEEE 1990 Int. Radar Conf.*, 217–222, 1990.
35. Caves, R.G. and Quegan, S., Segmentation based change detection in ERS-1 SAR images, *Proc. IGARSS*, 4, 2149–2151, 1994.
36. Inglada, J., Change detection on SAR images by using a parametric estimation of the Kullback–Leibler divergence, *Proc. IGARSS*, 6, 4104–4106, 2003.
37. Rignot, E. and Chellappa, R., A Bayes classifier for change detection in synthetic aperture radar imagery, *Proc. IEEE Int. Conf. Acoust., Speech, Signal Processing*, 3, 25–28, 1992.
38. Brown. L., A survey of image registration techniques, *ACM Computing Surveys*, 24, 1992.
39. Homer, J. and Longstaff, I.D., Minimizing the tie patch window size for SAR image co-registration, *Electron. Lett.*, 39, 122–124, 2003.
40. Frost, V.S. et al., A model for radar images and its application to adaptive digital filtering of multiplicative noise, *IEEE Trans. Pattern Anal. Machine Intell.*, PAMI-4, 157–165, 1982.
41. Lee, J.S., Digital image enhancement and noise filtering by use of local statistics, *IEEE Trans. Pattern Anal. Machine Intell.*, PAMI-2, 165–168, 1980.
42. Kuan, D.T. et al., Adaptive noise smoothing filter for images with signal-dependent noise, *IEEE Trans. Pattern Anal. Machine Intell.*, PAMI-7, 165–177, 1985.
43. Lopes, A. et al., Structure detection and statistical adaptive filtering in SAR images, *Int. J. Rem. Sens.* 41, 1735–1758, 1993.
44. Lopes, A. et al., Maximum *a posteriori* speckle filtering and first order texture models in SAR images, *Proc. Geosci. Rem. Sens. Symp., IGARSS90*, Collage Park, MD, 2409–2412, 1990.
45. Solbø, S. and Eltoft, T., Γ-WMAP: a statistical speckle filter operating in the wavelet domain, *Int. J. Rem. Sens.*, 25, 1019–1036, 2004.
46. Cihlar, J., Pultz, T.J., and Gray, A.L., Change detection with synthetic aperture radar, *Int. J. Rem. Sens.*, 13, 401–414, 1992.
47. Dekker, R.J., Speckle filtering in satellite SAR change detection imagery, *Int. J. Rem. Sens.*, 19, 1133–1146, 1998.
48. Bovolo, F. and Bruzzone, L., A detail-preserving scale-driven approach to change detection in multitemporal SAR images *IEEE Trans. Geosci. Rem. Sens.*, 43, 2963–2972, 2005.
49. Bazi, Y., Bruzzone, L. and Melgani, F., Change detection in multitemporal SAR images based on generalized Gaussian distribution and EM algorithm, *Proc. SPIE Conf. Image Signal Processing Rem. Sens. X*, 364–375, 2004.
50. Bruzzone, L. and Fernández Prieto, D., An adaptive semiparametric and context-based approach to unsupervised change detection in multitemporal remote-sensing images, *IEEE Trans. Image Processing*, 11, 452–446, 2002.
51. Fukunaga, K., *Introduction to Statistical Pattern Recognition*, 2nd ed., Academic Press, London, 1990.
52. Melgani, F., Moser, G., and Serpico, S.B., Unsupervised change-detection methods for remote-sensing data, *Optical Engineering*, 41, 3288–3297, 2002.
53. Kittler, J. and Illingworth, J., Minimum error thresholding, *Pattern Recognition*, 19, 41–47, 1986.
54. Rosin, P.L. and Ioannidis, E., Evaluation of global image thresholding for change detection, *Pattern Recognition Lett.*, 24, 2345–2356, 2003.
55. Lee, S.U., Chung, S.Y., and Park, R.H., A comparative performance study of several global thresholding techniques for segmentation, *Comput. Vision, Graph. Image Processing*, 52, 171–190, 1990.
56. Dempster, A.P., Laird, N.M., and Rubin, D.B., Maximum likelihood from incomplete data via the EM algorithm, *J. Roy. Statist. Soc.*, 39, 1–38, 1977.
57. Moon, T.K., The expectation–maximization algorithm, *Sig. Process. Mag.*, 13, 47–60, 1996.
58. Redner, A.P. and Walker, H.F., Mixture densities, maximum likelihood and the EM algorithm, *SIAM Review*, 26, 195–239, 1984.
59. Duda, R.O., Hart, P.E., and Stork, D.G., *Pattern Classification*, John Wiley & Sons, New York, 2001.

60. Bazi, Y., Bruzzone, L. and Melgani, F., Change detection in multitemporal SAR images based on the EM–GA algorithm and Markov Random Fields, *Proc. IEEE Third Int. Workshop Anal. Multi-Temporal Rem. Sens. Images* (*Multi-Temp 2005*), 2005.
61. Bazi, Y., Bruzzone, L., and Melgani, F., Automatic identification of the number and values of decision thresholds in the log-ratio image for change detection in SAR images, *IEEE Lett. Geosci. Rem. Sens.*, 43, 874–887, 2005.
62. Burt, P.J. and Adelson, E.H., The Laplacian pyramid as a compact image code *IEEE Trans. Comm.*, 31, 532–540, 1983.
63. Mallat, S.G., A theory for multiresolution signal decomposition: the wavelet representation, *IEEE Trans. Pattern Anal. Machine Intell.*, PAMI-11, 674–693, 1989.
64. Mallat, S.G., *A Wavelet Tour of Signal Processing*, Academic Press, San Diego, 1998.
65. Aiazzi, B. et al., A Laplacian pyramid with rational scale factor for multisensor image data fusion, in *Proc. Int. Conf. Sampling Theory and Applications*, SampTA 97, 55–60, 1997.
66. Aiazzi, B., Alparone, L. and Baronti, S., Multiresolution local-statistics speckle filtering based on a ratio Laplacian pyramid, *IEEE Trans. Geosci. Rem. Sens.*, 36, 1466–1476, 1998.
67. Xie, H., Pierce, L.E. and Ulaby, F.T., SAR speckle reduction using wavelet denoising and Markov random field modeling, *IEEE Trans. Geosci. Rem. Sens.*, 40, 2196–2212, 2002.
68. Zeng, Z. and Cumming, I., Bayesian speckle noise reduction using the discrete wavelet transform, *IEEE Int. Geosci. Rem. Sens. Symp. Proc.*, *IGARSS*-1998, 7–9, 1998.
69. Guo, H. et al., Wavelet based speckle reduction with application to SAR based ATD/R, *IEEE Int. Conf. Image Processing Proc.*, *ICIP*-1994, 1, 75–79, 1994.
70. Argenti, F. and Alparone, L., Speckle removal for SAR images in the undecimated wavelet domain, *IEEE Trans. Geosci. Rem. Sens.*, 40, 2363–2374, 2002.
71. Nason, G.P. and Silverman, B.W., The stationary wavelet transform and some statistical applications, in *Wavelets and Statistics*, Eds. Antoniadis, A. and Oppenheim, G., Springer-Verlag, New York, *Lect. Notes Statist.*, 103, 281–300, 1995.
72. Manian, V. and Vàsquez, R., On the use of transform for SAR image classification, *IEEE Int. Geosci. Rem. Sens. Symp. Proc.*, 2, 1068–1070, 1998.
73. Strang, G. and Nguyen, T, *Wavelets and Filter Banks*, Cambridge Press, Wellesley, USA, 1996.
74. Lopes, A., Touzi, R., and Nerzy, E., Adaptive speckle filters and scene heterogeneity, *IEEE Trans. Geosci. Rem. Sens.*, 28, 992–1000, 1990.
75. Donoho, D.L. and Johnstone, I.M., Ideal spatial adaptation via wavelet shrinkage, *Biometrika*, 81, 425–455, 1994.

6

Change-Detection Methods for Location of Mines in SAR Imagery

Maria Tates, Nasser Nasrabadi, Heesung Kwon, and Carl White

CONTENTS

6.1 Introduction

Using multi-temporal SAR images of the same scene, analysts employ several methods to determine changes among the set [1]. Changes may be abrupt in which case only two images are required or it may be gradual in which case several images of a given scene are compared to identify changes. The former case is considered here, where SAR images are analyzed to detect land mines. Much of the change-detection activity has been focused on optical data for specific applications, for example, several change-detection methods are implemented and compared to detect land-cover changes in multi-spectral imagery [2]. However, due to the natural limitations of optical sensors, such as sensitivity to weather and illumination conditions, SAR sensors may constitute a superior sensor for change detection as images for this purpose are obtained at various times of the day under varying conditions.

SAR systems are capable of collecting data in all weather conditions and are unaffected by cloud cover or changing sunlight conditions. Exploiting the long-range propagation qualities of radar signals and utilizing the complex processing capabilities of current digital technology, SAR produces high-resolution imagery providing a wealth of information in many civilian and military applications. One major characteristic of SAR data is the presence of speckle noise, which poses a challenge in classification. SAR data taken from different platforms also shows high variability.

The accuracy of preprocessing tasks, such as image registration, also affects the accuracy of the postprocessing task of change detection. Image registration, the process of aligning images into the same coordinate frame, must be done prior to change detection. Image registration can be an especially nontrivial task in instances where the images are acquired from nonstationary sources in the presence of sensor motion. In this paper we do not address the issue of image registration, but note that there is a plethora of information available on this topic and a survey is provided in Ref. [3].

In Ref. [4], Mayer and Schaum implemented several transforms on multi-temporal hyperspectral target signatures to improve target detection in a matched filter. The transforms implemented in their paper were either model-dependent, based solely on atmospheric corrections, or image-based techniques. The image-based transforms, including one that used Wiener filtering, yielded improvements in target detection using the matched filter and proved robust to various atmospheric conditions.

In Ref. [5], Rignot and van Zyl compared various change-detection techniques for SAR imagery. In one method they implemented image ratioing, which proves more appropriate for SAR than simple differencing. However, the speckle decorrelation techniques they implemented outperform the ratio method under certain conditions. The results lead to the conclusion that the ratio method and speckle decorrelation method have complimentary characteristics for detecting gradual changes in the structural and dielectric properties of remotely sensed surfaces.

In Ref. [6], Ranney and Soumekh evaluated change detection using the so-called signal subspace processing method, which is similar to the subspace projection technique [7], to detect mines in averaged multi-look SAR imagery. They posed the problem as a trinary hypothesis-testing problem to determine whether no change has occurred, a target has entered a scene, or whether a target has left the scene. In the signal subspace processing method, one image is modeled as the convolution of the other image by a transform that accounts for variances in the image-point response between the reference and test images due to varying conditions present during the different times of data collection.

For the detection of abrupt changes in soil characteristics following volcanic eruptions, in Ref. [8] a two-stage change-detection process was utilized using SAR data. In the first stage, the probability distribution functions of the classes present in the images were estimated and then changes in the images via paradoxical and evidential reasoning were detected. In Ref. [9], difference images are automatically analyzed using two different methods based on Bayes theory in an unsupervised change-detection paradigm, which estimates the statistical distributions of the changed and unchanged pixels using an iterative method based on the expectation maximization algorithm. In Ref. [10], the performance of the minimum error solution based on the matrix Wiener filter was compared to the covariance equalization method (CE) to detect changes in hyperspectral imagery. They found that CE, which has relaxed operational requirements, is more robust to imperfect image registration than the matrix Wiener filter method.

In this chapter, we have implemented image differencing, ratio, Euclidean distance, Mahalanobis distance, subspace projection-based, and Wiener filter-based change detection and compared their performances to one another. We demonstrate all algorithms on co-registered SAR images obtained from a high resolution, VV-polarized SAR system. In Section 6.2 we discuss difference-based change detection, Euclidean distance, and ratio change-detection methods, which comprise the simpler change-detection methods implemented in this paper. In Section 6.3 through Section 6.5, we discuss more complex methods such as Mahalanobis distance, subspace projection–based change detection, and Wiener filter–based change detection, respectively. We consider specific implementation issues in Section 6.6 and present results in Section 6.7. Finally, a conclusion is provided in Section 6.8.

6.2 Difference, Euclidean Distance, and Image Ratioing Change-Detection Methods

In simple difference–based change detection, a pixel from one image (the reference image) is subtracted from the pixel in the corresponding location of another image (the test image), which has changed with respect to the reference image. If the difference is greater than a threshold, then a change is said to have occurred. One can also subtract a block of pixels from one image from the corresponding block from a test image. This is referred to as Euclidean difference if the blocks of pixels are arranged into vectors and the L-2 norm of the difference of these vectors is computed. We implemented the Euclidean distance as in Equation 6.1, where we computed the difference between a block of pixels (arranged into a one-dimensional vector) from reference image, $\mathbf{f}_X$, and the corresponding block in the test image, $\mathbf{f}_Y$, to obtain an error, e_E:

$$e_E = (\mathbf{y} - \mathbf{x})^T \tag{6.1}$$

where $\mathbf{x}$ and $\mathbf{y}$ are vectors of pixels taken from $\mathbf{f}_X$ and $\mathbf{f}_Y$, respectively. We display this error, e_E, as an image. When no change has occurred this error is expected to be low and the error will be high when a change has occurred. While simple differencing considers only two pixels in making a change decision, the Euclidean distance takes into account pixels within the neighborhood of the pixel in question. This regional decision may have a smoothing effect on the change image at the expense of additional computational complexity.

Closely related to the Euclidean distance metric for change detection is image ratioing. In many change-detection applications, ratioing proved more robust to illumination effects than simple differencing. It is implemented as follows:

$$e_R = \frac{y}{x} \tag{6.2}$$

where y and x are pixels from the same locations in the test and reference images, respectively.

6.3 Mahalanobis Distance–Based Change Detection

Simple techniques like differencing and image ratioing suffer from sensitivity to noise and illumination intensity variations in the images. Therefore, these methods may prove adequate only for applications to images with low noise and which show little illumination variation. These methods may be inadequate when applied to SAR images due to the presence of highly correlated speckle noise, misregistration errors, and other unimportant changes in the images. There may be differences in the reference and test images caused by unknown fluctuations in the amplitude-phase in the radiation pattern of the physical radar between the two data sets and subtle inconsistencies in the data acquisition circuitry [5]; consequently, more robust methods are sought.

One such method uses the Mahalanobis distance to detect changes in SAR imagery. In this change-detection application, we obtain an error (change) image by computing the Mahalanobis distance between $\mathbf{x}$ and $\mathbf{y}$ to obtain e_{MD} as follows:

$$e_{\mathrm{MD}} = (\mathbf{x} - \mathbf{y})^{\mathrm{T}} \mathbf{C}_{\mathrm{X}}^{-1} (\mathbf{x} - \mathbf{y}) \tag{6.3}$$

The $\mathbf{C}_{\mathrm{X}}^{-1}$ term in the Mahalanobis distance is the inverse covariance matrix computed from vectors of pixels in $\mathbf{f}_{\mathrm{X}}$. By considering the effects of other pixels in making a change decision with the inclusion of second-order statistics, the Mahalanobis distance method is expected to reduce false alarms. The $\mathbf{C}_{\mathrm{X}}^{-1}$ term should improve the estimate by reducing the effects of background clutter variance, which, for the purpose of detecting mines in SAR imagery, does not constitute a significant change.

6.4 Subspace Projection–Based Change Detection

To apply subspace projection [11] to change detection a subspace must be defined for either the reference or the test image. One may implement Gram–Schmidt orthogonalization as in [6] or one can make use of eigen-decomposition to define a suitable subspace for the data. We computed the covariance of a sample from $\mathbf{f}_{\mathrm{X}}$ and its eigenvectors and eigenvalues as follows:

$$\mathbf{C}_{\mathrm{X}} = (\mathbf{X} - \boldsymbol{\mu}_{X})(\mathbf{X} - \boldsymbol{\mu}_{X})^{\mathrm{T}} \tag{6.4}$$

where $\mathbf{X}$ is a matrix of pixels whose columns represent a block of pixels from $\mathbf{f}_{\mathrm{X}}$, arranged as described in Section 6.6.1, having mean $\boldsymbol{\mu}_{X} = \mathbf{X}\mathbf{1}_{N \times N}$. $\mathbf{1}_{N \times N}$ is a square matrix of size $N \times N$ whose elements are $1/N$, where N is the number of columns of $\mathbf{X}$. We define the subspace of the reference data, which we can express using eigen-decomposition in terms of its eigenvectors, $\mathbf{Z}$, and eigenvalues, $\mathbf{V}$:

$$\mathbf{C}_{\mathrm{X}} = \mathbf{Z}\mathbf{V}\mathbf{Z}^{\mathrm{T}} \tag{6.5}$$

We truncate the number of eigenvectors in $\mathbf{Z}$, denoted by $\widetilde{\mathbf{Z}}$, to develop a subspace projection operator, $\mathbf{P}_{\mathrm{X}}$:

$$\mathbf{P}_{\mathrm{X}} = \widetilde{\mathbf{Z}}\widetilde{\mathbf{Z}}^{\mathrm{T}} \tag{6.6}$$

The projection of the test image onto the subspace of the reference image will provide a measure of how much of the test sample is represented by the reference image. Therefore, by computing the squared difference between the test image and its projection onto the subspace of the reference image we obtain an estimate of the difference between the two images:

$$e_{\mathrm{SP}}(\mathbf{y}) = [\mathbf{y}^{\mathrm{T}}(\mathbf{I} - \mathbf{P}_{\mathrm{X}})\mathbf{y}] \tag{6.7}$$

We evaluated the effects of various levels of truncation and display the best results achieved. In our implementations we include the $\mathbf{C}_X^{-1}$ term in the subspace projection error term as follows:

$$e_{\text{SP}}(\mathbf{y}) = \mathbf{y}^{\text{T}}(\mathbf{I} - \mathbf{P}_X)^{\text{T}}\mathbf{C}_X^{-1}(\mathbf{I} - \mathbf{P}_X)\mathbf{y} \tag{6.8}$$

We expect it will play a similar role as it does in the Mahalanobis prediction and further diminish false alarms by suppressing the background clutter.

Expanding the terms in (6.8) we get

$$e_{\text{SP}}(\mathbf{y}) = \mathbf{y}^{\text{T}}\mathbf{C}_X^{-1}\mathbf{y} - \mathbf{y}^{\text{T}}\mathbf{P}_X^{\text{T}}\mathbf{C}_X^{-1}\mathbf{y} - \mathbf{y}^{\text{T}}\mathbf{C}_X^{-1}\mathbf{P}_X\mathbf{y} + \mathbf{y}^{\text{T}}\mathbf{P}_X^{\text{T}}\mathbf{C}_X^{-1}\mathbf{P}_X\mathbf{y} \tag{6.9}$$

It can be shown that $\mathbf{y}^{\text{T}}\mathbf{C}_X^{-1}\mathbf{P}_X\mathbf{y} = \mathbf{y}^{\text{T}}\mathbf{P}_X^{\text{T}}\mathbf{C}_X^{-1}\mathbf{y} = \mathbf{y}^{\text{T}}\mathbf{P}_X^{\text{T}}\mathbf{C}_X^{-1}\mathbf{P}_X\mathbf{y}$, so we can rewrite (6.9) as

$$e_{\text{SP}}(\mathbf{y}) = \mathbf{y}^{\text{T}}\mathbf{C}_X^{-1}\mathbf{y} - \mathbf{y}\mathbf{C}_X^{-1}\mathbf{P}_X\mathbf{y}^{\text{T}} \tag{6.10}$$

6.5 Wiener Prediction–Based Change Detection

We propose a Wiener filter–based change-detection algorithm to overcome some of the limitations of simple differencing, namely to exploit the highly correlated nature of speckle noise, thereby reducing false alarms. Given two stochastic, jointly wide-sense stationary signals, the Wiener equation can be implemented for prediction of one data set from the other using the auto-and cross-correlations of the data. By minimizing the variance of the error committed in this prediction, the Wiener predictor provides the optimal solution in the linear minimum mean-squared error sense.

If the data are well represented by the second-order statistics, the Wiener prediction will be very close to the actual data. In applying Wiener prediction theory to change detection the number of pixels where abrupt change has occurred is very small relative to the total number of pixels in the image; therefore, these changes will not be adequately represented by the correlation matrix. Consequently, there will be a low error in the prediction where the reference and test images are the same and no change exists between the two, and there will be a large error in the prediction of pixels where a change has occurred.

The Wiener filter [11] is the linear minimum mean-squared error filter for second-order stationary data. Consider the signal $\hat{\mathbf{y}}_W = \mathbf{Wx}$. The goal of Wiener filtering is to find $\mathbf{W}$, which minimizes the error, $e_W = \| \mathbf{y} - \hat{\mathbf{y}}_W \|$, where $\mathbf{y}$ represents a desired signal. In the case of change detection $\mathbf{y}$ is taken from the test image, $\mathbf{f}_Y$, which contains changes as compared to the reference image, $\mathbf{f}_X$. The Wiener filter seeks the value of $\mathbf{W}$, which minimizes the mean-squared error, e_W. If the linear minimum mean-squared error estimator of $\mathbf{Y}$ satisfies the orthogonality condition, the following expressions hold:

$$\begin{aligned} E\{(\mathbf{Y} - \mathbf{WX})\mathbf{X}^{\text{T}}\} &= 0 \\ E\{\mathbf{YX}^{\text{T}}\} - E\{\mathbf{WXX}^{\text{T}}\} &= 0 \\ \mathbf{R}_{YX} - \mathbf{WR}_{XX} &= 0 \end{aligned} \tag{6.11}$$

where $E\{\cdot\}$ represents the expectation operator, $\mathbf{X}$ is a matrix of pixels whose columns represent a block of pixels from $\mathbf{f}_X$, $\mathbf{Y}$ is a matrix of pixels from $\mathbf{f}_Y$ whose columns represent a block of pixels at the same locations as those in $\mathbf{X}$, $\mathbf{R}_{YX}$ is the cross-correlation matrix of $\mathbf{X}$ and $\mathbf{Y}$, and $\mathbf{R}_{YX}^{-1}$ is the autocorrelation matrix of $\mathbf{X}$. For a complete description of the construction of $\mathbf{X}$ and $\mathbf{Y}$ see Section 6.6.1. Equations 6.11 imply that $\mathbf{W}$ satisfies the Wiener–Hopf equation, which has the following solution:

$$\mathbf{W} = \mathbf{R}_{YX}\mathbf{R}_{XX}^{-1} \tag{6.12}$$

Therefore, we have:

$$\hat{\mathbf{y}}_W = \mathbf{R}_{YX}\mathbf{R}_{XX}^{-1}\mathbf{x} \tag{6.13}$$

where $\mathrm{R}_{YX} = \mathrm{YX}^{\mathrm{T}}$ and $\mathrm{R}_{XX}^{-1} = (\mathrm{XX}^{\mathrm{T}})^{-1}$. So the error can be computed as follows:

$$e_W = \left(\mathbf{y} - \hat{\mathbf{y}}_W\right)^{\mathbf{T}} \tag{6.14}$$

In a modified implementation of the Wiener filter–based change-detection method, we insert a normalization (or whitening) term, $\mathbf{C}_X^{-1}$, into the error equation as follows:

$$e_{W'} = (\mathbf{y} - \hat{\mathbf{y}}_W)^{\mathrm{T}}\mathbf{C}_X^{-1}(\mathbf{y} - \hat{\mathbf{y}}_W) \tag{6.15}$$

We expect it will play a similar role as it does in the Mahalanobis prediction and further diminish false alarms by suppressing the background clutter. In our work we implemented Equation 6.15 to detect changes in pixels occurring from one image to another. The error at each pixel is $e_{W'}$, which we display as an image.

6.6 Implementation Considerations

6.6.1 Local Implementation Problem Formulation

In implementing the Mahalanobis distance, subspace projection–based and the Wiener filter–based change-detection methods, we constructed our data matrices and vectors locally to compute the local statistics and generate a change image. We generate $\mathbf{X}$, $\mathbf{Y}$, $\mathbf{x}$, and $\mathbf{y}$ using dual concentric sliding windows. $\mathbf{X}$ and $\mathbf{Y}$ are matrices of pixels contained within an outer moving window (size $m \times m$), whose columns are obtained from overlapping smaller blocks ($\mathbf{x}_1, \mathbf{x}_2, \ldots, \mathbf{x}_M$ each of size $n \times n$) within the outer windows of the reference image and test image, respectively, as demonstrated in Figure 6.1 and Figure 6.2 for $\mathbf{X}$, where $\mathbf{X} = [\mathbf{x}_1\ \mathbf{x}_2 \cdots \mathbf{x}_M]$. Accordingly, the dimensions of $\mathbf{X}$ and $\mathbf{Y}$ will be $N \times M$, where $N = n^2$ and $M = m^2$. Note, for clarity the windows that generate the columns of $\mathbf{X}$ and $\mathbf{Y}$ are not shown as overlapping in Figure 6.2. However, in actual implementation the $\mathbf{X}$ and $\mathbf{Y}$ matrices are constructed from overlapping windows within $\mathbf{X}$ and $\mathbf{Y}$. Figure 6.1 shows $\mathbf{x}$ and $\mathbf{y}$, which are $N \times 1$ vectors composed of pixels within the inner window (size $n \times n$) of the reference image, $\mathbf{f}_X$, and test image, $\mathbf{f}_Y$, respectively. In this local implementation, we must compute the necessary statistics at each iteration as we move pixel-by-pixel through the image, thereby generating new $\mathbf{X}$ and $\mathbf{Y}$ matrices and $\mathbf{x}$ and $\mathbf{y}$ vectors.

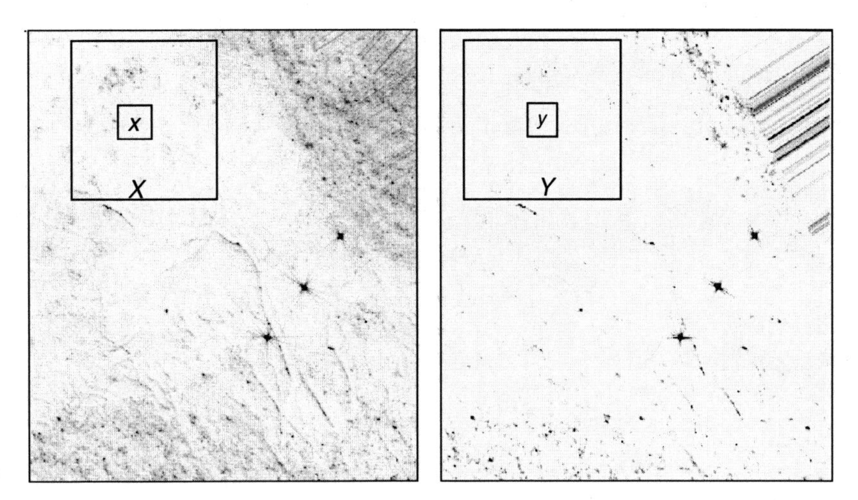

FIGURE 6.1
Dual concentric sliding windows of the test and reference images.

6.6.2 Computing Matrix Inversion for $\mathbf{C}_X^{-1}$ and $\mathbf{R}_{XX}^{-1}$

Because of numerical errors, computation of the inverse of these matrices may become unstable. We compute the pseudo-inverse using eigenvalue decomposition as shown below:

$$\mathbf{C}_X^{-1} = \mathbf{U}\mathbf{V}^{-1}\mathbf{U}^{\mathrm{T}}, \tag{6.16}$$

where **U** and **V** are the matrices of eigenvectors and eigenvalues of $\mathbf{C}_X$, respectively. However, since the eigenvalues are obtained using a numerical tool, some values are very small such that its inverse blows up and causes the matrix to be unstable. To avoid this problem, we discard eigenvalues that are below a certain threshold. We have evaluated the performance for various thresholds.

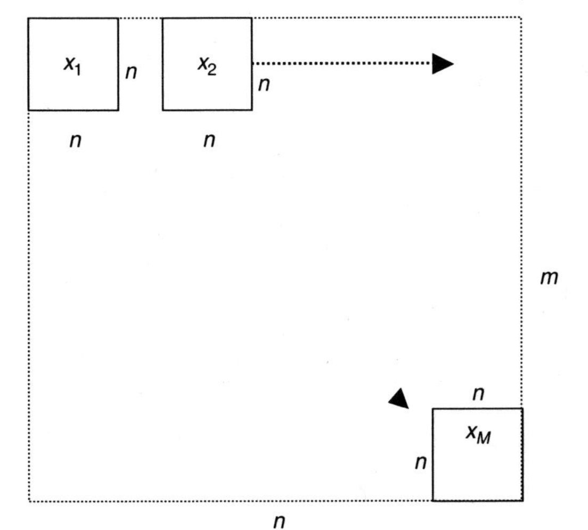

FIGURE 6.2
Construction of **X** matrix.

6.7 Experimental Results

Results from all methods implemented were normalized between zero and one before producing ROC curves. The threshold to generate the ROC curves also ranged from zero to one in increments of 10^{-3}. We used multi-look data collected at X band from a high resolution, VV-polarized SAR system (Figure 6.3a and Figure 6.3b). Small mines, comprising about a 5×5 pixel area, appear in the test image (Figure 6.3b). The size of both the reference and test images is 950×950 pixels. Ground truth was provided, which indicates the approximate center location of the known mines; we defined the mine area as an area slightly bigger than the average mine, centered at the locations indicated by ground-truth information. To determine the performance, a hit was tabulated if at least a single pixel within the mine area surpassed the threshold. Any pixel outside of a mine area, which surpassed the threshold, was counted as a false alarm.

Figure 6.4a shows the pixel differencing and Figure 6.4b shows the Euclidean distance change images. The ratio change image obtained from Equation 6.2 is represented in Figure 6.5. The local implementation results shown in Figure 6.6 through Figure 6.8 are from implementations where the inner window was of size 3×3; the outer window, from which we constructed $\mathbf{X}$ and $\mathbf{Y}$ matrices to compute the second-order statistics, was of size 13×13. We gathered 169 overlapping 3×3 blocks from the 13×13 area, arranged each into a 9×1 vector, and placed them as columns in a matrix of size 9×169. We re-computed $\mathbf{X}$ and $\mathbf{Y}$ at each iteration to determine the change at each pixel.

The three very bright spots in the original images, Figure 6.3, are used to register the images and were identified as changes in the simple change-detection methods. The local methods that incorporated image statistics were able to eliminate these bright areas from the change image (Figure 6.6 through Figure 6.8), as they are present in both the reference and test images at different intensities. Another area of the image that results in many false alarms, after processing by the simpler methods (results shown in Figure 6.4a,b, and Figure 6.5), is the area in the upper right quadrant of the image. In the test image, Figure 6.3b, this area is comprised of several slanted lines, while in the reference

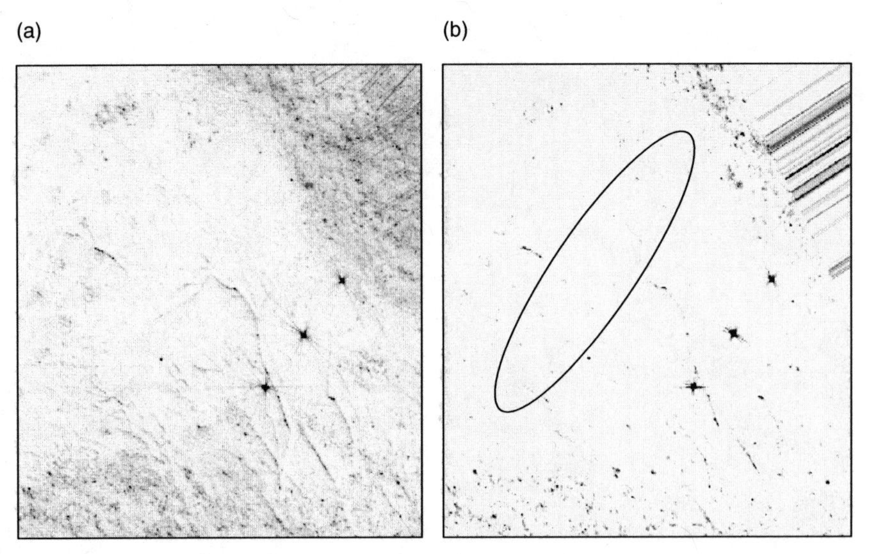

FIGURE 6.3
Original SAR Images: (a) reference image, (b) test image with mines contained in elliptical area.

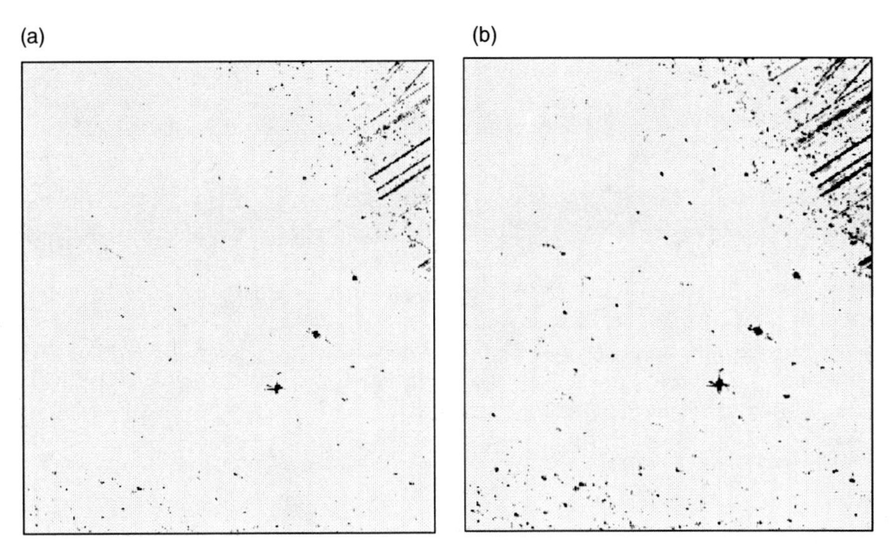

FIGURE 6.4
(a) Simple difference change image, (b) Euclidean distance change image.

image, Figure 6.3a, there is a more random, grainy pattern. The Wiener filter–based method with $\mathbf{C}_X^{-1}$, Mahalanobis distance, and the subspace projection method with $\mathbf{C}_X^{-1}$ methods all show far fewer false alarms in this area than is shown in the results from the simpler change-detection methods. ROC curves for the simple change-detection methods can be seen in Figure 6.9a and ROC curves for the more complicated methods that have the $\mathbf{C}_X^{-1}$ term are shown in Figure 6.9b. A plot of the ROC curves for all the methods implemented is displayed in Figure 6.10 for comparison.

FIGURE 6.5
Ratio change image.

FIGURE 6.6
Local Mahalanobis distance change image.

Local Wiener with $\mathbf{C}_X^{-1}$, local Mahalanobis, local subspace projection with $\mathbf{C}_X^{-1}$, and ratio methods show similar performance. However, the local Mahalanobis distance method begins to perform worse than the others from about .4 to .65 probability of detection (Pd) rate, then surpasses all the other methods above Pd $=.65$. The simple difference method has the worst performance over all Pd rates.

FIGURE 6.7
Local subspace projection (with $\mathbf{C}_X^{-1}$ term) change image.

FIGURE 6.8
Local Wiener filter–based (with $\mathbf{C}_X^{-1}$ term) change image.

6.8 Conclusions

As we have only demonstrated the algorithms on a limited data set of one scene, it is not prudent to come to broad conclusions. However there are some trends indicated by the

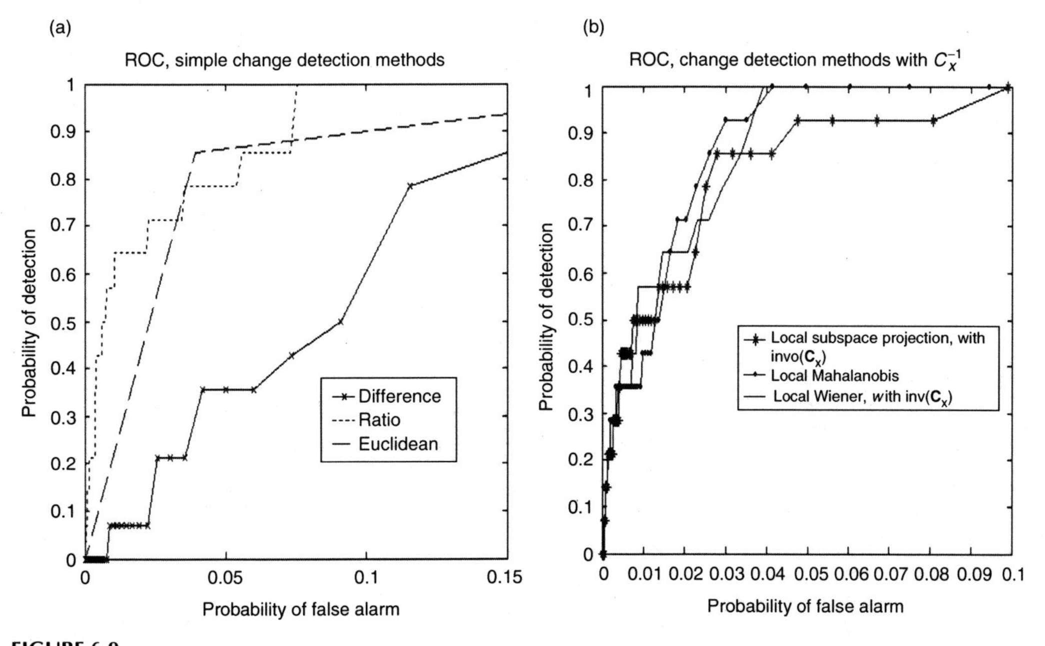

FIGURE 6.9
ROC curves displaying performance of (a) simple change-detection methods and (b) methods incorporating $\mathbf{C}_X^{-1}$ term.

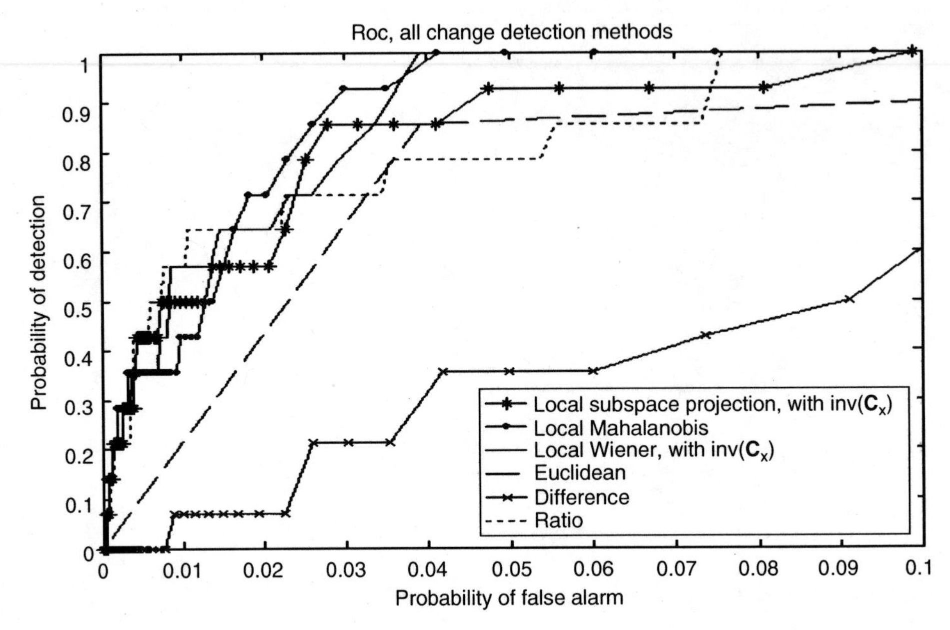

FIGURE 6.10
ROC curves displaying performance of the various change-detection methods implemented in this paper.

results above. The results above indicate that in general the more complex algorithms exhibited superior performance as compared to the simple methods; the exception seems to be the ratio method. The ratio method, although a simple implementation, has a performance that is competitive with the more complex change-detection methods implemented up to a Pd rate approximately equal to .65. Its performance degrades at higher detection rates, where it suffers from more false alarms than the methods that have the $\mathbf{C}_X^{-1}$.

Results indicate that taking into account the local spatial and statistical properties of the image improves the change-detection performance. Local characteristics are used in computing the inverse covariance matrix, resulting in fewer false alarms as compared to the simple change detection methods. The ROC curves in Figure 6.9 and Figure 6.10 also show that that addition of the $\mathbf{C}_X^{-1}$ term may serve to mitigate false alarms that arise from speckle, which is a hindrance to achieving low false alarm rates with the Euclidean distance and simple difference methods. As expected the simple difference method exhibits the worst performance, having a high occurrence of false alarms.

References

1. R.J. Radke, S. Andra, O. Al-Kofahi, and B. Roysam, Image change detection algorithms: A systematic survey, *IEEE Transactions on Image Processing*, 14(3), 294–307, 2005.
2. J.F. Mas, Monitoring land-cover changes: A comparison of change detection techniques, *International Journal of Remote Sensing*, 20(1), 139–152, 1999.
3. B. Zitova and J. Flusser, Image registration methods: A survey, *Image and Vision Computing*, 21, 977–1000, 2003.

4. R. Mayer and A. Schaum, Target detection using temporal signature propagation, *Proceedings of SPIE Conference Algorithms for Multispectral, Hyperspectral and Ultraspectral Imagery*, 4049, 64–74, 2000.
5. E.J.M. Rignot and J. van Zyl, Change detection techniques for ERS-1 SAR data, *IEEE Transactions on Geoscience and Remote Sensing*, 31(4), 896–906, 1993.
6. K. Ranney and M. Soumekh, Adaptive change detection in coherent and noncoherent SAR imagery, *Proceedings of IEEE International Radar Conference*, pp. 195–200, 2005.
7. H. Kwon, S.Z. Der, and N.M. Nasrabadi, Projection-based adaptive anomaly detection for hyperspectral imagery, *Proceedings of IEEE International Conference on Image Processing*, pp. 1001–1004, 2003.
8. G. Mercier and S. Derrode, SAR image change detection using distance between distributions of classes, *Proceedings of IEEE International Geoscience and Remote Sensing Symposium*, 6, 3872–3875, 2004.
9. L. Bruzzone and D.F. Prieto, Automatic analysis of the difference image for unsupervised change detection, *IEEE Transactions on Geoscience and Remote Sensing*, 38(3), 1171–1182, 2000.
10. A. Schaum and A. Stocker, Advanced algorithms for autonomous hyperspectral change detection, *Proceedings of IEEE, Applied Imagery Pattern Recognition Workshop*, pp. 39–44, 2004.
11. L. Scharf, *Statistical Signal Processing: Detection Estimation, and Time Series Analysis*, Addison Wesley, Colorado, 1991.

7

*Vertex Component Analysis: A Geometric-Based Approach to Unmix Hyperspectral Data**

José M.B. Dias and José M.P. Nascimento

CONTENTS

7.1 Introduction

Hyperspectral remote sensing exploits the electromagnetic scattering patterns of the different materials at specific wavelengths [2,3]. Hyperspectral sensors have been developed to sample the scattered portion of the electromagnetic spectrum extending from the visible region through the near-infrared and mid-infrared, in hundreds of narrow contiguous bands [4,5]. The number and variety of potential civilian and military applications of hyperspectral remote sensing is enormous [6,7].

Very often, the resolution cell corresponding to a single pixel in an image contains several substances (endmembers) [4]. In this situation, the scattered energy is a mixing of the endmember spectra. A challenging task underlying many hyperspectral imagery applications is then decomposing a mixed pixel into a collection of reflectance spectra, called endmember signatures, and the corresponding abundance fractions [8–10].

Depending on the mixing scales at each pixel, the observed mixture is either linear or nonlinear [11,12]. Linear mixing model holds approximately when the mixing scale is macroscopic [13] and there is negligible interaction among distinct endmembers [3,14]. If, however, the mixing scale is microscopic (or intimate mixtures) [15,16] and the incident solar radiation is scattered by the scene through multiple bounces involving several endmembers [17], the linear model is no longer accurate.

*Work partially based on [1].

Linear spectral unmixing has been intensively researched in the last years [9,10,12, 18–21]. It considers that a mixed pixel is a linear combination of endmember signatures weighted by the correspondent abundance fractions. Under this model, and assuming that the number of substances and their reflectance spectra are known, hyperspectral unmixing is a linear problem for which many solutions have been proposed (e.g., maximum likelihood estimation [8], spectral signature matching [22], spectral angle mapper [23], subspace projection methods [24,25], and constrained least squares [26]).

In most cases, the number of substances and their reflectances are not known and, then, hyperspectral unmixing falls into the class of blind source separation problems [27]. Independent component analysis (ICA) has recently been proposed as a tool to blindly unmix hyperspectral data [28–31]. ICA is based on the assumption of mutually independent sources (abundance fractions), which is not the case of hyperspectral data, since the sum of abundance fractions is constant, implying statistical dependence among them. This dependence compromises ICA applicability to hyperspectral images as shown in Refs. [21,32]. In fact, ICA finds the endmember signatures by multiplying the spectral vectors with an unmixing matrix, which minimizes the mutual information among sources. If sources are independent, ICA provides the correct unmixing, since the minimum of the mutual information is obtained only when sources are independent. This is no longer true for dependent abundance fractions. Nevertheless, some endmembers may be approximately unmixed. These aspects are addressed in Ref. [33].

Under the linear mixing model, the observations from a scene are in a simplex whose vertices correspond to the endmembers. Several approaches [34–36] have exploited this geometric feature of hyperspectral mixtures [35]. Minimum volume transform (MVT) algorithm [36] determines the simplex of minimum volume containing the data. The method presented in Ref. [37] is also of MVT type but, by introducing the notion of bundles, it takes into account the endmember variability usually present in hyperspectral mixtures.

The MVT type approaches are complex from the computational point of view. Usually, these algorithms find in the first place the convex hull defined by the observed data and then fit a minimum volume simplex to it. For example, the *gift wrapping algorithm* [38] computes the convex hull of n data points in a d-dimensional space with a computational complexity of $O(n^{\lfloor d/2 \rfloor + 1})$, where $\lfloor x \rfloor$ is the highest integer lower or equal than x and n is the number of samples. The complexity of the method presented in Ref. [37] is even higher, since the temperature of the simulated annealing algorithm used shall follow a $\log(\cdot)$ law [39] to assure convergence (in probability) to the desired solution.

Aiming at a lower computational complexity, some algorithms such as the pixel purity index (PPI) [35] and the N-FINDR [40] still find the minimum volume simplex containing the data cloud, but they assume the presence of at least one pure pixel of each endmember in the data. This is a strong requisite that may not hold in some data sets. In any case, these algorithms find the set of *most pure pixels* in the data.

PPI algorithm uses the minimum noise fraction (MNF) [41] as a preprocessing step to reduce dimensionality and to improve the signal-to-noise ratio (*SNR*). The algorithm then projects every spectral vector onto *skewers* (large number of random vectors) [35,42,43]. The points corresponding to extremes, for each *skewer* direction, are stored. A cumulative account records the number of times each pixel (i.e., a given spectral vector) is found to be an extreme. The pixels with the highest scores are the purest ones.

N-FINDR algorithm [40] is based on the fact that in p spectral dimensions, the p-volume defined by a simplex formed by the purest pixels is larger than any other volume defined by any other combination of pixels. This algorithm finds the set of pixels defining the largest volume by *inflating* a simplex inside the data.

ORASIS [44,45] is a hyperspectral framework developed by the U.S. Naval Research Laboratory consisting of several algorithms organized in six modules: exemplar selector, adaptative learner, demixer, knowledge base or spectral library, and spatial postprocessor. The first step consists in flat-fielding the spectra. Next, the exemplar selection module is used to select spectral vectors that best represent the smaller convex cone containing the data. The other pixels are rejected when the spectral angle distance (SAD) is less than a given threshold. The procedure finds the basis for a subspace of a lower dimension using a modified Gram–Schmidt orthogonalization. The selected vectors are then projected onto this subspace and a simplex is found by an MVT process. ORASIS is oriented to real-time target detection from uncrewed air vehicles using hyperspectral data [46].

In this chapter we develop a new algorithm to unmix linear mixtures of endmember spectra. First, the algorithm determines the number of endmembers and the signal subspace using a newly developed concept [47,48]. Second, the algorithm extracts the most pure pixels present in the data. Unlike other methods, this algorithm is completely automatic and unsupervised.

To estimate the number of endmembers and the signal subspace in hyperspectral linear mixtures, the proposed scheme begins by estimating signal and noise correlation matrices. The latter is based on multiple regression theory. The signal subspace is then identified by selecting the set of signal eigenvalues that best represents the data, in the least-square sense [48,49], we note, however, that VCA works with projected and with unprojected data.

The extraction of the endmembers exploits two facts: (1) the endmembers are the vertices of a simplex and (2) the affine transformation of a simplex is also a simplex. As PPI and N-FINDR algorithms, VCA also assumes the presence of pure pixels in the data. The algorithm iteratively projects data onto a direction orthogonal to the subspace spanned by the endmembers already determined. The new endmember signature corresponds to the extreme of the projection. The algorithm iterates until all endmembers are exhausted. VCA performs much better than PPI and better than or comparable to N-FINDR; yet it has a computational complexity between one and two orders of magnitude lower than N-FINDR.

The chapter is structured as follows. Section 7.2 describes the fundamentals of the proposed method. Section 7.3 and Section 7.4 evaluate the proposed algorithm using simulated and real data, respectively. Section 7.5 presents some concluding remarks.

7.2 Vertex Component Analysis Algorithm

Assuming the linear mixing scenario, each observed spectral vector is given by

$$\begin{aligned}\mathbf{r} &= \mathbf{x} + \mathbf{n} \\ &= \mathbf{M}\underbrace{\gamma\boldsymbol{\alpha}}_{\mathbf{s}} + \mathbf{n},\end{aligned} \tag{7.1}$$

where $\mathbf{r}$ is an L-vector (L is the number of bands), $\mathbf{M} \equiv [\mathbf{m}_1, \mathbf{m}_2, \ldots, \mathbf{m}_p]$ is the mixing matrix ($\mathbf{m}_i$ denotes the ith endmember signature and p is the number of endmembers present in the covered area), $\mathbf{s} \equiv \gamma\boldsymbol{\alpha}$ (γ is a scale factor modeling illumination variability due to surface topography), $\boldsymbol{\alpha} = [\alpha_1, \alpha_2, \ldots, \alpha_p]^T$ is the vector containing the abundance fraction of each endmember (the notation $(\cdot)^T$ stands for vector transposed), and $\mathbf{n}$ models system additive noise.

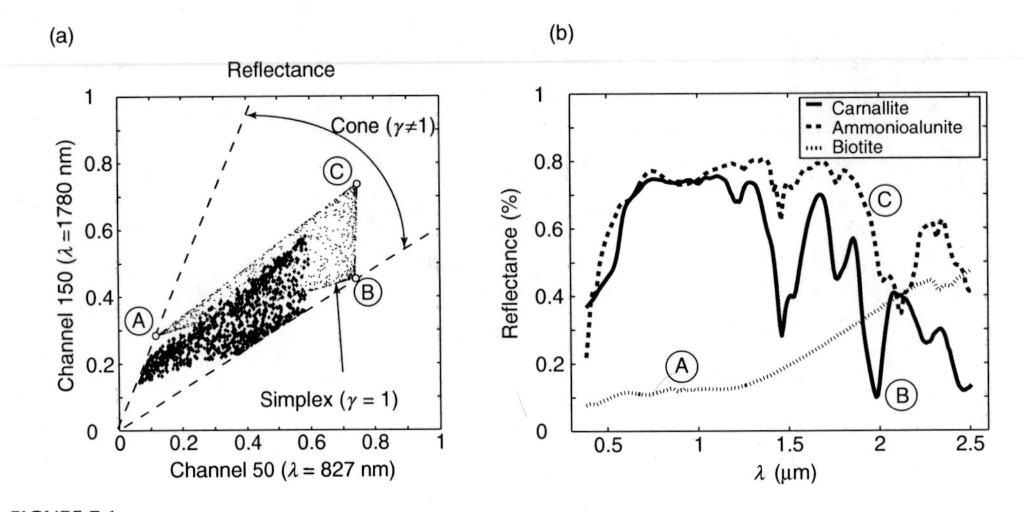

FIGURE 7.1
(a) 2D scatterplot of mixtures of the three endmembers shown in Figure 7.2; circles denote pure materials. (b) Reflectances of Carnallite, Ammonioalunite, and Biotite. (© 2005 IEEE. With permission.)

Owing to physical constraints [20], abundance fractions are non-negative ($\boldsymbol{\alpha} \geq 0$) and satisfy the so-called positivity constraint $\mathbf{1}^T\, \boldsymbol{\alpha} = 1$, where $\mathbf{1}$ is a $p \times 1$ vector of ones. Each pixel can be viewed as a vector in an L-dimensional Euclidean space, where each channel is assigned to one axis of space. Since the set $\{\boldsymbol{\alpha} \in \Re^p : \mathbf{1}^T\boldsymbol{\alpha} = 1, \boldsymbol{\alpha} \geq \mathbf{0}\}$ is a simplex, the set $S_x = \{\mathbf{x} \in \Re^L : \mathbf{x} = \mathbf{M}\boldsymbol{\alpha}, \mathbf{1}^T\boldsymbol{\alpha} = 1, \boldsymbol{\alpha} \geq 0\}$ is also a simplex. However, even assuming $\mathbf{n} = 0$, the observed vector set belongs to $C_p = \{\mathbf{r} \in \Re^L : \mathbf{r} = \mathbf{M}\gamma\boldsymbol{\alpha},\ \mathbf{1}^T\ \boldsymbol{\alpha} = 1,\ \boldsymbol{\alpha} \geq \mathbf{0},\ \gamma \geq 0\}$ that is a convex cone, owing to scale factor γ. For illustration purposes, a simulated scene was generated according to the expression in Equation 7.1. Figure 7.1a illustrates a simplex and a cone, projected on a 2D subspace, defined by a mixture of three endmembers. These spectral signatures (AM Biotite, BM Carnallite, and CM Ammonioalunite) were selected from the U.S. Geological Survey (USGS) digital spectral library [50]. The simplex boundary is a triangle whose vertices correspond to these endmembers shown in Figure 7.1b. Small and medium dots are simulated mixed spectra belonging to the simplex S_x ($\gamma = 1$) and to the cone C_p ($\gamma \rangle\ 0$), respectively.

The projective projection of the convex cone C_p onto a properly chosen hyperplane is a simplex with vertices corresponding to the vertices of the simplex S_x. This is illustrated in Figure 7.2. The simplex $S_p = \{\mathbf{y} \in \Re^L : \mathbf{y} = \mathbf{r}/(\mathbf{r}^T\ \mathbf{u}), \mathbf{r} \in C_p\}$ is the projective projection of the convex cone C_p onto the plane $\mathbf{r}^T\mathbf{u} = 1$, where the choice of $\mathbf{u}$ assures that there is no observed vectors orthogonal to it.

After identifying S_p, the VCA algorithm iteratively projects data onto a direction orthogonal to the subspace spanned by the endmembers already determined. The new endmember signature corresponds to the extreme of the projection. Figure 7.2 illustrates the two iterations of VCA algorithm applied to the simplex S_p defined by the mixture of two endmembers. In the first iteration, data is projected onto the first direction $\mathbf{f}_1$. The extreme of the projection corresponds to endmember $\mathbf{m}_a$. In the next iteration, endmember $\mathbf{m}_b$ is found by projecting data onto direction $\mathbf{f}_2$, which is orthogonal to $\mathbf{m}_a$. The algorithm iterates until the number of endmembers is exhausted.

7.2.1 Subspace Estimation

As mentioned before each spectral vector, also called pixel, of a hyperspectral image can be represented as a vector in the space $\Re^L$. Under the linear mixing scenario, those

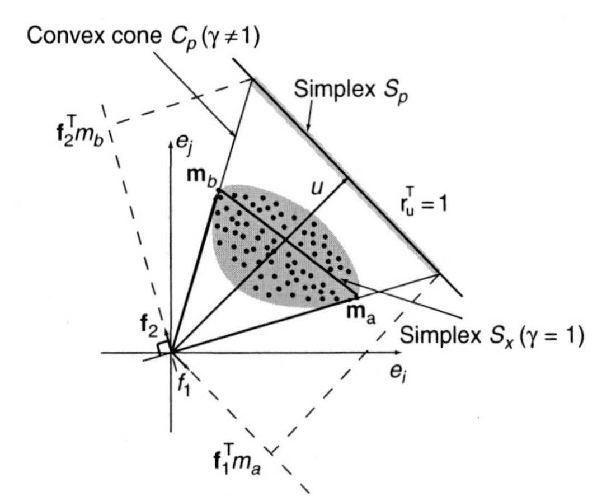

FIGURE 7.2
Illustration of the VCA algorithm. (© 2005 IEEE. With permission.)

spectral vectors are a linear combination of a few vectors, the so-called endmember signatures. Therefore, the dimensionality of data (number of endmembers) is usually much lower than the number of bands, that is, those pixels are in a subspace of dimension p. If $p \ll L$, it is worthy to project the observed spectral vectors onto the subspace signal. This leads to significant savings in computational complexity and to *SNR* improvements.

Principal component analysis (PCA) [51], maximum noise fraction (MNF) [52], and singular value decomposition (SVD) [49] are three well-known projection techniques widely used in remote sensing. PCA, also known as Karhunen–Loéve transform, seeks the projection that best represents data in a least-square sense; MNF or noise adjusted principal components (NAPC) [53] seeks the projection that optimizes *SNR*; and SVD provides the projection that best represents data in the maximum power sense. PCA and MNF are equal in the case of white noise. SVD and PCA are also equal in the case of zero-mean data.

A key problem in dimensionality reduction of hyperspectral data is the determination of the number of endmembers present in the data set. Recently, Harsanyi, Farrand, and Chang developed a Neyman–Pearson detection theory-based thresholding method (HFC), to determine the number of spectral endmembers in hyperspectral data (referred in Ref. [18] as virtual dimensionality—VD). HFC method uses the eigenvalues to measure signal [54]. A modified version of this method, energy in a detection frame work called noise-whitened HFC (NWHFC), includes a noise-whitening process as a preprocessing step to remove the second-order statistical correlation [18]. In the NWHFC method a noise estimation is required.

By specifying the false alarm probability, P_f, a Neyman–Pearson detector is derived to determine whether or not a distinct signature is present in each of spectral bands. The number of times the Neyman–Pearson detector fails the test is exactly the number of endmembers assumed to be present in the data.

A new mean-squared error-based approach is next presented to determine the signal subspace in hyperspectral imagery [47,48]. The method firstly estimates the signal and noise correlations matrices, then it selects the subset of eigenvalues that best represents the signal subspace in the least-square sense.

7.2.1.1 Subspace Estimation Method Description

Let $\mathbf{R} = [\mathbf{r}_1, \mathbf{r}_2, \ldots, \mathbf{r}_N]$ be an $L \times N$ matrix of spectral vectors, one per pixel, where N is the number of pixels and L is the number of bands. Assuming a linear mixing scenario, each observed spectral vector is given by Equation 7.1.

The correlation matrix of vector $\mathbf{r}$ is $\mathbf{K}_r = \mathbf{K}_x + \mathbf{K}_n$, where $\mathbf{K}_x = \mathbf{M}\mathbf{K}_s\,\mathbf{M}^T$ is the signal correlation matrix, $\mathbf{K}_n$ is the noise correlation matrix, and $\mathbf{K}_s$ is the abundance correlation matrix. An estimate of the signal correlation matrix is given by

$$\hat{\mathbf{K}}_x = \hat{\mathbf{K}}_r - \hat{\mathbf{K}}_n, \tag{7.2}$$

where $\hat{\mathbf{K}}_r = \mathbf{R}\mathbf{R}^T/N$ is the sample correlation matrix of $\mathbf{R}$, and $\hat{\mathbf{K}}_n$ is an estimate of noise correlation matrix.

Define $\mathbf{R}^i = ([\mathbf{R}]_{i,:})^T$, where symbol $[\mathbf{R}]_{i,:}$ stands for the ith row of $\mathbf{R}$, that is, $\mathbf{R}^i$ is the transpose of the ith line of matrix $\mathbf{R}$, thus containing the data read by the hyperspectral sensor at the ith band for all image pixels. Define also the matrix $\mathbf{R}^{\partial_i} = [\mathbf{R}^1, \ldots, \mathbf{R}^{i-1}, \mathbf{R}^{i+1}, \ldots, \mathbf{R}^L]$.

Assuming that the dimension of the signal subspace is much lower than the number of bands, the noise correlation matrix $\hat{\mathbf{K}}_n$ can be inferred based on multiple regression theory [55]. This consists in assuming that

$$\mathbf{R}^i = \mathbf{R}^{\partial_i}\mathbf{B}_i + \varepsilon_i, \tag{7.3}$$

where $\mathbf{R}^{\partial_i}$ is the explanatory data matrix, $\mathbf{B}_i = [B_1, \ldots, B_{L-1}]^T$ is the regression vector, and ε_i are modeling errors. For each $i \in \{1, \ldots, L\}$, the regression vector is given by $\mathbf{B}_i = [\mathbf{R}^{\partial_i}]^{\#}\,\mathbf{R}^i$, where $(\cdot)^{\#}$ denotes pseudo-inverse matrix. Finally, we compute $\hat{\varepsilon}_i = \mathbf{R}^i - \mathbf{R}^{\partial_i}\,\hat{B}_i$ and its sample correlation matrix $\hat{\mathbf{K}}_n$.

Let the singular value decomposition (SVD) of $\hat{\mathbf{K}}_x$ be

$$\hat{\mathbf{K}}_x = \mathbf{E}\Sigma\mathbf{E}^T, \tag{7.4}$$

where $\mathbf{E} = [\mathbf{e}_1, \ldots, \mathbf{e}_k, \mathbf{e}_{k+1}, \ldots, \mathbf{e}_L]$ is a matrix with the singular vectors ordered by the descendent magnitude of the respective singular values. The space $\Re^L$ can be split into two orthogonal subspaces: $\langle E_k \rangle$ spanned by $\mathbf{E}_k = [\mathbf{e}_1, \ldots, \mathbf{e}_k]$ and $\langle E_k^{\perp} \rangle$ spanned by $\mathbf{E}_k^{\perp} = [\mathbf{e}_{k+1}, \ldots, \mathbf{e}_L]$, where k is the order of the signal subspace.

Because hyperspectral mixtures have non-negative components, the projection of the mean value of $\mathbf{R}$ onto any eigenvector $\mathbf{e}_i$, $1 \leq i \leq k$, is always nonzero. Therefore, the signal subspace can be identified by finding the subset of eigenvalues that best represents, in the least-square sense, the mean value of data set.

The sample mean value of $\mathbf{R}$ is

$$\bar{\mathbf{r}} = \frac{1}{N}\sum_{i=1}^{N}\mathbf{r}_i = \frac{1}{N}\mathbf{M}\sum_{i=1}^{N}\mathbf{s}_i + \frac{1}{N}\sum_{i=1}^{N}\mathbf{n}_i = \mathbf{c} + \mathbf{w}, \tag{7.5}$$

where $\mathbf{c}$ is in the signal subspace and $\mathbf{w} \sim N(0, \mathbf{K}_n/N)$ [the notation $N(\mu, \mathbf{C})$ stands for normal density function with mean μ and covariance $\mathbf{C}$]. Let $\mathbf{c}_k$ be the projection of $\mathbf{c}$ onto $\langle E_k \rangle$. The estimation of $\mathbf{c}_k$ can be obtained by projecting $\bar{\mathbf{r}}$ onto the signal subspace $\langle E_k \rangle$, i.e., $\hat{\mathbf{c}}_k = \mathbf{U}_k\bar{\mathbf{r}}$, where $\mathbf{U}_k = \mathbf{E}_k\,\mathbf{E}_k^T$ is the projection matrix onto $\langle E_k \rangle$.

The first and the second-order moments of the estimated error $\mathbf{c} - \hat{\mathbf{c}}_k$ are

$$E[\mathbf{c} - \hat{\mathbf{c}}_k] = \mathbf{c} - E[\hat{\mathbf{c}}_k] = \mathbf{c} - E[\mathbf{U}_k\bar{r}] = \mathbf{c} - \mathbf{U}_k\mathbf{c} = \mathbf{c} - \mathbf{c}_k \equiv \mathbf{b}_k, \tag{7.6}$$

$$E[(\mathbf{c} - \hat{\mathbf{c}}_k)(\mathbf{c} - \hat{\mathbf{c}}_k)^{\mathrm{T}}] = \mathbf{b}_k\mathbf{b}_k^{\mathrm{T}} + \mathbf{U}_k\mathbf{K}_n\mathbf{U}_k^{\mathrm{T}}/N, \tag{7.7}$$

where $E[\cdot]$ denotes the expectation operator and the bias $\mathbf{b}_k = \mathbf{U}_k^{\perp}\mathbf{c}$ is the projection of $\mathbf{c}$ onto the space $\langle \mathbf{E}_k^{\perp} \rangle$. Therefore, the density of the estimated error $\mathbf{c} - \hat{\mathbf{c}}_k$ is $N(\mathbf{b}_k, \mathbf{b}_k^{\mathrm{T}}\,\mathbf{b}_k + \mathbf{U}_k\,\mathbf{K}_n\,\mathbf{U}_k^{\mathrm{T}}/\mathrm{N})$.

The mean-squared error between $\mathbf{c}$ and $\hat{\mathbf{c}}_k$ is

$$\begin{aligned}\mathrm{mse}(k) &= E[(\mathbf{c} - \hat{\mathbf{c}}_k)^{\mathrm{T}}(\mathbf{c} - \hat{\mathbf{c}}_k)]\\ &= \mathrm{tr}\{E[(\mathbf{c} - \hat{\mathbf{c}}_k)(\mathbf{c} - \hat{\mathbf{c}}_k)^{\mathrm{T}}]\}\\ &= \mathbf{b}_k^{\mathrm{T}}\mathbf{b}_k + \mathrm{tr}(\mathbf{U}_k\mathbf{K}_n\mathbf{U}_k^{\mathrm{T}}/N),\end{aligned} \tag{7.8}$$

where $\mathrm{tr}(\cdot)$ denotes the trace operator. As we do not know the bias $\mathbf{b}_k$, an approximation of Equation 7.8 can be achieved by using the bias estimate $\hat{\mathbf{b}}_k = \mathbf{U}_k^{\perp}\hat{\mathbf{r}}$. However, $E[\hat{\mathbf{b}}_k] = \mathbf{b}_k$ and $E[\hat{\mathbf{b}}_k^{T}\hat{\mathbf{b}}_k] = \mathbf{b}_k^{\mathrm{T}}\mathbf{b}_k + \mathrm{tr}(\mathbf{U}_k^{\perp}\mathbf{K}_n\mathbf{U}_k^{\perp\mathrm{T}}/N)$, that is, an unbiased estimate of $\mathbf{b}_k^{\mathrm{T}}\mathbf{b}_k$ is $\hat{\mathbf{b}}_k^{\mathrm{T}}\hat{\mathbf{b}}_k - \mathrm{tr}(\mathbf{U}_k^{\perp}\mathbf{K}_n\mathbf{U}_k^{\perp\mathrm{T}}/N)$. The criteria for the signal subspace order determination are then

$$\begin{aligned}\hat{k} &= \arg\min_k\,(\hat{\mathbf{b}}_k^{T}\hat{\mathbf{b}}_k + \mathrm{tr}(\mathbf{U}_k\mathbf{K}_n\mathbf{U}_k^{T}/N) - \mathrm{tr}(\mathbf{U}_k^{\perp}\mathbf{K}_n\mathbf{U}_k^{\perp^{T}}/N))\\ &= \arg\min_k\,(\bar{\mathbf{r}}^{T}\mathbf{U}_k^{\perp^{T}}\mathbf{U}_k^{\perp}\bar{\mathbf{r}} + 2\mathrm{tr}(\mathbf{U}_k\mathbf{K}_n/N) - \mathrm{tr}(\mathbf{K}_n/N))\\ &= \arg\min_k\,(\bar{\mathbf{r}}^{T}\mathbf{U}_k^{\perp}\bar{\mathbf{r}} + 2\mathrm{tr}(\mathbf{U}_k\mathbf{K}_n/N)),\end{aligned} \tag{7.9}$$

where we have used $\mathbf{U} = \mathbf{U}^{\mathrm{T}}$ and $\mathbf{U}^2 = \mathbf{U}$ for any projection matrix.

Each term of Equation 7.9 has a clear meaning: the first accounts for projection error power and it is a decreasing function of k; the second accounts for noise power and it is an increasing function of k.

7.2.1.2 Computer Simulations

In this section we test the proposed method in simulated scenes. The spectral signatures are selected from the USGS digital spectral library. Abundance fractions are generated according to a Dirichlet distribution given by

$$p(\alpha_1, \alpha_2, \ldots, \alpha_p) = \frac{\Gamma(\mu_1 + \mu_2 + \cdots + \mu_p)}{\Gamma(\mu_1)\Gamma(\mu_2)\cdots\Gamma(\mu_p)}\alpha_1^{\mu_1-1}\alpha_2^{\mu_2-1}\cdots\alpha_p^{\mu_p-1}, \tag{7.10}$$

where $0 \leq \alpha_i \leq 1$, $\sum_{i=1}^{p}\alpha_i = 1$, and $\Gamma(\cdot)$ denotes the Gamma function. The mean value of the ith endmember fraction α_i is $E[\alpha_i] = \mu_i / \sum_{k=1}^{p}\mu_k$.

The results next presented are organized into two experiments: in the first experiment the method is evaluated with respect to the *SNR* and to the number of endmembers p. We define *SNR* as

$$SNR \equiv 10\log_{10}\frac{E[\mathbf{x}^{\mathrm{T}}\mathbf{x}]}{E[\mathbf{n}^{\mathrm{T}}\mathbf{n}]} \tag{7.11}$$

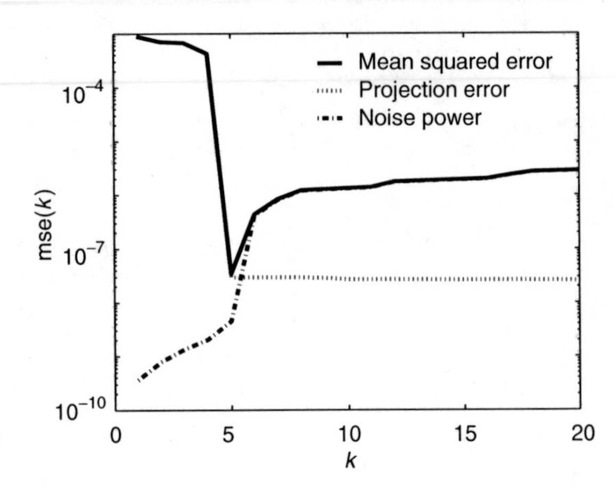

FIGURE 7.3
Mean-squared error versus k, with $SNR = 35$ dB, $p = 5$ (first experiment).

In the second experiment, the method is evaluated when a subset of the endmembers are present only in a few pixels of the scene.

In the first experiment, the hyperspectral scene has 10^4 pixels and the number of endmembers vary from 3 to 21. The abundance fractions are Dirichlet distributed with mean value $\mu_i = 1/p$, for $i = 1, \ldots, p$.

Figure 7.3 shows the evolution of the mean-squared error, that is, of $\bar{\mathbf{r}}^T \mathbf{U}_k^{\perp} \bar{\mathbf{r}} + 2 \text{ tr}(\mathbf{U}_k \mathbf{K}_n/N)$ as a function of the parameter k, for $SNR = 35$ dB and $p = 5$. The minimum of the mean-squared error occurs for $k = 5$, which is exactly the number of endmembers present in the image.

Table 7.1 presents the signal subspace order estimate as a function of the SNR and of p. In this table we compare the proposed method and the VD, recently proposed in Ref. [18]. The VD was estimated by the NWHFC-based eigen-thresholding method using the Neyman–Pearson test with the false-alarm probability set to $P_f = 10^{-4}$. The proposed method finds the correct ID for SNR larger than 25 dB, and underestimates number of endmembers as the SNR decreases. In comparison with the NWHFC algorithm, the proposed approach yields systematically equal or better results.

In the second experiment $SNR = 35$ dB and $p = 8$. The first five endmembers have a Dirichlet distribution as in the previous experiment and Figure 7.4 shows the mean-squared error versus k, when $p = 8$. The minimum of mse(k) is achieved for $k = 8$. This means that the method is able to detect rare endmembers in the image. However, this ability degrades as SNR decreases, as expected.

TABLE 7.1

Signal Subspace Dimension $\hat{k}$ as a Function of SNR and of p (Bold: Proposed method; In Brackets: VD Estimation with NWHFC Method and $P_f = 10^{-4}$)

Method *SNR*	**New (VD)** **50 dB**	**New (VD)** **35 dB**	**New (VD)** **25 dB**	**New (VD)** **15 dB**	**New (VD)** **5 dB**
$p = 3$	**3** (3)	**3** (3)	**3** (3)	**3** (3)	**3** (2)
$p = 5$	**5** (5)	**5** (5)	**5** (5)	**5** (3)	**4** (2)
$p = 10$	**10** (10)	**10** (10)	**10** (8)	**8** (5)	**6** (2)
$p = 15$	**15** (15)	**15** (15)	**13** (12)	**9** (9)	**5** (2)
$p = 21$	**21** (21)	**21** (19)	**14** (12)	**10** (6)	**5** (1)

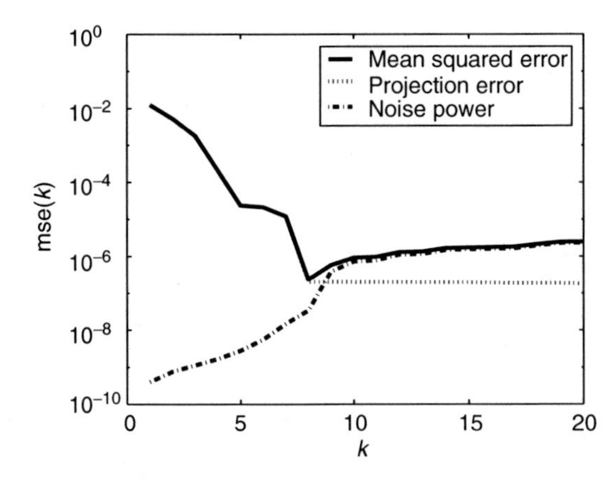

FIGURE 7.4
Mean-squared error versus k, with $SNR = 35$ dB, $p = 8$ (3 spectral vectors occur only on 4 pixels each (second experiment).

7.2.2 Dimensionality Reduction

As discussed before, in the absence of noise, observed vectors $\mathbf{r}$ lie in a convex cone C_p contained in a subspace $\langle \boldsymbol{E}_p \rangle$ of dimension p. The VCA algorithm starts by identifying $\langle \boldsymbol{E}_p \rangle$ and then projects points in C_p onto a simplex S_p by computing $\mathbf{y} = \mathbf{r}/(\mathbf{r}^T \mathbf{u})$ (see Figure 7.2). This simplex is contained in an affine set of dimension $p - 1$. We note that the rational underlying the VCA algorithm is still valid if the observed data set is projected onto any subspace $\langle \boldsymbol{E}_d \rangle \supset \langle \boldsymbol{E}_p \rangle$ of dimension d, for $p \leq d \leq L$; that is, the projection of the cone C_p onto $\langle \boldsymbol{E}_d \rangle$ followed by a projective projection is also a simplex with the same vertices. Of course, the *SNR* decreases as d increases.

To illustrate the point above-mentioned, a simulated scene was generated according to the expression in Equation 7.1 with three spectral signatures (A, Biotite; B, Carnallite; and C, Ammonioalunite) (see Figure 7.1b). The abundance fractions follow a Dirichlet distribution, parameter γ is set to 1, and the noise is zero-mean white Gaussian with covariance matrix $\sigma^2 \mathbf{I}$, where $\mathbf{I}$ is the identity matrix and $\sigma = 0.045$, leading to an $SNR = 20$ dB. Figure 7.5a presents a scatterplot of the simulated spectral mixtures without projection (bands $\lambda = 827$ nm and $\lambda = 1780$ nm). Two triangles whose vertices represent the true endmembers (solid line) and the estimated endmembers (dashed line) by the VCA algorithm are also plotted. Figure 7.5b presents a scatterplot (same bands) of projected data onto the estimated affine set of dimension 2 inferred by SVD. Noise is clearly reduced, leading to a visible improvement on the VCA results.

As referred before, we apply the rescaling $\mathbf{r}/(\mathbf{r}^T \mathbf{u})$ to get rid of the topographic modulation factor. As the *SNR* decreases, this rescaling amplifies noise and is preferable to identify directly the affine space of dimension $p - 1$ by using only PCA. This phenomenon is illustrated in Figure 7.6, where data clouds (noiseless and noisy) generated by two signatures are shown. Affine spaces $\langle \boldsymbol{A}_{p-1} \rangle$ and $\langle \boldsymbol{A}_{p-1} \rangle$ identified by PCA of dimension $p - 1$ and SVD of dimension p, respectively, followed by projective projection are schematized by straight lines. In the absence of noise, the direction of $\mathbf{m}_a$ is better identified by projective projection onto $\langle \boldsymbol{A}_{p-1} \rangle$ ($\hat{m}_a$ better than $\hat{\hat{m}}_a$); in the presence of strong noise, the direction of $\mathbf{m}_a$ is better identified by orthogonal projection onto $\langle \boldsymbol{A}_{p-1} \rangle$ ($\hat{\hat{m}}_a$ better than $\hat{m}_a$). As a conclusion, when the *SNR* is higher than a given threshold SNR_{th}, data is projected onto $\langle \boldsymbol{E}_p \rangle$ followed by the rescaling $\mathbf{r}/(\mathbf{r}^T \mathbf{u})$; otherwise data is projected onto $\langle \boldsymbol{A}_{p-1} \rangle$. Based on experimental results, we propose the threshold $SNR_{\text{th}} = 15 + 10 \log_{10}(p)$ dB. Since for zero-mean white noise $SNR = E[\mathbf{x}^T \mathbf{x}]/(L\sigma^2)$, we conclude that at

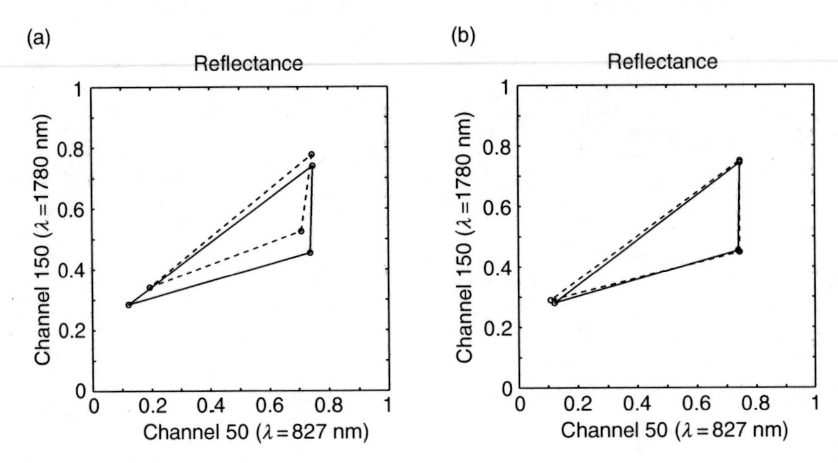

FIGURE 7.5
Scatterplot (bands $\lambda = 827$ nm and $\lambda = 1780$ nm) of the three endmembers mixture. (a) Unprojected data. (b) Projected data using SVD. Solid and dashed lines represent simplexes computed from original and estimated endmembers (using VCA), respectively. (© 2005 IEEE. With permission.)

SNR_{th}, $E[\mathbf{x}^T\mathbf{x}]/(p\sigma^2) = 10^{1.5}\,L$; that is, the SNR_{th} corresponds to the fixed value $L \times 10^{1.5}$ of the SNR measured with respect to the signal subspace.

7.2.3 VCA Algorithm

The pseudo-code for the VCA method is shown in Algorithm 1. Symbols $[\widehat{\mathbf{M}}]_{:,j}$ and $[\widehat{\mathbf{M}}]_{:,j:k}$ stand for the jth column of $\widehat{\mathbf{M}}$ and for the ith to k th columns of $\widehat{\mathbf{M}}$, respectively. Symbol $\widehat{\mathbf{M}}$ stands for the estimated mixing matrix.

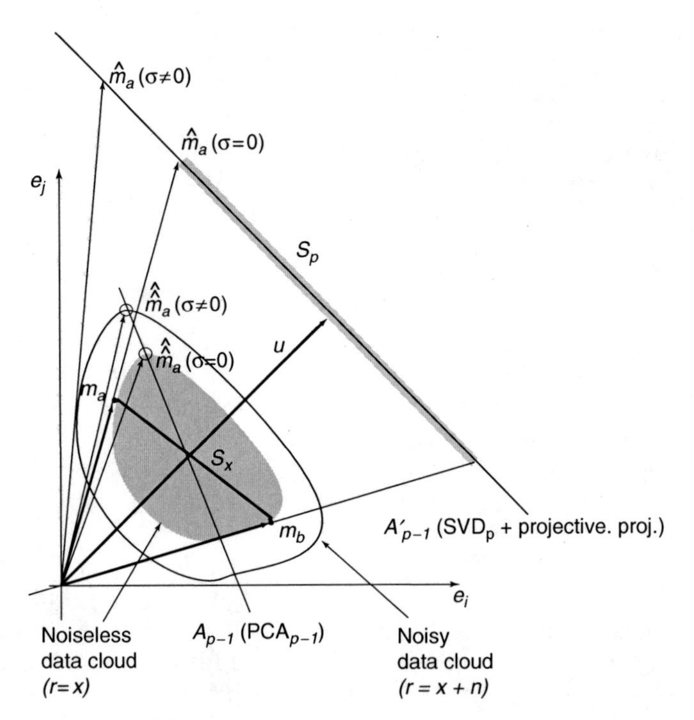

FIGURE 7.6
Illustration of the noise effect on the dimensionality reduction. (© 2005 IEEE. With permission.)

Algorithm 1: Vertex Component Analysis (VCA)

INPUT: p, $\mathbf{R} \equiv [\mathbf{r}_1, \mathbf{r}_2, \ldots, \mathbf{r}_N]$

1: $p := \text{argmin}_k\ (\bar{\mathbf{r}}^T\ \mathbf{U}_k^{\perp}\ \bar{\mathbf{r}} + 2\ \text{tr}(\mathbf{U}_k\ \mathbf{K}_n/N))$; $\mathbf{U}_k$ obtained by SVD, and $\mathbf{K}_n$ estimated by multiple regression

2: $SNR_{\text{th}} = 15 + 10\ \log_{10}(p)$ dB

3: **if** $SNR > SNR_{\text{th}}$ then

  4: $d := p$;

  5: $\mathbf{X} := \mathbf{U}_\mathbf{d}^T\ \mathbf{R}$; $\mathbf{U}_\mathbf{d}$ obtained by SVD

  6: $\mathbf{u} := \text{mean}\ (\mathbf{X})$; $\mathbf{u}$ is a $1 \times d$ vector

  7: $[\mathbf{Y}]_{:,j} := [\mathbf{X}]_{:,j}/([\mathbf{X}]_{:,j}^T\ \mathbf{u})$; projective projection

8: else

  9: $d := p - 1$;

  10: $[\mathbf{X}]_{:,j} := \mathbf{U}_\mathbf{d}^T\ ([\mathbf{R}]_{:,j} - \bar{r})$; $\{\mathbf{U}_\mathbf{d}$ obtained by PCA$\}$

  11: $\kappa := \arg\max_{j=1\ldots N} \|\ [\mathbf{X}]_{:,j}\ \|$; $\kappa := [\kappa \mid \kappa \mid \ldots \mid \kappa]$;

  12: κ is a $1 \times N$ vector

  13: $\mathbf{Y} := \begin{bmatrix} \mathbf{X} \\ \kappa \end{bmatrix}$;

14: end if

15: $\mathbf{A} := [\varepsilon_u \mid 0 \mid \cdots \mid 0]$; $\{\varepsilon_u := [0, \ldots, 0, 1]^T$ and $\mathbf{A}$ is a $p \times p$ auxiliary matrix$\}$

16: for $i := 1$ to p **do**

  17: $\mathbf{w} := \text{randn}\ (0, \mathbf{I}_p)$; $\{\mathbf{w}$ is a zero-mean random Gaussian vector of covariance $\mathbf{I}_p\}$.

  18: $\mathbf{f} := \frac{(\mathbf{I}-\mathbf{A}\mathbf{A}^{\#})\mathbf{w}}{\|(\mathbf{I}-\mathbf{A}\mathbf{A}^{\#})\mathbf{w}\|}$; $\{\mathbf{f}$ is a vector orthonormal to the subspace spanned by $[\mathbf{A}]_{:,1:i}$.$\}$

  19: $\mathbf{v} := \mathbf{f}^T\ \mathbf{Y}$;

  20: $k := \arg\max_{j=1,\ldots,N} |[\mathbf{v}]_{:,j}|$; $\{$find the projection extreme$\}$

  21: $[\mathbf{A}]_{:,i} := [\mathbf{Y}]_{:,k}$;

  22: $[indice]_i := k$; $\{$stores the pixel index$\}$

23: end for

24: if $SNR \rangle SNR_{\text{th}}$ then

  25: $\widehat{\mathbf{M}} := \mathbf{U}_\mathbf{d}[\mathbf{X}]_{:,\text{indice}}$; $\{\widehat{\mathbf{M}}$ is a $L \times p$ estimated mixing matrix$\}$

26: else

  27: $\widehat{\mathbf{M}} := \mathbf{U}_\mathbf{d}[\mathbf{X}]_{:,\text{indice}} + \bar{\mathbf{r}}$; $\{\widehat{\mathbf{M}}$ is a $L \times p$ estimated mixing matrix$\}$

28: end if

Step 3: Test if the SNR is higher than SNR_{th} to decide whether the data is to be projected onto a subspace of dimension p or $p - 1$. In the first case the projection matrix $\mathbf{U}_\mathbf{d}$ is obtained by SVD from $\mathbf{R}\mathbf{R}^T/N$. In the second case the projection is obtained by PCA from $(\mathbf{R}-\bar{\mathbf{r}})(\mathbf{R}-\bar{\mathbf{r}})^T/N$; recall that $\bar{\mathbf{r}}$ is the sample mean of $[\mathbf{R}]_{:,i}$, for $i = 1, \ldots, N$.

Step 5 and Step 10: Assure that the inner product between any vector $[\mathbf{X}]_{:,j}$ and vector $\mathbf{u}$ is non-negative—a crucial condition for the VCA algorithm to work correctly. The chosen value of $\kappa = \text{argmax}_{j=1\cdots N} \|[\mathbf{X}]_{:,j}\|$ assures that the colatitude angle between $\mathbf{u}$ and any vector $[\mathbf{X}]_{:,j}$ is between 0° and 45°, avoiding numerical errors that otherwise would occur for angles near 90°.

Step 15: Initializes the auxiliary matrix $\mathbf{A}$, which stores the projection of the estimated endmembers signatures. Assume that there exists at least one pure pixel of each endmember in the input sample **R** (see Figure 7.2). Each time the loop *for* is executed, a vector **f** orthonormal to the space spanned by the columns of the auxiliary matrix **A** is randomly generated and **y** is projected onto **f**. Because we assume that pure endmembers occupy the vertices of a simplex, $a \leq \mathbf{f}^T [\mathbf{Y}]_{:,i} \leq b$, for $i = 1, \ldots, N$, where values a and b correspond to only pure pixels. We store the endmember signature corresponding to $\max(|a|,|b|)$. The next time loop *for* is executed, **f** is orthogonal to the space spanned by the signatures already determined. Since **f** is the projection of a zero-mean Gaussian independent random vector onto the orthogonal space spanned by the columns of $[\mathbf{A}]_{:,1:i}$, then the probability of **f** being null is zero. Notice that the underlying reason for generating a random vector is only to get a non-null projection onto the orthogonal space generated by the columns of **A**. Figure 7.2 shows the input samples and the chosen pixels, after the projection $\mathbf{v} = \mathbf{f}^T \mathbf{Y}$. Then a second vector **f** orthonormal to the endmember a is generated and the second endmember is stored. Finally, Step 25 and Step 27 compute the columns of matrix $\widehat{\mathbf{M}}$, which contain the estimated endmembers signatures in the L-dimensional space.

7.3 Evaluation of the VCA Algorithm

In this section, we compare VCA, PPI, and N-FINDR algorithms. N-FINDR and PPI were coded according to Refs. [40] and [35], respectively. Regarding PPI, the number of *skewers* must be *large* [41,42,56–58]. On the basis of Monte Carlo runs, we concluded that the minimum number of *skewers* beyond which there is no unmixing improvements is about 1000. All experiments are based on simulated scenes from which we know the number of endmembers, their signatures, and their abundance fractions. Estimated endmembers are the columns of $\widehat{\mathbf{M}} \equiv [\widehat{\mathbf{m}}_1, \widehat{\mathbf{m}}_2, \ldots, \widehat{\mathbf{m}}_p]$. We also compare estimated abundance fractions given by $\hat{\mathbf{S}} = \widehat{\mathbf{M}}^{\#} [\mathbf{r}_1, \mathbf{r}_2, \ldots, \mathbf{r}_N]$, ($\widehat{\mathbf{M}}^{\#}$ stands for pseudo-inverse of $\widehat{\mathbf{M}}$) with the true abundance fractions.

To evaluate the performance of the three algorithms, we compute vectors of angles $\theta \equiv [\theta_1, \theta_2, \ldots, \theta_p]^T$ and $\phi \equiv [\phi_1, \phi_2, \ldots, \phi_p]^T$ with[1]

$$\theta_i \equiv \left(\arccos \frac{< \mathbf{m}_i, \hat{\mathbf{m}}_i >}{\|\mathbf{m}_i\| \|\hat{\mathbf{m}}_i\|} \right), \tag{7.12}$$

$$\phi_i \equiv \left(\arccos \frac{< [\mathbf{S}]_{i,:}, [\hat{\mathbf{S}}]_{i,:} >}{\|[\mathbf{S}]_{i,:}\| \|[\hat{\mathbf{S}}]_{i,:}\|} \right), \tag{7.13}$$

where θ_i is the angle between vectors $\mathbf{m}_i$ and $\widehat{\mathbf{m}}_i$ (ith endmember signature estimate) and ϕ_i is the angle between vectors $[\mathbf{S}]_{i,:}$ and $[\hat{\mathbf{S}}]_{i,:}$ (vectors of $\Re^N$ formed by the ith lines of matrices $\hat{\mathbf{S}}$ and $\mathbf{S} \equiv [\mathbf{s}_1, \mathbf{s}_2, \ldots, \mathbf{s}_N]$, respectively).

Based on θ and ϕ, we estimate the following root mean square error distances

$$\varepsilon_\theta = \left(\frac{1}{p} E\left[\|\theta\|_2^2 \right] \right)^{1/2}, \tag{7.14}$$

$$\varepsilon_\phi = \left(\frac{1}{p} E\left[\|\phi\|_2^2 \right] \right)^{1/2}. \tag{7.15}$$

The first quantity measures distances between $\hat{\mathbf{m}}_i$ and $\mathbf{m}_i$, for $i = 1, \ldots, p$; the second is similar to the first, but for the estimated abundance fractions. Here we name ε_θ and ε_ϕ as

[1]Notation $\langle \mathbf{x}, \mathbf{y} \rangle$ stands for the inner product $\mathbf{x}^T \mathbf{y}$.

rmsSAE and rmsFAAE, respectively (SAE stands for signature angle error and FAAE stands for fractional abundance angle error). Mean values in Equation 7.14 and Equation 7.15 are approximated by sample means based on one hundred Monte Carlo runs.

In all experiments, the spectral signatures are selected from the USGS digital spectral library (Figure 7.1b shows three of these endmember signatures). Abundance fractions are generated according to a Dirichlet distribution given by Equation 7.10. Parameter γ is Beta (β_1, β_2) distributed, that is,

$$p(\gamma) = \frac{\Gamma(\beta_1 + \beta_2)}{\Gamma(\beta_1)\Gamma(\beta_2)}\gamma^{\beta_1 - 1}(\gamma - 1)^{\beta_2 - 1},$$

which is also a Dirichlet distribution*. The Dirichlet density, besides enforcing positivity and full additivity constraints, displays a wide range of shapes depending on the parameters $\mu_1, \ldots, \mu_i$. This flexibility influences its choice in our simulations.

The results presented next are organized into five experiments: in the first experiment, the algorithms are evaluated with respect to the *SNR* and to the absence of pure pixels. As mentioned before, we define

$$SNR \equiv 10 \log_{10} \frac{E[\mathbf{x}^T\mathbf{x}]}{E[\mathbf{n}^T\mathbf{n}]}. \tag{7.16}$$

In the case of zero-mean noise with covariance $\sigma^2\mathbf{I}$ and Dirichlet abundance fractions, one obtains

$$SNR = 10 \log_{10} \frac{\text{tr}[\mathbf{M}\mathbf{K}_s\mathbf{M}^T]}{L\sigma^2}, \tag{7.17}$$

where

$$\mathbf{K}_s \equiv \sigma_\gamma^2 E[\boldsymbol{\alpha}\boldsymbol{\alpha}^T] = \sigma_\gamma^2 \frac{\boldsymbol{\mu}\boldsymbol{\mu}^T + \text{diag}(\boldsymbol{\mu})}{(\sum_{i=1}^{p} \mu_i)(1 + \sum_{i=1}^{p} \mu_i)}, \tag{7.18}$$

$\boldsymbol{\mu} = [\mu_1 \cdots \mu_p]^T$, and σ_γ^2 is the variance of parameter γ. For example, assuming abundance fractions equally distributed, we have, after some algebra, $SNR \simeq 10 \log_{10} {\sigma_\gamma}^2 \sum_i^p = 1 \left(\sum_{j=1}^{p} {m_{ij}}^2/p\right)/(L\sigma^2)$ for $\mu p \ll 1$ and $SNR \simeq 10 \log_{10} \sigma_\gamma^2 \left(\sum_{i=1}^{p}(\sum_{j=1}^{p} m_{ij})^2/p^2\right)/(L\sigma^2)$ for $\mu\, p \gg 1$.

In the second experiment, the performance is measured as function of the parameter γ, which models fluctuations on the illumination due to surface topography. In the third experiment, to illustrate the algorithm performance, the number of pixels of the scene varies with the size of the covered area—as the number of pixels increases, the likelihood of having pure pixels also increases, improving the performance of the unmixing algorithms. In the fourth experiment, the algorithms are evaluated as a function of the number of endmembers present in the scene. Finally, in the fifth experiment, the number of floating point operations (flops) is measured, to compare the computational complexity of the VCA, N-FINDR, and the PPI algorithms.

In the first experiment, the hyperspectral scene has 1000 pixels and the abundance fractions are Dirichlet distributed with $\mu_i = 1/3$, for $i = 1,2,3$; parameter γ is Beta-distributed with $\beta_1 = 20$ and $\beta_2 = 1$ implying $E[\gamma] = 0.952$ and $\sigma_\gamma = 0.05$.

Figure 7.7 shows performance results as a function of the *SNR*. As expected, the presence of noise degrades the performance of all algorithms. In terms of rmsSAE and rmsFAAE (Figure 7.7a and Figure 7.7b), we can see that when *SNR* is less than 20 dB

*With one component.

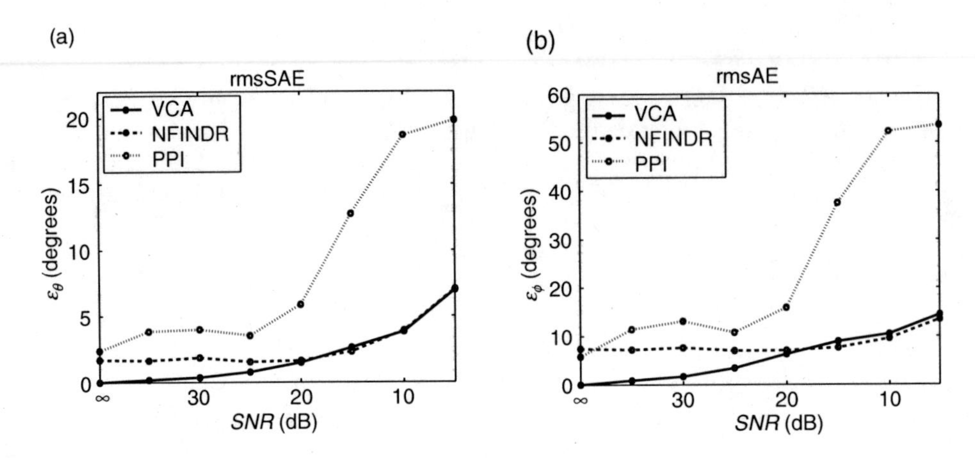

FIGURE 7.7
First scenario: ($N=1000$, $p=3$, $L=224$, $\mu_1=\mu_2=\mu_3=1/3$, $\beta_1=20$, $\beta_2=1$): (a) rmsSAE as function of *SNR*; (b) rmsFAAE as function of *SNR*. (© 2005 IEEE. With permission.)

the VCA algorithm exhibits the best performance. Note that for noiseless scenes, only the VCA algorithm has zero rmsSAE. PPI algorithm displays the worst result.

Figure 7.8 shows performance results as a function of the *SNR* in the absence of pure pixels. Spectral data without pure pixels was obtained by rejecting pixels with any abundance fraction smaller than 0.2. Figure 7.9 shows the scatterplot obtained. VCA and N-FINDR display similar results, and both are better than PPI. Notice that the performance is almost independent of the *SNR* and is uniformly worse than that displayed with pure pixels and $SNR=5$ dB in the first experiment. We conclude that this family of algorithms is more affected by the lack of pure pixels than by low *SNR*.

For economy of space and also because rmsSAE and rmsFAAE disclose similar pattern of behavior, we only present the rmsSAE in the remaining experiments.

In the second experiment, abundance fractions are generated as in the first one, *SNR* is set to 20 dB, and parameter γ is Beta-distributed with $\beta_2=2, \ldots, 28$. This corresponds to the variation of $E[\gamma]$ from 0.66 to 0.96 and σ_γ from 0.23 to 0.03. By varying parameter β_1, the severity of topographic modulation is also varied. Figure 7.10 illustrates the effect of topographic modulation on the performance of the three algorithms. When β_1 grows (σ_γ

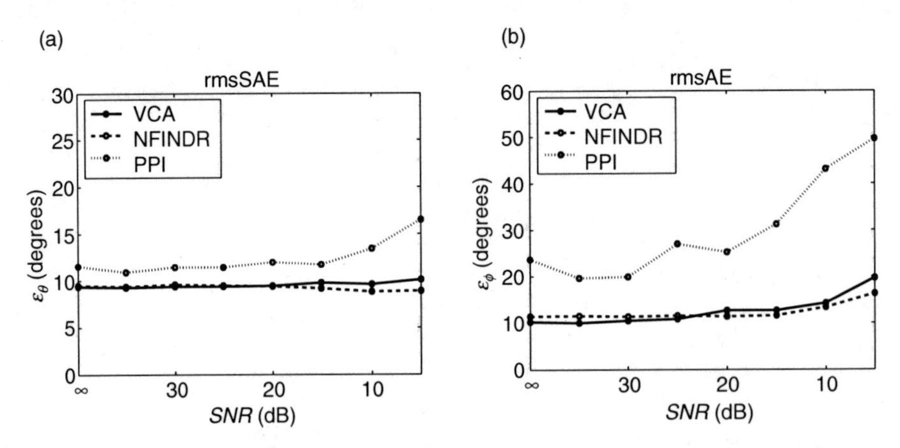

FIGURE 7.8
Robustness to the absence of pure pixels ($N=1000, p=3, L=224, \mu_1=\mu_2=\mu_3=1/3, \beta_1=20, \beta_2=1$): (a) rmsSAE as function of *SNR*; (b) rmsFAAE as function of *SNR*. (© 2005 IEEE. With permission.)

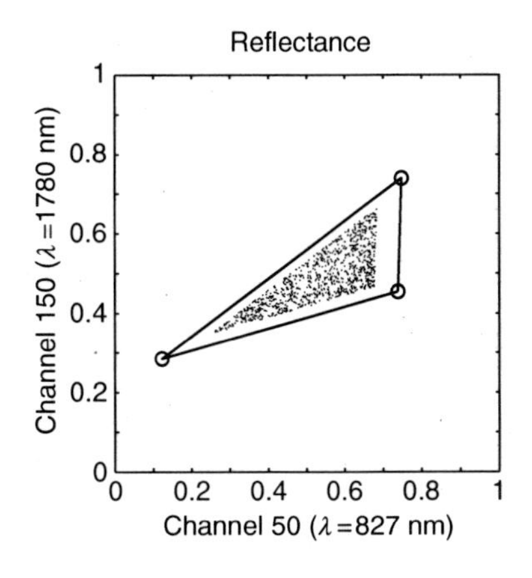

FIGURE 7.9
Illustration of the absence of pure pixels ($N=1000$, $p=3$, $L=224$, $\mu_1=\mu_2=\mu_3=1/3$, $\gamma=1$). Scatterplot (bands $\lambda=827$ nm and $\lambda=1780$ nm), with abundance fraction smaller than 0.2 rejected. (© 2005 IEEE. With permission.)

gets smaller) the performance improves. This is expected because the simplex identification is more accurate when the topographic modulation is smaller. The PPI algorithm displays the worst performance for $\sigma_\gamma \langle 0.1$. VCA and N-FINDR algorithms have identical performances when β_1 takes higher values ($\sigma_\gamma \langle 0.045$); otherwise VCA algorithm has the best performance. VCA is more robust to topographic modulation because it seeks for the extreme projections of the simplex, whereas N-FINDR seeks for the maximum volume, which is more sensitive to fluctuations on γ.

In the third experiment, the number of pixels is varied, the abundance fractions are generated as in the first one, and $SNR=20$ dB. Figure 7.11 shows that VCA and N-FINDR exhibit identical results, whereas the PPI algorithm displays the worst result. Note that the behavior of the three algorithms is quasi-independent of the number of pixels.

In the fourth experiment, we vary the number of signatures from $p=3$ to $p=21$, the scene has 1000 pixels, and $SNR=30$ dB. Figure 7.12a shows that VCA and N-FINDR performances are comparable, whereas PPI displays the worst result. The rmsSAE increase slightly as the number of endmembers present in the scene increases. The rmsSAE is also plotted as a function of the SNR with $p=10$ (see Figure 7.12b). Comparing with Figure 7.7a we conclude that when the number of endmembers increases, the performance of the algorithms slightly decreases.

In the fifth and last experiment, the number of flops is measured to compare the computational complexity of VCA, PPI, and N-FINDR algorithms. Here we use the scenarios of the second and third experiments. Table 7.2 presents approximated expressions for

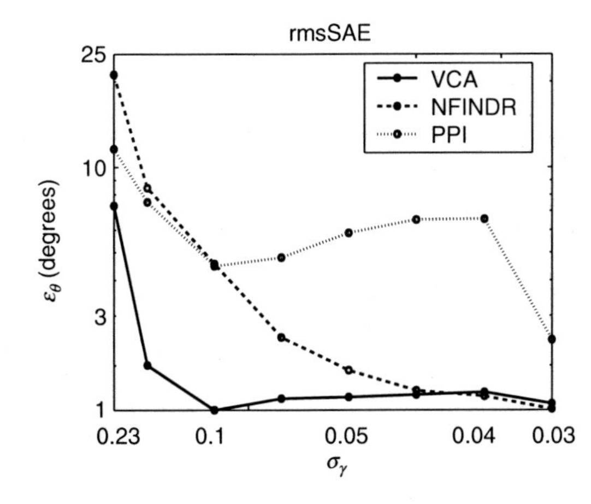

FIGURE 7.10
Robustness to the topographic modulation ($N=1000$, $p=3$, $L=224$, $\mu_1=\mu_2=\mu_3=1/3$, $SNR=20$ dB, $\beta_2=1$), rmsSEA as function of the σ_γ^2 (variance of γ). (© 2005 IEEE. With permission.)

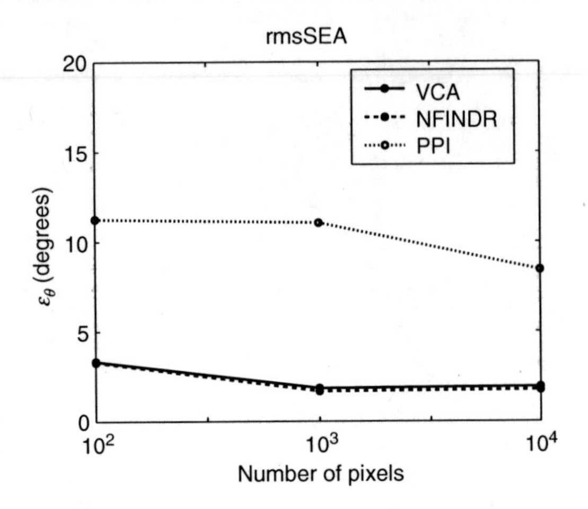

FIGURE 7.11
rmsSEA as function of the number of pixels in a scene ($p = 6$, $L = 224$, $\mu_1 = \mu_2 = \mu_3 = 1/3$, $SNR = 20$ dB, $\beta_2 = 20$, $\beta_2 = 1$). (© 2005 IEEE. With permission.)

the number of flops used by each algorithm. These expressions neither account for the computational complexities involved in the computations of the sample covariance $(\mathbf{R} - \bar{\mathbf{r}})(\mathbf{R} - \bar{\mathbf{r}})^T/N$ nor in the computations of the eigen decomposition. The reason is that these operations, compared with the VCA, PPI, and N-FINDR algorithms, have a negligible computational cost since:

- The computation of $(\mathbf{R} - \bar{\mathbf{r}})(\mathbf{R} - \bar{\mathbf{r}})^T/N$ has a complexity of $2NL^2$ flops. However, in practice one does not need to use the complete set of N hyperspectral vectors. If the scene is noiseless, only $p - 1$ linearly independent vectors would be enough to infer the exact subspace $\langle \mathbf{E}_{p-1} \rangle$. In the presence of noise, however, a larger set should be used. For example in a 1000×1000 hyperspectral image, we found out that only 1000 samples randomly sampled are enough to find a very good estimate of $\langle \mathbf{E}_{p-1} \rangle$. Even a sample size of 100 leads to good results in this respect.
- Concerning the eigen decomposition of $(\mathbf{R} - \bar{\mathbf{r}})(\mathbf{R} - \bar{\mathbf{r}})^T/N$ (or the SVD of $\mathbf{R}\mathbf{R}^T/N$), we only need to compute $p - 1$ (or p) eigenvectors corresponding to the largest $p - 1$ eigenvalues (or p single values). For these partial eigen decompositions, we

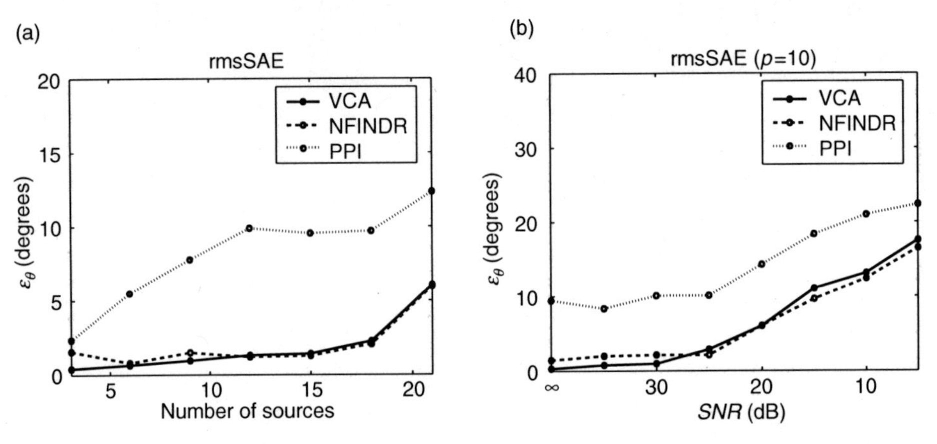

FIGURE 7.12
Impact of the number of endmembers ($N = 1000$, $L = 224$, $\mu_1 = \mu_2 = \mu_3 = 1/3$, $SNR = 30$ dB, $\beta_2 = 20$, $\beta_2 = 1$), (a) rmsSEA as function of the number of endmembers; (b) rmsSEA function of the SNR with $p = 10$. (© 2005 IEEE. With permission.)

TABLE 7.2

Computational Complexity of VCA, N-FINDR, and PPI Algorithms

Algorithm	Complexity (flops)
VCA	$2p^2N$
N-FINDR	$p^{\eta+1}N$
PPI	$2psN$

Source: © 2005 IEEE. With permission.

have used the PCA algorithm [51] (or SVD analysis [49]) whose complexity is negligible compared with the remaining operations.

The VCA algorithm projects all data (N vectors of size p) onto p orthogonal directions. N-FINDR computes pN times the determinant of a $p \times p$ matrix, whose complexity is p^η, with $2.3 \langle \eta \langle 2.9$ [59]. Assuming that $N \gg p \rangle 2$, VCA complexity is lower than that of N-FINDR. With regard to PPI, given that the number of *skewers* (s) is much higher than the usual number of endmembers, the PPI complexity is much higher than that of VCA. Hence, we conclude that the VCA algorithm has always the lowest complexity.

Figure 7.13 plots the flops for the three algorithms after data projection. In Figure 7.13a the abscissa is the number of endmembers in the scene, whereas in Figure 7.13b the abscissa is the number of pixels. Note that for five endmembers, VCA computational complexity is one order of magnitude lower than that of the N-FINDR algorithm. When the number of endmembers is higher than 15, the VCA computational complexity is, at least, two orders of magnitude lower than PPI and N-FINDR algorithms.

In the introduction, besides PPI and N-FINDR algorithms, we have also mentioned ORASIS. Nevertheless, no comparison whatsoever was made with this method. The reason is that there are no ORASIS implementation details published in the literature. We can, however, make a few considerations based on the results recently published in Ref. [58]. This work compares, among others, PPI, N-FINDR, and ORASIS algorithms. Although the relative performance of the three algorithms varies, depending on *SNR*, number of endmembers, spectral signatures, type of atmospheric correction, and so on,

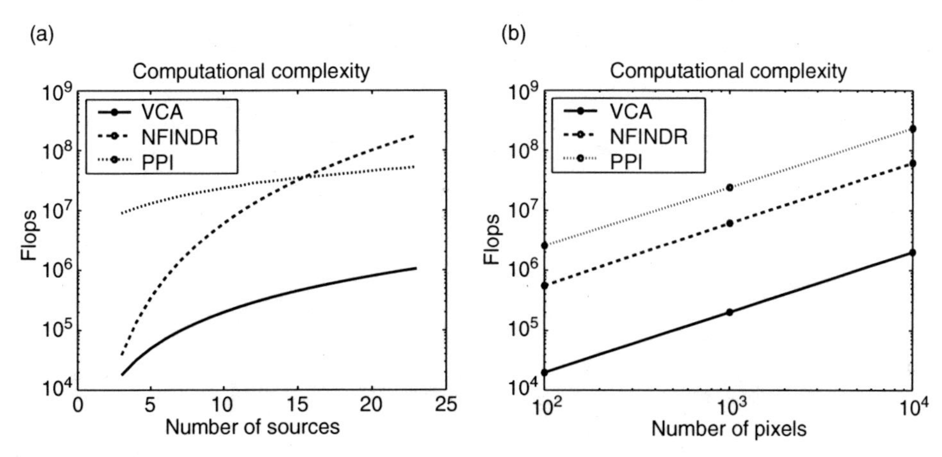

FIGURE 7.13
Computational complexity measured in the number of flops.

both PPI and N-FINDR generally perform better than ORASIS when *SNR* is low. Since in all comparisons conducted here the VCA performs better than or equal to PPI and N-FINDR, we expect that the proposed method performs better than or equal to ORASIS when low *SNR* dominates the data, although further experiments would be required to demonstrate the above remark.

7.4 Evaluation with Experimental Data

In this section, we apply the VCA algorithm to real hyperspectral data collected by the AVIRIS [5] sensor over Cuprite, Nevada. Cuprite is a mining area in southern Nevada with mineral and little vegetation [60], located approximately 200 km northwest of Las Vegas. The test site is a relatively undisturbed acid-sulphate hydrothermal system near highway 95. The geology and alteration were previously mapped in detail [61,62]. A geologic summary and a mineral map can be found in Refs. [60,63]. This site has been used extensively for remote sensing experiments over the past years [64,65].

Our study is based on a subimage (250×190 pixels and 224 bands) of a data set acquired on the AVIRIS flight 19 June 1997 (see Figure 7.14a). The AVIRIS instrument covers the spectral region from 0.41 to 2.45 μm in 224 bands with 10 nm bands. Flying at an altitude of 20 km, the AVIRIS flight has an instantaneous field of view (IFOV) of 20 m and views a swath over 10 km wide. To compare results with a signature library, we process the reflectance image after atmospheric correction.

The proposed method to estimate the number of endmembers when applied to this data set estimates $\hat{k} = 23$ (see Figure 7.15b). According to the truth data presented in Ref. [60], there are eight materials in this area. This difference is due to (1) the presence of rare pixels not accounted for in the truth data [60] and (2) spectral variability.

The bulk of spectral energy is explained with only a few eigenvectors. This can be observed from Figure 7.15a where the accumulated signal energy is plotted as a function of the eigenvalue index. The energy contained in the first eight eigenvalues is 99.94% of the total signal energy. This is further confirmed in Figure 7.14b where we show, in gray level and for each pixel, the percentage of energy contained in the subspace

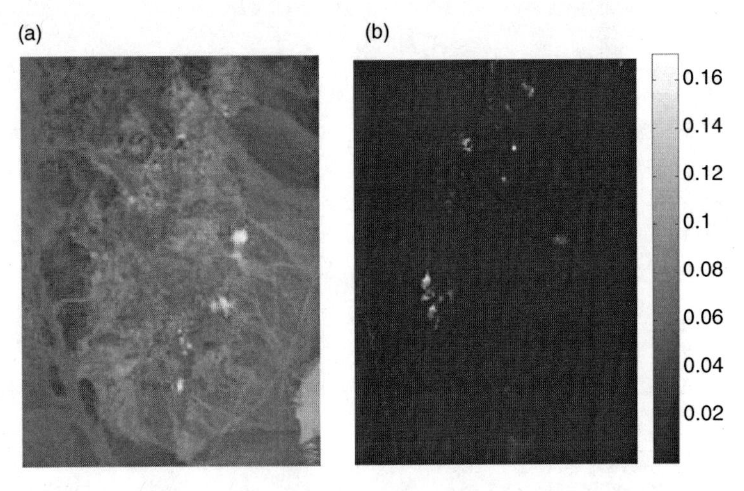

FIGURE 7.14
(a) Band 30 (wavelength $\lambda = 667.3$ nm) of the subimage of AVIRIS Cuprite Nevada data set; (b) percentage of energy in the subspace $E_{9:23}$.

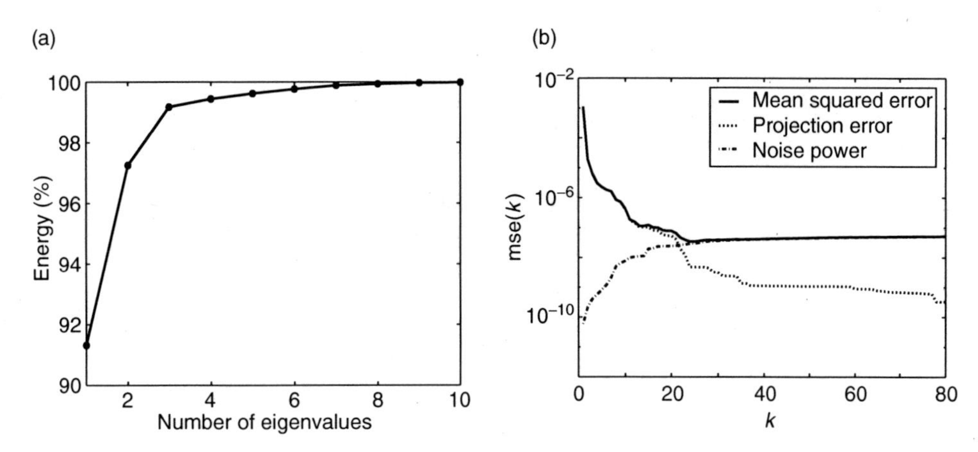

FIGURE 7.15
(a) Percentage of signal energy as a function of the number of eigenvalues; (b) mean-squared error versus k for cuprite data set.

$\langle \, \boldsymbol{E}_{9:23} \, \rangle = \langle \, [\mathbf{e}_9, \ldots, \mathbf{e}_{23}] \, \rangle$. Notice that only a few (rare) pixels contain energy in $\langle \, \boldsymbol{E}_{9:23} \, \rangle$. Furthermore, these energies are a very small percentage of the corresponding spectral vector energies (less than 0.16%) in this subspace.

The VD estimated by the HFC-based eigen-thresholding method [18] ($P_f = 10^{-3}$) on the same data set yields $\hat{k} = 20$. A lower value of P_f would lead to a lower number of endmembers. This result seems to indicate that the proposed method performs better than the HFC with respect to rare materials.

To determine the type of projection applied by VCA, we compute

$$SNR \simeq 10 \log_{10} \frac{P_{R_p} - (p/L)P_R}{P_R - P_{R_p}}, \tag{7.19}$$

where $P_R \equiv E[\mathbf{r}^{\mathrm{T}} \, \mathbf{r}]$ and $P_{R_p} = E[\mathbf{r}^{\mathrm{T}} \, \mathbf{U}_d \, \mathbf{U}_d{}^{\mathrm{T}} \, \mathbf{r}]$ in the case of SVD and $P_{R_p} \equiv E[\mathbf{r}^{\mathrm{T}} \, \mathbf{U}_d \, \mathbf{U}_d{}^{\mathrm{T}} \, \mathbf{r}] + \bar{\mathbf{r}}^{\mathrm{T}} \bar{\mathbf{r}}$ in the case of PCA.

A visual comparison between VCA results on the Cuprite data set and the ground truth presented in Ref. [63] shows that the first component (see Figure 7.16a) is predominantly Alunite, the second component (see Figure 7.16b) is Sphene, the third component (see Figure 7.16c) is Buddingtonite, and the fourth component (see Figure 7.16d) is Montmorillonite. The fifth, seventh, and the eighth components (see Figure 7.16e, Figure 7.16g, and Figure 7.16h) are Kaolinite and the sixth component (see Figure 7.16f) is predominantly Nontronite.

To confirm the classification based on the estimated abundance fractions, a comparison of the estimated VCA endmember signatures with laboratory spectra [53] is presented in Figure 7.17. The signatures provided by VCA are scaled by a factor to minimize the mean square error between them and the respective library spectra. The estimated signatures are close to the laboratory spectra. The larger mismatches occur for buddingtonite and kaolinite (#1) signatures, but only on a small percentage of the total bands.

Table 7.3 compares the spectral angles between extracted endmembers and laboratory reflectances for the VCA, N-FINDR, and the PPI algorithms. The first column shows the laboratory substances with smaller spectral angle distance with respect to the signature extracted by VCA algorithm; the second column shows the respective angles. The third and the fourth columns are similar to the second one, except when the closest spectral substance is different from the corresponding VCA one. In these

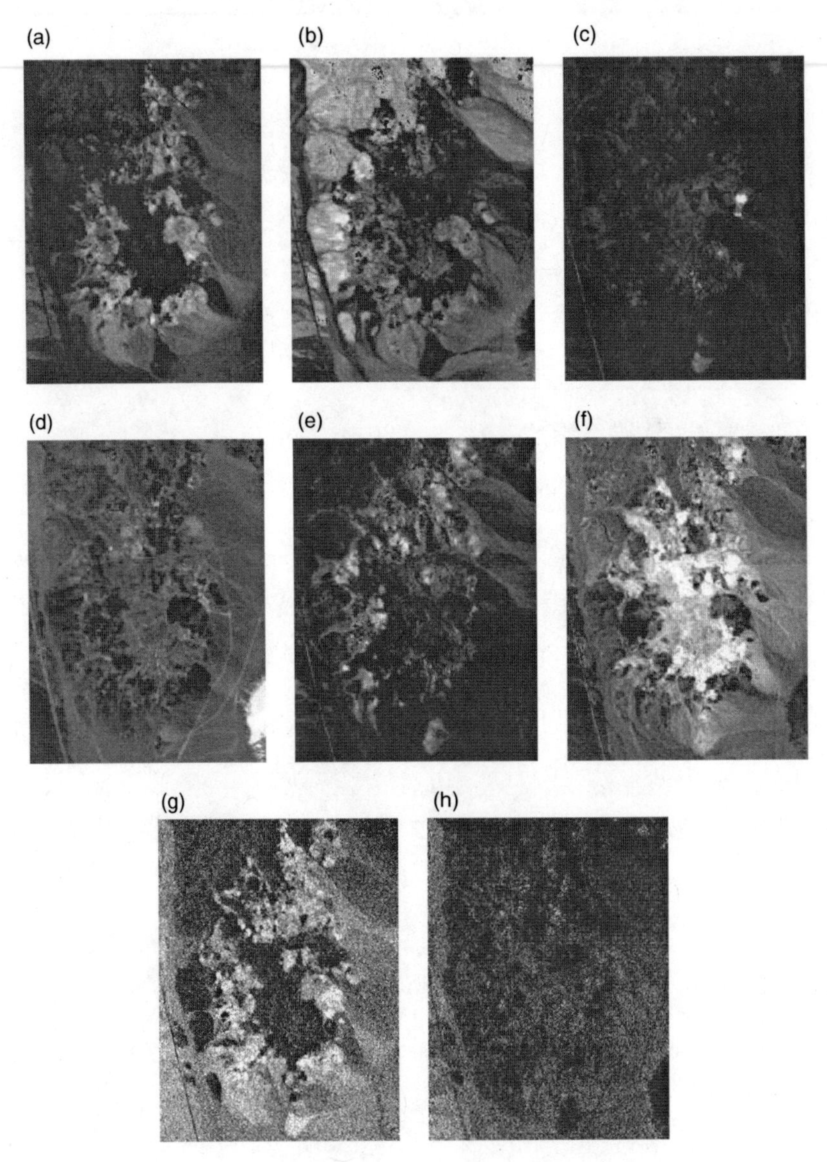

FIGURE 7.16
Eight abundance fractions estimated with VCA algorithm: (a) Alunite or Montmorillonite; (b) Sphene; (c) Buddingtonite; (d) Montmorillonite; (e) Kaolinite #1; (f) Nontronite or Kaolinite; (g) Kaolinite #2; (h) Kaolinite #3;

cases, we write the name of the substance. The displayed results follow the pattern of behavior shown in the simulations, where VCA performs better than PPI and better or similar to N-FINDR.

7.5 Conclusions

We have presented a new algorithm to unmix linear mixtures of hyperspectral sources, termed vertex component analysis. The VCA algorithm is unsupervised and is based on the geometry of convex sets. It exploits the fact that endmembers occupy the vertices

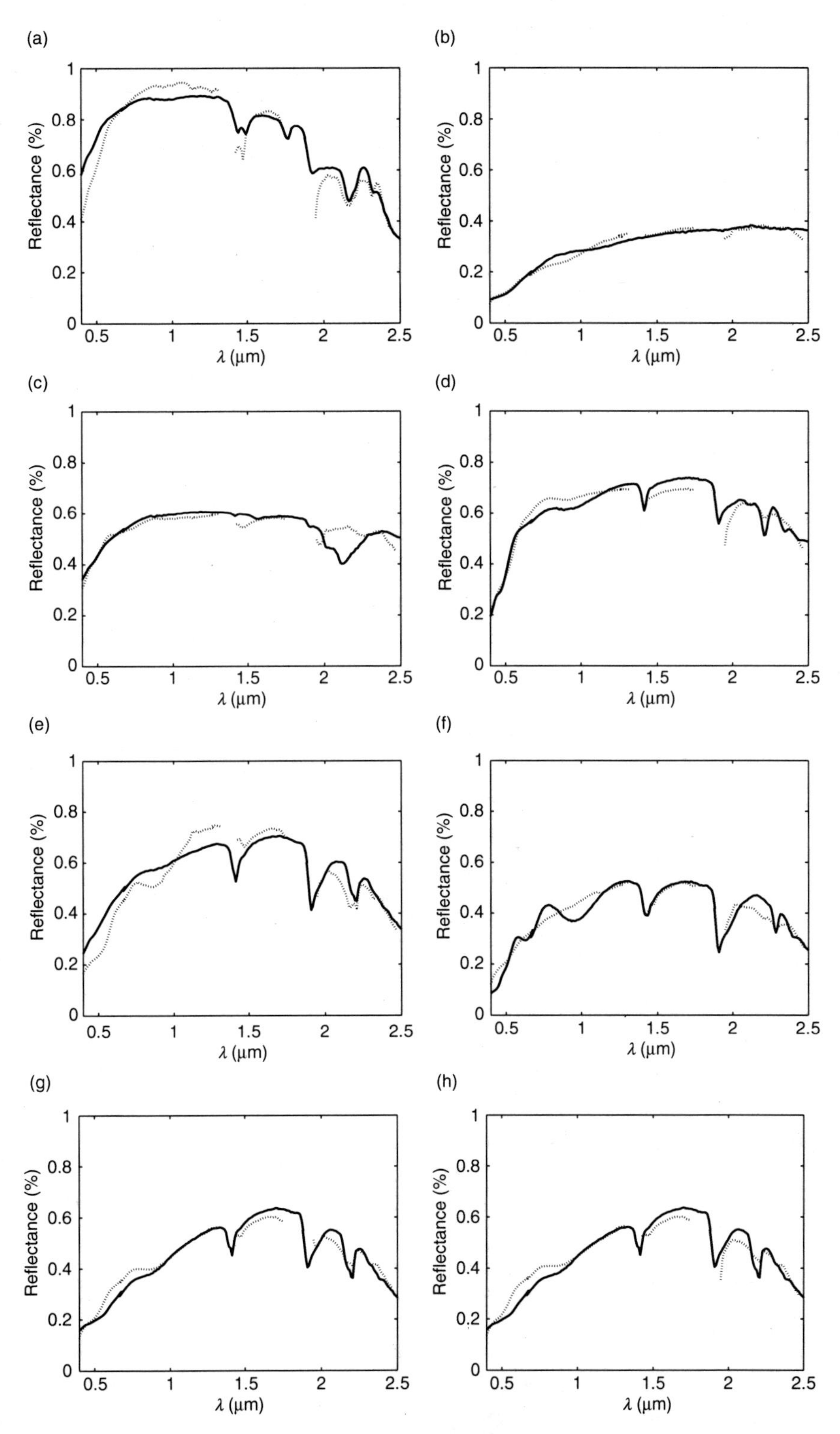

FIGURE 7.17
Comparison of the extracted signatures (dotted line) with the USGS spectral library (solid line): (a) Alunite or Montmorillonite; (b) Sphene; (c) Buddingtonite; (d) Montmorillonite; (e) Kaolinite #1; (f) Nontronite or Kaolinite; (g) Kaolinite #2; (h) Kaolinite #3;

TABLE 7.3

Spectral Angle Distance between Extracted Endmembers and Laboratory Reflectances for VCA, N-FINDR, and PPI Algorithms

Substance	VCA	N-FINDR	PPI
Alunite or montmorillonite	3.9	3.9	4.3
Sphene	3.1	Barite (2.7)	Pyrope (3.9)
Buddingtonite	4.2	4.1	3.9
Montmorillonite	3.1	3.0	2.9
Kaolinite #1	5.7	5.3	Dumortierite (5.3)
Nontronite or kaolinite	3.4	4.8	4.7
Kaolinite #2	3.5	Montmor. (4.2)	3.5
Kaolinite #3	4.2	4.3	5.0

of a simplex. This algorithm also estimates the dimensionality of hyperspectral linear mixtures.

To determine the signal subspace in hyperspectral data set the proposed method first estimates the signal and noise correlations matrices, and then it selects the subset of eigenvalues that best represents the signal subspace in the least-square sense. A comparison with HFC and NWHFC methods is conducted yielding comparable or better results than these methods.

The VCA algorithm assumes the presence of pure pixels in the data and iteratively projects data onto a direction orthogonal to the subspace spanned by the endmembers already determined. The new endmember signature corresponds to the extreme of the projection. The algorithm iterates until the number of endmembers is exhausted.

A comparison of VCA with PPI [35] and N-FINDR [40] algorithms is conducted. Several experiments with simulated data lead to the conclusion that VCA performs better than PPI and better than or similar to N-FINDR. However, VCA has the lowest computational complexity among these three algorithms. Savings in computational complexity ranges between one and two orders of magnitude. This conclusion has great impact when the data set has a large number of pixels. VCA was also applied to real hyperspectral data. The results achieved show that VCA is an effective tool to unmix hyperspectral data.

References

1. José M.P. Nascimonto and José M. Bioucas Dias, Vertex component analysis: A fast algorithm to unmix hyperspectral data, *IEEE Transactions on Geoscience and Remote Sensing*, vol. 43, no. 4, pp. 898–910, 2005.
2. B. Hapke, *Theory of Reflectance and Emmittance Spectroscopy*, Cambridge, U.K., Cambridge University Press, 1993.
3. R.N. Clark and T.L. Roush, Reflectance spectroscopy: Quantitative analysis techniques for remote sensing applications, *Journal of Geophysical Research*, vol. 89, no. B7, pp. 6329–6340, 1984.
4. T.M. Lillesand, R.W. Kiefer, and J.W. Chipman, *Remote Sensing and Image Interpretation*, 5th ed., John Wiley & Sons, Inc., New York, 2004.
5. G. Vane, R. Green, T. Chrien, H. Enmark, E. Hansen, and W. Porter, The airborne visible/infrared imaging spectrometer (AVIRIS), *Remote Sensing of the Environment*, vol. 44, pp. 127–143, 1993.

6. M.O. Smith, J.B. Adams, and D.E. Sabol, *Spectral mixture analysis-New strategies for the analysis of multispectral data*, Brussels and Luxemburg, Belgium, 1994, pp. 125–143.
7. A.R. Gillespie, M.O. Smith, J.B. Adams, S.C. Willis, A.F. Fisher, and D.E. Sabol, Interpretation of residual images: Spectral mixture analysis of AVIRIS images, Owens Valley, California, in *Proceedings of the 2nd AVIRIS Workshop*, R.O. Green, Ed., JPL Publications, vol. 90–54, pp. 243–270, 1990.
8. J.J. Settle, On the relationship between spectral unmixing and sub-space projection, *IEEE Transactions on Geoscience And Remote Sensing*, vol. 34, pp. 1045–1046, 1996.
9. Y.H. Hu, H.B. Lee, and F.L. Scarpace, Optimal linear spectral un-mixing, *IEEE Transactions on Geoscience and Remote Sensing*, vol. 37, pp. 639–644, 1999.
10. M. Petrou and P.G. Foschi, Confidence in linear spectral unmixing of single pixels, *IEEE Transactions on Geoscience and Remote Sensing*, vol. 37, pp. 624–626, 1999.
11. S. Liangrocapart and M. Petrou, Mixed pixels classification, in *Proceedings of the SPIE Conference on Image and Signal Processing for Remote Sensing IV*, vol. 3500, pp. 72–83, 1998.
12. N. Keshava and J. Mustard, Spectral unmixing, *IEEE Signal Processing Magazine*, vol. 19, no. 1, pp. 44–57, 2002.
13. R.B. Singer and T.B. McCord, Mars: Large scale mixing of bright and dark surface materials and implications for analysis of spectral reflectance, in *Proceedings of the 10th Lunar and Planetary Science Conference*, pp. 1835–1848, 1979.
14. B. Hapke, Bidirection reflectance spectroscopy. I. theory, *Journal of Geophysical Research*, vol. 86, pp. 3039–3054, 1981.
15. R. Singer, Near-infrared spectral reflectance of mineral mixtures: Systematic combinations of pyroxenes, olivine, and iron oxides, *Journal of Geophysical Research*, vol. 86, pp. 7967–7982, 1981.
16. B. Nash and J. Conel, Spectral reflectance systematics for mixtures of powdered hypersthene, labradoride, and ilmenite, *Journal of Geophysical Research*, vol. 79, pp. 1615–1621, 1974.
17. C.C. Borel and S.A. Gerstl, Nonlinear spectral mixing models for vegetative and soils surface, *Remote Sensing of the Environment*, vol. 47, no. 2, pp. 403–416, 1994.
18. C.-I. Chang, *Hyperspectral Imaging: Techniques for spectral detection and classification*, New York, Kluwer Academic, 2003.
19. G. Shaw and H. Burke, Spectral imaging for remote sensing, *Lincoln Laboratory Journal*, vol. 14, no. 1, pp. 3–28, 2003.
20. D. Manolakis, C. Siracusa, and G. Shaw, Hyperspectral subpixel target detection using linear mixing model, *IEEE Transactions on Geoscience and Remote Sensing*, vol. 39, no. 7, pp. 1392–1409, 2001.
21. N. Keshava, J. Kerekes, D. Manolakis, and G. Shaw, An algorithm taxonomy for hyperspectral unmixing, in *Proceedings of the SPIE AeroSense Conference on Algorithms for Multispectral and Hyperspectral Imagery VI*, vol. 4049, pp. 42–63, 2000.
22. A.S. Mazer, M. Martin, et al., Image processing software for imaging spectrometry data analysis, *Remote Sensing of the Environment*, vol. 24, no. 1, pp. 201–210, 1988.
23. R.H. Yuhas, A.F.H. Goetz, and J.W. Boardman, Discrimination among semi-arid landscape endmembres using the spectral angle mapper (SAM) algorithm, in *Summaries of the 3rd Annual JPL Airborne Geoscience Workshop*, R.O. Green, Ed., Publ., 92–14, vol. 1, pp. 147–149, 1992.
24. J.C. Harsanyi and C.-I. Chang, Hyperspectral image classification and dimensionality reduction: an orthogonal subspace projection approach, *IEEE Transactions on Geoscience and Remote Sensing*, vol. 32, no. 4, pp. 779–785, 1994.
25. C. Chang, X. Zhao, M.L.G. Althouse, and J.J. Pan, Least squares subspace projection approach to mixed pixel classification for hyperspectral images, *IEEE Transactions on Geoscience and Remote Sensing*, vol. 36, no. 3, pp. 898–912, 1998.
26. D.C. Heinz, C.-I. Chang, and M.L.G. Althouse, Fully constrained least squares-based linear unmixing, in *Proceedings of the IEEE International Geoscience and Remote Sensing Symposium*, pp. 1401–1403, 1999.
27. P. Common, C. Jutten, and J. Herault, Blind separation of sources, part II: Problem statement, *Signal Processing*, vol. 24, pp. 11–20, 1991.
28. J. Bayliss, J.A. Gualtieri, and R. Cromp, Analysing hyperspectral data with independent component analysis, in *Proceedings of SPIE*, vol. 3240, pp. 133–143, 1997.

29. C. Chen and X. Zhang, Independent component analysis for remote sensing study, in *Proceedings of the SPIE Symposium on Remote Sensing Conference on Image and Signal Processing for Remote Sensing V*, vol. 3871, pp. 150–158, 1999.
30. T.M. Tu, Unsupervised signature extraction and separation in hyperspectral images: A noise-adjusted fast independent component analysis approach, *Optical Engineering/SPIE*, vol. 39, no. 4, pp. 897–906, 2000.
31. S.-S. Chiang, C.-I. Chang, and I.W. Ginsberg, Unsupervised hyperspectral image analysis using independent component analysis, in *Proceedings of the IEEE International Geoscience and Remote Sensing Symposium*, 2000.
32. Jośe M.P. Nascimonto and José M. Bioucas Dias, Does independent component analysis play a role in unmixing hyperspectral data? in *Pattern Recognition and Image Analysis*, ser. Lecture Notes in Computer Science, F. j. Perales, A. Campilho, and N.P.B.A. Sanfeliu, Eds., vol. 2652. Springer-Verlag, pp. 616–625, 2003.
33. Jośe M.P. Nascimonto and José M. Bioucas Dias, Does independent component analysis play a role in unmixing hyperspectral data? *IEEE Transactions on Geoscience and Remote Sensing*, vol. 43, no. 1, pp. 175–187, 2005.
34. A. Ifarraguerri and C.-I. Chang, Multispectral and hyperspectral image analysis with convex cones, *IEEE Transactions on Geoscience and Remote Sensing*, vol. 37, no. 2, pp. 756–770, 1999.
35. J. Boardman, Automating spectral unmixing of AVIRIS data using convex geometry concepts, in *Summaries of the Fourth Annual JPL Airborne Geoscience Workshop*, JPL Publications, 93–26, AVIRIS Workshop, vol. 1, 1993, pp. 11–14.
36. M.D. Craig, Minimum-volume transforms for remotely sensed data, *IEEE Transactions on Geoscience and Remote Sensing*, vol. 32, pp. 99–109, 1994.
37. C. Bateson, G. Asner, and C. Wessman, Endmember bundles: A new approach to incorporating endmember variability into spectral mixture analysis, *IEEE Transactions on Geoscience and Remote Sensing*, vol. 38, pp. 1083–1094, 2000.
38. R. Seidel, *Convex Hull Computations*, Boca Raton, CRC Press, ch. 19, pp. 361–375, 1997.
39. S. Geman and D. Geman, Stochastic relaxation, Gibbs distribution and the Bayesian restoration of images, *IEEE Trans. Pattern Anal. Machine Intell.*, vol. 6, no. 6, pp. 721–741, 1984.
40. M.E. Winter, N-findr: an algorithm for fast autonomous spectral endmember determination in hyperspectral data, in *Proceedings of the SPIE Conference on Imaging Spectrometry V*, pp. 266–275, 1999.
41. J. Boardman, F.A. Kruse, and R.O. Green, Mapping target signatures via partial unmixing of AVIRIS data, in *Summaries of the V JPL Airborne Earth Science Workshop*, vol. 1, pp. 23–26, 1995.
42. J. Theiler, D. Lavenier, N. Harvey, S. Perkins, and J. Szymanski, Using blocks of skewers for faster computation of pixel purity index, in *Proceedings of the SPIE International Conference on Optical Science and Technology*, vol. 4132, pp. 61–71, 2000.
43. D. Lavenier, J. Theiler, J. Szymanski, M. Gokhale, and J. Frigo, Fpga implementation of the pixel purity index algorithm, in *Proceedings of the SPIE Photonics East, Workshop on Reconfigurable Architectures*, 2000.
44. J.H. Bowles, P.J. Palmadesso, J.A. Antoniades, M.M. Baumback, and L.J. Rickard, Use of filter vectors in hyperspectral data analysis, in *Proceedings of the SPIE Conference on Infrared Spaceborne Remote Sensing III*, vol. 2553, pp. 148–157, 1995.
45. J.H. Bowles, J.A. Antoniades, M.M. Baumback, J.M. Grossmann, D. Haas, P.J. Palmadesso, and J. Stracka, Real-time analysis of hyperspectral data sets using nrl's orasis algorithm, in *Proceedings of the SPIE Conference on Imaging Spectrometry III*, vol. 3118, pp. 38–45, 1997.
46. J.M. Grossmann, J. Bowles, D. Haas, J.A. Antoniades, M.R. Grunes, P. Palmadesso, D. Gillis, K. Y. Tsang, M. Baumback, M. Daniel, J. Fisher, and I. Triandaf, Hyperspectral analysis and target detection system for the adaptative spectral reconnaissance program (asrp), in *Proceedings of the SPIE Conference on Algorithms for Multispectral and Hyperspectral Imagery IV*, vol. 3372, pp. 2–13, 1998.
47. J.M.P. Nascimento and J.M.B. Dias, Signal subspace identification in hyperspectral linear mixtures, in *Pattern Recognition and Image Analysis*, ser. Lecture Notes in Computer Science, J. S. Marques, N.P. de la Blanca, and P. Pina, Eds., vol. 3523, no. 2., Heidelberg, Springer-Verlag, pp. 207–214, 2005.
48. J.M.B. Dias and J.M.P. Nascimento, Estimation of signal subspace on hyperspectral data, in *Proceedings of SPIE conference on Image and Signal Processing for Remote Sensing XI*, L. Bruzzone, Ed., vol. 5982, pp. 191–198, 2005.

49. L.L. Scharf, *Statistical Signal Processing, Detection Estimation and Time Series Analysis*, Reading, MA, Addison-Wesley, 1991.
50. R.N. Clark, G.A. Swayze, A. Gallagher, T.V. King, and W.M. Calvin, The U.S. geological survey digital spectral library: Version 1: 0.2 to 3.0 μm, U.S. Geological Survey, Open File Report 93–592, 1993.
51. I.T. Jolliffe, *Principal Component Analysis*, New York, Springer-Verlag, 1986.
52. A. Green, M. Berman, P. Switzer, and M.D. Craig, A transformation for ordering multispectral data in terms of image quality with implications for noise removal, *IEEE Transactions on Geoscience and Remote Sensing*, vol. 26, no. 1, pp. 65–74, 1994.
53. J.B. Lee, S. Woodyatt, and M. Berman, Enhancement of high spectral resolution remote-sensing data by noise-adjusted principal components transform, *IEEE Transactions on Geoscience and Remote Sensing*, vol. 28, pp. 295–304, 1990.
54. J. Harsanyi, W. Farrand, and C.-I. Chang, Determining the number and identity of spectral endmembers: An integrated approach using neymanpearson eigenthresholding and iterative constrained rms error minimization, in *Proceedings of the 9th Thematic Conference on Geologic Remote Sensing*, 1993.
55. R. Roger and J. Arnold, Reliably estimating the noise in aviris hyperspectral imagers, *International Journal of Remote Sensing*, vol. 17, no. 10, pp. 1951–1962, 1996.
56. J.H. Bowles, M. Daniel, J.M. Grossmann, J.A. Antoniades, M.M. Baumback, and P. J. Palmadesso, Comparison of output from orasis and pixel purity calculations, in *Proceedings of the SPIE Conference on Imaging Spectrometry IV*, vol. 3438, pp. 148–156, 1998.
57. A. Plaza, P. Martinez, R. Perez, and J. Plaza, Spatial/spectral endmember extraction by multidimensional morphological operations, *IEEE Transactions on Geoscience and Remote Sensing*, vol. 40, no. 9, pp. 2025–2041, 2002.
58. A. Plaza, P. Martinez, R. Perez, and J. Plaza, A quantitative and comparative analysis of endmember extraction algorithms from hyperspectral data, *IEEE Transactions on Geoscience and Remote Sensing*, vol. 42, no. 3, pp. 650–663, 2004.
59. E. Kaltofen and G. Villard, On the complexity of computing determinants, in *Proceedings of the Fifth Asian Symposium on Computer Mathematics*, Singapore, ser. Lecture Notes Series on Computing, K. Shirayanagi and K. Yokoyama, Eds., vol. 9, pp. 13–27, 2001.
60. G. Swayze, R. Clark, S. Sutley, and A. Gallagher, Ground-truthing aviris mineral mapping at Cuprite, Nevada, in *Summaries of the Third Annual JPL Airborne Geosciences Workshop*, vol. 1, pp. 47–49, 1992.
61. R. Ashley and M. Abrams, Alteration mapping using multispectral images—Cuprite mining district, Esmeralda county, U.S. Geological Survey, Open File Report 80-367, 1980.
62. M. Abrams, R. Ashley, L. Rowan, A. Goetz, and A. Kahle, Mapping of hydrothermal alteration in the Cuprite mining district, Nevada, using aircraft scanner images for the spectral region 0.46 to 2.36 mm, *Geology*, vol. 5, pp. 713–718, 1977.
63. G. Swayze, The hydrothermal and structural history of the Cuprite mining district, southwestern Nevada: An integrated geological and geo-physical approach, Ph.D. Dissertation, University of Colorado, 1997.
64. A. Goetz and V. Strivastava, Mineralogical mapping in the cuprite mining district, in *Proceedings of the Airborne Imaging Spectrometer Data Analysis Workshop*, JPL Publications 85-41, pp. 22–29, 1985.
65. F. Kruse, J. Boardman, and J. Huntington, Comparison of airborne and satellite hyperspectral data for geologic mapping, in *Proceedings of the SPIE Aerospace Conference*, vol. 4725, pp. 128–139, 2002.

8

Two ICA Approaches for SAR Image Enhancement

Chi Hau Chen, Xianju Wang, and Salim Chitroub

CONTENTS

8.1 Part 1: Subspace Approach of Speckle Reduction in SAR Images Using ICA

8.1.1 Introduction

The use of synthetic aperture radar (SAR) can provide images with good details under many environmental conditions. However, the main disadvantage of SAR imagery is the poor quality of images, which are degraded by multiplicative speckle noise. SAR image speckle noise appears to be randomly granular and results from phase variations of radar waves from unit reflectors within a resolution cell. Its existence is undesirable because it degrades quality of the image and affects the task of human interpretation and evaluation. Thus, speckle removal is a key preprocessing step for automatic interpretation of SAR images. A subspace method using independent component analysis (ICA) for speckle reduction is presented here.

8.1.2 Review of Speckle Reduction Techniques in SAR Images

Many adaptive filters for speckle reduction have been proposed in the past. Earlier approaches include Frost filter, Lee filter, Kuan filter, etc. The Frost filter was designed as an adaptive Wiener filter based on the assumption that the scene reflectivity is an autoregressive (AR) exponential model [1]. The Lee filter is a linear approximation filter based on the minimum mean-square error (MMSE) criterion [2]. The Kuan filter is the generalized case of the Lee filter. It is an MMSE linear filter based on the multiplicative speckle model and is optimal when both the scene and the detected intensities are Gaussian distributed [3].

Recently, there has been considerable interest in using the ICA as an effective tool for signal blind separation and deconvolution. In the field of image processing, ICA has strong adaptability for representing different kinds of images and is very suitable for tasks like compression and denoising. Since the mid-1990s its applications have been extended to more practical fields, such as signal and image denoising and pattern recognition. Zhang [4] presented a new ICA algorithm by working directly with high-order statistics and demonstrated its better performance on SAR image speckle reduction problem. Malladi [5] developed a speckle filtering technique using Holder regularity analysis of the Sparse coding. Other approaches [6–8] employ multi-scale and wavelet analysis.

8.1.3 The Subspace Approach to ICA Speckle Reduction

In this approach, we assume that the speckle noise in SAR images comes from a different signal source, which accompanies but is independent of the ''true signal source'' (image details). Thus the speckle removal problem can also be described as ''signal source separation'' problem. The steps taken are illustrated by the nine-channel SAR images considered in Chapter 2 of the companion volume (*Signal Processing for Remote Sensing*), which are reproduced here as shown in Figure 8.1.

8.1.3.1 Estimating ICA Bases from the Image

One of the important problems in ICA is how to estimate the transform from the given data. It has been shown that the estimation of the ICA data model can be reduced to the search for uncorrelated directions in which the components are as non-Gaussian as possible [9]. In addition, we note that ICA usually gives one component (DC component) representing the local mean image intensity, which is noise-free. Thus we should treat it separately from the other components in image denoising applications. Therefore, in all experiments we first subtract the local mean, and then estimate a suitable basis for the rest of the components.

The original image is first linearly normalized so that it has zero mean and unit variance. A set of overlapped image windows of 16×16 pixels are taken from it and the local mean of each patch is subtracted. The choice of window size can be critical in this application. For smaller sizes, the reconstructed separated sources can still be very correlated. To overcome the difficulties related to the high dimensionality of vectors, their dimensionality has been reduced to 64 by PCA. (Experiments prove that for SAR images that have few image details, 64 components can make image reconstruction nearly error-free.) The preprocessed data set is used as the input to FastICA algorithm, using the tanh nonlinearity.

Figure 8.2 shows the estimated basis vectors after convergence of the FastICA algorithm.

8.1.3.2 Basis Image Classification

As alluded earlier, we believe that ''speckle pattern'' (speckle noise) in the SAR image comes from another kind of signal source, which is independent of true signal source; hence our problem can be considered as signal source separation. However, for the

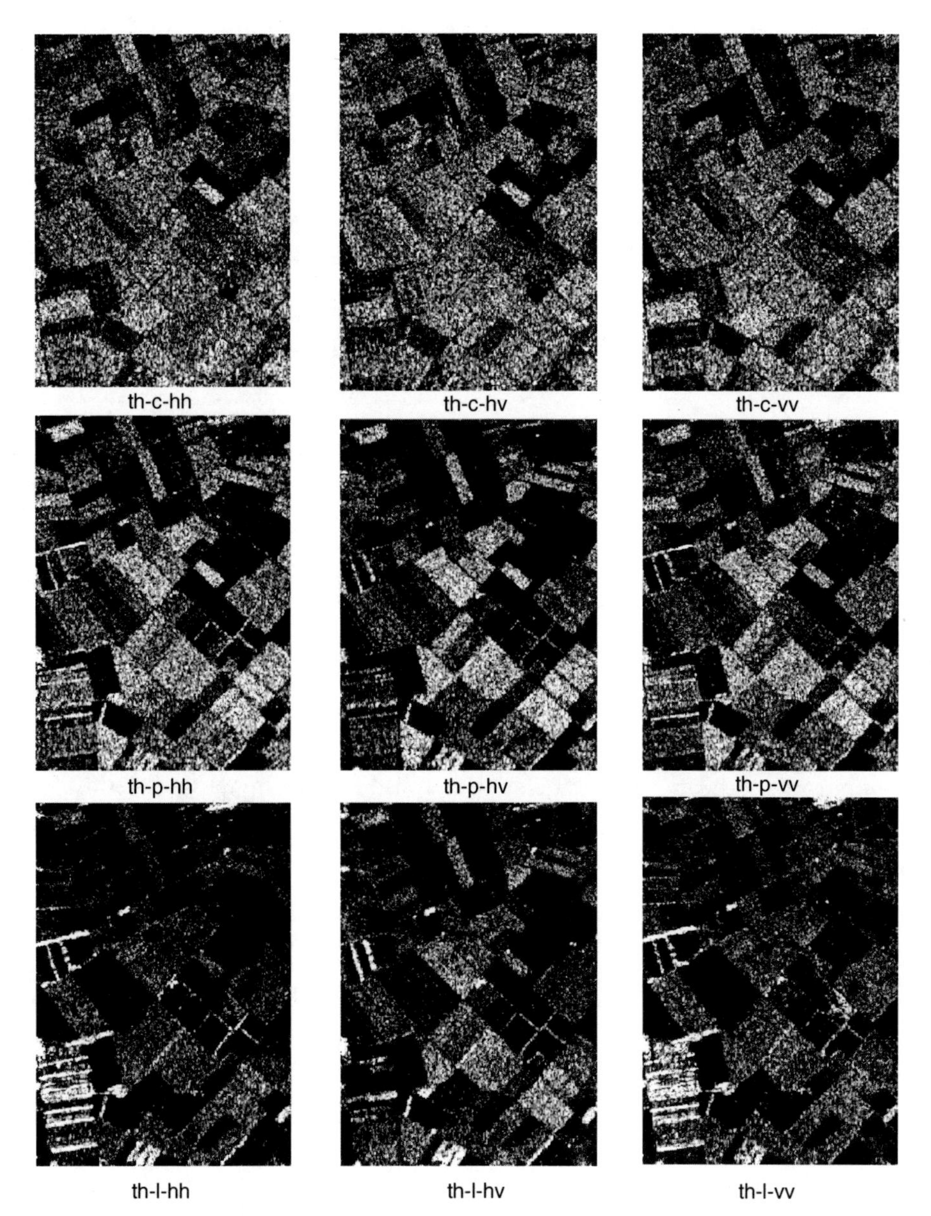

FIGURE 8.1
The nine-channel polarimetric synthetic aperture radar (POLSAR) images.

image signal separation, we first need to classify the basis images; that is, we denote basis images that span speckle pattern space by S2 and the basis images that span "true signal" space by S1. Then we have S1 + S2 = V. The whole signal space that is spanned by all the basis images is denoted by V. Here, we sample in the main noise regions, which we denote by *P*. From the above discussion, S1 and S2 are essentially nonoverlapping or "orthogonal." Then our classification rule is

$$\begin{cases} \frac{1}{N}\sum_{j\in P}\left|s_{ij}\right| > T & i\text{th component } \in \text{S2} \\ \frac{1}{N}\sum_{j\in P}\left|s_{ij}\right| < T & i\text{th component } \in \text{S1} \end{cases}$$

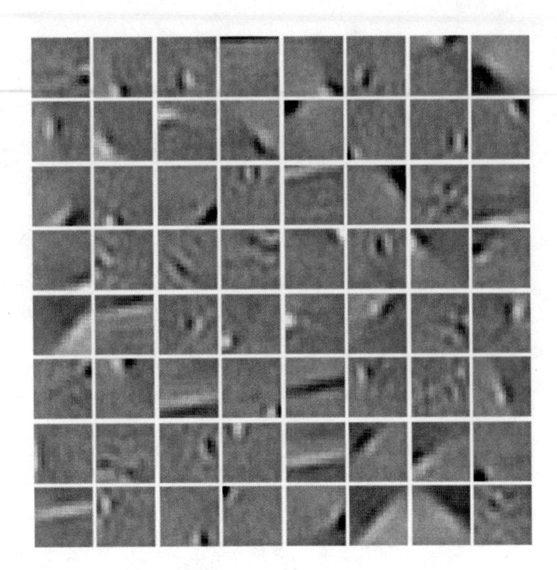

FIGURE 8.2
ICA basis images of the images in Figure 8.1.

where T is a selected threshold.

Figure 8.3 shows the classification result.

The processing results of the first five channels are shown in Figure 8.4. We further calculate the ratio of local standard deviation to mean (SD/mean) for each image and use it as a criterion for image quality. Both visual quality and performance criterion demonstrate that our method can remove the speckle noise in SAR images efficiently.

8.1.3.3 Feature Emphasis by Generalized Adaptive Gain

We now apply nonlinear contrast stretching in each component to enhance the image features. Here, adaptive gain [6] through nonlinear processing, denoted as $f(\cdot)$, is generalized to incorporate hard thresholding to avoid amplifying noise and remove small noise perturbations.

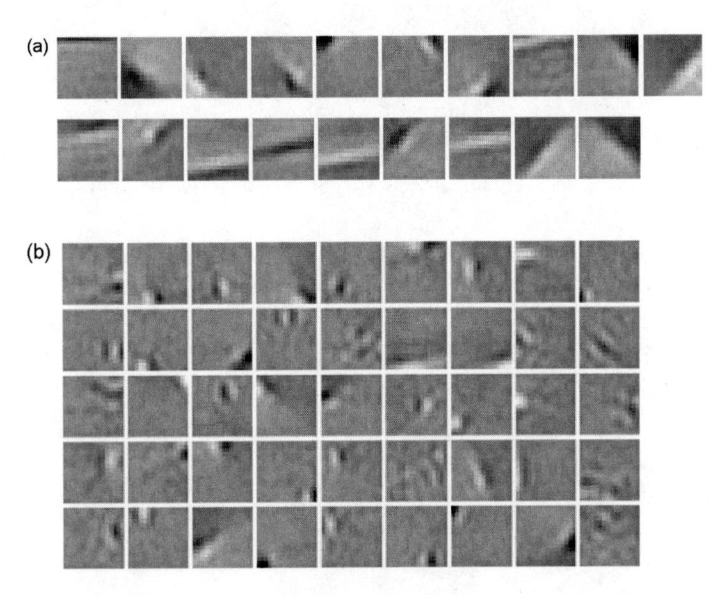

FIGURE 8.3
(a) The basis images $\in$S1 (19 components). (b) The basis images $\in$S2 (45 components).

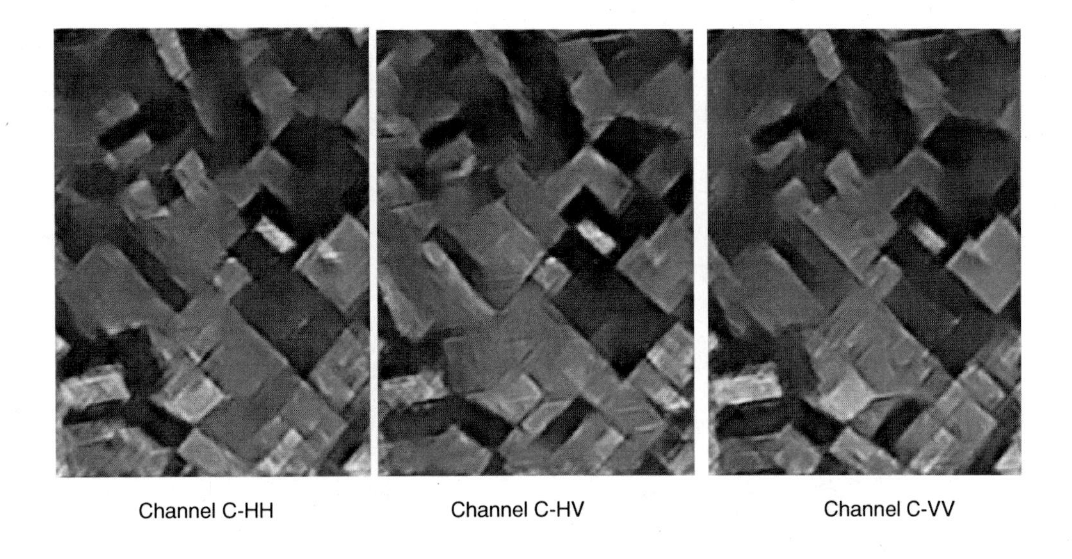

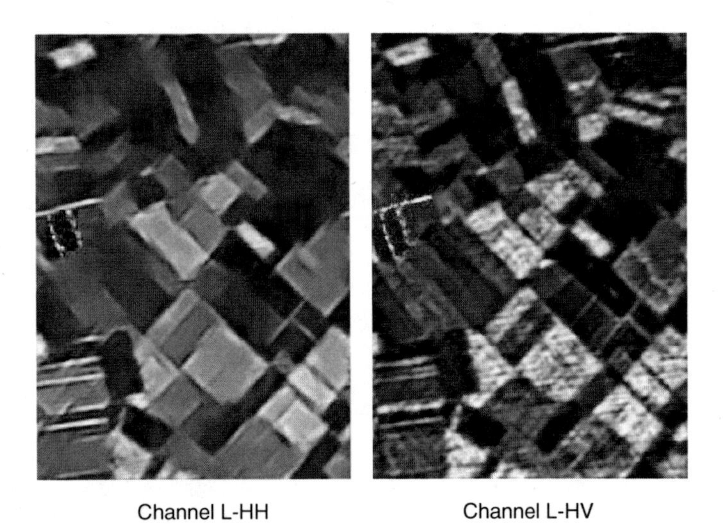

FIGURE 8.4
Recovered images with our method.

8.1.3.4 Nonlinear Filtering for Each Component

Our nonlinear filtering is simple to realize. For the components that belong to S2, we simply set them to zero, but we apply our GAG operator to other components that belong to S1, to enhance the image feature. Then the recovered S_{ij} can be calculated by the following equation:

$$\hat{s}_{ij} = \begin{cases} 0 & i\text{th component} \in \text{S2} \\ f(s_{ij}) & i\text{th component} \in \text{S1} \end{cases}$$

Finally the restored image can be obtained after a mixing transform.

A comparison is made with other methods including the Wiener filter, the Lee filter, and Kuan filter. The result of using Lee filter is shown in Figure 8.5. The ratio comparison is shown in Table 8.1. The smaller the ratio, the better the image quality. Our method has the smallest ratios in most cases.

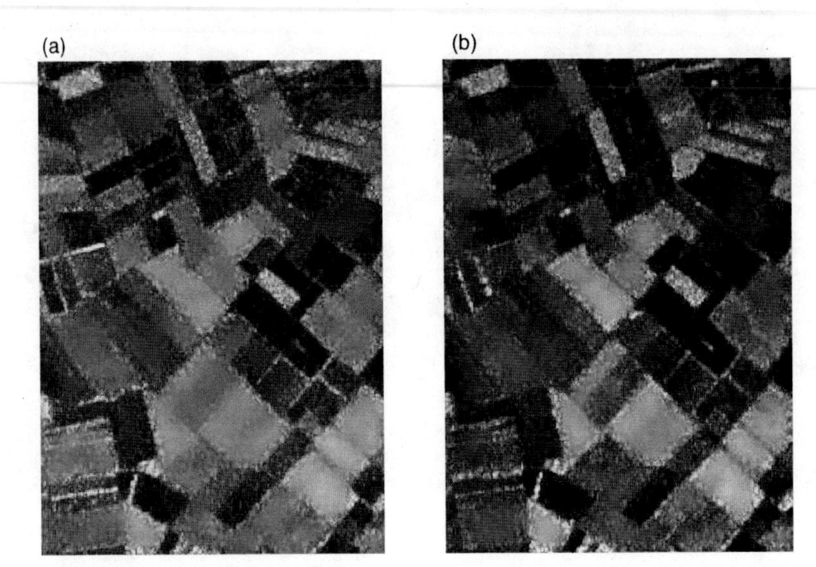

FIGURE 8.5
Recovered images using Lee filter.

As a concluding remark the subspace approach as presented allows quite a flexibility to adjust parameters such that significant improvement in speckle reduction with the SAR images can be achieved.

8.2 Part 2: A Bayesian Approach to ICA of SAR Images

8.2.1 Introduction

We present a PCA–ICA neural network for analyzing the SAR images. With this model, the correlation between the images is eliminated and the speckle noise is largely reduced in only the first independent component (IC) image. We have used, as input data for the ICA parts, only the first principal component (PC) image. The IC images obtained are of very high quality and better contrasted than the first PC image. However, when the second and third PC images are also used as input images with the first PC image, the results are less impressive and the first IC images become less contrasted and more affected by the noise. This can be justified by the fact that the ICA parts of the models

TABLE 8.1
Ratio Comparison

	Original	Our Method	Wiener Filter	Lee Filter	Kuan Filter
Channel 1	0.1298	0.1086	0.1273	0.1191	0.1141
Channel 2	0.1009	0.0526	0.0852	0.1133	0.0770
Channel 3	0.1446	0.0938	0.1042	0.1277	0.1016
Channel 4	0.1259	0.0371	0.0531	0.0983	0.0515
Channel 5	0.1263	0.1010	0.0858	0.1933	0.0685

are essentially based on the principle of the Infomax algorithm for the model proposed in Ref. [10]. The Informax algorithm, however, is efficient only in the case where the input data have low additive noise.

The purpose of Part 2 is to propose a Bayesian approach of the ICA method that performs well for analyzing images and that presents some advantages compared to the previous model. The Bayesian approach ICA method is based on the so-called ensemble learning algorithm [11,12]. The purpose is to overcome the disadvantages of the method proposed in Ref. [10]. Before detailing the present method in Section 8.2.4, we present in Section 8.2.2 the SAR image model and the statistics to be used later. Section 8.2.3 is devoted to the whitening phase of the proposed method. This step of processing is based on the so-called simultaneous diagonalization transform for performing the PCA method of SAR images [13]. Experimental results based on real SAR images shown in Figure 8.1 are discussed in Section 8.2.5. To prove the effectiveness of the proposed method, the FastICA-based method [9] is used for comparison. The conclusion for Part 2 is in Section 8.2.6.

8.2.2 Model and Statistics

We adopt the same model used in Ref. [10]. Speckle has the characteristics of a multiplicative noise in the sense that its intensity is proportional to the value of the pixel content and is dependent on the target nature [13]. Let x_i be the content of the pixel in the ith image, s_i the noise-free signal response of the target, and n_i the speckle. Then, we have the following multiplicative model:

$$x_i = s_i n_i \tag{8.1}$$

By supposing that the speckle has unity mean, standard deviation of σ_i, and is statistically independent of the observed signal x_i [14], the multiplicative model can be rewritten as

$$x_i = s_i + s_i(n_i - 1) \tag{8.2}$$

The term $s_i\,(n_i - 1)$ represents the zero mean signal-dependent noise and characterizes the speckle noise variation. Now, let X be the stationary random vector of input SAR images. The covariance matrix of X, Σ_X, can be written as

$$\Sigma_X = \Sigma_s + \Sigma_n \tag{8.3}$$

where Σ_s and Σ_n are the covariance matrices of the noise-free signal vector and the signal-dependent noise vector, respectively. The two matrices, Σ_X and Σ_n, are used in constructing the linear transformation matrix of the whitening phase of the proposed method.

8.2.3 Whitening Phase

The whitening phase is ensured by the PCA part of the proposed model (Figure 8.6). The PCA-based part (Figure 8.7) is devoted to the extraction of the PC images. It is based on the simultaneous diagonalization concept of the two matrices Σ_X and Σ_n, via one orthogonal matrix A. This means that the PC images (vector Y) are uncorrelated and have an additive noise that has a unit variance. This step of processing allows us to make our application coherent with the theoretical development of ICA. In fact, the constraint to

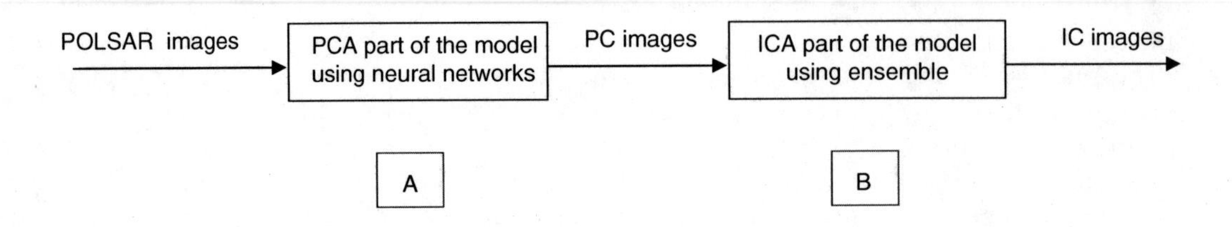

FIGURE 8.6
The proposed PCA–ICA model for SAR image analysis.

have whitening uncorrelated inputs is desirable in ICA algorithms because it simplifies the computations considerably [11,12]. These inputs are assumed non-Gaussian, centered, and have unit variance. It is ordinarily assumed that X is zero-mean, which in turn means that Y is also zero-mean, where the condition of unit variance can be achieved by standardizing Y. For the non-Gaussianity of Y, it is clear that the speckle, which has non-Gaussianity properties, is not affected by this step of processing because only the second-order statistics are used to compute the matrix A.

The criterion for determining A is: "Finding A such as the matrix Σ_n becomes an identity matrix and the matrix Σ_X is transformed, at the same time, to a diagonal matrix." This criterion can be formulated in the constrained optimization framework as

$$\text{Maximize } A^T\Sigma_X A \text{ subject to } A^T\Sigma_n A = I \tag{8.4}$$

where I is the identity matrix. Based on the well-developed aspects of the matrix theories and computations, the existence of A is proved in Ref. [12] and a statistical algorithm for obtaining it is also proposed. Here, we propose a neuronal implementation of this algorithm [15] with some modifications (Figure 8.7). It is composed of two PCA neural networks that have the same topology. The lateral weights c_j^1 and c_j^2, forming the vectors C_1 and C_2, respectively, connect all the first $m-1$ neurons with the mth one. These connections play a very important role in the model because they work toward the orthogonalization of the synaptic vector of the mth neuron with the vectors of the previous $m-1$ neurons. The solid lines denote the weights w_i^1, c_j^1 and w_i^2, c_j^2, respectively,

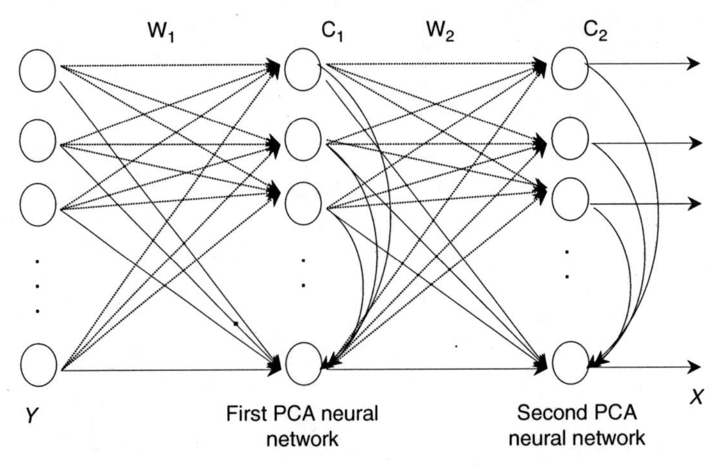

FIGURE 8.7
The PCA part of the proposed model for SAR image analysis.

which are trained at the mth stage, while the dashed lines correspond to the weights of the already trained neurons. Note that the lateral weights asymptotically converge to zero, so they do not appear among the already trained neurons.

The first network of Figure 8.7 is devoted to whitening the noise in Equation 8.2, while the second one is for maximizing the variance given that the noise is already whitened. Let X_1 be the input vector of the first network. The noise is whitened, through the feed-forward weights $\{w_{ij}^1\}$, where i and j correspond to the input and output neurons, respectively, and the superscript 1 designates the weighted matrix of the first network. After convergence, the vector X is transformed to the new vector X' via the matrix $U = W_1\Lambda^{-1/2}$, where W_1 is the weighted matrix of the first network, Λ is the diagonal matrix of eigenvalues of Σ_n (variances of the output neurons) and $\Lambda^{-1/2}$ is the inverse of its square root. Next, X' be the input vector of the second network. It is connected to M outputs, with $M \leq N$, corresponding to the intermediate output vector noted X_2, through the feed-forward weights $\{w_{ij}^2\}$. Once this network is converged, the PC images to be extracted (vector Y) are obtained as

$$Y = A^{\mathrm{T}}X = UW_2X \tag{8.5}$$

where W_2 is the weighted matrix of the second network. The activation of each neuron in the two parts of the network is a linear function of their inputs. The kth iteration of the learning algorithm, for both networks, is given as:

$$w(k+1) = w(k) + \beta(k)(q_m(k)P - q_m^2(k)w(k)) \tag{8.6}$$

$$c(k+1) = c(k) + \beta(k)(q_m(k)Q - q_m^2(k)c(k)) \tag{8.7}$$

Here P and Q are the input and output vectors of the network, respectively. $\beta(k)$ is a positive sequence of the learning parameter. The global convergence of the PCA-based part of the model is strongly dependent on the parameter β. The optimal choice of this parameter is well studied in Ref. [15].

8.2.4 ICA of SAR Images by Ensemble Learning

Ensemble learning is a computationally efficient approximation for exact Bayesian analysis. With Bayesian learning, all information is taken into account in the posterior probabilities. However, the posterior probability density function (pdf) is a complex high-dimensional function whose exact treatment is often difficult, if not impossible. Thus some suitable approximation method must be used. One solution is to find the maximum A posterior (MAP) parameters. But this method can overfit because it is sensitive to probability density rather than probability mass. The correct way to perform the inference would be to average over all possible parameter values by drawing samples from the posterior density. Rather than performing a Markov chain Monte Carlo (MCMC) approach to sample from the true posterior, we use the ensemble learning approximation [11].

Ensemble learning [11,12] which is a special case of variational learning, is a recently developed method for parametric approximation of posterior pdfs where the search takes into account the probability mass of the models. Therefore, it solves the tradeoff between under- and overfitting. The basic idea in ensemble learning is to minimize the misfit between the posterior pdf and its parametric approximation by choosing a computationally tractable parametric approximation—an ensemble—for the posterior pdf.

In fact, all the relevant information needed in choosing an appropriate model is contained in the posterior pdfs of hidden sources and parameters.

Let us denote the set of available data, which are the PC output images of the PCA part of Figure 8.7, by X, and the respective source vectors by S. Given the observed data X, the unknown variables of the model are the sources S, the mixing matrix B, the parameters of the noise and source distributions, and the hyperparameters. For notational simplicity, we shall denote the ensemble of these variables and parameters by θ. The posterior $P(S, \theta|X)$ is thus a pdf of all these unknown variables and parameters.

We wish to infer the set pdf parameters θ given the observed data matrix X. We approximate the exact posterior probability density, $P(S, \theta|X)$, by a more tractable parametric approximation, $Q(S, \theta|X)$, for which it is easy to perform inferences by integration rather than by sampling. We optimize the approximate distribution by minimizing the Kullback–Leibler divergence between the approximate and the true posterior distribution. If we choose a separable distribution for $Q(S, \theta|X)$, the Kullback–Leibler divergence will split into a sum of simpler terms. An ensemble learning model can approximate the full posterior of the sources by a more tractable separable distribution.

The Kullback–Leibler divergence C_{KL}, between $P(S, \theta|X)$ and $Q(S, \theta|X)$, is defined by the following cost function:

$$C_{KL} = \int Q(S, \theta|X) \log\left(\frac{Q(S, \theta|X)}{P(S, \theta|X)}\right) d\theta \, dS \tag{8.8}$$

C_{KL} measures the difference in the probability mass between the densities $P(S, \theta|X)$ and $Q(S, \theta|X)$. Its minimum value 0 is achieved when the two densities are the same. For approximating and then minimizing C_{KL}, we need the exact posterior density $P(S, \theta|X)$ and its parametric approximation $Q(S, \theta|X)$. According to the Bayes rule, the posterior pdf of the unknown variables S and θ is such as:

$$P(S, \theta|X) = \frac{P(X|S, \theta)P(S|\theta)P(\theta)}{P(X)} \tag{8.9}$$

The term $P(X|S, \theta)$ is obtained from the model that relates the observed data and the sources. The terms $P(S|\theta)$ and $P(\theta)$ are products of simple Gaussian distributions and they are obtained directly from the definition of the model structure [16]. The term $P(X)$ does not depend on the model parameters and can be neglected. The approximation $Q(S, \theta|X)$ must be simple for mathematical tractability and computational efficiency. Here, both the posterior density $P(S, \theta|X)$ and its approximation $Q(S, \theta|X)$ are products of simple Gaussian terms, which simplify the cost function given by Equation 8.8 considerably: it splits into expectations of many simple terms. In fact, to make the approximation of the posterior pdf computationally tractable, we shall choose the ensemble $Q(S, \theta|X)$ to be a Gaussian pdf with diagonal covariance. The independent sources are assumed to have mixtures of Gaussian as distributions. The observed data are also assumed to have additive Gaussian noise with diagonal covariance. This hypothesis is verified by performing the whitening step using the simultaneous diagonalization transform as it is given in Section 8.2.3. The model structure and all the parameters of the distributions are estimated from the data. First, we assume that the sources S are independent of the other parameters θ, so that $Q(S, \theta|X)$ decouples into

$$Q(S, \theta|X) = Q(S|X)Q(\theta|X) \tag{8.10}$$

For the parameters θ, a Gaussian density with a diagonal covariance matrix is supposed. This implies that the approximation is a product of independent distributions:

$$Q(\theta|X) = \prod_i Q_i(\theta_i|X) \tag{8.11}$$

The parameters of each Gaussian component density $Q_i(\theta_i|X)$ are its means $\bar{\theta}_i$ and variances $\tilde{\theta}_i$. The $Q(\theta|X)$ is similar. The cost function C_{KL} is a function of the posterior means $\bar{\theta}_i$ and variances $\tilde{\theta}_i$ of the sources and the parameters of the network. This is because instead of finding a point estimate, the joint posterior pdf of the sources and parameters is estimated in ensemble learning. The variances give information about the reliability of the estimates.

Let us now denote the two parts of the cost function given by Equation 8.8 arising in the denominator and numerator of the logarithm respectively by $C_p = -E_p(\log(P))$ and $C_q = E_q(\log(Q))$. The variances $\tilde{\theta}_i$ are obtained by differentiating Equation 8.8 with respect to $\tilde{\theta}_i$ [16]:

$$\frac{\partial C_{KL}}{\partial \tilde{\theta}_i} = \frac{\partial C_p}{\partial \tilde{\theta}_i} + \frac{\partial C_q}{\partial \tilde{\theta}_i} = \frac{\partial C_p}{\partial \tilde{\theta}_i} - \frac{1}{2\tilde{\theta}_i} \tag{8.12}$$

Equating this to zero yields a fixed-point iteration for updating the variances:

$$\tilde{\theta}_i = \left(2\frac{\partial C_p}{\partial \tilde{\theta}_i}\right)^{-1} \tag{8.13}$$

The means $\bar{\theta}_i$ can be estimated from the approximate Newton iteration [16]:

$$\bar{\theta}_i \leftarrow \bar{\theta}_i - \frac{\partial C_p}{\partial \bar{\theta}_i}\left(\frac{\partial^2 C_p}{\partial \bar{\theta}_i^2}\right)^{-1} \approx \bar{\theta}_i - \frac{\partial C_p}{\partial \bar{\theta}_i}\tilde{\theta}_i \tag{8.14}$$

The algorithm solves Equation 8.13 and Equation 8.14 iteratively until convergence is achieved. The practical learning procedure consists of applying the PCA part of the model. The output PC images are used to find sensible initial values for the posterior means of the sources. The PCA part of the model yields clearly better initial values than a random choice. The posterior variances of the sources are initialized to small values.

8.2.5 Experimental Results

The SAR images used are shown in Figure 8.1. To prove the effectiveness of the proposed method, the FastICA-based method [13,14] is used for comparison. The extracted IC images using the proposed method are given in Figure 8.8. The extracted IC images using the FastICA-based method are presented in Figure 8.9. We note that the FastICA-based method gives inadequate results because the IC images obtained are contrasted too much. It is clear that the proposed method gives the IC images that are better than the original SAR images. Also, the results of ICA by ensemble learning exceed largely the results of the FastICA-based method. We observe that the effect of the speckle noise is largely reduced in the images based on ensemble learning especially in the sixth image, which is an image of high quality. It appears that the low quality of some of the images by the FastICA method is caused by being trapped in local minimum while the ensemble learning–based method is much more robust.

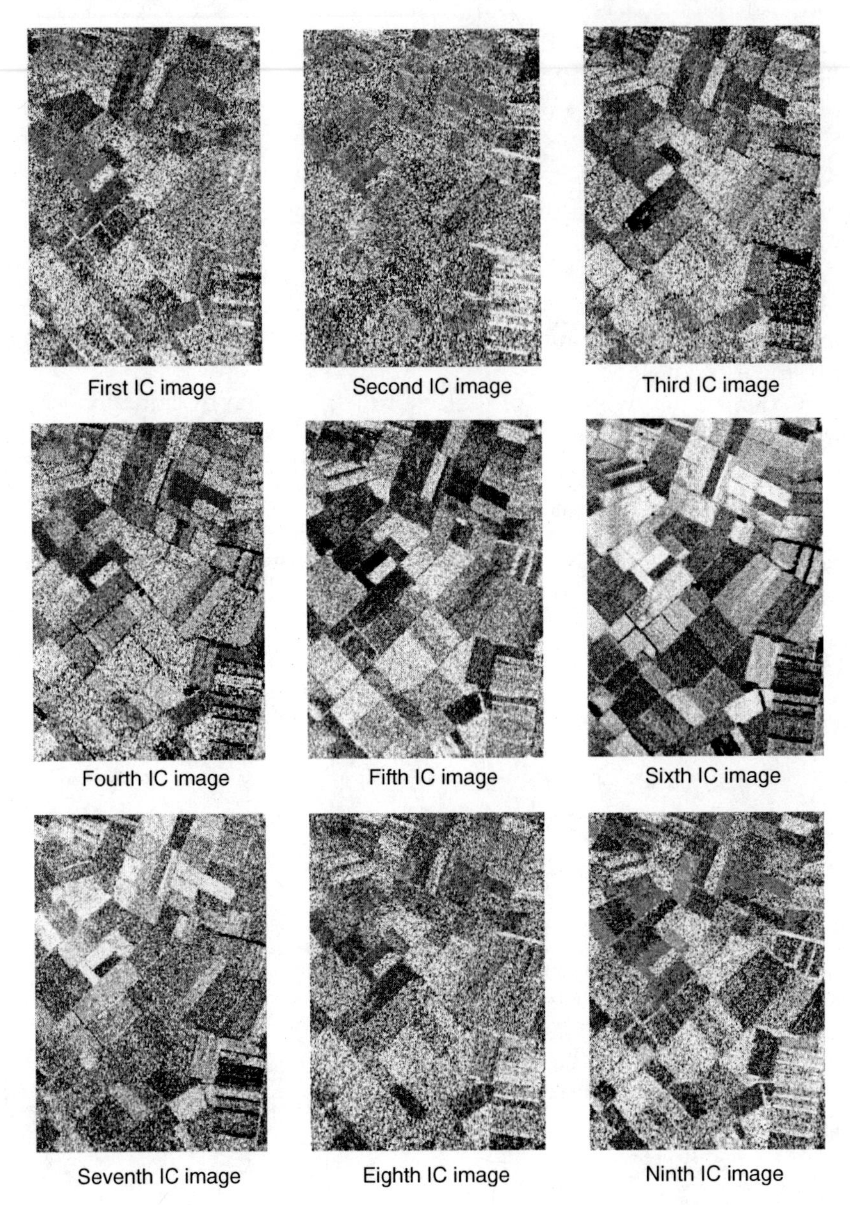

FIGURE 8.8
The results of ICA by ensemble learning.

Table 8.2 shows a comparison of the computation time between the FastICA method and the proposed method. It is evident that the FastICA method has significant advantage in computation time.

8.2.6 Conclusions

We have suggested a Bayesian approach of ICA applied to SAR image analysis. This consists of using the ensemble learning, which is a computationally efficient approximation for exact Bayesian analysis. Before performing the ICA by ensemble learning, a PCA neural network model that performs the simultaneous diagonalization of the noise

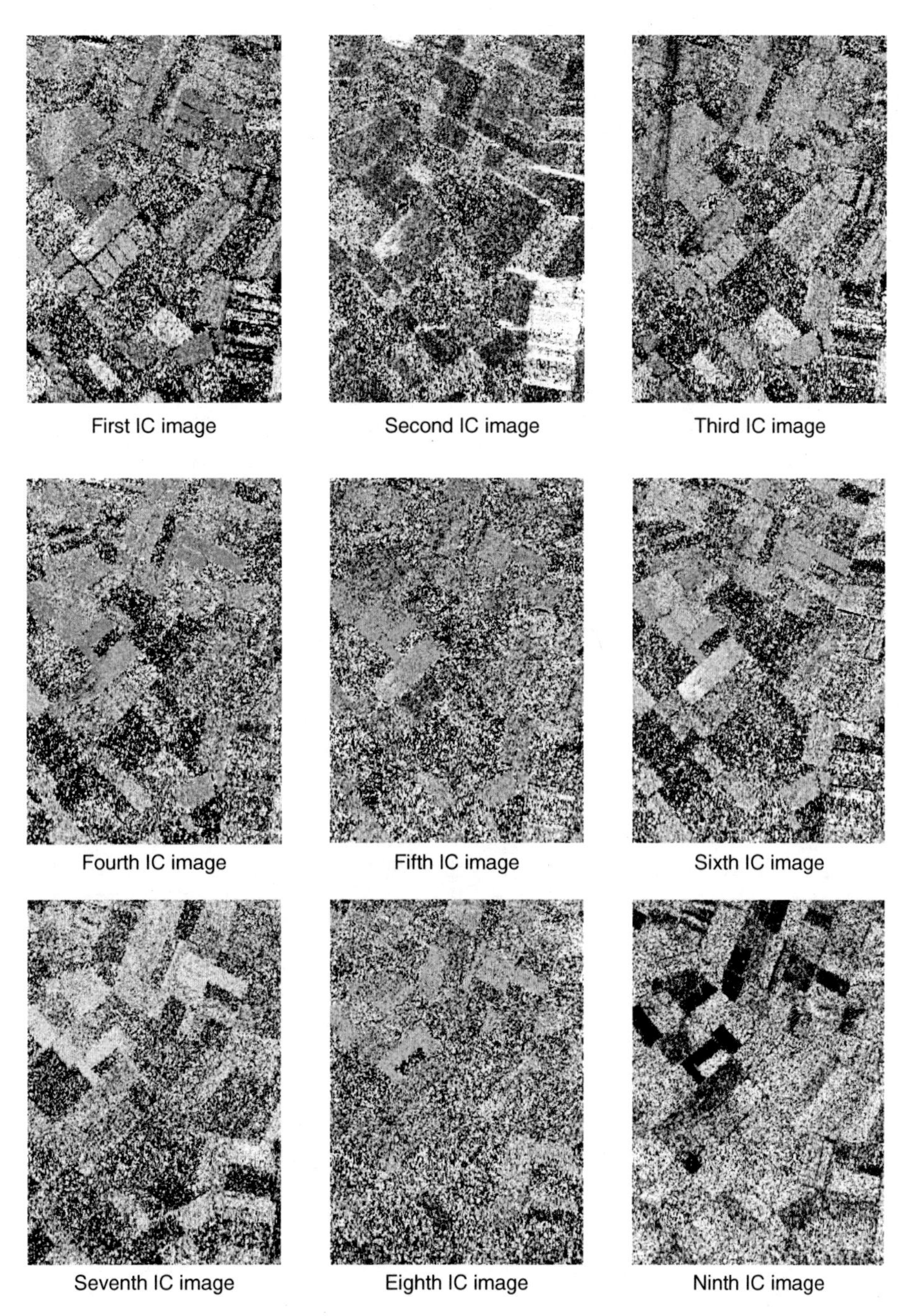

FIGURE 8.9
The results of the FastICA-based method.

TABLE 8.2

Computation Time of FastICA-Based Method and ICA by Ensemble Learning

Method	Computation Time (sec)	Number of Iterations
FastICA-based method	23.92	270
ICA by ensemble learning	2819.53	130

covariance matrix and the observed data covariance matrix is applied to SAR images. The PC images are used as input data of ICA by ensemble learning. The obtained results are satisfactory. The comparative study with FastICA-based method shows that ICA by ensemble learning is a robust technique that has an ability to avoid the local minimal and so reaching the global minimal in contrast to the FastICA-based method, which does not have this ability. However, the drawback of ICA by ensemble learning is the prohibitive computation time compared to that of the FastICA-based method. This can be justified by the fact that ICA by ensemble learning requires many parameter estimations during its learning process. Further investigation is needed to reduce the computational requirement.

References

1. Frost, V.S., Stiles, J.A., Shanmugan, K.S., and Holtzman, J.C., A model for radar images and its application to adaptive digital filtering of multiplicative noise, *IEEE Trans. Pattern Anal. Mach. Intell.*, 4, 157–166, 1982.
2. Lee, J.S., Digital image enhancement and noise filtering by use of local statistics, *IEEE Trans. Pattern Anal. Mach. Intell.*, 2(2), 165–168, 1980.
3. Kuan, D.R., Sawchuk, A.A., Strand, T.C., and Chavel, P., Adaptive noise smoothing filter for images with signal-dependent noise, *IEEE Trans. Pattern Anal. Mach. Intell.*, 7, 165–177, 1985.
4. Zhang, X. and Chen, C.H., Independent component analysis by using joint cumulations and its application to remote sensing images, *J. VLSI Signal Process. Syst.*, 37(2/3), 2004.
5. Malladi, R.K., Speckle filtering of SAR images using Holder Regularity Analysis of the Sparse Code, Master Dissertation of ECE Department, University of Massachusetts, Dartmouth, September 2003.
6. Zong, X., Laine, A.F., and Geiser, E.A., Speckle reduction and contrast enhancement of echocardiograms via multiscale nonlinear processing, *IEEE Trans. Med. Imag.*, 17, 532–540, 1998.
7. Fukuda, S. and Hirosawa, H., Suppression of speckle in synthetic aperture radar images using wavelet, *Int. J. Rem. Sens.*, 19(3), 507–519, 1998.
8. Achim, A., Tsakalides, P., and Bezerianos, A., SAR image denoising via Bayesian, *IEEE Trans. Geosci. Rem. Sens.*, 41(8), 1773–1784, 2003.
9. Hyvarinen, A., Karhunen, J., and Oja, E., *Independent Component Analysis*, Wiley Interscience, New York, 2001.
10. Chitroub, S., PCA–ICA neural network model for POLSAR images analysis, in *Proc. IEEE Int. Conf. Acoustics, Speech Signal Process. (ICASSP'04)*, Montreal, Canada, May 17–21, 2004, pp. 757–760.
11. Lappalainen, H. and Miskin, J., Ensemble learning, in M. Girolami, Ed., *Advances in Independent Component Analysis*, Springer-Verlag, Berlin, 2000, pp. 75–92.
12. Mackay, D.J.C., Developments in probabilistic modeling with neural networks—ensemble learning, in *Proc. 3rd Annu. Symp. Neural Networks*, Springer-Verlag, Berlin, 1995, pp. 191–198.
13. Chitroub, S., Houacine, A., and Sansal, B., Statistical characterisation and modelling of SAR images, *Signal Processing*, 82(1), 69–92, 2002.
14. Chitroub, S., Houacine, A., and Sansal, B., A new PCA-based method for data compression and enhancement of multi-frequency polarimetric SAR imagery, *Intell. Data Anal. Int. J.*, 6(2), 187–207, 2002.
15. Chitroub, S., Houacine, A., and Sansal, B., Neuronal principal component analysis for an optimal representation of multispectral images, *Intell. Data Anal. Int. J.*, 5(5), 385–403, 2001.
16. Lappalainen, H. and Honkela, A., Bayesian non-linear independent component analysis by multilayer perceptrons, in M. Girolami, Ed., *Advances in Independent Component Analysis*, Springer-Verlag, Berlin, 2000, pp. 93–121.

9

Long-Range Dependence Models for the Analysis and Discrimination of Sea-Surface Anomalies in Sea SAR Imagery

Massimo Bertacca, Fabrizio Berizzi, and Enzo Dalle Mese

CONTENTS

9.1 Introduction

In this chapter, by employing long-memory spectral analysis techniques, the discrimination between oil spill and low-wind areas in sea synthetic aperture radar (SAR) images and the simulation of spectral densities of sea SAR images are described. Oil on the sea surface dampens capillary waves, reduces Bragg's electromagnetic backscattering effect and, therefore, generates darker zones in the SAR image. A low surface wind speed, which reduces the amplitudes of all the wave components (not just capillary waves), and the presence of phytoplankton, algae, or natural films can also cause analogous effects.

Some current recognition and classification techniques span from different algorithms for fractal analysis [1] (i.e., spectral algorithms, wavelet, and box-counting algorithms for the estimation of the fractal dimension D) to algorithms for the calculation of the normalized intensity moments (NIM) of the sea SAR image [2]. The problems faced when estimating the value of D include small variations due to oil slick and weak-wind areas and the effect of the edges between two anomaly regions with different physical characteristics. There are also computational problems that arise when the calculation of NIM is related to real (i.e., not simulated) sea SAR images.

In recent years, the analysis of natural clutter in high-resolution SAR images has improved by the utilization of self-similar random process models. Many natural surfaces, like terrain, grass, trees, and also sea surfaces, correspond to SAR precision images (PRI) that exhibit long-term dependence behavior and scale-limited fractal properties. Specifically, the long-term dependence or long-range dependence (LRD) property describes the high-order correlation structure of a process. Suppose that $Y(m,n)$ is a discrete two-dimensional (2D) process whose realizations are digital images. If $Y(m,n)$ exhibits long memory, persistent spatial (linear) dependence exists even between distant observations.

On the contrary, the short memory or short-range dependence (SRD) property describes the low-order correlation structure of a process. If $Y(m,n)$ is a short-memory process, observations separated by a long spatial span are nearly independent.

Among the possible self-similar models, two classes have been used in the literature to describe the spatial correlation properties of the scattering from natural surfaces: fractional Brownian motion (fBm) models and fractionally integrated autoregressive moving average (FARIMA) models. In particular, fBm provides a mathematical framework for the description of scale-invariant random textures and amorphous clutter of natural settings. Datcu [3] used an fBm model for synthesizing SAR imagery. Stewart et al. [4] proposed an analysis technique for natural background clutter in high-resolution SAR imagery. They employed fBm models to discriminate among three clutter types: grass, trees, and radar shadows.

If the fBm model provides a good fit with the periodogram of the data, it means that the power spectral density (PSD), as a function of the frequency, is approximately a straight line with negative slope in a log–log plot.

For particular data sets, the estimated PSD cannot be correctly represented by an fBm model. There are different slopes that characterize the plot of the logarithm of the periodogram versus the logarithm of the frequency. They reveal a greater complexity of the analyzed phenomenon. Therefore, we can utilize FARIMA models that preserve the negative slope of the long-memory data PSD near the origin and, through the so-called SRD functions, modify the shape and the slope of the PSD with increasing frequency.

The SRD part of a FARIMA model is an autoregressive moving average (ARMA) process. Ilow and Leung [5] used the FARIMA model as a texture model for sea SAR images to capture the long-range and short-range spatial dependence structures of some sea SAR images collected by the RADARSAT sensor. Their work was limited to the analysis of isotropic and homogeneous random fields, and only to AR or MA models (ARMA models were not considered). They observed that, for a statistically isotropic and homogeneous field, it is a common practice to derive a 2D model from a one-dimensional (1D) model by replacing the argument K in the PSD of a 1D process, $S(K)$, with $\|K\| = \sqrt{K_x^2 + K_y^2}$ to get the radial PSD: $S(\|K\|)$. When such properties hold, the PSD of the correspondent image can be completely described by using the radial PSD.

Unfortunately, sea SAR images cannot be considered simply in terms of a homogeneous, isotropic, or amorphous clutter. The action of the wind contributes to the anisotropy of the sea surfaces and the particular self-similar behavior of sea surfaces and spectra, correctly described by means of the Weierstrass-like fractal model [6], strongly complicates the self-similar representation of sea SAR imagery.

Bertacca et al. [7,8] extended the work of Ilow and Leung to the analysis of nonisotropic sea surfaces. The authors made use of ARMA processes to model the SRD part of the mean radial PSD (MRPSD) of sea European remote sensing 1 and 2 (ERS-1 and ERS-2) SAR PRI. They utilized a FARIMA analysis technique of the spectral densities to discriminate low-wind from oil slick areas on the sea surface.

A limitation to the applicability of FARIMA models is the high number of parameters required for the ARMA part of the PSD. Using an excessive number of parameters is undesirable because it increases the uncertainty of the statistical inference and the parameters become difficult to interpret. Using fractionally exponential (FEXP) models allows the representation of the logarithm of the SRD part of the long-memory PSD to be obtained, and greatly reduces the number of parameters to be estimated. FEXP models provide the same goodness of fit as FARIMA models at lower computational costs.

We have experimentally determined that three parameters are sufficient to characterize the SRD part of the PSD of sea SAR images corresponding to absence of wind, low surface wind speeds, or to oil slicks (or spills) on the sea surface [9].

The first step in all the methods presented in this chapter is the calculation of the directional spectrum of a sea SAR image by using the 2D periodogram of an $N \times N$ image. To decrease the variance of the spectral estimation, we average spectral estimates obtained from nonoverlapping squared blocks of data. The characterization of isotropic or anisotropic 2D random fields is done first using a rectangular to polar coordinates transformation of the 2D PSD, and then considering, as radial PSD, the average of the radial spectral densities for ϑ ranging from 0 to 2π radians. This estimated MRPSD is finally modeled using a FARIMA or an FEXP model independently of the anisotropy of sea SAR images. As the MRPSD is a 1D signal, we define these techniques as *1D PSD modeling techniques.*

It is observed that sea SAR images, in the presence of a high or moderate wind, do not have statistical isotropic properties [7,8]. In these cases, MRPSD modeling permits discrimination between different sea surface anomalies, but it is not sufficient to completely represent anisotropic and nonhomogeneous fields in the spectral domain. For instance, to characterize the sea wave directional spectrum of a sea surface, we can use its MRPSD together with an apposite spreading function. Spreading functions describe the anisotropy of sea surfaces and depend on the directions of the waves. The assumption of spatial isotropy and nondirectionality for sea SAR images is valid when the sea is calm, as the sea wave energy is spread in all directions and the SAR image PSD shows a circular symmetry. However, with surface wind speeds over 7 m/sec, and, in particular, when the wind and the radar directions are orthogonal [10], the anisotropy of the PSD of sea SAR images starts to be perceptible. Using a 2D model allows the information on the shape of the SAR image PSD to be preserved and provides a better representation of sea SAR images. In this chapter, LRD models are used in addition to the fractal sea surface spectral model [6] to obtain a suitable representation of the spectral densities of sea SAR images. We define this technique as the *2D PSD modeling technique*. These 2D spectral models (FARIMA-fractal or FEXP-fractal models) can be used to simulate sea SAR image spectra in different sea states and wind conditions—and with oil slicks—at a very low computational cost. All the presented methods demonstrated reliable results when applied to ERS-2 SAR PRI and to ERS-2 SAR Ellipsoid Geocoded Images.

9.2 Methods of Estimating the PSD of Images

The problem of spectral estimation can be faced in two ways: applying classical methods, which consist of estimating the spectrum directly from the observed data, or by a parametrical approach, which consists of hypothesizing a model, estimating its parameters from the data, and verifying the validity of the adopted model *a posteriori*.

The classical methods of spectrum estimation are based on the calculation of the Fourier transform of the observed data or of their autocorrelation function [11]. These techniques of estimation ensure good performances in case the available samples are numerous and require the sole hypothesis of stationarity of the observed data. The methods that depend on the choice of a model ensure a better estimation than the ones obtainable with the classical methods in case the available data are less (provided the adopted model is correct).

The classical methods are preferable for the study of SAR images. In these applications, an elevated number of pixels are available and one cannot use models that describe the process of generation of the samples and that turn out to be simple and accurate at the same time.

9.2.1 The Periodogram

This method of estimation, in the 1D case, requires the calculation of the Fourier transform of the sequence of the observed data. When working with bidimensional stochastic processes, whose sample functions are images [12], in place of a sequence $x[n]$, we consider a data matrix $x[m, n]$, $m = 0, 1, \ldots, (M-1)$, $n = 0, 1, \ldots, (N-1)$. In these cases, one uses the bidimensional version of the periodogram as defined by the equation

$$\hat{P}_{\mathrm{PER}}(f_1, f_2) = \frac{1}{MN}\left|\sum_{m=0}^{M-1}\sum_{n=0}^{N-1} x[m,n]e^{-j2\pi(f_1 m + f_2 n)}\right|^2 \tag{9.1}$$

Observing that

$$X(f_1, f_2) = \sum_{m=0}^{M-1}\sum_{n=0}^{N-1} x[m,n]e^{-j2\pi(f_1 m + f_2 n)} \tag{9.2}$$

is the Fourier bidimensional discrete transform of the data sequence, Equation (9.1), can thus be rewritten as

$$\hat{P}_{\mathrm{PER}}(f_1, f_2) = \frac{1}{MN}|X(f_1, f_2)|^2 \tag{9.3}$$

It can be demonstrated that the estimator in the above equation is not only asymptotically unbiased (the average value of the estimator tends, at the limit for $N \to \infty$ and $M \to \infty$, to the PSD of the data) but also inconsistent (the variance of the estimator does not tend to zero in the limit for $N \to \infty$ and $M \to \infty$). The technique of estimation through the periodogram remains, however, of great practical interest: from Equation 9.3, we perceive the computational simplicity achievable through an implementation of the calculation of the fast Fourier transform (FFT).

9.2.2 Bartlett Method: Average of the Periodograms

A simple strategy adopted to reduce the variance of the estimator (Equation 9.3) consists of calculating the average of several independent estimations. As the variance of the estimator does not decrease with the increasing of the dimensions of the matrix of the data, one can subdivide this matrix in disconnected subsets, calculate the periodogram of each subset and execute the average of all the periodograms. Figure 9.1 shows an image of $N \times M$ pixels (a matrix of $N \times M$ elements) subdivided into K^2 subwindows that are not superimposed by each of the $R \times S$ elements

$$x_{l_x l_y}[m, n] = x[m + l_x K, n + l_y K] \quad \begin{cases} m = 0, 1, \ldots, (R-1) \\ n = 0, 1, \ldots, (S-1) \end{cases} \tag{9.4}$$

with

$$\begin{matrix} M = R \times K \\ N = S \times K, \end{matrix} \begin{cases} l_x = 0, 1, \ldots, (K-1) \\ l_y = 0, 1, \ldots, (K-1) \end{cases} \tag{9.5}$$

The estimation according to Bartlett's procedure gives

$$\hat{P}_{\mathrm{BART}}(f_1, f_2) = \frac{1}{K^2} \sum_{l_x=0}^{K-1} \sum_{l_y=0}^{K-1} \hat{P}_{\mathrm{PER}}^{(l_x l_y)}(f_1, f_2) \tag{9.6}$$

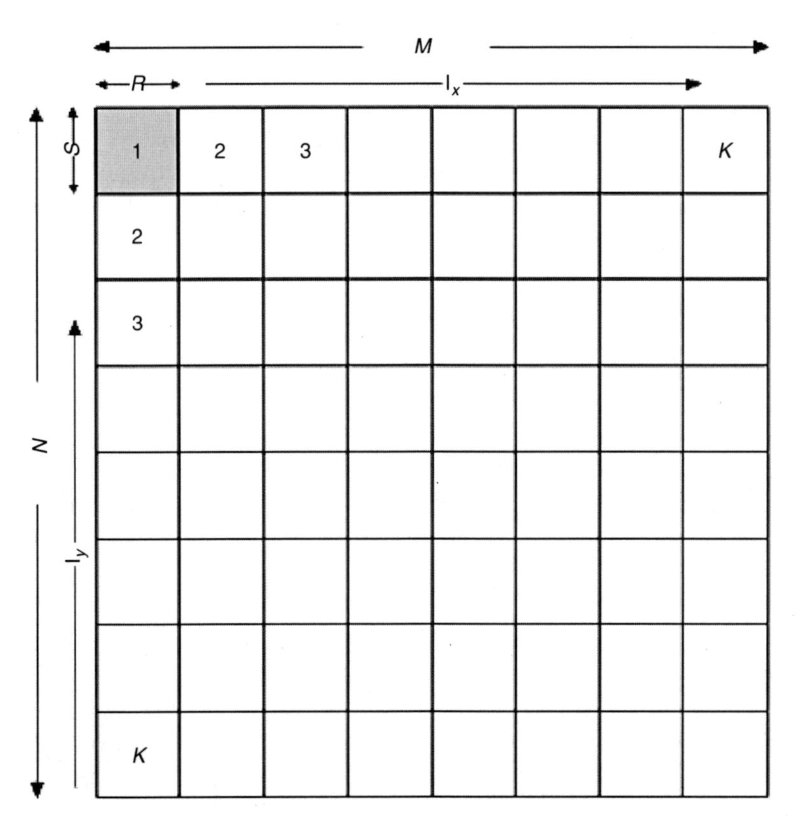

FIGURE 9.1
Calculation of the periodogram in an image (a bidimensional sequence) with Bartlett's method.

In the above equation, $\hat{P}_{\mathrm{PER}}^{(l_x l_y)}(f_1, f_2)$ represents the periodogram calculated on the subwindows identified in the couple (l_x, l_y) and defined by the equation

$$\hat{P}_{\mathrm{PER}}^{(l_x l_y)}(f_1, f_2) = \frac{1}{RS}\left|\sum_{m=0}^{R-1}\sum_{n=0}^{S-1} x_{l_x l_y}[m, n]\mathrm{e}^{-j2\pi(f_1 m + f_2 n)}\right|^2 \tag{9.7}$$

A reduction in the dimensions of the windows containing the data that are analyzed corresponds to a loss of resolution for the estimated spectrum. Consider the average value of the estimator modified according to Bartlett:

$$E\left\{\hat{P}_{\mathrm{BART}}(f_1, f_2)\right\} = \frac{1}{K^2}\sum_{l_x=0}^{K-1}\sum_{l_y=0}^{K-1} E\left\{\hat{P}_{\mathrm{PER}}^{(l_x l_y)}(f_1, f_2)\right\} = E\left\{\hat{P}_{\mathrm{PER}}^{(l_x l_y)}(f_1, f_2)\right\} \tag{9.8}$$

From the above equation, we obtain the equation

$$\mathrm{E}\left\{\hat{P}_{\mathrm{BART}}(f_1, f_2)\right\} = W_B(f_1, f_2) \otimes \otimes S_x(f_1, f_2) \tag{9.9}$$

From the above equation, we have that the average value of the estimator is the result of the double periodical convolution between the function $W_B(f_1, f_2)$ and the spectral density of power $S_x(f_1, f_2)$ relative to the data matrix. Equation 9.6 thus defines a biased estimator. By a direct extension of the 1D theory of the spectral estimate, it is possible to interpret the function $W_B(f_1, f_2)$ as a 2D Fourier transform of the window

$$w(k, l) = \begin{cases} \left(\frac{1-|k|}{R}\right)\left(1 - \frac{|l|}{S}\right) & \text{if } \begin{cases} |k| \leq R \\ |l| \leq S \end{cases} \\ 0 & \text{otherwise} \end{cases} \tag{9.10}$$

Given that the window (Equation 9.10) is separable, the Fourier transform is the product of the 1D transforms:

$$W_B(f_1, f_2) = \frac{1}{R}\frac{1}{S}\left(\frac{\sin(\pi f_1 R)}{\sin(\pi f_1)}\right)^2\left(\frac{\sin(\pi f_2 S)}{\sin(\pi f_2)}\right)^2 \tag{9.11}$$

The application of Bartlett's method determines the smearing of the estimated spectrum. Such a phenomenon is scarcely relevant only if $W_B(f_1, f_2)$ is very narrow in relation to $X(f_1, f_2)$, that is, if the window used is sufficiently long. For example, for $R = S = 256$, we have that the band at -3 dB of the principal lobe of $W_B(f_1, f_2)$ is equal to around $\frac{1}{R} = \frac{1}{256}$. The resolution in frequency is equal to this value. Bartlett's method permits a reduction in the variance of the estimator by a factor proportional to K^2 [12]:

$$\mathrm{var}\left\{\hat{P}_{\mathrm{BART}}(f_1, f_2)\right\} = \frac{1}{K^2}\mathrm{var}\left\{\hat{P}_{\mathrm{PER}}^{(l)}(f_1, f_2)\right\} \tag{9.12}$$

Such a result is correct only if the various periodograms are independent estimates. When the subsets of data are chosen as the contiguous blocks of the same realization, as shown in Figure 9.1, the windows of the data do not turn out to be uncorrelated among themselves, and a reduction in the variance by a factor inferior to K^2 must be accepted.

In conclusion, the periodogram is a nonconsistent and asymptotically unbiased estimator of the PSD of a bidimensional sequence. The bias of the estimator can be mathematically described in the form of a double convolution of the true spectrum with a spectral window of the type $W_{Btot}(f_1, f_2) = \frac{1}{M}\frac{1}{N}\left(\frac{\sin(\pi f_1 M)}{\sin(\pi f_1)}\right)^2\left(\frac{\sin(\pi f_2 N)}{\sin(\pi f_2)}\right)^2$, where with M and N we indicate the dimensions of the bidimensional sequence considered. If we carry out an average of the periodograms calculated on several adjacent subsequences (Bartlett's method), each of $R \times S$ samples, as in Figure 9.1, the bias of the estimator can still be represented as the double convolution (see Equation 9.9) of the true spectrum with the spectral window (Equation 9.11):

$$W_B(f_1, f_2) = \frac{1}{R}\frac{1}{S}\left(\frac{\sin(\pi f_1 R)}{\sin(\pi f_1)}\right)^2\left(\frac{\sin(\pi f_2 S)}{\sin(\pi f_2)}\right)^2, \text{ where } R = \frac{M}{K}, S = \frac{N}{K} \tag{9.13}$$

The bias of the estimator $\hat{P}_{BART}(f_1, f_2)$ is greater than that of $\hat{P}_{PER}(f_1, f_2)$, because of the greater width of the principal lobe of the corresponding spectral window. The bias can hence be interpreted in relation to its effects on the resolution of the spectrum. For a fixed dimension of the sequence to be analyzed, M rows and N columns, the variance of the estimator diminishes with the increase in the number of the periodograms, but R and S also diminish and thus the resolution of the estimated spectrum.

Therefore, in Bartlett's method, a compromise needs to be reached between the bias or resolution of the spectrum on one side and the variance of the estimator on the other. The actual choice of the parameters M, N, R, and S in a real situation is orientated by the *a priori* knowledge of the signal to be analyzed. For example, if we know that the spectrum has a very narrow peak and if it is important to resolve it, we must choose sufficiently large R and S values to obtain the desired resolution in frequency. It is then necessary to use a pair of sufficiently high M and N values to obtain a conveniently low variance for Bartlett's estimator.

9.3 Self-Similar Stochastic Processes

In this section, we recall the definitions of self-similar and long-memory stochastic processes.

Definition 1: *Let $Y(u)$, $u \in R$ be a continuous random process. It is called self-similar with self-similarity parameter H, if for any positive constant β, the following relation holds:*

$$\beta^{-H} Y(\beta u) \stackrel{d}{=} Y(u) \tag{9.14}$$

In Definition 1, $\beta^{-H}\ Y(\beta u)$ is called *rescaled process* with scale factor β, and $\stackrel{d}{=}$ means the equality of all the finite dimensional probability distributions.

For any sequence of points $u_1, \ldots, u_N$ and all $\beta > 0$, we have that $\beta^{-H}\ [Y(\beta u_1), \ldots, Y(\beta u_N)]$ has the same distribution as $[Y(u_1), \ldots, Y(u_N)]$

In this chapter, we analyze the correlation and PSD structure of discrete stochastic processes. Then we can impose a condition on the correlation of stochastic stationary processes introducing the definition of second-order self-similarity.

Let $Y(n)$, $n \in N$ be a covariance stationary discrete random process with mean $\eta = E\{Y(n)\}$, variance σ^2, and autocorrelation function $R(m) = E\{Y(n+m)Y(n)\}$, $m \geq 0$.

The spectral density of the process is defined as [13]:

$$S(K) = \frac{\sigma^2}{2\pi} \sum_{m=-\infty}^{\infty} R(m) e^{imK} \tag{9.15}$$

Assume that [14,15]

$$\lim_{m \to \infty} R(m) = c_R \cdot m^{-\gamma} \tag{9.16}$$

where c_R is a constant and $0 < \gamma < 1$. For each $l = 1, 2, \ldots$, we indicate with the symbols $Y^l(n) = \{Y^l(n),\ n = 1, 2, \ldots\}$ the series obtained by averaging $Y(n)$ over nonoverlapping blocks of size l:

$$Y^l(n) = \frac{Y[(n-1)l] \cdots + Y[(nl-1)]}{l}, \quad n > 1 \tag{9.17}$$

Definition 2: *A process is called exactly second-order self-similar with self-similarity parameter* $H = 1 - \frac{\gamma}{2}$ *if, for each* $l = 1, 2, \ldots$, *we have that*

$$\begin{aligned} &\mathrm{Var}\{Y^l\} = \sigma^2 l^{-\gamma} \\ &R^l(m) = R(m) = \frac{1}{2}\left[(m+1)^{2H} - 2m^{2H} + |m-1|^{2H}\right], \quad m \geq 0 \end{aligned} \tag{9.18}$$

where $R^l(m)$ *denotes the autocorrelation function of* $Y^l(n)$

Definition 3: *A process is called asymptotically second-order self-similar with self-similarity parameter* $H = 1 - \frac{\gamma}{2}$ *if*

$$\lim_{l \to \infty} R^l(m) = R(m) \tag{9.19}$$

Thus, if the autocorrelation functions of the processes $Y^l(n)$ are the same as or become indistinguishable from $R(m)$ as $m \to \infty$, the covariance stationary discrete process $Y(n)$ is second-order self-similar.

Lamperti [16] demonstrated that self-similarity is produced as a consequence of limit theorems for sums of stochastic variables.

Definition 4: *Let* $Y(u)$, $u \in R$ *be a continuous random process. Suppose that for any* $n \geq 1$, $\nu \in R$ *and any n points* $(u_1, \ldots, u_n)$, *the random vectors* $\{Y(u_1 + \nu) - Y(u_1 + \nu - 1), \ldots, Y(u_n + \nu) - Y(u_n + \nu - 1)\}$ *show the same distribution. Then the process* $Y(u)$ *has stationary increments.*

Theorem 1: *Let* $Y(u)$, $u \in R$ *be a continuous random process. Suppose that:*

1. $P\{Y(1) \neq 0\} > 0$
2. $X_1, X_2, \cdots$ *is a stationary sequence of stochastic variables*
3. $b_1, b_2, \cdots$ *are real, positive normalizing constants for which* $\lim_{n \to \infty} \{\log(b_n)\} = \infty$
4. $Y(u)$ *is the limit in distribution of the sequence of the normalized partial sums*

$$b_n^{-1} S_{nu} = b_n^{-1} \sum_{j=1}^{\lfloor nu \rfloor} X_j, \quad n = 1, 2, \ldots \tag{9.20}$$

Then, for each $t > 0$ *there exists an* $H > 0$ *such that*

1. $Y(u)$ *is self-similar with self-similarity parameter* H
2. $Y(u)$ *has stationary increments*

Furthermore, all self-similar processes with stationary increments and $H > 0$ can be obtained as sequences of normalized partial sums.

Let $Y(u)$, $u \in R$ be a continuous self-similar random process with self-similarity parameter H such that

$$Y(u) \stackrel{d}{=} u^H Y(1) \tag{9.21}$$

for any $u > 0$.

Then, indicating with $\stackrel{d}{\rightarrow}$ the convergence in distribution, we have the following behavior of $Y(u)$ as u tends to infinity [17]:

- If $H < 0$, then $Y(u) \stackrel{d}{\rightarrow} 0$
- If $H = 0$, then $Y(u) \stackrel{d}{=} Y(1)$
- If $H > 0$ and $Y(u) \neq 0$, then $|Y(u)| \stackrel{d}{\rightarrow} \infty$

If u tends to zero, we have the following:

- If $H < 0$ and $Y(u) \neq 0$, then $|Y(u)| \stackrel{d}{\rightarrow} \infty$
- If $H = 0$, then $Y(u) \stackrel{d}{=} Y(1)$
- If $H > 0$, then $Y(u) \stackrel{d}{\rightarrow} 0$

We notice that:

- $Y(u)$ is not stationary unless $Y(u) \equiv 0$ or $H = 0$
- If $H = 0$, then $P\{Y(u) = Y(1)\} = 1$ for any $u > 0$
- If $H < 0$, then $Y(u)$ is not a measurable process unless $P\{Y(u) = Y(1) = 0\} = 1$ for any $u > 0$ [18]
- As stationary data models, we use self-similar processes, $Y(u)$, with stationary increments, self-similarity parameter $H > 0$ and $P\{Y(0) = 0\} = 1$

9.3.1 Covariance and Correlation Functions for Self-Similar Processes with Stationary Increments

Let $Y(u)$, $u \in R$ be a continuous self-similar random process with self-similarity parameter H. Assume that $Y(u)$ has stationary increments and that $E\{Y(u)\} = 0$. Indicate with $\sigma^2 = E\{Y(u) - Y(u-1)\} = E\{Y^2(u)\}$ the variance of the stationary increment process $X(u)$ with $X(u) = Y(u) - Y(u-1)$. We have that

$$E\left\{[Y(u) - Y(v)]^2\right\} = E\left\{[Y(u-v) - Y(0)]^2\right\} = \sigma^2 (u-v)^{2H} \tag{9.22}$$

where $u, v \in R$ and $u > v$.

In addition

$$\begin{aligned} E\left\{[Y(u) - Y(v)]^2\right\} &= E\left\{[Y(u)]^2\right\} + E\left\{[Y(v)]^2\right\} - 2E\{Y(u)Y(v)\} = \\ &= \sigma^2 u^{2H} + \sigma^2 v^{2H} - 2C_Y(u,v) \end{aligned} \tag{9.23}$$

where $C_Y(u, v)$ denotes the covariance function of the nonstationary process $Y(u)$.

Thus, we obtain that

$$C_Y(u,v) = \frac{1}{2}\sigma^2\left[u^{2H} - (u-v)^{2H} + v^{2H}\right] \tag{9.24}$$

The covariance function of the stationary increment sequence $X(j) = Y(j) - Y(j-1), j = 1, 2, \ldots$ is

$$\begin{aligned} C(m) &= Cov\{X(j), X(j+m)\} = Cov\{X(1), X(1+m)\} = \\ &= \frac{1}{2}E\left\{\left[\sum_{p=1}^{m+1} X(p)\right]^2 + \left[\sum_{p=2}^{m} X(p)\right]^2 - \left[\sum_{p=1}^{m} X(p)\right]^2 - \left[\sum_{p=2}^{m+1} X(p)\right]^2\right\} \\ &= \frac{1}{2}\left\{E\left\{[Y(1+m) - Y(0)]^2\right\} + E\left\{[Y(m-1) - Y(0)]^2\right\} - 2E\left\{[Y(m) - Y(0)]^2\right\}\right\} \end{aligned} \tag{9.25}$$

After some algebra, we obtain

$$\begin{aligned} C(m) &= \frac{1}{2}\sigma^2\left[(m+1)^{2H} - 2m^{2H} + (m-1)^{2H}\right], \quad m \geq 0 \\ C(m) &= C(-m), \quad m < 0 \end{aligned} \tag{9.26}$$

Then, the correlation function, $R(m) = \frac{C(m)}{\sigma^2}$, is

$$\begin{aligned} R(m) &= \frac{1}{2}\left[(m+1)^{2H} - 2m^{2H} + (m-1)^{2H}\right], \quad m \geq 0 \\ R(m) &= R(-m), \quad m < 0 \end{aligned} \tag{9.27}$$

If $0 < H < 1$ and $H \neq 0$, we have that [19]

$$\lim_{m\to\infty}\{R(m)\} = H(2H-1)m^{2H-2} \tag{9.28}$$

We notice that:

- If $\frac{1}{2} < H < 1$, then the correlations decay very slowly and sum to infinity: $\sum_{m=-\infty}^{\infty} R(m) = \infty$. The process has long memory (it has LRD behavior).
- If $H = \frac{1}{2}$, then the observations are uncorrelated: $R(m) = 0$ for each m.
- If $0 < H < \frac{1}{2}$, then the correlations sum to zero: $\sum_{m=-\infty}^{\infty} R(m) = 0$. In reality, the last condition is very unstable. In the presence of arbitrarily small disturbances [19], the series sums to a finite number: $\sum_{m=-\infty}^{\infty} R(m) = c, c \neq 0$.
- If $H = 1$, from Equation 9.7 we obtain

$$Y(m) \stackrel{d}{=} u^H Y(1) \stackrel{d}{=} Y(m), \quad m = 1, 2, \ldots \tag{9.29}$$

and $R(m) = 1$ for each m.

- If $H > 1$, then $R(m)$ can become greater than 1 or less than -1 when m tends to infinity.

The last two points, corresponding to $H \geq 1$, are not of any importance. Thus, if correlations exist and $\lim_{m \to \infty} \{R(m)\} = 0$, then $0 < H < 1$.

We can conclude by observing that:

- A self-similar process for which Equation 9.7 holds is nonstationary.
- Stationary data can be modeled using self-similar processes with stationary increments.
- Analyzing the autocorrelation function of the stationary increment process, we obtain

  – If $\frac{1}{2} < H < 1$, then the increment process has long memory (LRD)
  – If $H = \frac{1}{2}$, then the observations are uncorrelated
  – If $0 < H < \frac{1}{2}$, then the process has short memory and its correlations sum to zero

9.3.2 Power Spectral Density of Self-Similar Processes with Stationary Increments

Let $Y(u)$ be a self-similar process with stationary increments, finite second-order moments, $0 < H < 1$ and $\lim_{m \to \infty} R(m) = 0$.

Under these hypotheses, the PSD of the stationary increment sequence $X(j)$ is [20]:

$$S(k) = 2c_S[1 - \cos(k)] \sum_{p=-\infty}^{\infty} |2\pi p + k|^{-2H-1}, \quad k \in [-\pi, \pi] \tag{9.30}$$

where $c_S = \frac{\sigma^2}{2\pi} \sin(\pi H)\Gamma(2H + 1)$ and $\sigma^2 = \text{Var}\,\{X(j)\}$.

Calculating the Taylor expansion of $S(K)$ in zero, we obtain

$$S(K) = c_S|k|^{1-2H} + o\left(|k|^{\min(3-2H,2)}\right) \tag{9.31}$$

We note that if $\frac{1}{2} < H < 1$, then the logarithm of the PSD, $\log[S(k)]$, plotted against the logarithm of frequency, $\log(k)$, diverges when k tends to zero. In other words, the PSD of long-memory data tends to infinity at the origin.

9.4 Long-Memory Stochastic Processes

Intuitively, long-memory or LRD can be considered as a phenomenon in which current observations are strongly correlated to observations that are far away in time or space.

In Section 9.3, the concept of self-similar LRD processes was introduced and was shown to be related to the shape of the autocorrelation function of the stationary increment sequence $X(j)$. If the correlations $R(m)$ decay asymptotically as a hyperbolic function, their sum over all lags diverges and the self-similar process exhibits an LRD behavior.

For the correlations and the PSD of a stationary LRD process, the following properties hold [19]:

- The correlations $R(m)$ are asymptotically equal to $c_R|m|^{-\delta}$ for some $0 < \delta < 1$.

- The PSD $S(K)$ has a pole at zero that is equal to a constant $c_S k^{-\beta}$ for some $0 < \beta < 1$.
- Near the origin, the logarithm of the periodogram $I(k)$ plotted versus the logarithm of the frequency is randomly scattered around a straight line with negative slope.

Definition 5: *Let $X(v)$, $v \in R$ be a continuous stationary random process. Assume that there exists a real number $0 < \delta < 1$ and a constant c_R such that*

$$\lim_{m \to \infty} \frac{R(m)}{c_R m^{-\delta}} = 1 \tag{9.32}$$

Then $X(v)$ is called a stationary process with long-memory or LRD

In Equation 9.32, the Hurst parameter $H = 1 - \dfrac{\delta}{2}$ is often used instead of δ.

On the contrary, stationary processes with exponentially decaying correlations

$$R(m) \leq cb^m \tag{9.33}$$

where c and b are real constants and $0 < c < \infty$, $0 < b < 1$ are called stationary processes with short-memory or SRD.

We can also define LRD by imposing a condition on the PSD of a stationary process.

Definition 6: *Let $X(v)$, $v \in R$ be a continuous stationary random process. Assume that there exists a real number $0 < \beta < 1$ and a constant c_S such that*

$$\lim_{k \to \infty} \frac{S(k)}{c_S |k|^{-\beta}} = 1 \tag{9.34}$$

Then $X(v)$ is called a stationary process with long-memory or LRD.

Such spectra occur frequently in engineering, geophysics, and physics [21,22]. In particular, studies on sea spectra using long-memory processes have been carried out by Sarpkaya and Isaacson [23] and Bretschneider [24].

We notice that the definition of LRD by Equation 9.33 or Equation 9.34 is an asymptotic definition. It depends on the behavior of the spectral density as the frequency tends to zero and behavior of the correlations as the lag tends to infinity.

9.5 Long-Memory Stochastic Fractal Models

Examples of LRD processes include fractional Gaussian noise (fGn), FARIMA, and FEXP.

fGn is the stationary first-order increment of the well-known fractionally fBm model. fBm was defined by Kolmogorov and studied by Mandelbrot and Van Ness [25]. It is a Gaussian, zero mean, nonstationary self-similar process with stationary increments. In one dimension, it is the only self-similar Gaussian process with stationary increments. Its covariance function is given by Equation 9.24.

A particular case of an fBm model is the Wiener process (Brownian motion). It is a zero-mean Gaussian process whose covariance function is equal to Equation 9.24 with $H = \frac{1}{2}$ (the observations are uncorrelated). In fact, one of the most important properties of Brownian motion is the independence of its increments.

As fBm is nonstationary, its PSD cannot be defined. Therefore, we can study the characteristics of the process by analyzing the autocorrelation function and the PSD of the fGn process.

Fractional Gaussian noise is a Gaussian, null mean, and stationary discrete process. Its autocorrelation function is given by Equation 9.27) and is proportional to $|m|^{2H-2}$ as m tends to infinity (Equation 9.28). Therefore, the discrete process exhibits SRD for $0 < H < \frac{1}{2}$, independence for $H = \frac{1}{2}$ (Brownian motion), and LRD for $\frac{1}{2} < H < 1$.

As all second-order self-similar processes with stationary increments have the same second-order statistics as fBm, their increments have all the same correlation functions as fGn. FGn processes can be completely specified by three parameters: mean, variance, and the Hurst parameter.

Some data sets in diverse fields of statistical applications, such as hydrology, broadband network traffic, and sea SAR images analysis, can exhibit a complex mixture of SRD and LRD. It means that the corresponding autocorrelation function behaves similar to that of LRD processes at large lags, and to that of SRD processes at small lags [8,26]. Models such as fGn can capture LRD but not SRD behavior. In these cases, we can use models specifically developed to characterize both LRD and SRD, like FARIMA and FEXP.

9.5.1 FARIMA Models

FARIMA models can be introduced as an extension of the classic ARMA and ARIMA models.

To simplify the notation, we consider only zero-mean stochastic processes. An ARMA(p, q) process model is a stationary discrete random process. It is defined to be the stationary solution of

$$\Phi(B)Y(n) = \Psi(B)W(n) \tag{9.35}$$

where:

- B denotes the backshift operator defined by

$$\begin{aligned} &Y(n) - Y(n-1) = (1-B)Y(n) \\ &(Y(n) - Y(n-1)) - (Y(n-1) - Y(n-2)) = (1-B)^2 Y(n) \end{aligned} \tag{9.36}$$

- $\Phi(x)$ and $\Psi(x)$ are polynomials of order p and q, respectively:

$$\Phi(x) = 1 - \sum_{m=1}^{p} \Phi_m x^m$$

$$\Psi(x) = 1 - \sum_{m=1}^{q} \Psi_m x^m$$

- It is assumed that all solutions of $\Phi(x) = 0$ and $\Psi(x) = 0$ are outside the unit circle.
- $W(n)$, $n = 1, 2, \ldots$ are i.i.d. Gaussian random variables with zero mean and variance σ_W^2.

An ARIMA process is the stationary solution of

$$\Phi(B)(1-B)^d Y(n) = \Psi(B)W(n) \tag{9.37}$$

where d is an integer.

If $d \geq 0$, then $(1 - B)^d$ is given by

$$(1 - B)^d = \sum_{m=0}^{d} \binom{d}{m} (-1)^m B^m \tag{9.38}$$

with the binomial coefficients

$$\binom{d}{m} = \frac{d!}{m!(d-m)!} = \frac{\Gamma(d+1)}{\Gamma(m+1)\Gamma(d-m+1)} \tag{9.39}$$

As the gamma function $\Gamma(x)$ is defined for all real numbers, the definition of binomial coefficients can be extended to all real numbers d.

The extended definition is given by

$$(1 - B)^d = \sum_{m=0}^{\infty} \binom{d}{m} (-1)^m B^m \tag{9.40}$$

In Equation 9.26, if d is an integer, all the terms for $m > d$ are zero. On the contrary, if d is a real number, we have a summation over an infinite number of indices.

Definition 7: *Let $Y(n)$ be a discrete stationary stochastic process. If it is the solution of*

$$\Phi(B)(1 - B)^d Y(n) = \Psi(B)W(n) \tag{9.41}$$

for some $-\frac{1}{2} < d < \frac{1}{2}$, then $Y(n)$ is a FARIMA(p, d, q) process [27,28].

LRD occurs for $0 < d < \frac{1}{2}$, and for $d \geq \frac{1}{2}$ the process is not stationary. The coefficient d is called the *fractional differencing parameter*.

There are four special cases of a FARIMA(p, d, q) model:

- Fractional differencing (FD) $=$ FARIMA$(0, d, 0)$
- Fractionally autoregressive (FAR) $=$ FARIMA$(p, d, 0)$
- Fractionally moving average (FMA) $=$ FARIMA$(0, d, q)$
- FARIMA(p, d, q)

In Ref. [5], Ilow and Leung used FMA and FAR models to represent some sea SAR images collected by the RADARSAT sensor as 2D isotropic and homogeneous random fields. Bertacca et al. extended Ilow's approach to characterize nonhomogeneous high-resolution sea SAR images [7,8]. In these papers, the modeling of the SRD part of the PSD model (MA, AR, or ARMA) required from 8 to more than 30 parameters.

FARIMA(p, d, q) has $p + q + 3$ parameters, it is much more flexible than fGn in terms of the simultaneous modeling of both LRD and SRD, but it is known to require a large number of model parameters and to be not computationally efficient [26]. Using FEXP models, Bertacca et al. defined a simplified analysis technique of sea SAR images PSD [9].

9.5.2 FEXP Models

FEXP models were introduced by Beran in Ref. [29] to reduce the numerical complexity of Whittle's approximate maximum likelihood estimator (time domain MLE estimation of long memory) [30] principally for large sample size, and to decrease the CPU times required for the approximate frequency domain MLE of long-memory, in particular, for high-dimensional parameter vectors to estimate. Operating with FEXP models leads to the estimation of the parameters in a generalized linear model. This methodology permits the valuation of the whole vector of parameters independently of their particular character (i.e., LRD or SRD parameters). In a generalized linear model, we observe a random response y with mean μ and distribution function F [31]. The mean μ can be expressed as

$$g(\mu) = \eta_0 + \eta_1 v_1 + \cdots + \eta_n v_n \tag{9.42}$$

where $v_1, v_2, \ldots, v_n$ are called explanatory variables and are related to μ through the link function $g(\mu)$.

Let $X(l)$, $l = 1, 2, \ldots, n$ be samples of a stationary sequence at points $l = 1, 2, \ldots, n$. The periodogram ordinates, $I(k_{j,n})$, $j = 1, 2, \ldots, n$, are given by [19]:

$$I(k_{j,n}) = \frac{1}{2\pi n}\left|\sum_{l=1}^{n}\left[X(l) - \overline{X}_n\right]e^{ilk_{j,n}}\right|^2 = \frac{1}{2\pi}\sum_{m=-(n-1)}^{n-1}\hat{C}_n(m)e^{imk_{j,n}} \tag{9.43}$$

where $k_{j,n} = \frac{2\pi j}{n}$, $j = 1, 2, \ldots, n^*$ are the Fourier frequencies, $n^* = \left\lfloor \frac{n-1}{2} \right\rfloor$, $\lfloor x \rfloor$ denotes the integer part of x, $\overline{X}_n$ is the sample mean, and $\hat{C}_n(m)$ are the sample covariances:

$$\hat{C}_n(m) = \frac{1}{n}\sum_{l=1}^{n-|m|}\left[X(l) - \overline{X}_n\right]\left[X(l + |m|) - \overline{X}_n\right] \tag{9.44}$$

It is known that the periodogram $I(k)$, calculated on an n-size data vector, is an asymptotically unbiased estimate of the power spectral density $S(k)$:

$$\lim_{n\to\infty} E\{I(k)\} = S(k) \tag{9.45}$$

Further, for short-memory processes and a finite number of frequencies $k_1, \ldots, k_N \in [0, \pi]$, the corresponding periodogram ordinates $I(k_1), \ldots, I(k_N)$ are approximately independent exponential random variables with means $S(k_1), \ldots, S(k_N)$. For long-memory processes, this result continues to be valid under certain mild regularity conditions [32]. The periodogram of the data is usually calculated at the Fourier frequencies. Thus, the samples $I(k_{j,n})$ are independent exponential random variables with means $S(k_{j,n})$.

When we estimate the MRPSD, we can employ the central limit theorem and consider the mean radial periodogram ordinates $I_{mr}(k_{j,n})$ as approximately independent Gaussian random variables with means $S_{mr}(k_{j,n})$ (the MRPSD at the frequencies $k_{j,n}$).

Assume that

$$y_{j,n} = I_{mr}(k_{j,n}) \tag{9.46}$$

Then the expected value of $y_{j,n}$ is given by

$$\mu = S_{mr}(k_{j,n}) \tag{9.47}$$

Suppose that there exists a link function

$$\nu(\mu) = \eta_1 f_1(k) + \eta_2 f_2(k) + \cdots + \eta_M f_M(k) \tag{9.48}$$

Equation 9.48 defines a generalized linear model where y is the vector of the mean radial periodogram ordinates with a Gaussian distribution F, the functions $f_i(k)$, $i = 1,\ldots,M$ are called *explanatory variables*, and ν is the link function.

It is observed that, near the origin, the spectral densities of long-memory processes are proportional to

$$k^{-2d} = e^{-2d\log(k)} \tag{9.49}$$

so that a convenient choice of the link function is

$$\nu(\mu) = \log(\mu) \tag{9.50}$$

Definition 8: *Let* $r(k)$*:*$[-\pi, \pi] \rightarrow R^+$ *be an even function* [29] *for which* $\lim_{k\rightarrow 0} \frac{r(k)}{k} = 1$.

Define $z_0 = 1$. *Let* $z_1,\ldots, z_q$ *be smooth even functions in* $[-\pi,\pi]$ *such that the* $n^* \times (q+1)$ *matrix A, with column vectors* $\left[z_l\left(\frac{2\pi}{n}\right), z_l\left(\frac{4\pi}{n}\right), z_l\left(\frac{6\pi}{n}\right),\ldots, z_l\left(\frac{2n^*\pi}{n}\right)\right]^{T}$, $l = 1,\ldots, q$, *is nonsingular for any n. Define a real vector* $\phi = [\eta_0, H, \eta_1,\ldots, \eta_q]$ *with* $\frac{1}{2} \leq H < 1$.

A stationary discrete process $Y(m)$ *whose spectral density is given by*

$$S(k;\phi) = r(k)^{1-2H}\exp\left\{\sum_{l=0}^{q}\eta_l z_l(k)\right\} \tag{9.51}$$

is an FEXP process. The functions $z_1,\ldots, z_q$ *are called short-memory components, whereas r is the long-memory component of the process.*

We can choose different sets of short-memory components. Each set $z_1,\ldots, z_q$ corresponds to a class of FEXP models. Beran observed that two classes of FEXP models are particularly convenient. These classes are characterized by the same LRD component

$$r(k) = |1 - e^{ik}| = \left|2\sin\left(\frac{k}{2}\right)\right| \tag{9.52}$$

If we define the short-memory functions as

$$z_l(k) = \cos(lk), \quad l = 1,2,\ldots,q \tag{9.53}$$

then the short-memory part of the PSD can be expressed as a Fourier series [33].

If the short-memory components are equal to

$$z_l(k) = k^l, \quad l = 1, 2,\ldots,q \tag{9.54}$$

then the logarithm of the SRD component of the spectral density is assumed to be a finite-order polynomial [34]. In all that follows, this class of FEXP models is referred to as *polynomial FEXP models*.

9.5.3 Spectral Densities of FARIMA and FEXP Processes

It is observed that a FARIMA process can be obtained by passing an FD process through an ARMA filter [19]. Therefore, in deriving the PSD of a FARIMA process $S_{\mathrm{FARIMA}}(k)$, we can refer to the spectral density of an ARMA model $S_{\mathrm{ARMA}}(k)$.

Following the notation of Section 9.5.1, we have that

$$S_{\mathrm{ARMA}}(k) = \frac{\sigma_W^2 |\Psi(e^{ik})|^2}{2\pi |\Psi(e^{ik})|^2} \tag{9.55}$$

Then we can write the expression of $S_{\mathrm{FARIMA}}(k)$ as

$$S_{\mathrm{FARIMA}}(k) = |1 - e^{ik}|^{-2d} S_{\mathrm{ARMA}}(k) = \left| 2\sin\left(\frac{k}{2}\right) \right|^{-2d} S_{\mathrm{ARMA}}(k) \tag{9.56}$$

If k tends to zero, $S_{\mathrm{FARIMA}}(k)$ is asymptotically equal to

$$S_{\mathrm{FARIMA}}(k) = S_{\mathrm{ARMA}}(0)|k|^{-2d} = \frac{\sigma_W^2 |\psi(1)|^2}{2\pi |\Psi(1)|^2} |k|^{-2d} \tag{9.57}$$

and it diverges for $d > 0$.

Comparing Equation 9.57 and Equation 9.31, we see that

$$d = H - \frac{1}{2} \tag{9.58}$$

As mentioned in the preceding sections, LRD occurs for $\frac{1}{2} < H < 1$ or $0 < d < \frac{1}{2}$. It has been demonstrated that LRD occurs for $0 < d < \frac{3}{2}$ for 2D self-similar processes, and, in particular, for a fractal image [35].

Let us rewrite the expression of the polynomial FEXP PSD as follows:

$$S(k;\phi) = |1 - e^{-ik}|^{1-2H} \exp\left\{ \sum_{l=0}^{q} \eta_l z_l(k) \right\} = \left| 2\sin\left(\frac{k}{2}\right) \right|^{-2d} S_{\mathrm{SRD}}(k;\phi) \tag{9.59}$$

where $S_{\mathrm{SRD}}(k;\phi)$ denotes the SRD part of the PSD.

Modeling of sea SAR images PSD is concerned with handling functions of both LRD and SRD behaviors. If we compare Equation 9.56 with Equation 9.59, the result is that FARIMA and FEXP models provide an equivalent description for the LRD behavior of the estimated sea SAR image MRPSD. To have a better understanding of the gain obtained by using polynomial FEXP PSD, we must analyze the expressions of its SRD component.

It goes without saying that the exponential SRD of FEXP is more suitable than a ratio of polynomials to represent rapid SRD variability of the estimated MRPSD. In the next section, we employ FARIMA and FEXP models to fit some MRPSD obtained from high-resolution ERS sea SAR images.

9.6 LRD Modeling of Mean Radial Spectral Densities of Sea SAR Images

As described in Section 9.1, the MRPSD of a sea SAR image is obtained by using a rectangular to polar coordinates transformation of the 2D PSD and by calculating the average of the radial spectral densities for ϑ ranging from 0 to 2π radians. This MRPSD can assume different shapes corresponding to low-wind areas, oil slicks, and the sea in the presence of strong winds. In any case, it diverges at the origin independently of the surface wind speeds, sea states, and the presence of oily substances on the sea surface [8].

LRD was first used by Ilow and Leung in the context of sea SAR images modeling and simulations. Analyzing some images collected by the RADARSAT sensor and utilizing FAR and FMA processes, they extended the applicability of long-memory models to characterize the sea clutter texture in high-resolution SAR imagery. An accurate representation of clutter in satellite images is important in sea traffic monitoring as well as in search and rescue operations [5], because it can lead to the development of automatic target recognition algorithms with improved performance.

The work of Ilow and Leung was limited to isotropic processes. In fact, in moderate weather conditions, the ocean surface exhibits homogeneous behavior. The anisotropy of sea SAR images depends, in a complex way, on the SAR system (frequency, polarization, look-angle, sensor velocity, pulse repetition frequency, pulse duration, chirp bandwidth) and on the ocean (directional spectra of waves), slick (concentration, surface tension, and then ocean spectral dampening), and environment (wind history) parameters.

It is observed [36] that the backscattered signal intensity depends on the horizontal angle φ between the radar look direction and the directions of the waves on the sea surface. The maximum signal occurs when the radar looks in the upwind direction, a smaller signal when the radar looks downwind, and the minimum signal is measured when the radar looks normal to the wind direction.

In employing the direct transformation defined by Hasselmann and Hasselmann [10], we observed that the anisotropy of sea SAR image spectra tends to decrease as φ increases from 0 to $\pi/2$ radians [37]. We can consider the sea SAR images corresponding to sea surfaces with no swells and with surface winds with speeds lower than 5 m/sec as isotropic. When the surface wind speed increases, the anisotropy of the estimated directional spectra starts to be noticeable. For surface wind speeds higher than 10 m/sec, evident main lobes appear in the directional spectra of sea SAR images, and the hypothesis of isotropy is no longer appropriate.

It is important to underline that a low surface wind speed is not always associated with an isotropic sea surface. Sometimes, wind waves generated far from the considered area can spread to a sea surface with little wind (with low amplitudes) and determine a reduction in the spatial correlation in the context of long-memory models.

In addition, the land and some man-made structures can act as a shelter from the wind in the proximity of the coast. Also, an island or the particular conformation of the shore can represent a barrier and partially reduce the surface wind speed.

Nevertheless, a low surface wind speed measured far from the coast, which is responsible for the attenuation of the sea wave spectra in the region of the wind waves, frequently corresponds to almost isotropic sea SAR images.

The spectral density and autocorrelation properties, introduced in the previous sections for 1D self-similar and LRD series, can be extended to 2D processes $Y(x, y)$, where $(x, y) \in R^2$, whose realizations are images [35]. This process is a collection of matrices whose elements are random variables. We also refer to $Y(x, y)$ as a random field.

If the autocorrelation function of the image is invariant under all Euclidean motions, then it depends only on the Euclidean distance between the points

$$R_Y(u, v) = E\{Y(x, y)Y(x + u, y + v)\} \tag{9.60}$$

and the field is referred to as homogeneous and isotropic.

The PSD of the homogeneous field, $S_Y(k_x, k_y)$, can be calculated by taking the 2D Fourier transform of $R_Y(u, v)$. For sampled data, this is commonly implemented using the 2D FFT algorithm.

As observed in Ref. [5], for a statistically isotropic and homogeneous field, we can derive a 2D model from a 1D model by replacing $k \in R$ in the PSD of a 1D process with the radial frequency $k_R = \sqrt{(k_x^2 + k_y^2)}$ to get the radial PSD.

Isotropic and homogeneous random fields can be completely represented in the spectral domain by means of their radial PSD. On the contrary, the representation of nonisotropic fields requires additional information. Our experimental results show that FARIMA models always provide a good fit with the MRPSD of isotropic and nonisotropic sea SAR images.

In the next section, the results obtained in Refs. [5,8] are compared with those we get by employing a modified FARIMA-based technique or using FEXP models.

In particular, the proposed method utilizes a noniterative technique to estimate the fractional differencing parameter d [9].

As in Ref. [8], we use the LRD parameter d, which depends on the SAR image roughness, and a SRD parameter, which depends on the amplitude of the backscattered signal to discriminate between low-wind and oil slick areas in sea SAR imagery.

Furthermore, we will demonstrate that FEXP models allow the computational efficiency to be maximized.

In this chapter, with the general term "SAR images," we refer to amplitude multi-look ground range SAR images, such as either ERS-2 PRI or ERS-2 GEC.

ERS-2 PRI are corrected for antenna elevation gain and range-spreading loss. The terrain-induced radiometric effects and terrain distortion are not removed.

ERS-2 SAR GEC are multi-look (speckle-reduced), ground range, system-corrected images. They are precisely located and rectified onto a map projection, but not corrected for terrain distortion. These images are high-level products, calibrated, and corrected for the SAR antenna pattern and range-spreading loss.

9.6.1 Estimation of the Fractional Differencing Parameter *d*

The LRD estimation algorithms set includes the *R/S* statistic, first proposed by Hurst in a hydrological context, the log–log correlogram, the log–log plot of the sample mean, the semivariogram, and the least-squares regression in the spectral domain.

Least-squares regression in the spectral domain is concerned with the analysis of the asymptotic behavior of the estimated MRPSD $\lfloor x \rfloor$ as k_R tends to zero.

As mentioned in Section 9.3.5, the mean radial periodogram is usually sampled at the Fourier frequencies $k_{j,n} = \frac{2\pi j}{n}$, $j = 1, 2, \ldots n^*$, where $n^* = \lfloor \frac{n-1}{2} \rfloor$ and $\lfloor x \rfloor$ denotes the integer part of x.

To obtain the LRD parameter d, we use a linear regression algorithm and estimate the slope of the mean radial periodogram, in a log–log plot, when the radial frequency tends to zero. It is observed that the LRD is an asymptotic property and that the negative slope of $\log\{I_{mr}(k_R)\}$ is usually proportional to the fractional differencing parameter only in a restricted neighborhood of the origin. This means that we must perform the linear regression using only a certain number of the smallest Fourier frequencies [38,39].

Least-squares regression in the spectral domain was employed by Geweke and Porter-Hudak in Ref. [38]. Ilow and Leung [5] extended their techniques and made use of a two-step procedure to estimate the FARIMA parameters of homogeneous and isotropic sea SAR images.

An iterative two-step analogous approach was introduced by Bertacca et al. [7,8]. In all these papers, the d parameter was repeatedly estimated inside a loop for increasing Fourier frequencies. The loop was terminated by a logical expression, which compared the logarithms of the long-memory and short-memory spectral density components. In the proximity of the origin, the logarithm of the spectral density was dominated by the long-memory component, so that only the Fourier frequencies for which the SRD contribution was negligible were chosen.

There is an important difference between the 1D spectral densities introduced by Ilow and Bertacca. Ilow calculated the radial PSD through the conversion of frequency points on a square lattice from the 2D FFT to the discrete radial frequency points. The radial PSD was obtained at frequency points, which were not equally spaced, by an angular integration and averaging of the spectral density.

Bertacca first introduced a rectangular to polar coordinates transformation of the 2D PSD. This was obtained through a cubic interpolation of the spectral density on a polar grid. Thus, the rapid fluctuations of the 2D periodogram were reduced before the mean radial periodogram computation.

It is assumed (see Section 9.2.2) that the mean radial periodogram ordinates $I_{mr}(k_{j,n})$ are approximately independent Gaussian random variables with means $S_{mr}(k_{j,n})$ (the MRPSD at the Fourier frequencies $k_{j,n}$). Then the average of the radial spectral densities for ϑ ranging from 0 to 2π radians allows the variance of $I_{mr}(k_{j,n})$ to be significantly reduced.

For example, if we calculate the 2D spectral density using nonoverlapping 256×256 size squared blocks of data, after going into polar coordinates (with the cubic interpolation), the mean radial periodogram is obtained by averaging over 256 different radial spectral densities.

On the contrary, in Ilow's technique, there is no interpolation and the average is calculated over a smaller number of frequency points.

This means that the variance of the mean radial periodogram ordinates $I_{mr}(k_{j,n})$ introduced by Bertacca is significantly smaller than that obtained by Ilow. Furthermore, supposing that the scattering of the random variables $I_{mr}(k_{j,n})$ around their means $S_{mr}(k_{j,n})$ is the effect of an additive noise, then the variance reduction is not affected by the anisotropy of sea SAR images. Let $\{k_{j,n}\}_{j=1}^{m_R}$ denote the set of the smallest Fourier frequencies to be used for the fractional differencing parameter estimation. Experimental results (shown in Section 9.7) demonstrate that, near the origin, the ordinates $\log\{I_{mr}(k_{j,n})\}$, represented versus $\log\{k_{j,n}\}$, are approximately the samples of a straight line with a negative slope. In addition, the small variance of all the samples $\log\{I_{mr}(k_{j,n})\}$, $k_{j,n} = 1, 2, \ldots, \lfloor \frac{n-1}{2} \rfloor$ allows the set of the Fourier frequencies $\{k_{j,n}\}_{j=1}^{m_R} = 1$ to be easily distinguished from this log–log plot. In all that follows, we will limit the set $\{k_{j,n}\}_{j=1}^{m_R}$ to the last Fourier frequency before the lobe (the slope increment) of the plot.

In Ref. [9], Bertacca et al. obtained good results by determining the number m_R of the Fourier frequencies to be used for the parameter d estimation from the log–log plot of the mean radial periodogram. This method prevents the estimation of the set $\{k_{j,n}\}_{j=1}^{m_R}$ inside a loop and increases the computational efficiency of the proposed technique.

The FARIMA MRPSD ordinates are equal to

$$S_{mr}(k_{j,n}) = \left[2\sin\left(\frac{k_{j,n}}{2}\right)\right]^{-2d} S_{\mathrm{ARMA}}(k_{j,n}) \tag{9.61}$$

Calculating the logarithm of the above equation, we obtain

$$\log\{S_{mr}(k_{j,n})\} = -d\log\left[4\sin^2\left(\frac{k_{j,n}}{2}\right)\right] + \log\{S_{\mathrm{ARMA}}(k_{j,n})\} \tag{9.62}$$

After determining $\{k_{j,n}\}_{j=1}^{m_R}$ and substituting the ordinates $S_{mr}(k_{j,n})$ with their estimates $I_{mr}\ (k_{j,n})$, we have

$$\log\{I_{mr}(k_{j,n})\} \approx -d \log\left[4 \sin^2\left(\frac{k_{j,n}}{2}\right)\right] \tag{9.63}$$

The least-squares estimator of d is given by

$$\hat{d} = -\frac{\sum_{j=1}^{m_R} (u_j - \bar{u})(v_j - \bar{v})}{\sum_{j=1}^{m_R} (u_j - \bar{u})^2} \tag{9.64}$$

where $v_j = \log\{I_{mr}(k_{j,n})\}$, $u_j = \log\left[4 \sin^2\left(\frac{k_{j,n}}{2}\right)\right]$, $\bar{v} = \sum_{j=1}^{m_R} \frac{v_j}{m_R}$, and $\bar{u} = \sum_{j=1}^{m_R} \frac{u_j}{m_R}$.

In Section 9.7, we show the experimental results corresponding to high-resolution sea SAR images of clean and windy sea areas, oil spill, and low-wind areas.

It is observed that considering the SAR image of a rough sea surface that shows clearly identifiable darker zones corresponding to an oil slick (or to an oil spill) and a low-wind area, we have [8] the following:

- These darker areas are both caused by the amplitude attenuation for the tiny capillary and short-gravity waves contributing to the Bragg resonance phenomenon.
- In the oil slick (or spill), the low amplitude of the backscattered signal is due to the concentration of the surfactant in the water that affects only the short-wavelength waves.
- In the low-wind area, the amplitudes of all the wind-generated waves are reduced. In other words, the low-wind area tends to a flat sea surface.

Bertacca et al. observed that the spatial correlation of SAR subimages of low-wind areas always decays to zero at a slower rate than that related to the oil slicks (or spills) in the same SAR image [8]. Since lower decaying correlations are related to higher values of the fractional differencing parameter, low-wind areas are characterized by the greatest d value in the whole image. It is observed that oily substances determine the attenuation of only tiny capillary and short-gravity waves contributing to the Bragg resonance phenomenon [36]. Out of the Bragg wavelength range, the same waves are present in oil spill and clean water areas, and experimental results show that the shapes (and the fractional differencing parameter d) of the MRPSD of oil slick (or spill) and clean water areas on the SAR image are very similar. On the contrary, the shapes of the mean radial spectra differ for low-wind and windy clean water areas [8].

9.6.2 ARMA Parameter Estimation

After obtaining the fractional differencing parameter estimate $\hat{d}$, we can write the SRD component of the mean radial periodogram as

$$I_{\mathrm{SRD}}(k_{j,n}) = \hat{S}_{\mathrm{SRD}}(k_{j,n}) = I_{mr}(k_{j,n})\left[2 \sin\left(\frac{k_{j,n}}{2}\right)\right]^{2\hat{d}} \tag{9.65}$$

The square root of this function,

$$h_{SRD}(k_{j,n}) = \sqrt{I_{SRD}(k_{j,n})} \tag{9.66}$$

can be defined as the short-memory frequency response vector. The vector of the ARMA parameter estimates that fits the data, $h_{SRD}(k_{j,n})$, can be obtained using different algorithms.

Ilow considered either MA or AR representation of the ARMA part. With his method, a 2D MA model, with a circular symmetric impulse response, is obtained from a 1D MA model with a symmetric impulse response in a similar way as the 2D filter is designed in the frequency transformation technique [40].

Bertacca considered ARMA models. He estimated the real numerator and denominator coefficients of the transfer function by employing the classical Levi algorithm [41], whose output was used to initialize the Gaussian Newton method that directly minimized the mean square error. The stability of the system was ensured by using the damped Gauss–Newton algorithm for iterative search [42].

9.6.3 FEXP Parameter Estimation

FEXP models for sea SAR images modeling and representation were first introduced by Bertacca et al. in Ref. [9]. As observed in Ref. [19], using FEXP models leads to the estimation of parameters in a generalized linear model.

A different approach consists of estimating the fractional differencing parameter d as described in Section 9.6.1. The SRD component of the mean radial periodogram, $I_{SRD}(k_{j,n})$, can be calculated as in Equation 9.65.

After determining the logarithm of the data

$$\bar{y},\ y(j) = \log\left\{\hat{S}_{SRD}(k_{j,n})\right\} = \log\{I_{SRD}(k_{j,n})\} \tag{9.67}$$

we define the vector $\bar{x} = x(j) = k_{j,n}$, and compute the coefficients vector η of the polynomial $p(x)$ that fits the data, $p(x(j))$ to $y(j)$, in a least-squares sense. The SRD part of the FEXP model is equal to

$$S_{FEXP\_srd}(k_{j,n}) = \exp\left\{\sum_{i=0}^{m} \eta(i) k_{j,n}^{i}\right\} \tag{9.68}$$

where m denotes the order of the polynomial $p(x)$.

It is worth noting the difference between the FMA and the polynomial FEXP models for considering the computational costs and the number of the parameters to estimate. Using FMA models, we calculate the square root of $I_{SRD}(k_{j,n}) = \hat{S}_{SRD}(k_{j,n})$ and estimate the best-fit polynomial regression of $f_1(k_{j,n}) = \sqrt{I_{SRD}(k_{j,n})}$, $j = 1, \ldots, m_R$ on $\{k_{j,n}\}_{j=1}^{m_R}$.

Employing FEXP models, we perform a polynomial regression of $f_2(k_{j,n}) = \log\{\sqrt{I_{SRD}(k_{j,n})}\}$, $j = 1, \ldots, m_R$ on $\{k_{j,n}\}_{j=1}^{m_R}$.

Since the logarithm generates functions much smoother than those obtained by calculating the square root of $I_{SRD}(k_{j,n})$, FEXP models give the same goodness of fit of FARIMA models and require a reduced number of parameters.

9.7 Analysis of Sea SAR Images

In this section, we show the results obtained for an ERS-2 GEC and an ERS-2 SAR PRI (Figure 9.2 and Figure 9.3). In fact, the results show that the geocoding process does not

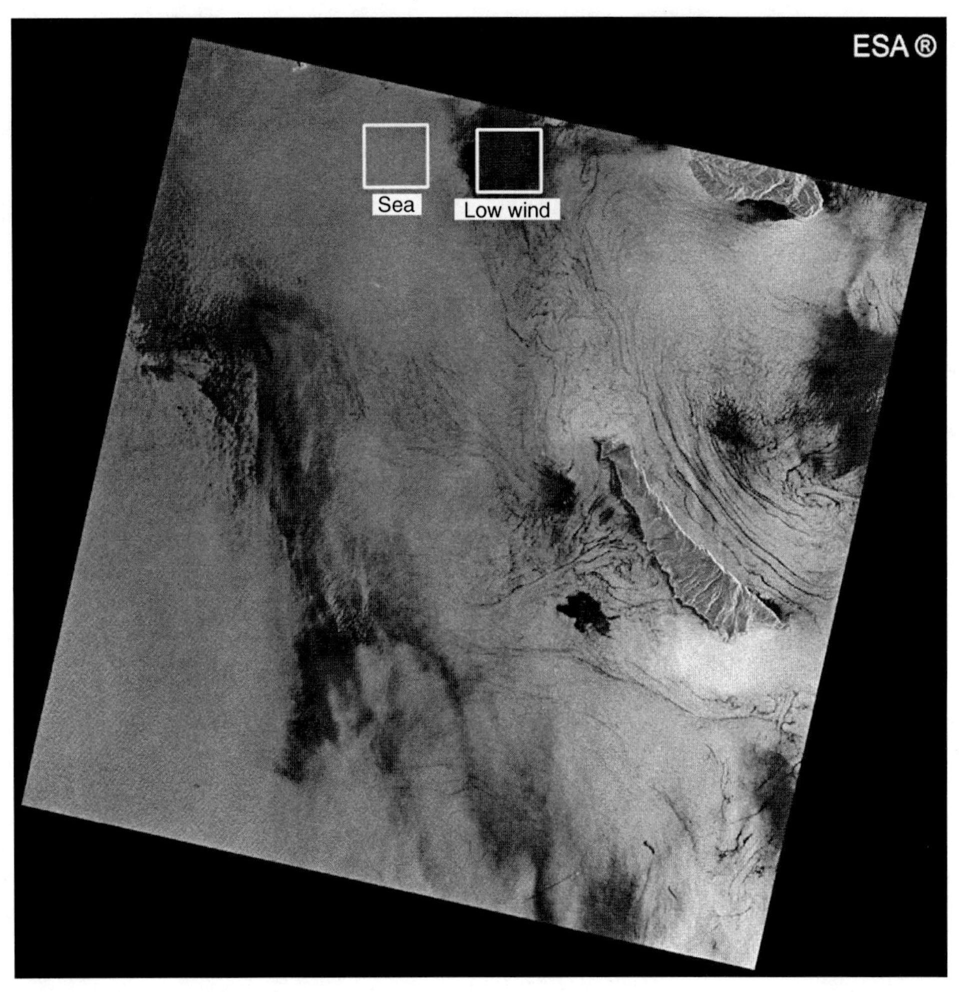

FIGURE 9.2
Sea ERS-2 SAR GEC acquired near the Santa Catalina and San Clemente Islands (Los Angeles, CA). (Data provided by the European Space Agency © ESA (2002). With permission.)

affect the LRD behavior of sea SAR image spectral densities. In all that follows, figures are represented with respect to the spatial frequency *ks* (in linear or logarithmic scale). The spatial resolution of ERS-2 SAR PRI is 12.5 m for both the azimuth and the (ground) range coordinates. The North and East axes pixel spacing is 12.5 m for ERS-2 SAR GEC.

In the calculation of the periodogram, we average spectral estimates obtained from squared blocks of data containing 256 × 256 pixels. This square window size permits the representation of the low-frequency spectral components that provide good estimates of the fractional differencing parameter d. Thus, we have

$$\begin{aligned}
&\text{Spatial resolution} = 12.5(m) \\
&f_{\max} = \frac{1}{\text{Resolution}} = 0.08(m^{-1}) \\
&k_{j,n} = \frac{2\pi j}{n},\ j = 1, \ldots, 127,\ n = 256 \\
&ks_{j,n} = k_{j,n} f_{\max}
\end{aligned} \tag{9.69}$$

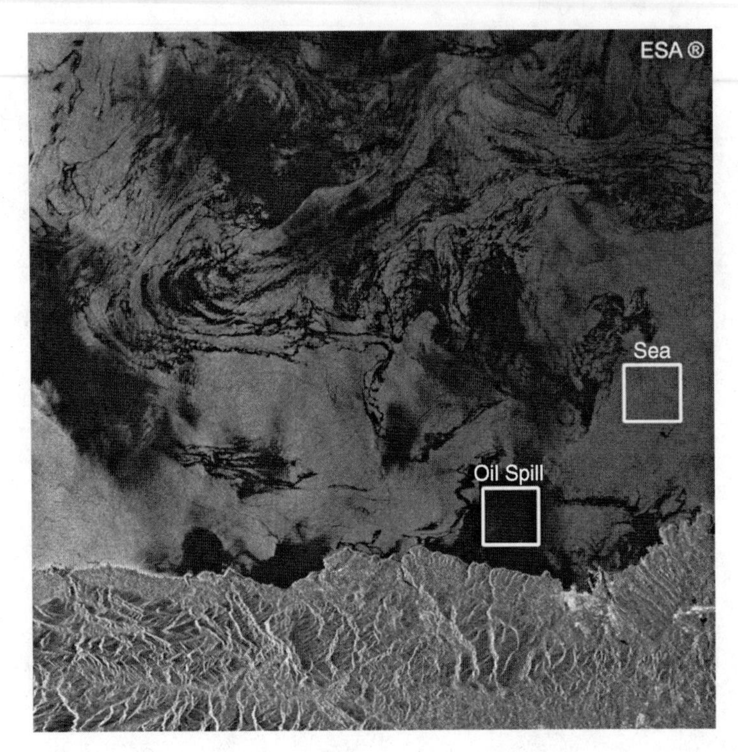

FIGURE 9.3
Sea ERS-2 SAR PRI acquired near the coast of Asturias region, Spain. (Data provided by the European Space Agency © ESA (2002). With permission.)

The first image (Figure 9.2) represents a rough sea area near Santa Catalina and San Clemente Islands (Los Angeles, USA), on the Pacific Ocean, containing a low-wind area (ERS-2 SAR GEC orbit 34364 frame 2943). The second image (Figure 9.3) covers the coast of the Asturias region on the Atlantic Ocean (ERS-2 SAR PRI orbit 40071 frame 2727), and represents a rough sea surface with extended oil spill areas.

Four subimages from Figure 9.2 and Figure 9.3 are considered for this analysis. The two subimages marked in Figure 9.2 represent a rough sea area and a low-wind area, respectively. The two subimages marked in Figure 9.3 correspond to a rough sea area and to an oil spill. Note that not all the considered subimages have the same dimension. Subimages of greater dimension provide lower variance of the 2D periodogram. The two clean water areas contain 1024 × 1024 pixels. The low-wind area in Figure 9.2 and the oil spill in Figure 9.3 contain 512 × 512 pixels.

Using FARIMA models, we estimate the parameter d as explained in Section 9.6.1. The estimation of the ARMA part of the spectrum is made as in Ref. [8]. First we choose the maximum orders, $m_{\max}$ and $n_{\max}$, of the numerator and denominator polynomials of the ARMA transfer function. Then we estimate the real numerator and denominator coefficients in vectors **b** and **a**, for m and n indexes ranging from 0 to $m_{\max}$ and $n_{\max}$, respectively.

The estimated frequency response vector of the short memory PSD, $h_{\mathrm{SRD}}(k_{j,n})$, has been defined in Equation 9.66 as the square root of $I_{\mathrm{SRD}}(k_{j,n})$. The ARMA transfer function is the one that corresponds to the estimated frequency response $\hat{h}_{\mathrm{ARMA}}(k_{j,n})$ for which the RMS error between $h_{\mathrm{SRD}}(k_{j,n})$ and $\hat{h}_{\mathrm{ARMA}}(k_{j,n})$ is the minimum among all values of (m, n).

The estimated ARMA part of the spectrum then becomes

TABLE 9.1

Estimated FARIMA Parameters

Analyzed Image	m_R	m	n	d	$\bar{A}_{\mathrm{SRD}}$
Asturias region sea	11	11	18	0.471	4.87
Asturias region oil spill	15	16	11	0.568	0.24
Santa Catalina sea	7	6	6	0.096	218.88
Santa Catalina low wind	11	3	14	0.423	0.47

$$\hat{S}_{\mathrm{ARMA}}(k_{j,n}) = |\hat{h}_{\mathrm{ARMA}}(k_{j,n})|^2 \tag{9.70}$$

In the data processing performed using FARIMA models, we have used $m_{\mathrm{max}} = 20$, $n_{\mathrm{max}} = 20$. The FARIMA parameters estimated from the four considered subimages are shown in Table 9.1. The parameter $\bar{A}_{\mathrm{SRD}}$, introduced by Bertacca et al. in Ref. [8], is given by

$$\bar{A}_{\mathrm{SRD}} = \sum_{j=1}^{n^*} I_{\mathrm{SRD}}(k_{j,n})/n^* \tag{9.71}$$

It depends on the strength of the backscattered signal and thus on the roughness of the sea surface. It is observed [8] that the ratio of these parameters for the clean sea and the oil spill SAR images is always lower than those corresponding to the clean sea and the low-wind subimages

$$\frac{\bar{A}_{\mathrm{SRD\_S}}}{\bar{A}_{\mathrm{SRD\_OS}}} < \frac{\bar{A}_{\mathrm{SRD\_S}}}{\bar{A}_{\mathrm{SRD\_LW}}} \tag{9.72}$$

From Table 9.1, we have $\frac{\bar{A}_{\mathrm{SRD\_S}}}{\bar{A}_{\mathrm{SRD\_OS}}} = 20.29$ and $\frac{\bar{A}_{\mathrm{SRD\_S}}}{\bar{A}_{\mathrm{SRD\_LW}}} = 465.7$. The parameter $\bar{A}_{\mathrm{SRD\_S}}$ assumes low values for clean sea areas with moderate wind velocities, as we can see in Table 9.1, for the subimage of Figure 9.3. However, the attenuation of the Bragg resonance phenomenon due to the oil on the sea surface leads to the measurement of a high ratio between $\bar{A}_{\mathrm{SRD\_S}}$ and $\bar{A}_{\mathrm{SRD\_OS}}$. In any case, for rough sea surfaces, the discrimination between oil spill and low wind is very easy using either the LRD parameter d or the SRD parameter $\bar{A}_{\mathrm{SRD}}$. The parameter d is used together with $\bar{A}_{\mathrm{SRD}}$ to resolve different darker areas corresponding to smooth sea surfaces [8]. Figure 9.4 and Figure 9.5 show the estimated mean radial spectral densities and their FARIMA models for all the subimages of Figure 9.2 and Figure 9.3.

Table 9.2 displays the results obtained by using an FEXP model, and estimating its parameters in a generalized linear model of order $m = 2$, for the almost isotropic sea SAR subimage shown in Figure 9.3. In Figure 9.6, we examine visually the fit of the estimated model to the MRPSD of the sea clutter. Figure 9.7 and Figure 9.8 show the estimated mean radial spectral densities of all the subimages of Figure 9.2 and Figure 9.3 and their FEXP models obtained by using the two-step technique described in Section 9.6.3. We notice that increasing the polynomial FEXP order allows better fits to be obtained. To illustrate this, FEXP models of order $m = 2$ and $m = 20$ are shown in Figure 9.7 and Figure 9.8.

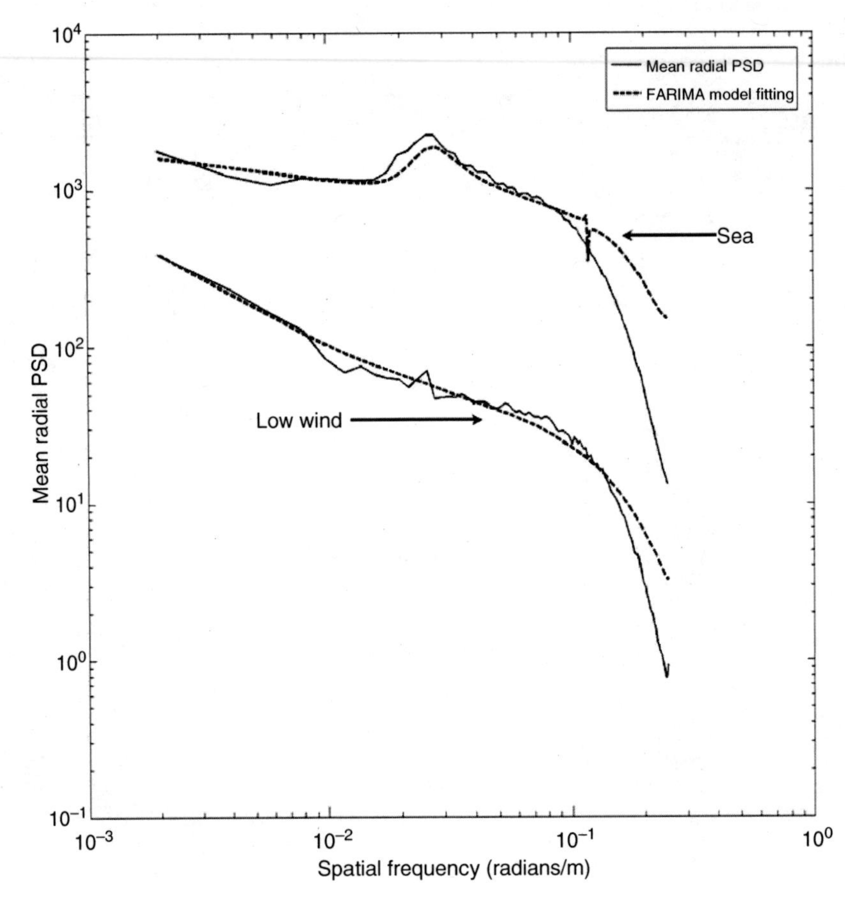

FIGURE 9.4
Mean radial power spectral densities and FARIMA models corresponding to Figure 9.2.

9.7.1 Two-Dimensional Long-Memory Models for Sea SAR Image Spectra

As mentioned in Section 9.6, the 2D periodograms estimated for sea SAR images can exhibit either isotropic or anisotropic spatial structure [8]. This depends on the SAR system, on the directional spectra of waves, on the possible presence of oil spills or slicks, and on the wind history parameters. fBm and FARIMA models have been used to represent isotropic and homogeneous self-similar random fields [5,8,35,43]. Owing to the complex nature of the surface scattering phenomenon involved in the problem and the highly nonlinear techniques of SAR image generation, the definition of anisotropic spectral densities models is not merely related to the sea state or wind conditions. In the context of self-similar random fields, anisotropy can be introduced by linear spatial transformations of isotropic fractal fields, by spatial filtering of fields with desired characteristics, or by building intrinsically anisotropic fields as in the case of the so-called fractional Brownian sheet [43]. Anisotropic 2D FARIMA has been introduced in Ref. [43] as the discrete-space equivalent of a 2D fractionally differenced Gaussian noise, taking into account the long-memory property and the directionality of the image. In that work, Pesquet-Popescu and Lévy Véhel defined the anisotropic extension of 2D FARIMA by multiplying its isotropic spectral density by an anisotropic part $A_{\alpha,\varphi}(k_x, k_y)$. The function $A_{\alpha,\varphi}(k_x, k_y)$ depended on two parameters: α and φ. The parameter $\varphi \in (0,2\pi]$ provided the

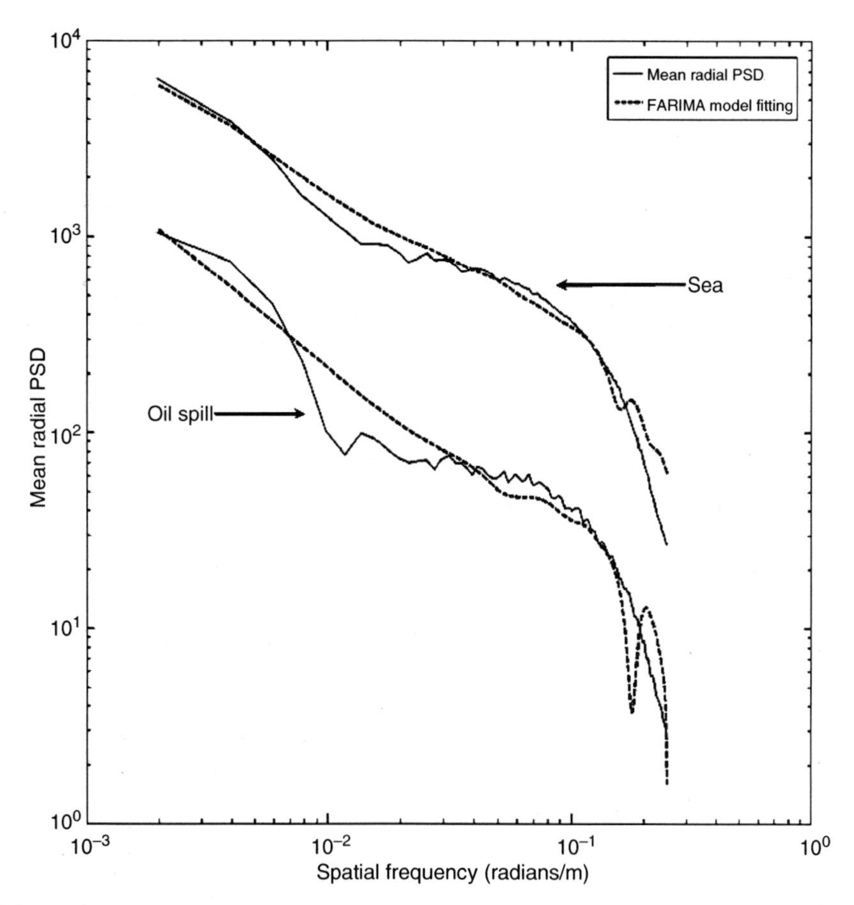

FIGURE 9.5
Mean radial power spectral densities and FARIMA models corresponding to Figure 9.3.

orientation of the field. The anisotropy coefficient $\alpha \in [0,1)$ determined the dispersion of the PSD around its main direction φ.

This anisotropic model allows the directionality and the shape of the 2D PSD to be modified, but the maximum of the PSD main lobe depends on the values assumed by the anisotropy coefficient. Furthermore, the anisotropy of this model depends on only one parameter (α).

Experimental results show that this is not sufficient to correctly represent the anisotropy of the PSD of sea SAR images. This means the model must be changed both to provide a better fit with the spectral densities of sea SAR images and allow the shape and the position of the spectral density main lobes to be independently modified.

TABLE 9.2

Parameter Estimates for the FEXP Model of Order $m = 2$ of Figure 9.6.

Parameter	Coefficient Estimates	Standard Errors	*t*-Statistics	*p*-Values
$b(0) = -2d$	−1.0379	0.018609	−55.775	3.5171e−089
$b(1) = \eta(0)$	2.2743	0.11393	19.962	2.1482e−040
$b(2) = \eta(1)$	25.651	1.7359	14.777	4.881e−029
$b(3) = \eta(2)$	−126.49	9.7813	−12.932	1.1051e−024

Note: *p*-Values are given for testing $b(i) = 0$ against the two-sided alternative $b(i) \neq 0$.

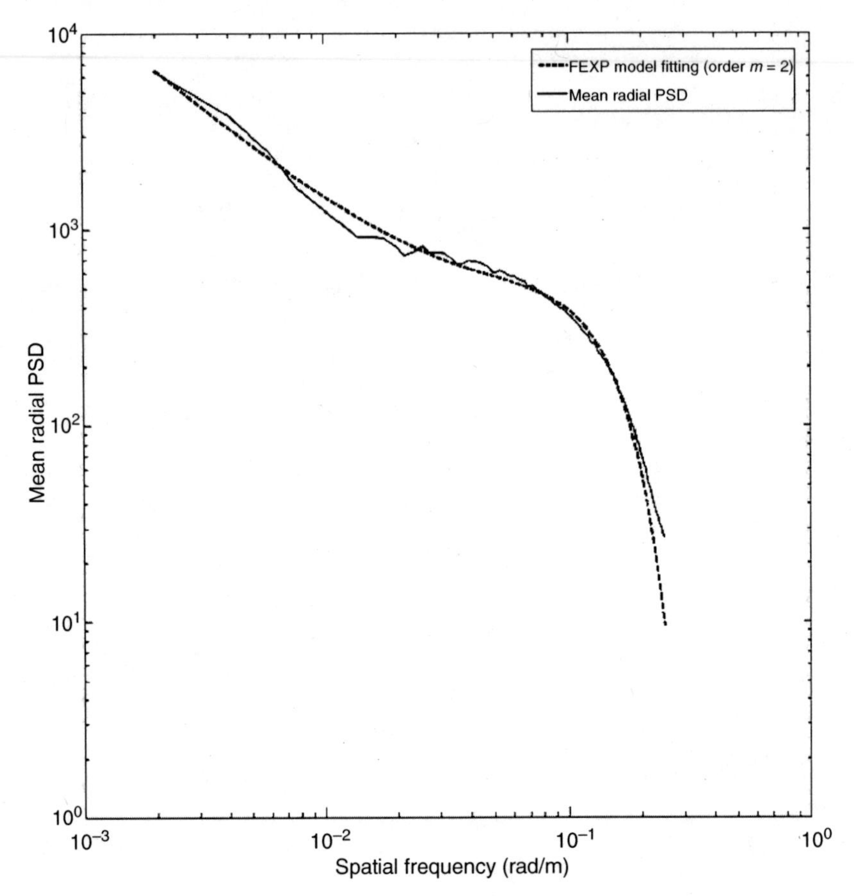

FIGURE 9.6
Mean radial power spectral densities and FEXP model corresponding to the sea subimage of Figure 9.3.

Bertacca et al. [44] showed that the PSD of either isotropic or anisotropic sea SAR images can be modeled by multiplying the isotropic spectral density of a 2D FARIMA by an anisotropic part derived from the fractal model of sea surface spectra [6]. This anisotropic component depends on seven different parameters and can be broken down into a radial component, the omnidirectional spectrum $S(ks_{j,n})$, and a spreading function $G(\vartheta_{i,m})$, $\vartheta_{i.m} = \frac{2\pi i}{m}$, $i = 1, 2, \ldots, m$. This allows the shape, the slope, and the anisotropic characteristics of the 2D PSD to be independently modified and a good fit with the estimated periodogram to be obtained.

With a different approach [37], a reliable model of anisotropic PSD has been defined, for sea SAR intensity images, by adding a 2D isotropic FEXP to an anisotropic term newly defined starting from the fractal model of sea surface spectra [6]. However, the additive model cannot be interpreted as an anisotropic extension of the LRD spectral models as in Refs. [5,44].

Analyzing homogeneous and isotropic random fields leads to the estimation of isotropic spectral densities. In such cases, it is better to utilize FEXP instead of FARIMA models [9]. To illustrate this, in Figure 9.9 and Figure 9.10, we show the isotropic PSD of the low-wind area in Figure 9.3 and the result obtained by employing a 2D isotropic FEXP model to represent it. Figure 9.11 and Figure 9.12 compare the anisotropic PSD estimated from a directional sea SAR image (ERS-2 SAR PRI orbit 40801 frame 2727, data provided by the European Space Agency © ESA (2002)) with its additive model [37].

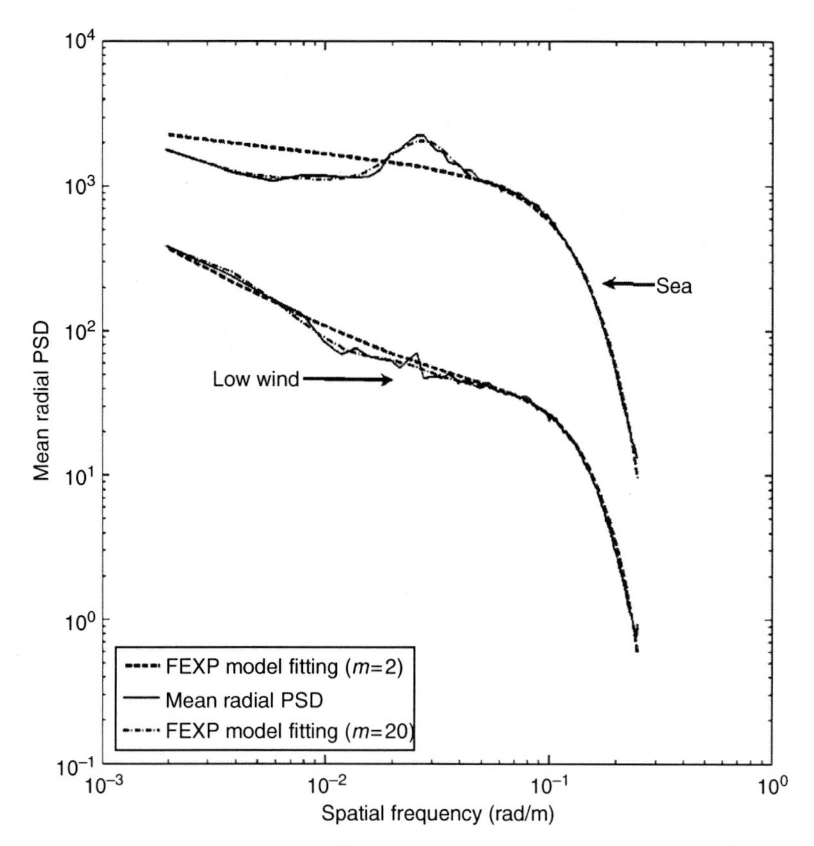

FIGURE 9.7
Mean radial PSD and FEXP models corresponding to Figure 9.2.

In Figure 9.9 and Figure 9.10, we present results using the log scale for the z-axis. This emphasizes the divergence of the long-memory PSD. In Figure 9.11 and Figure 9.12, to better examine the anisotropy characteristics of the spectral densities at medium and high frequencies, the value of the PSD at the origin has been set to zero.

9.8 Conclusions

Some self-similar and LRD processes, such as fBm, fGn, FD, and FMA or FAR models, have been used in the literature to model the spatial correlation properties of the scattering from natural surfaces. These models have demonstrated reliable results in the analysis and modeling of high-definition sea SAR images under the assumption of homogeneity and isotropy.

Unfortunately, a wind speed greater than 7 m/sec is often present on sea surfaces and this can produce directionality and nonhomogeneity of the corresponding SAR images. Furthermore, the mean radial spectral density (MRPSD) of sea SAR images always shows an LRD behavior and a slope that changes with increasing frequency. This means that:

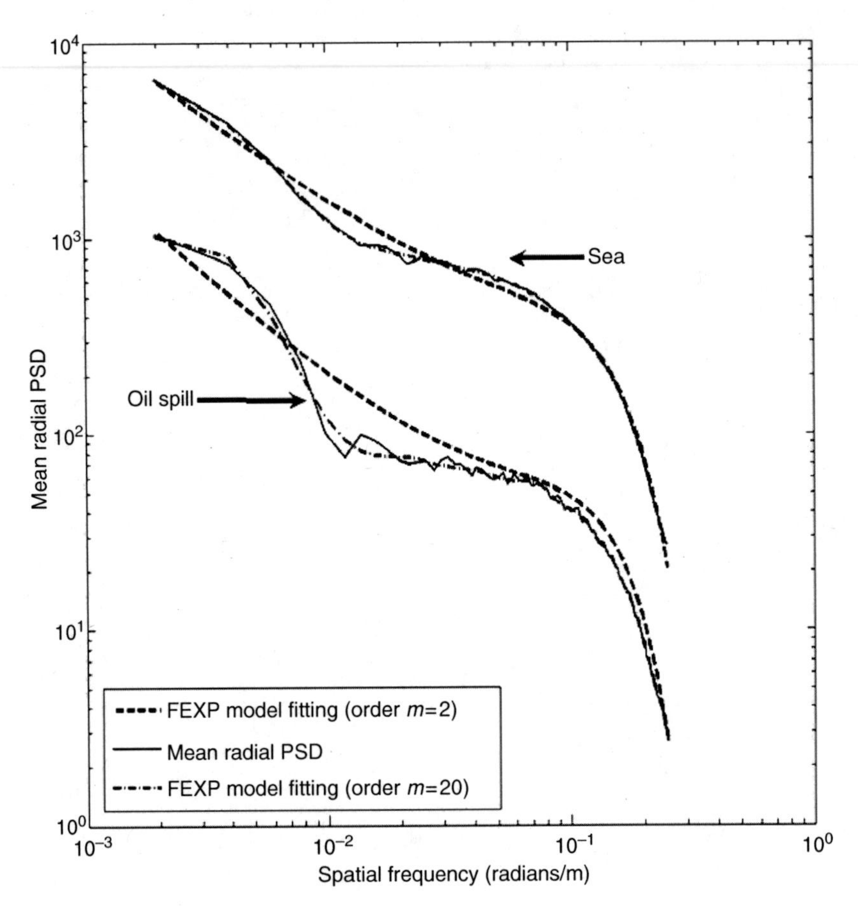

FIGURE 9.8
Mean radial PSD and FEXP models corresponding to Figure 9.3.

- When analyzing mean radial spectral densities (1D), fBm and FD models, which do not represent both the long and short-memory behaviors, cannot provide a good fit.
- When analyzing nonhomogeneous and nonisotropic sea SAR images, due to their directionality, suitable 2D LRD spectral models, corresponding to nonhomogeneous and anisotropic random fields, must be defined to better fit their power spectral densities.

Here we have presented a brief review of the most important LRD models as well as an explanation of the techniques of analysis and discrimination of sea SAR images corresponding to oil slicks (or spills) and low-wind areas.

In the context of MRPSD modeling, we have shown that FEXP models improve the goodness of fit of FARIMA models and ensure lower computational costs. In particular, we have used Bertacca's noniterative technique for estimating the LRD parameter d.

The result obtained by using this algorithm has been compared with that corresponding to the estimation of parameters in a generalized linear model. The two estimates of the parameter d, corresponding to the sea SAR subimage of Figure 9.3, are $d_{\mathrm{I}} = 0.471$, $d_{\mathrm{II}} = -\frac{b(0)}{2} = 0.518$, from Table 9.1 and Table 9.2, respectively. This demon-

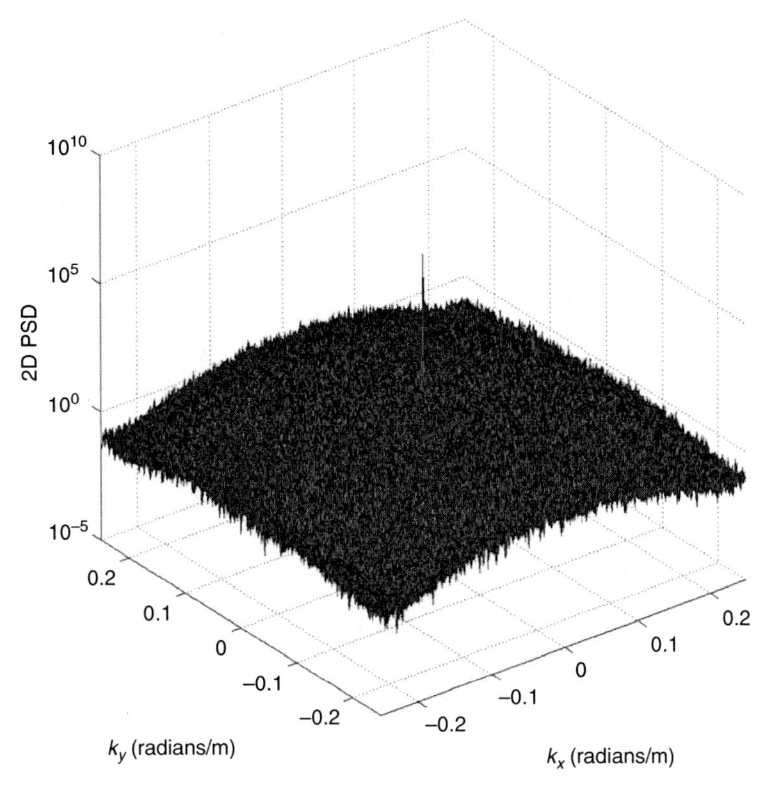

FIGURE 9.9
The isotropic spectral density estimated from the low-wind area of Figure 9.2.

strates the reliability of the noniterative algorithm: it prevents the estimation of the set $\{k_j,n\}_{j=1}^{m_R}$ inside a loop and increases the computational efficiency.

We notice that the ratios of the FD parameters for oil spill and clean sea subimages are always lower than those corresponding to low-wind and clean sea areas. From Table 9.1, we have that

$$\frac{d_{\text{OS}}}{d_{\text{S}}} = \frac{0.568}{0.471} = 1.206$$
$$\frac{d_{\text{LW}}}{d_{\text{S}}} = \frac{0.423}{0.096} = 4.406 \tag{9.73}$$

As observed by Bertacca et al. [8], the discrimination among clean water, oil spill, and low-wind areas is very easy by using either the long-memory parameter d or the short-memory parameter $\bar{A}_{\text{SRD}}$, in the case of rough sea surfaces. The two parameters must be used together only to resolve different darker areas corresponding to smooth sea surfaces.

Furthermore, using large square window sizes in the estimation of the SAR image PSD allows low-frequency spectral components to be represented and good estimates of the fractional differencing parameter to be obtained. Therefore, the proposed technique can be better used to distinguish oil spills or slicks with large extent from low-wind areas.

In the context of modeling of 2D spectral densities, we have shown that the anisotropic models introduced in the literature by linear spatial transformations of isotropic fractal fields, by spatial filtering of fields with desired characteristics, or by building intrinsically

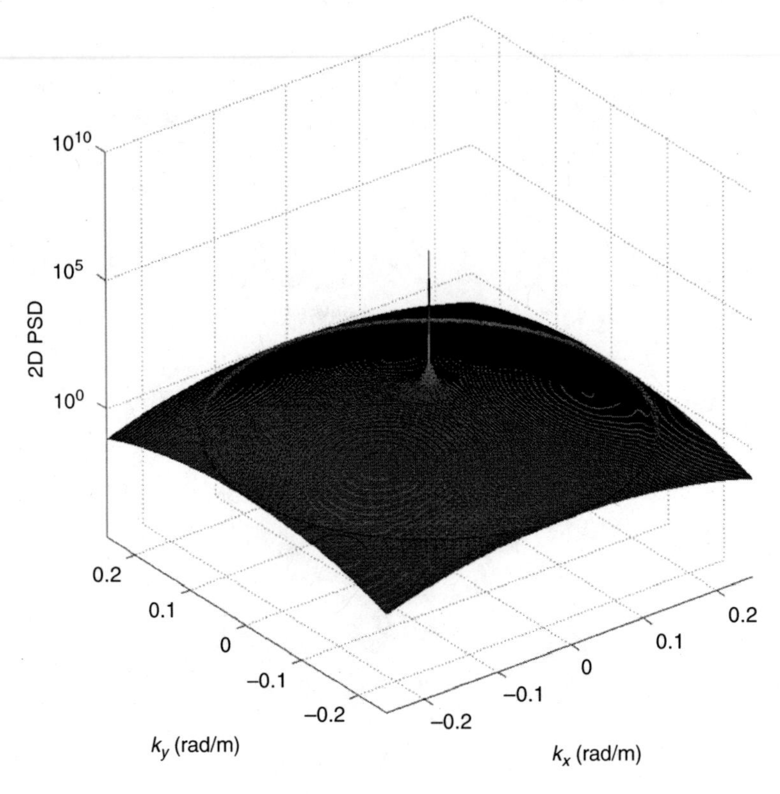

FIGURE 9.10
The 2D isotropic FEXP model ($m = 20$) of the PSD in Figure 9.9.

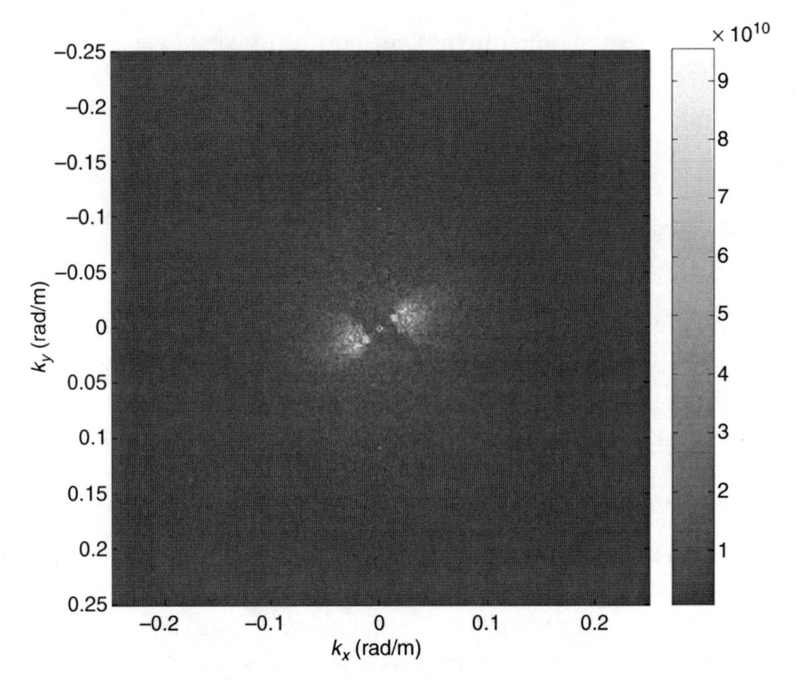

FIGURE 9.11
Anisotropic PSD estimate from a directional sea SAR image.

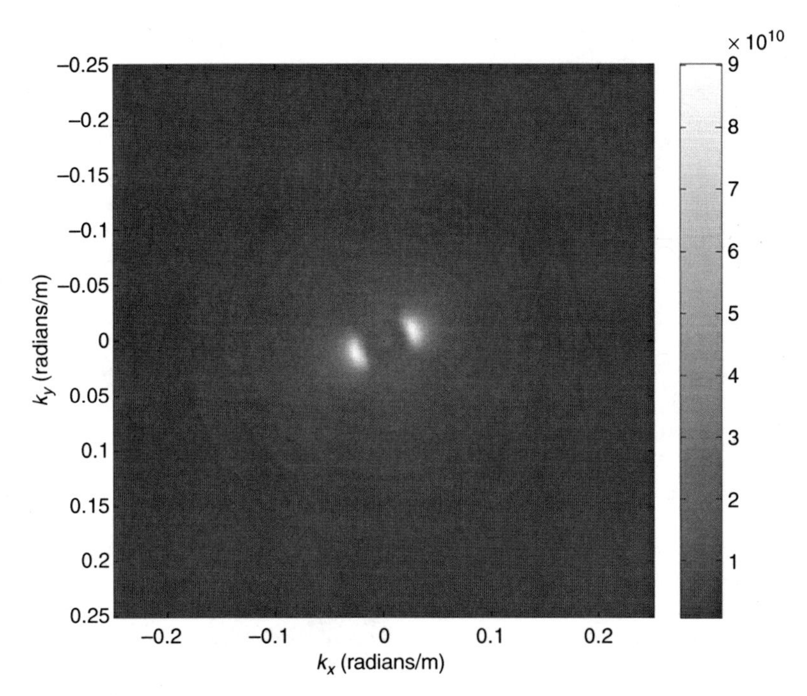

FIGURE 9.12
Anisotropic 2D model of the PSD of Figure 9.11.

anisotropic fields, are not suitable for representing the spectral densities of sea SAR images. In this chapter, two anisotropic spectral models have been presented. They have been obtained by either adding or multiplying a 2D isotropic FEXP to an anisotropic term defined starting from the fractal model of sea surface spectra [6,37]. To illustrate this, we have shown an anisotropic sea SAR image PSD in Figure 9.11, and its spectral model, in Figure 9.12, obtained by adding a 2D FEXP model of order $m = 20$ to an anisotropic fractal spectral component.

References

1. Berizzi, F. et al., Fractal mapping for sea surface anomalies recognition, *IEEE Proc. IGARSS 2003*, Toulouse, France, 4, 2665–2667, 2003.
2. Franceschetti, G. et al., SAR raw signal simulation of oil slicks in ocean environments, *IEEE Trans. Geosci. Rem. Sens.*, 40, 1935, 2002.
3. Datcu, M., Model for SAR images, *Int. Symp. Opt. Eng. Photonics Aerospace Sens.*, SPIE, Orlando, FL, Apr 1992.
4. Stewart, C.V. et al., Fractional Brownian motion models for synthetic aperture radar imagery scene segmentation, *Proc. IEEE*, 81, 1511, 1993.
5. Ilow, J. and Leung, H., Self-similar texture modeling using FARIMA processes with applications to satellite images, *IEEE Trans. Image Process.*, 10, 792, 2001.
6. Berizzi, F. and Dalle Mese, E., Sea-wave fractal spectrum for SAR remote sensing, *IEEE Proc. Radar Sonar Navigat.*, 148, 56, 2001.
7. Bertacca, M. et al., A FARIMA-based analysis for wind falls and oil slicks discrimination in sea SAR imagery, *Proc. IGARSS 2004*, Alaska, 7, 4703, 2004.

8. Bertacca, M., Berizzi, F., and Dalle Mese, E., A FARIMA-based technique for oil slick and low-wind areas discrimination in sea SAR imagery, *IEEE Trans. Geosci. Rem. Sens.*, 43, 2484, 2005.
9. Bertacca, M., Berizzi, F., and Dalle Mese, E., FEXP models for oil slick and low-wind areas analysis and discrimination in sea SAR images, SEASAR 2006: Advances in SAR oceanophy from ENVISAT and ERS mission, ESA ESRIN, Prascati (Rome), Italy, 2006 to *ESA-SEASAR 2006 Advances in SAR oceanography from ENVISAT and ERS missions.*
10. Hasselmann, K. and Hasselmann, S., The nonlinear mapping of an ocean wave spectrum into a synthetic aperture radar image spectrum and its inversion, *J. Geophys. Res.*, 96, 713, 1991.
11. Therrien, C.W., *Discrete Random Signals and Statistical Signal Processing*, Prentice Hall, Englewood Cliffs, NJ, 1992.
12. Oppenheim, A.V. and Schafer, R.W. *Digital Signal Processing*, Prentice-Hall, Englewood Cliffs, NJ, 1975.
13. Priestley, M.B., *Spectral Analysis of Time Series*, Accdemic Press, London, 1981.
14. Willinger, W., et al., Self-similarity in high-speed packet traffic: analysis and modeling of ethernet traffic measurements, *Stat. Sci.*, 10, 67, 1995.
15. Rose, O., Estimation of the Hurst parameter of long-range dependent time series, Institute Of Computer Science, University of Würzburg, Würzburg, Germany, Res. Rep. 137, 1996.
16. Lamperti, J.W., Semi-stable stochastic processes, *Trans. Am. Math. Soc.*, 104, 62, 1962.
17. Veervaat, W., Properties of general self-similar processes, *Bull. Int. Statist. Inst.*, 52, 199, 1987.
18. Veervaat, W., Sample path properties of self-similar processes with stationary increments, *Ann. Probab.*, 13, 1, 1985.
19. Beran, J., *Statistics for Long-Memory Processes*, Chapman & Hall, London, U.K., 1994, chap.2.
20. Sinai, Ya.G., Self-similar probability distributions, *Theory Probab. Appl.*, 21, 64, 1976.
21. Akaike, H., A limiting process which asymptotically produces f^{-2} spectral density, *Ann. Inst. Statist. Math.*, 12, 7, 1960.
22. Sayles, R.S. and Thomas, T.R., Surface topography as a nonstationary random process, *Nature*, 271, 431, 1978.
23. Sarpkaya, T. and Isaacson, M., *Mechanics of Wave Forces on Offshore Structures*, Van Nostrand Reinhold, New York, 1981.
24. Bretschneider, C.L., Wave variability and wave spectra for wind-generated gravity waves, Beach Erosion Board, U.S. Army Corps Eng., U.S. Gov. Printing Off., Washington, DC, Tech. Memo., pp. 118, 1959.
25. Mandelbrot, B.B. and Van Ness, J.W., Fractional Brownian motions, fractional noises and applications, *SIAM Rev.*, 10, 422, 1968.
26. Ma, S. and Chuanyi, J., Modeling heterogeneous network traffic in wavelet domain, *IEEE/ACM Trans. Networking*, 9, 634, 2001.
27. Granger, C.W.J. and Joyeux, R., An introduction to long-range time series models and fractional differencing, *J. Time Ser. Anal.*, 1, 15, 1980.
28. Hosking, J.R.M., Fractional differencing, *Biometrika*, 68, 165, 1981.
29. Beran, J., Fitting long-memory models by generalized linear regression, *Biometrika*, 80, 785, 1993.
30. Whittle, P., Estimation and information in stationary time series, *Ark. Mat.*, 2, 423, 1953.
31. McCullagh, P. and Nelder, J.A., *Generalized Linear Models*, Chapman and Hall, London, 1983.
32. Yajima, Y., A central limit theorem of Fourier transforms of strongly dependent stationary processes, *J. Time Ser. Anal.*, 10, 375, 1989.
33. Bloomfield, P., An exponential model for the spectrum of a scalar time series, *Biometrika*, 60, 217, 1973.
34. Diggle, P., *Time Series: A Biostatistical Introduction*, Oxford University Press, Oxford, 1990.
35. Reed, I., Lee, P., and Truong, T., Spectral representation of fractional Brownian motion in n dimension and its properties, *IEEE Trans. Inf. Theory*, 41, 1439, 1995.
36. Ulaby, F.T., Moore, R.K., and Fung, A.K., *Microwave Remote Sensing*, Artech House Inc., Vol. II, Norwood, 1982, chap. 11.
37. Bertacca, M., *Analisi spettrale delle immagini SAR della superficie marina per la discriminazione delle anomalie di superficie*, Ph.D. thesis, University of Pisa, Pisa, 2005.
38. Geweke, J. and Porter-Hudak, S., The estimation and application of long memory time series models, *J. Time Ser. Anal.*, 4, 221,1983.

39. Robinson, P.M. and Hidalgo, F.J. Time series regression with long-range dependence, *Ann. Statist.*, 25, 77, 1997.
40. Lim, J., *Two-Dimensional Signal and Image Processing*, Prentice-Hall, Englewood Cliffs, NJ, 1990.
41. Levi, E.C., Complex-curve fitting, *IRE Trans. Autom. Control*, AC-4, 37, 1959.
42. Dennis, J.E. Jr. and Schnabel, R.B., *Numerical Methods for Unconstrained Optimization and Nonlinear Equations*, Prentice-Hall, Upper Saddle River, NJ, 1983.
43. Pesquet-Popescu, B. and Véhel, J.L., Stochastic fractal models for image processing, *IEEE Signal Process. Mag.*, 19, 48, 2002.
44. Berizzi, F., Bertacca, M., et al., Development and validation of a sea surface fractal model: project results and new perspectives, in *Proc. 2004 Envisat & ERS Symp.*, Salzburg, Austria, 2004.

10

*Spatial Techniques for Image Classification**

Selim Aksoy

CONTENTS

10.1 Introduction

The amount of image data that is received from satellites is constantly increasing. For example, nearly 3 terabytes of data are being sent to Earth by NASA's satellites every day [1]. Advances in satellite technology and computing power have enabled the study of multi-modal, multi-spectral, multi-resolution, and multi-temporal data sets for applications such as urban land-use monitoring and management, GIS and mapping, environmental change, site suitability, and agricultural and ecological studies. Automatic content extraction, classification, and content-based retrieval have become highly desired goals for developing intelligent systems for effective and efficient processing of remotely sensed data sets.

There is extensive literature on classification of remotely sensed imagery using parametric or nonparametric statistical or structural techniques with many different features [2]. Most of the previous approaches try to solve the content extraction problem by building pixel-based classification and retrieval models using spectral and textural features. However, a recent study [3] that investigated classification accuracies reported in the last 15 years showed that there has not been any significant improvement in the

*This work was supported by the TUBITAK CAREER Grant 104E074 and European Commission Sixth Framework Programme Marie Curie International Reintegration Grant MIRG-CT-2005-017504.

performance of classification methodologies over this period. The reason behind this problem is the large semantic gap between the low-level features used for classification and the high-level expectations and scenarios required by the users. This semantic gap makes a human expert's involvement and interpretation in the final analysis inevitable, and this makes processing of data in large remote-sensing archives practically impossible. Therefore, practical accessibility of large remotely sensed data archives is currently limited to queries on geographical coordinates, time of acquisition, sensor type, and acquisition mode [4].

The commonly used statistical classifiers model image content using distributions of pixels in spectral or other feature domains by assuming that similar land-cover and land-use structures will cluster together and behave similarly in these feature spaces. However, the assumptions for distribution models often do not hold for different kinds of data. Even when nonlinear tools such as neural networks or multi-classifier systems are used, the use of only pixel-based data often fails expectations.

An important element of understanding an image is the spatial information because complex land structures usually contain many pixels that have different feature characteristics. Remote-sensing experts also use spatial information to interpret the land-cover because pixels alone do not give much information about image content. Image segmentation techniques [5] automatically group neighboring pixels into contiguous regions based on similarity criteria on the pixels' properties. Even though image segmentation has been heavily studied in image processing and computer vision fields, and despite the early efforts [6] that use spatial information for classification of remotely sensed imagery, segmentation algorithms have only recently started receiving emphasis in remote-sensing image analysis. Examples of image segmentation in the remote-sensing literature include region growing [7] and Markov random field models [8] for segmentation of natural scenes, hierarchical segmentation for image mining [9], region growing for object-level change detection [10] and fuzzy rule–based classification [11], and boundary delineation of agricultural fields [12].

We model spatial information by segmenting images into spatially contiguous regions and classifying these regions according to the statistics of their spectral and textural properties and shape features. To develop segmentation algorithms that group pixels into regions, first, we use nonparametric Bayesian classifiers that create probabilistic links between low-level image features and high-level user-defined semantic land-cover and land-use labels. Pixel-level characterization provides classification details for each pixel with automatic fusion of its spectral, textural, and other ancillary attributes [13]. Then, each resulting pixel-level classification map is converted into a set of contiguous regions using an iterative split-and-merge algorithm [13,14] and mathematical morphology. Following this segmentation process, resulting regions are modeled using the statistical summaries of their spectral and textural properties along with shape features that are computed from region polygon boundaries [14,15]. Finally, nonparametric Bayesian classifiers are used with these region-level features that describe properties shared by groups of pixels to classify these groups into land-cover and land-use categories defined by the user.

The rest of the chapter is organized as follows. An overview of feature data used for modeling pixels is given in Section 10.2. Bayesian classifiers used for classifying these pixels are described in Section 10.3. Algorithms for segmentation of regions are presented in Section 10.4. Feature data used for modeling resulting regions are described in Section 10.5. Application of the Bayesian classifiers to region-level classification is described in Section 10.6. Experiments are presented in Section 10.7 and conclusions are provided in Section 10.8.

10.2 Pixel Feature Extraction

The algorithms presented in this chapter will be illustrated using three different data sets:

- *DC Mall*: Hyperspectral digital image collection experiment (HYDICE) image with 1,280 × 307 pixels and 191 spectral bands corresponding to an airborne data flightline over the Washington DC Mall area.

  The *DC Mall* data set includes seven land-cover and land-use classes: roof, street, path, grass, trees, water, and shadow. A thematic map with ground-truth labels for 8,079 pixels was supplied with the original data [2]. We used this ground truth for testing and separately labeled 35,289 pixels for training. Details are given in Figure 10.1.

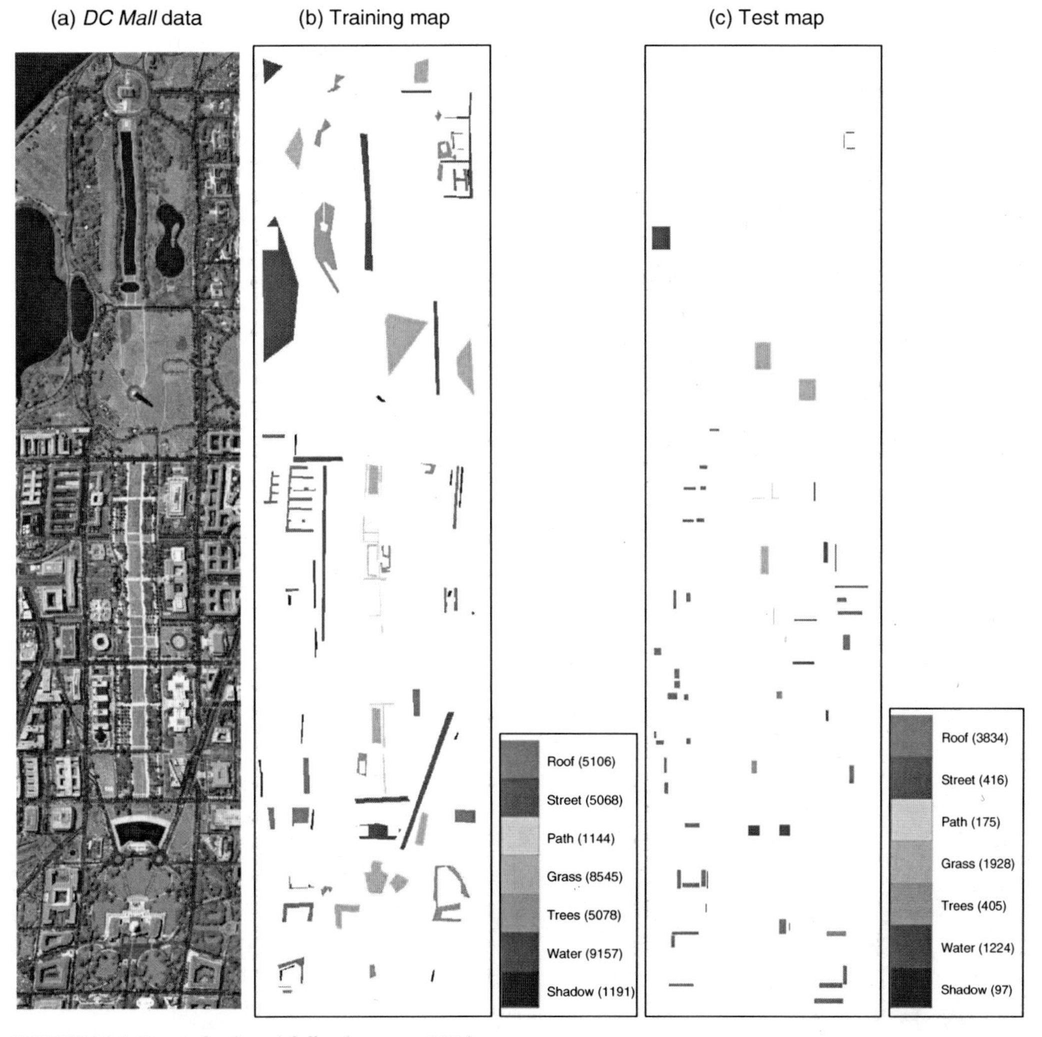

FIGURE 10.1 (See color insert following page 240.)
False color image of the *DC Mall* data set (generated using the bands 63, 52, and 36) and the corresponding ground-truth maps for training and testing. The number of pixels for each class is shown in parenthesis in the legend.

- *Centre*: Digital airborne imaging spectrometer (DAIS) and reflective optics system imaging spectrometer (ROSIS) data with 1,096 × 715 pixels and 102 spectral bands corresponding to the city center in Pavia, Italy.

The *Centre* data set includes nine land-cover and land-use classes: water, trees, meadows, self-blocking bricks, bare soil, asphalt, bitumen, tiles, and shadow. The thematic maps for ground truth contain 7,456 pixels for training and 148,152 pixels for testing. Details are given in Figure 10.2.

- *University*: DAIS and ROSIS data with 610 × 340 pixels and 103 spectral bands corresponding to a scene over the University of Pavia, Italy.

The *University* data set also includes nine land-cover and land-use classes: asphalt, meadows, gravel, trees, (painted) metal sheets, bare soil, bitumen, self-blocking bricks, and shadow. The thematic maps for ground truth contain 3,921 pixels for training and 42,776 pixels for testing. Details are given in Figure 10.3.

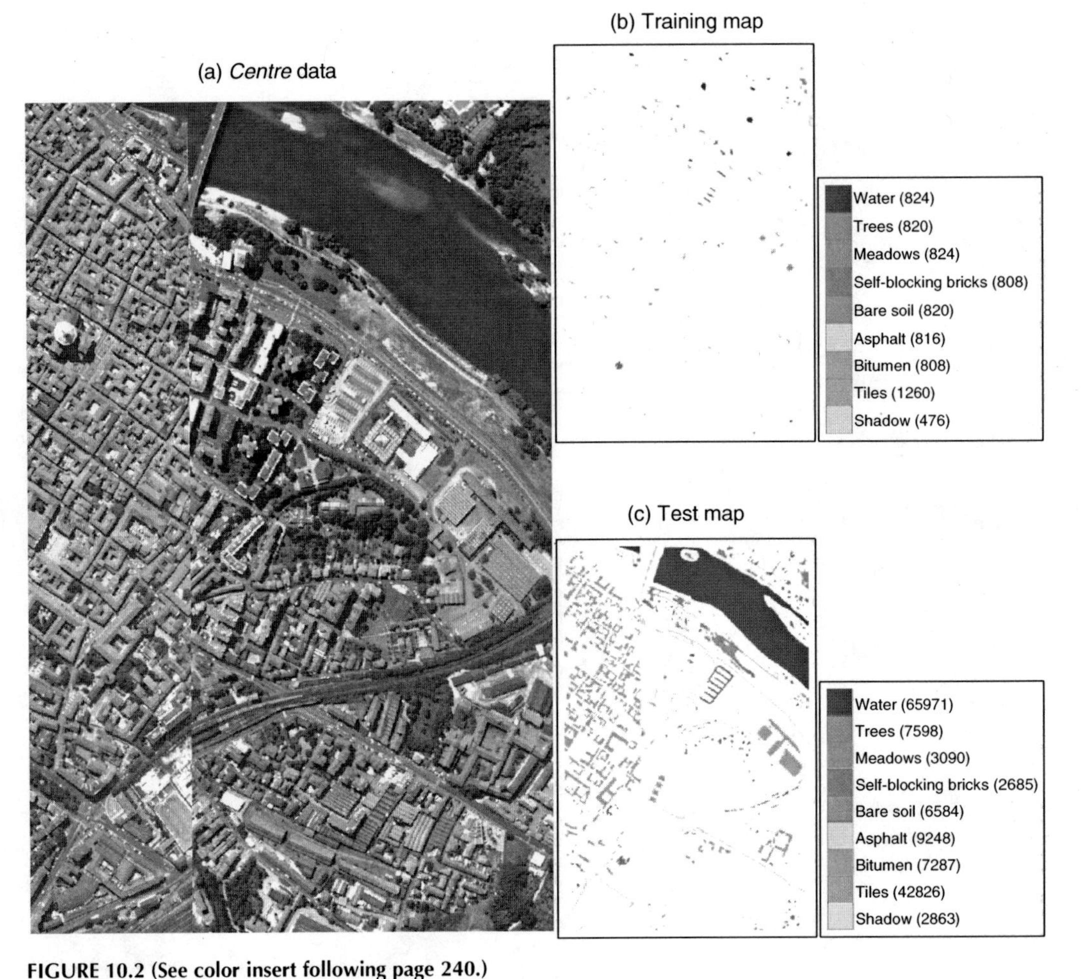

FIGURE 10.2 (See color insert following page 240.)
False color image of the *Centre* data set (generated using the bands 68, 30, and 2) and the corresponding ground-truth maps for training and testing. The number of pixels for each class is shown in parenthesis in the legend. (A missing vertical section in the middle was removed.)

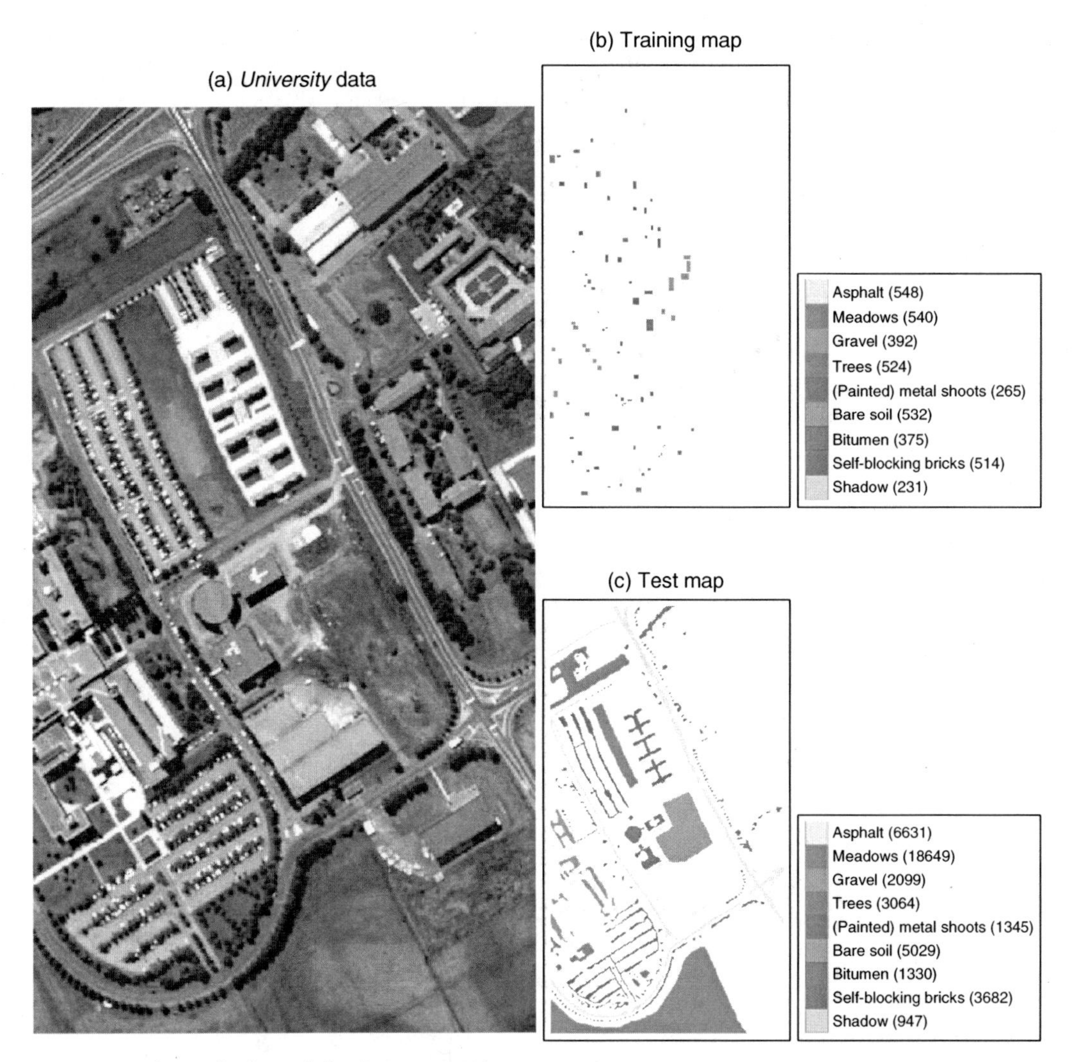

FIGURE 10.3 (See color insert following page 240.)
False color image of the *University* data set (generated using the bands 68, 30, and 2) and the corresponding ground-truth maps for training and testing. The number of pixels for each class is shown in parenthesis in the legend.

The Bayesian classification framework that will be described in the rest of the chapter supports fusion of multiple feature representations such as spectral values, textural features, and ancillary data such as elevation from DEM. In the rest of the chapter, pixel-level characterization consists of spectral and textural properties of pixels that are extracted as described below.

To simplify computations and to avoid the curse of dimensionality during the analysis of hyperspectral data, we apply Fisher's linear discriminant analysis (LDA) [16] that finds a projection to a new set of bases that best separate the data in a least-square sense. The resulting number of bands for each data set is one less than the number of classes in the ground truth.

We also apply principal components analysis (PCA) [16] that finds a projection to a new set of bases that best represent the data in a least-square sense. Then, we retain the top ten principal components instead of the large number of hyperspectral bands. In addition, we extract Gabor texture features [17] by filtering the first principal component image with Gabor kernels at different scales and orientations shown in Figure 10.4. We use kernels rotated by $n\pi/4$, $n = 0, \ldots, 3$, at four scales resulting in feature vectors of length 16. In previous work [13], we observed that, in general, microtexture analysis algorithms like Gabor features smooth noisy areas and become useful for modeling neighborhoods of pixels by distinguishing areas that may have similar spectral responses but have different spatial structures.

Finally, each feature component is normalized by linear scaling to unit variance [18] as

$$\tilde{x} = \frac{x - \mu}{\sigma} \tag{10.1}$$

where x is the original feature value, $\tilde{x}$ is the normalized value, μ is the sample mean, and σ is the sample standard deviation of that feature, so that the features with larger

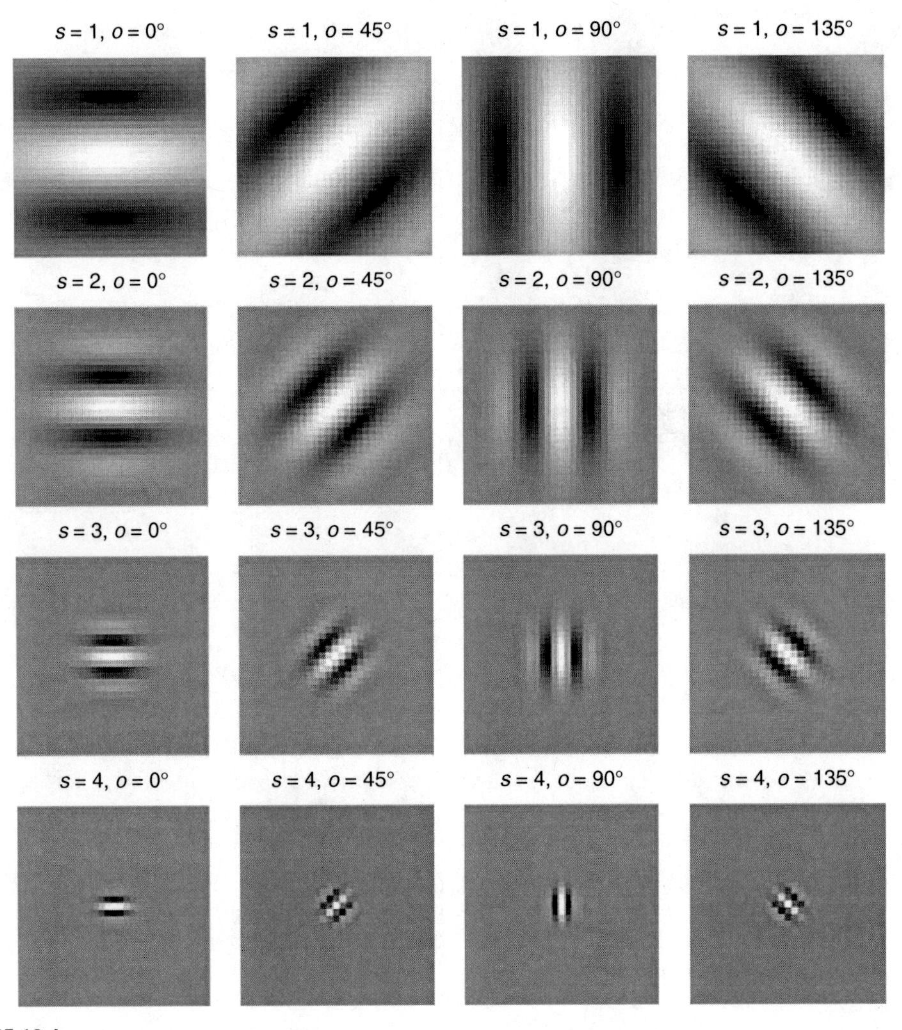

FIGURE 10.4
Gabor texture filters at different scales ($s = 1, \ldots, 4$) and orientations ($o \in \{0°, 45°, 90°, 135°\}$). Each filter is approximated using 31×31 pixels.

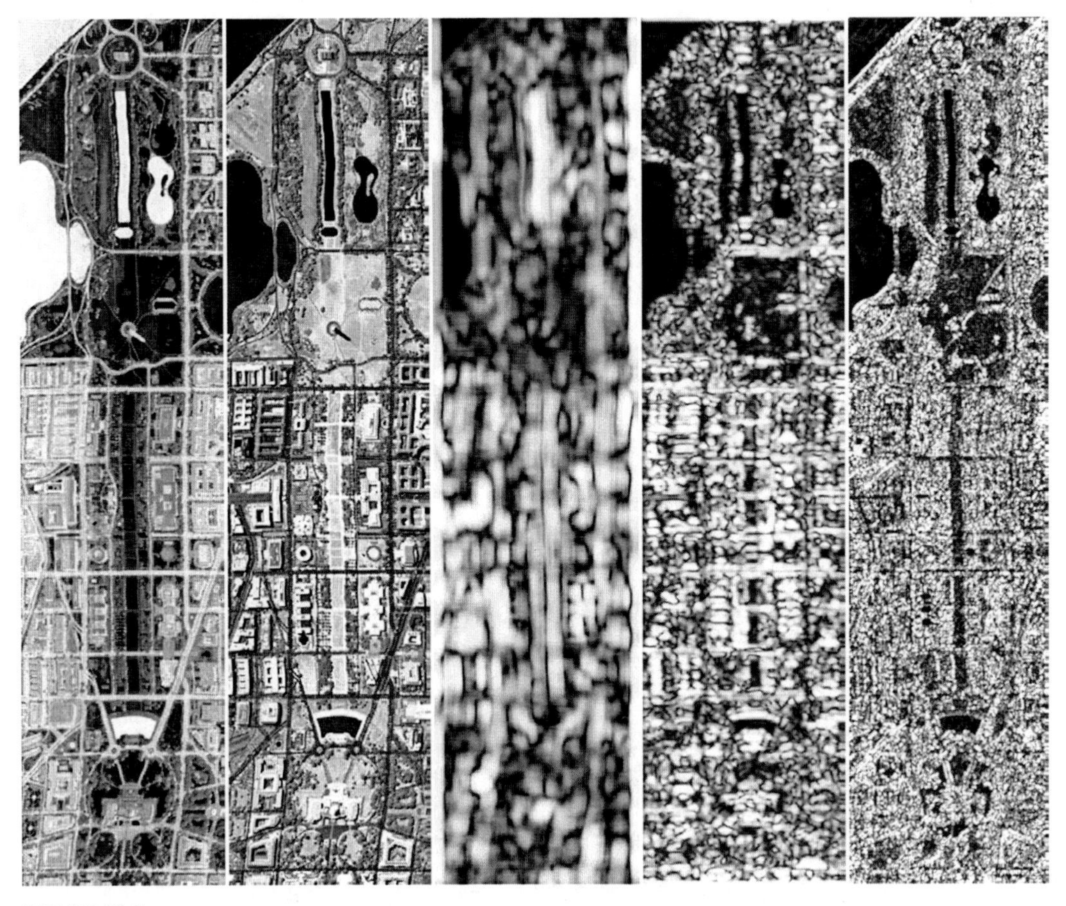

FIGURE 10.5
Pixel feature examples for the *DC Mall* data set. From left to right: the first LDA band, the first PCA band, Gabor features for 90° orientation at the first scale, Gabor features for 0° orientation at the third scale, and Gabor features for 45° orientation at the fourth scale. Histogram equalization was applied to all images for better visualization.

ranges do not bias the results. Examples for pixel-level features are shown in Figure 10.5 through Figure 10.7.

10.3 Pixel Classification

We use Bayesian classifiers to create subjective class definitions that are described in terms of easily computable objective attributes such as spectral values, texture, and ancillary data [13]. The Bayesian framework is a probabilistic tool to combine information from multiple sources in terms of conditional and prior probabilities. Assume there are k class labels, $w_1, \ldots, w_k$, defined by the user. Let $x_1, \ldots, x_m$ be the attributes computed for a pixel. The goal is to find the most probable label for that pixel given a particular set of values of these attributes. The degree of association between the pixel and class w_j can be computed using the posterior probability

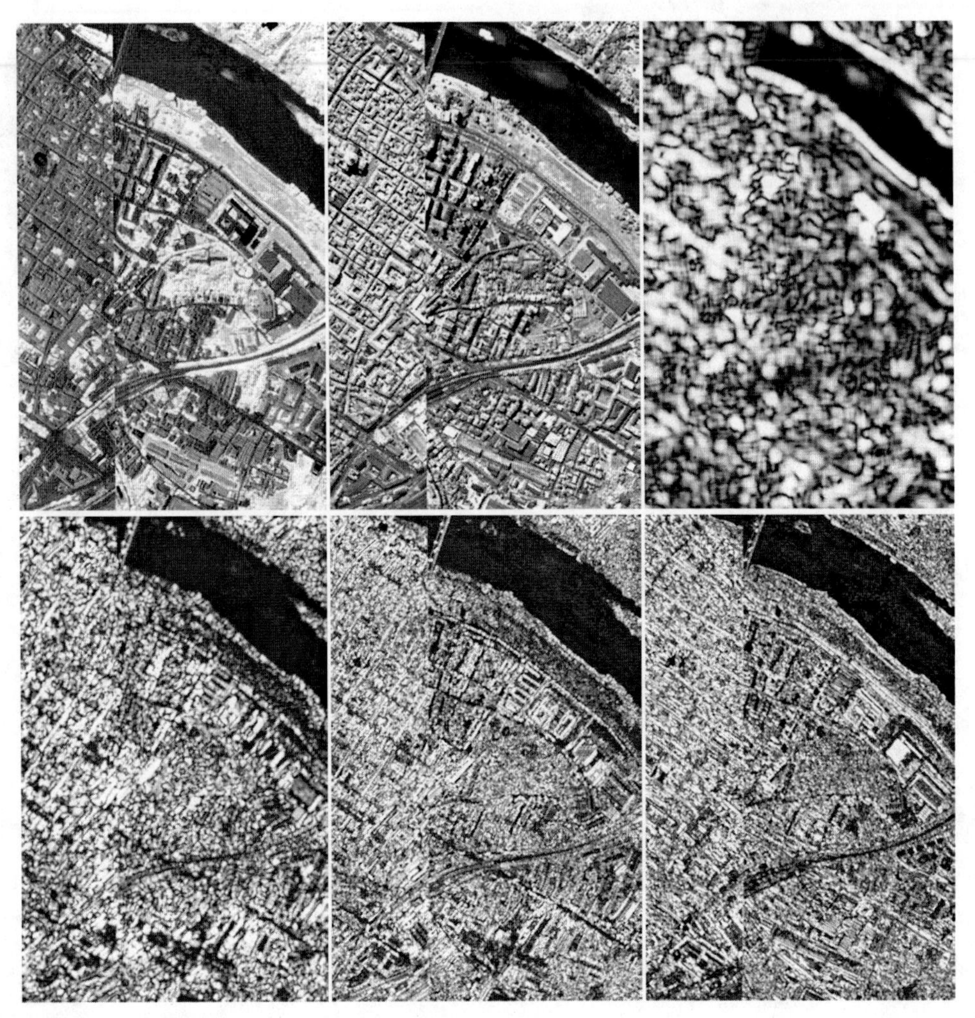

FIGURE 10.6
Pixel feature examples for the *Centre* data set. From left to right, first row: the first LDA band, the first PCA band, and Gabor features for 135° orientation at the first scale; second row: Gabor features for 45° orientation at the third scale, Gabor features for 45° orientation at the fourth scale, and Gabor features for 135° orientation at the fourth scale. Histogram equalization was applied to all images for better visualization.

$$
\begin{aligned}
p(w_j \mid x_1, \ldots, x_m) \\
&= \frac{p(x_1, \ldots, x_m \mid w_j)p(w_j)}{p(x_1, \ldots, x_m)} \\
&= \frac{p(x_1, \ldots, x_m \mid w_j)p(w_j)}{p(x_1, \ldots, x_m \mid w_j)p(w_j) + p(x_1, \ldots, x_m \mid \neg w_j)p(\neg w_j)} \\
&= \frac{p(w_j)\prod_{i=1}^{m} p(x_i \mid w_j)}{p(w_j)\prod_{i=1}^{m} p(x_i \mid w_j) + p(\neg w_j)\prod_{i=1}^{m} p(x_i \mid \neg w_j)}
\end{aligned}
\tag{10.2}
$$

under the conditional independence assumption. The conditional independence assumption simplifies learning because the parameters for each attribute model $p(x_i \mid w_j)$ can be estimated separately. Therefore, user interaction is only required for the labeling of pixels as positive (w_j) or negative ($-w_j$) examples for a particular class under training. Models for

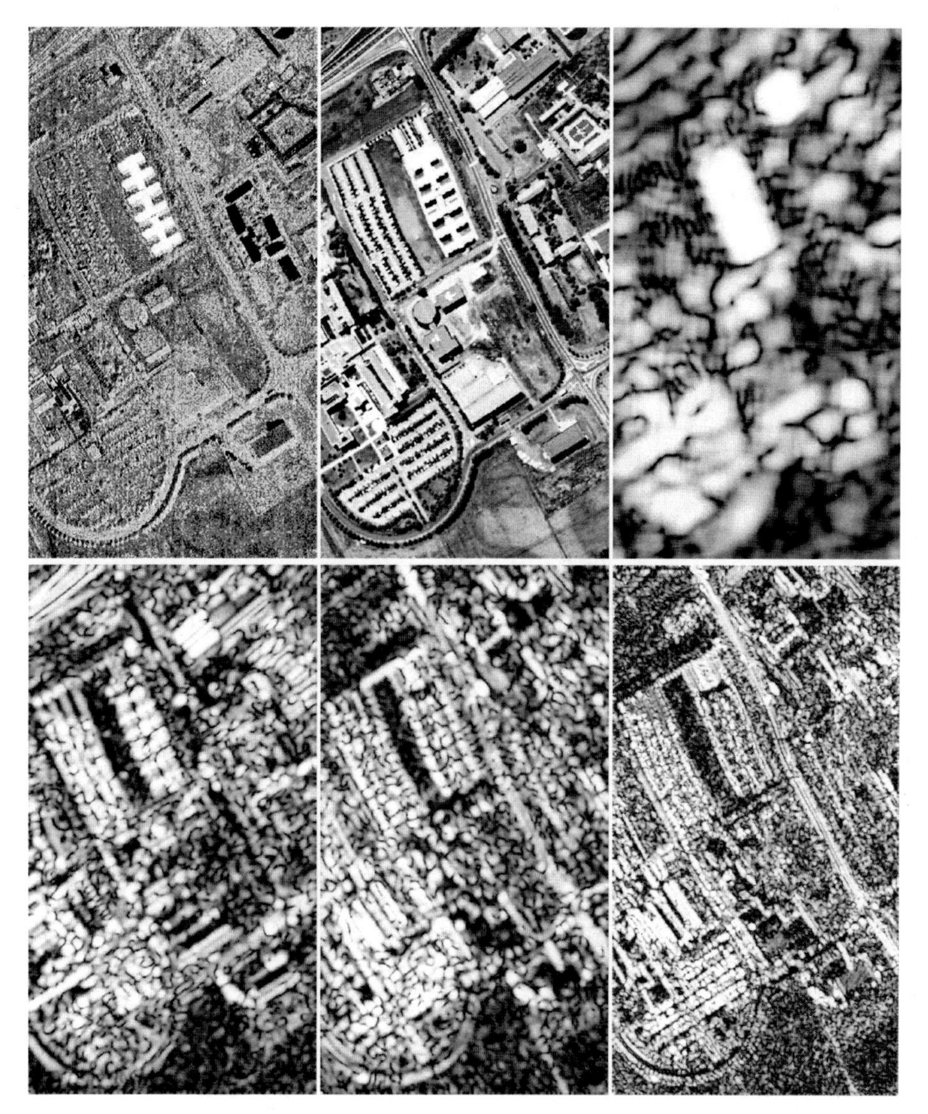

FIGURE 10.7
Pixel feature examples for the *University* data set. From left to right, first row: the first LDA band, the first PCA band, and Gabor features for 45° orientation at the first scale; second row: Gabor features for 45° orientation at the third scale, Gabor features for 135° orientation at the third scale, and Gabor features for 135° orientation at the fourth scale. Histogram equalization was applied to all images for better visualization.

different classes are learned separately from the corresponding positive and negative examples. Then, the predicted class becomes the one with the largest posterior probability and the pixel is assigned the class label

$$w_j^* = \arg \max_{j=1,\ldots,k} p(w_j \mid x_1, \ldots, x_m) \tag{10.3}$$

We use discrete variables and a nonparametric model in the Bayesian framework where continuous features are converted to discrete attribute values using the unsupervised k-means clustering algorithm for vector quantization. The number of clusters (quantization levels) is empirically chosen for each feature. (An alternative is to use a parametric distribution assumption, for example, Gaussian, for each individual continuous feature but these parametric assumptions do not always hold.) Schroder et al. [19] used similar classifiers to retrieve images from remote-sensing archives by approximating the probabilities of images belonging to different classes using pixel-level probabilities. In the following, we describe learning of the models for $p(x_i \mid w_j)$ using the positive training examples for the jth class label. Learning of $p(x_i \mid \neg\, w_j)$ is done the same way using the negative examples.

For a particular class, let each discrete variable x_i have r_i possible values (states) with probabilities

$$p(x_i = z \mid \theta_\mathrm{i}) = \theta_\mathrm{iz} > 0 \tag{10.4}$$

where $z \in \{1, \ldots, r_i\}$ and $\theta_\mathrm{i} = \{\theta_\mathrm{iz}\}_{z=1}^{r_i}$ is the set of parameters for the ith attribute model. This corresponds to a multinomial distribution. Since maximum likelihood estimates can give unreliable results when the sample is small and the number of parameters is large, we use the Bayes estimate of θ_iz that can be computed as the expected value of the posterior distribution.

We can choose any prior for θ_i in the computation of the posterior distribution but there is a big advantage in using conjugate priors. A conjugate prior is one which, when multiplied with the direct probability, gives a posterior probability having the same functional form as the prior, thus allowing the posterior to be used as a prior in further computations [20]. The conjugate prior for the multinomial distribution is the Dirichlet distribution [21]. Geiger and Heckerman [22] showed that if all allowed states of the variables are possible (i.e., $\theta_\mathrm{iz} > 0$) and if certain parameter independence assumptions hold, then a Dirichlet distribution is indeed the only possible choice for the prior.

Given the Dirichlet prior $p(\theta_\mathrm{i}) = \mathrm{Dir}(\theta_\mathrm{i} \mid \alpha_{i1}, \ldots, \alpha_{ir_i})$, where α_{iz} are positive constants, the posterior distribution of θ_i can be computed using the Bayes rule as

$$\begin{aligned} p(\theta_\mathrm{i} \mid D) &= \frac{p(D \mid \theta_\mathrm{i}) p(\theta_\mathrm{i})}{p(D)} \\ &= \mathrm{Dir}(\theta_\mathrm{i} \mid \alpha_{i1} + N_{i1}, \ldots, \alpha_{ir_i} + N_{ir_i}) \end{aligned} \tag{10.5}$$

where D is the training sample and N_{iz} is the number of cases in D in which $x_i = z$. Then, the Bayes estimate for θ_iz can be found by taking the conditional expected value

$$\hat{\theta}_\mathrm{iz} = E_{p(\theta_\mathrm{i} \mid D)}[\theta_\mathrm{iz}] = \frac{\alpha_{iz} + N_{iz}}{\alpha_i + N_i} \tag{10.6}$$

where $\alpha_i = \sum_{z=1}^{r_i} \alpha_{iz}$ and $N_i = \sum_{z=1}^{r_i} N_{iz}$.

An intuitive choice for the hyperparameters $\alpha_{i1}, \ldots, \alpha_{ir_i}$ of the Dirichlet distribution is Laplace's uniform prior [23] that assumes all r_i states to be equally probable ($\alpha_{iz} = 1\ \forall z \in \{1, \ldots, r_i\}$), which results in the Bayes estimate

$$\hat{\theta}_\mathrm{iz} = \frac{1 + N_{iz}}{r_i + N_i} \tag{10.7}$$

Laplace's prior is regarded to be a safe choice when the distribution of the source is unknown and the number of possible states r_i is fixed and known [24].

Given the current state of the classifier that was trained using the prior information and the sample D, we can easily update the parameters when new data D' is available. The new posterior distribution for θ_i becomes

$$p(\theta_i \mid D, D') = \frac{p(D' \mid \theta_i)p(\theta_i \mid D)}{p(D' \mid D)} \tag{10.8}$$

With the Dirichlet priors and the posterior distribution for $p(\theta_i \mid D)$ given in Equation 10.5, the updated posterior distribution becomes

$$p(\theta_i \mid D, D') = \text{Dir}(\theta_i \mid \alpha_{i1} + N_{i1} + N'_{i1}, \ldots, \alpha_{ir_i} + N_{ir_i} + N'_{ir_i}) \tag{10.9}$$

where N'_{iz} is the number of cases in D' in which $x_i = z$. Hence, updating the classifier parameters involves only updating the counts in the estimates for $\hat{\theta}_{iz}$.

The Bayesian classifiers that are learned from examples as described above are used to compute probability maps for all land-cover and land-use classes and assign each pixel to one of these classes using the maximum *a posteriori* probability (MAP) rule given in Equation 10.3. Example probability maps are shown in Figure 10.8 through Figure 10.10.

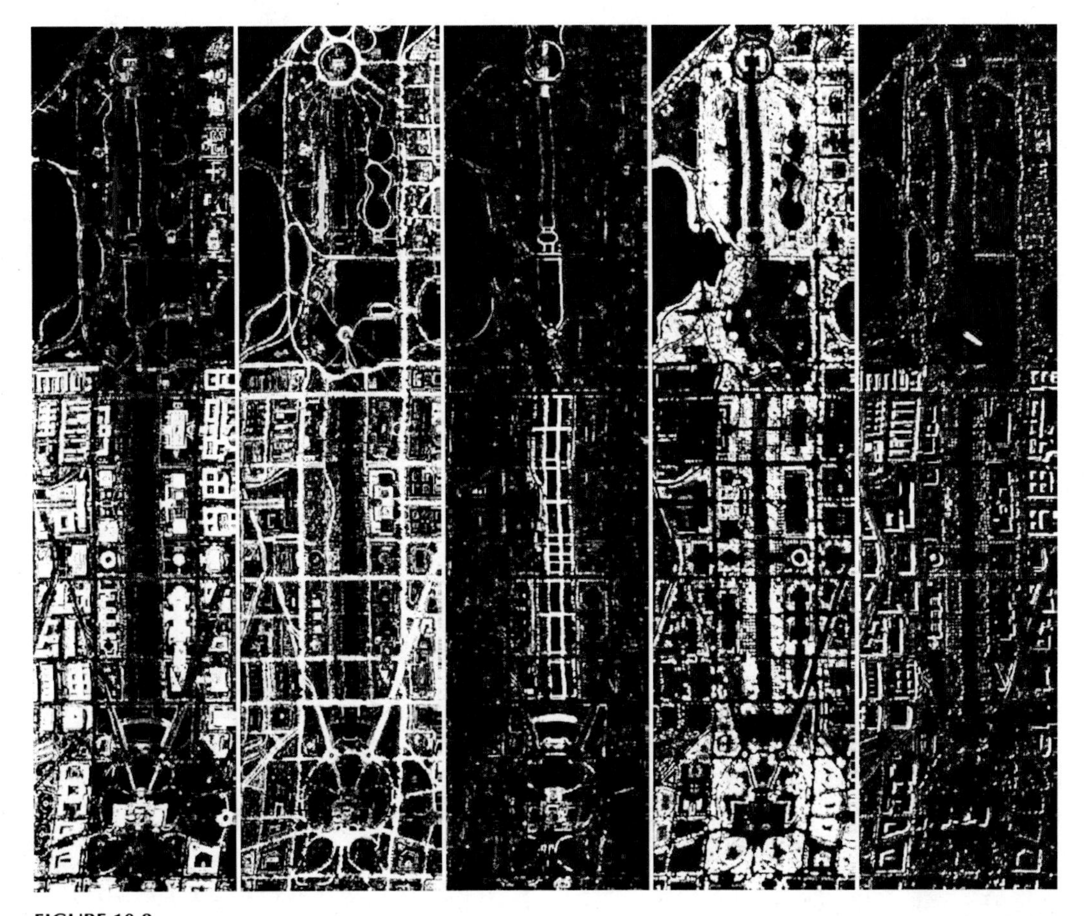

FIGURE 10.8
Pixel-level probability maps for different classes of the *DC Mall* data set. From left to right: roof, street, path, trees, shadow. Brighter values in the map show pixels with high probability of belonging to that class.

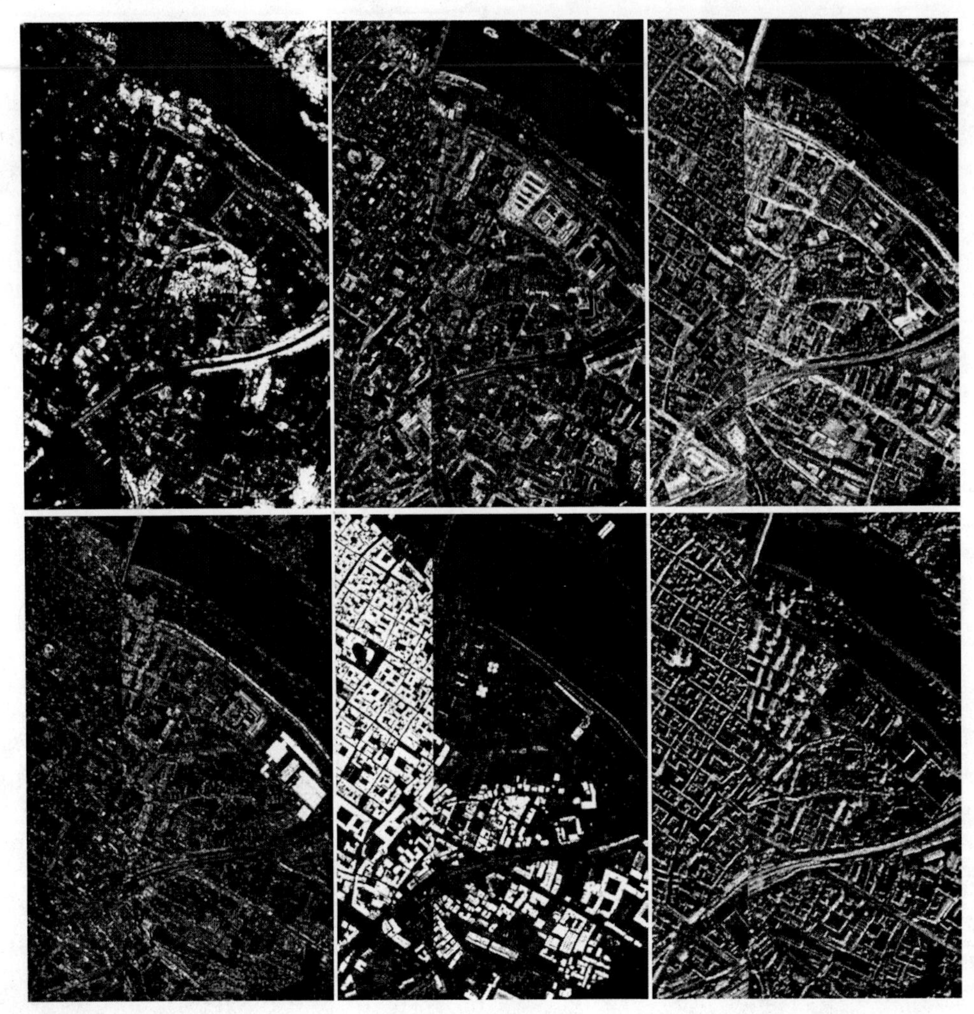

FIGURE 10.9
Pixel-level probability maps for different classes of the *Centre* data set. From left to right, first row: trees, self-blocking bricks, asphalt; second row: bitumen, tiles, shadow. Brighter values in the map show pixels with high probability of belonging to that class.

10.4 Region Segmentation

Image segmentation is used to group pixels that belong to the same structure with the goal of delineating each individual structure as an individual region. In previous work [25], we used an automatic segmentation algorithm that breaks an image into many small regions and merges them by minimizing an energy functional that trades off the similarity of regions against the length of their shared boundaries. We have also recently experimented with several segmentation algorithms from the computer vision literature. Algorithms that are based on graph clustering [26], mode seeking [27], and classification [28] have been reported to successful in moderately sized color images with relatively homogeneous structures. However, we could not apply these techniques successfully to our data sets because the huge amount of data in hyperspectral images made processing infeasible due to both memory and computational requirements, and the detailed

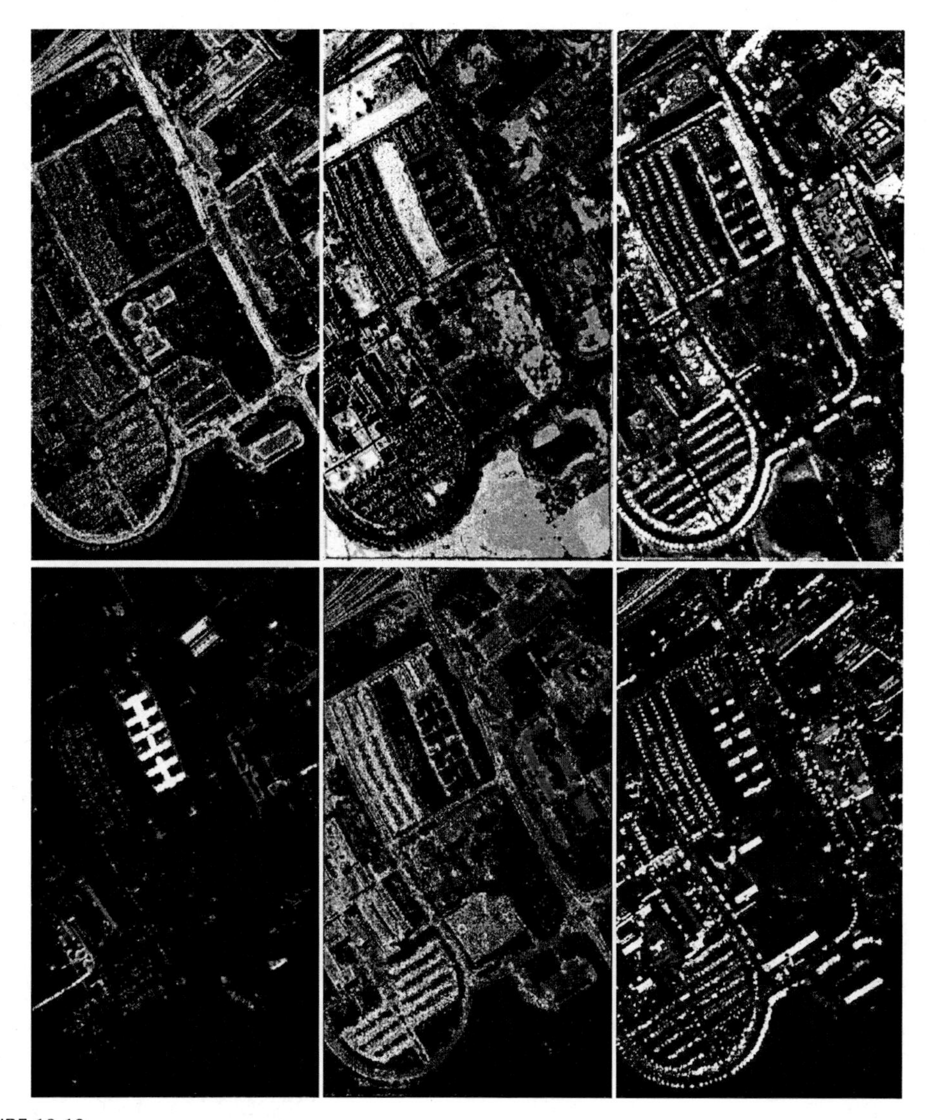

FIGURE 10.10
Pixel-level probability maps for different classes of the *University* data set. From left to right, first row: asphalt, meadows, trees; second row: metal sheets, self-blocking bricks, shadow. Brighter values in the map show pixels with high probability of belonging to that class.

structure in high-resolution remotely sensed imagery prevented the use of sampling that has been often used to reduce the computational requirements of these techniques.

The segmentation approach we have used in this work consists of smoothing filters and mathematical morphology. The input to the algorithm includes the probability maps for all classes, where each pixel is assigned either to one of these classes or to the reject class for probabilities smaller than a threshold (the latter type of pixels are initially marked as background). Because pixel-based classification ignores spatial correlations, the initial segmentation may contain isolated pixels with labels different from those of their neighbors. We use an iterative split-and-merge algorithm [13] to convert this intermediate step into contiguous regions as follows:

1. Merge pixels with identical class labels to find the initial set of regions and mark these regions as foreground.

2. Mark regions with areas smaller than a threshold as background using connected components analysis [5].
3. Use region growing to iteratively assign background pixels to the foreground regions by placing a window at each background pixel and assigning it to the class that occurs the most in its neighborhood.

This procedure corresponds to a spatial smoothing of the clustering results. We further process the resulting regions using mathematical morphology operators [5] to automatically divide large regions into more compact subregions as follows [13]:

1. Find individual regions using connected components analysis for each class.
2. For all regions, compute the erosion transform [5] and repeat:
   - Threshold erosion transform at steps of 3 pixels in every iteration
   - Find connected components of the thresholded image
   - Select subregions that have an area smaller than a threshold
   - Dilate these subregions to restore the effects of erosion
   - Mark these subregions in the output image by masking the dilation using the original image
   - Until no more sub-regions are found
3. Merge the residues of previous iterations to their smallest neighbors.

The merging and splitting process is illustrated in Figure 10.11. The probability of each region belonging to a land-cover or land-use class can be estimated by propagating class labels from pixels to regions. Let $X = \{x_1, \ldots, x_n\}$ be the set of pixels that are merged to form a region. Let w_j and $p(w_j \mid x_i)$ be the class label and its posterior probability, respectively, assigned to pixel x_i by the classifier. The probability $p(w_j \mid x \in X)$ that a pixel in the merged region belongs to the class w_j can be computed as

$$
\begin{aligned}
p(w_j \mid x \in X) &\\
&= \frac{p(w_j, x \in X)}{p(x \in X)} = \frac{p(w_j, x \in X)}{\sum_{t=1}^{k} p(w_t, x \in X)} \\
&= \frac{\sum_{x \in X} p(w_j, x)}{\sum_{t=1}^{k} \sum_{x \in X} p(w_t, x)} = \frac{\sum_{x \in X} p(w_j \mid x) p(x)}{\sum_{t=1}^{k} \sum_{x \in X} p(w_t \mid x) p(x)} \\
&= \frac{E_x\{I_{x \in X}(x) p(w_j \mid x)\}}{\sum_{t=1}^{k} E_x\{I_{x \in X}(x) p(w_t \mid x)\}} = \frac{1}{n} \sum_{i=1}^{n} p(w_j \mid x_i)
\end{aligned} \tag{10.10}
$$

where $I_A(\cdot)$ is the indicator function associated with the set A. Each region in the final segmentation is assigned labels with probabilities using Equation 10.10.

10.5 Region Feature Extraction

Region-level representations include properties shared by groups of pixels obtained through region segmentation. The regions are modeled using the statistical summaries of their spectral and textural properties along with shape features that are computed from

(a) A large connected region formed by merging pixels labeled as street in *DC Mall* data

(b) More compact sub-regions after splitting the region in (a)

(c) A large connected region formed by merging pixels labeled as tiled in *Centre* data

(d) More compact sub-regions after splitting the region in (c).

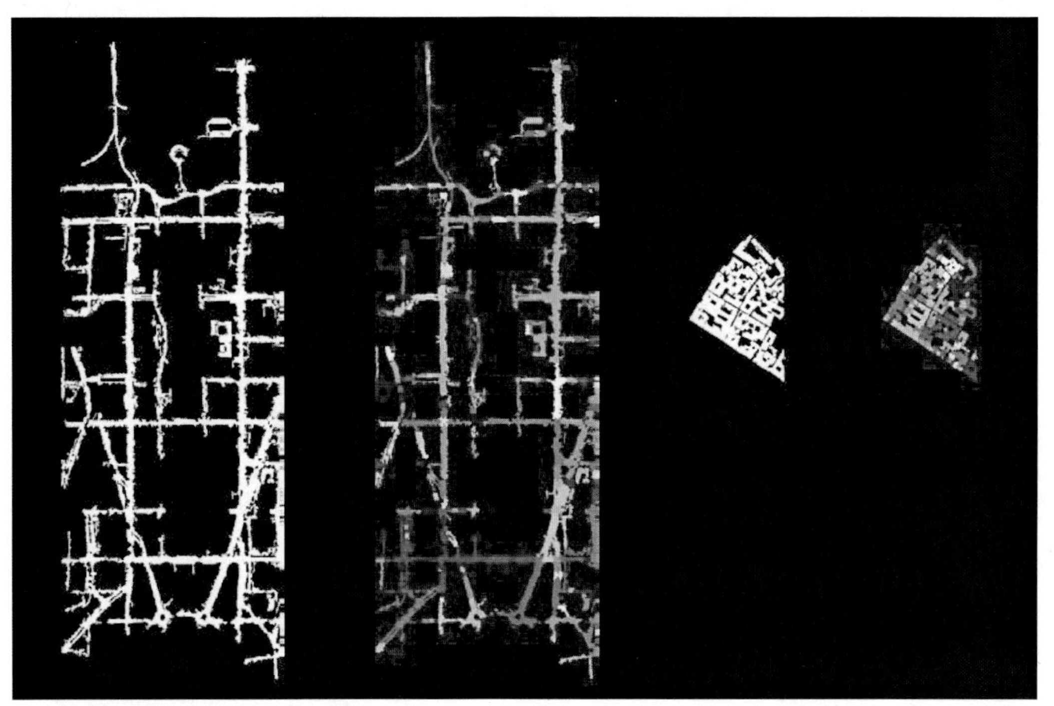

FIGURE 10.11 (See color insert following page 240.)
Examples for the region segmentation process. The iterative algorithm that uses mathematical morphology operators is used to split a large connected region into more compact subregions.

region polygon boundaries. The statistical summary for a region is computed as the means and standard deviations of features of the pixels in that region. Multi-dimensional histograms also provide pixel feature distributions within individual regions. The shape properties [5] of a region correspond to its

- Aarea
- Orientation of the region's major axis with respect to the x axis
- Eccentricity (ratio of the distance between the foci to the length of the major axis; for example, a circle is an ellipse with zero eccentricity)
- Euler number (1 minus the number of holes in the region)
- Solidity (ratio of the area to the convex area)
- Extent (ratio of the area to the area of the bounding box)
- Spatial variances along the x and y axes
- Spatial variances along the region's principal (major and minor) axes
- Resulting in a feature vector of length 10

10.6 Region Classification

In the remote-sensing literature, image classification is usually done by using pixel features as input to classifiers such as minimum distance, maximum likelihood, neural networks, or decision trees. However, large within-class variations and small between-class variations of these features at the pixel level and the lack of spatial information limit the accuracy of these classifiers.

In this work, we perform final classification using region-level information. To use the Bayesian classifiers that were described in Section 10.3, different region-based features such as statistics and shape features are independently converted to discrete random variables using the k-means algorithm for vector quantization. In particular, for each region, we obtain four values from

- Clustering of the statistics of the LDA bands (6 bands for *DC Mall* data, 8 bands for *Centre* and *University* data)
- Clustering of the statistics of the 10 PCA bands
- Clustering of the statistics of the 16 Gabor bands
- Clustering of the 10 shape features

In the next section, we evaluate the performance of these new features for classifying regions (and the corresponding pixels) into land-cover and land-use categories defined by the user.

10.7 Experiments

Performances of the features and the algorithms described in the previous sections were evaluated both quantitatively and qualitatively. First, pixel-level features (LDA, PCA, and Gabor) were extracted and normalized for all three data sets as described in Section 10.2. The ground-truth maps shown in Figure 10.1 through Figure 10.3 were used to divide the data into independent training and test sets. Then, the k-means algorithm was used to cluster (quantize) the continuous features and convert them to discrete attribute values, and Bayesian classifiers with discrete nonparametric models were trained using these attributes and the training examples as described in Section 10.3. The value of k was set to 25 empirically for all data sets. Example probability maps for some of the classes were given in Figure 10.8 through Figure 10.10. Confusion matrices, shown in Table 10.1 through Table 10.3, were computed using the test ground truth for all data sets.

Next, the iterative split-and-merge algorithm described in Section 10.4 was used to convert the pixel-level classification results into contiguous regions. The neighborhood size for region growing was set to 3×3. The minimum area threshold in the segmentation process was set to 5 pixels. After the region-level features (LDA, PCA, and Gabor statistics, and shape features) were computed and normalized for all resulting regions as described in Section 10.5, they were also clustered (quantized) and converted to discrete values. The value of k was set to 25 again for all data sets. Then, Bayesian classifiers were trained using the training ground truth as described in Section 10.6, and were applied to the test data to produce the confusion matrices shown in Table 10.4 through Table 10.6.

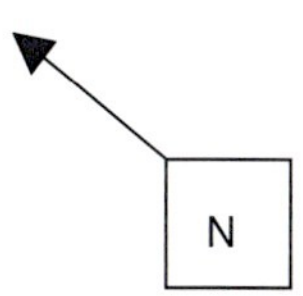

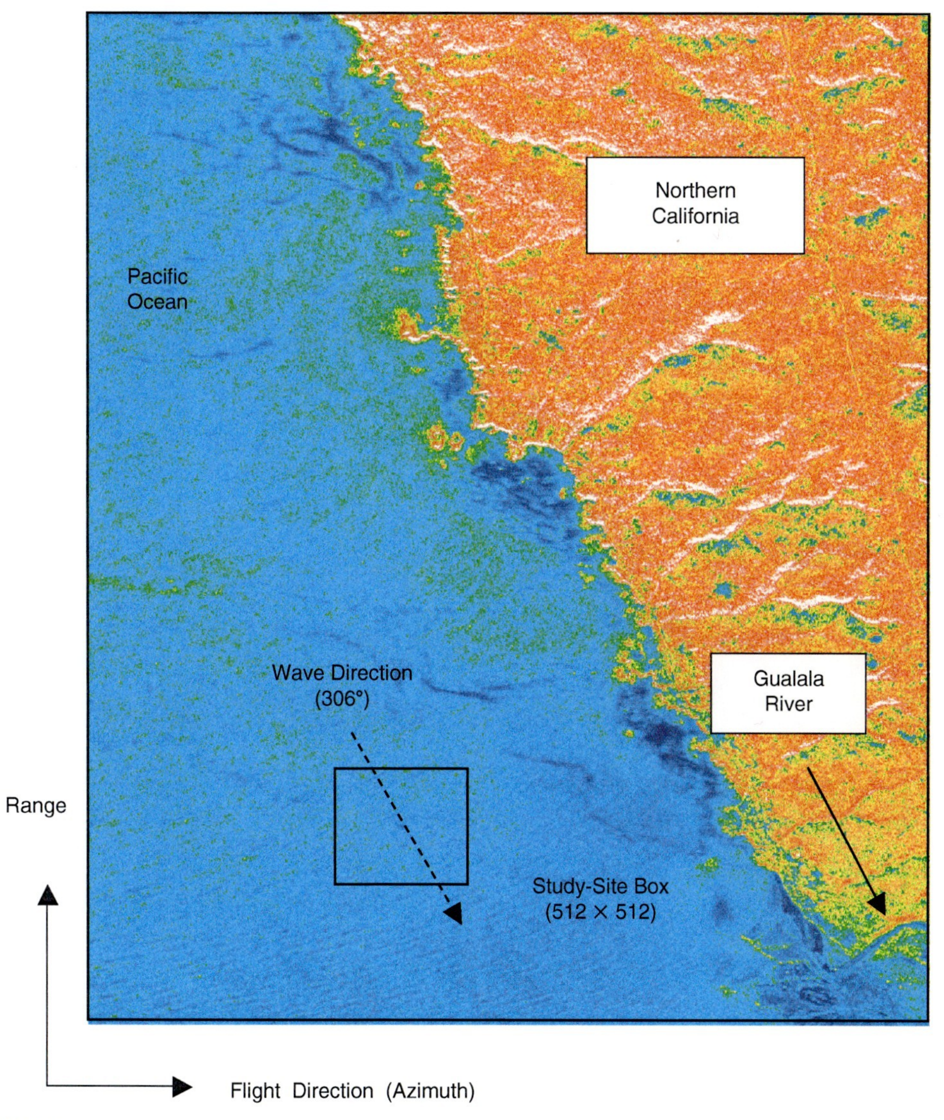

FIGURE 1.1
An L-band, VV-pol, AIRSAR image, of northern California coastal waters (Gualala River dataset), showing ocean waves propagating through a study-site box.

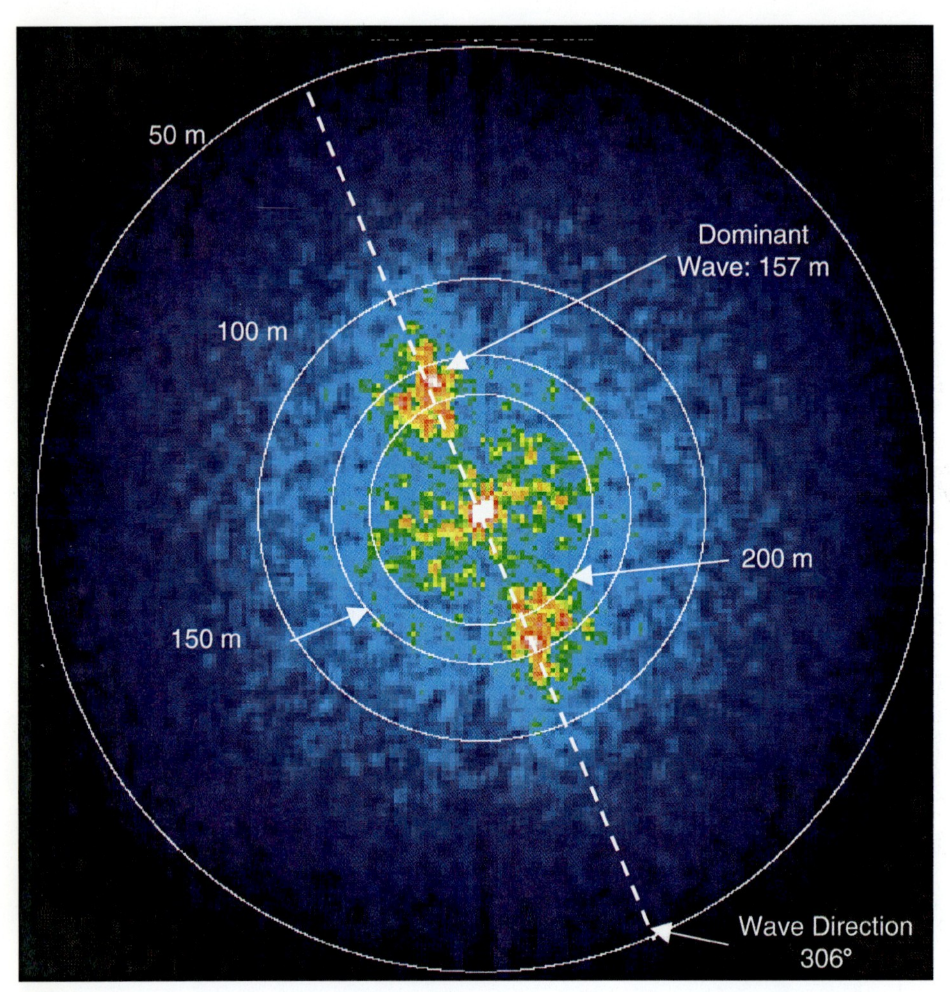

FIGURE 1.4
Orientation angle spectra versus wave number for azimuth direction waves propagating through the study site. The white rings correspond to 50 m, 100 m, 150 m, and 200 m. The dominant wave, of wavelength 157 m, is propagating at a heading of 306°.

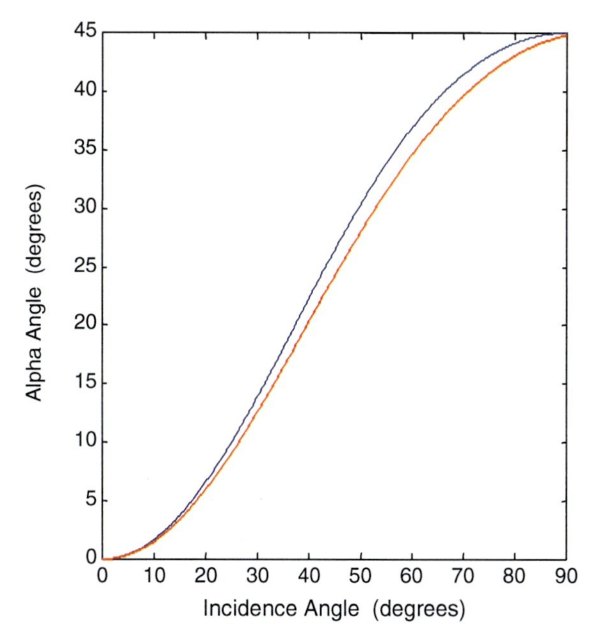

FIGURE 1.6
Small perturbation model dependence of alpha on the incidence angle. The red curve is for a dielectric constant representative of sea water (80–70j) and the blue curve is for a perfectly conducting surface.

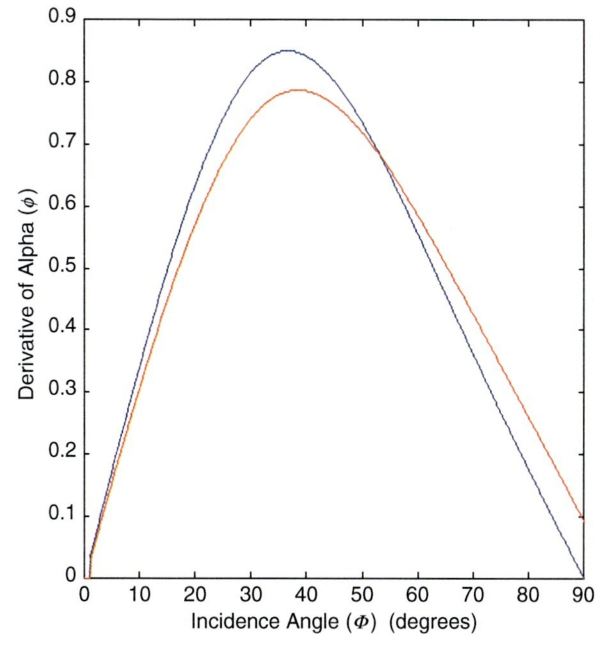

FIGURE 1.7
Derivative of alpha with respect to the incidence angle. The red curve is for a sea water dielectric and the blue curve is for a perfectly conducting surface.

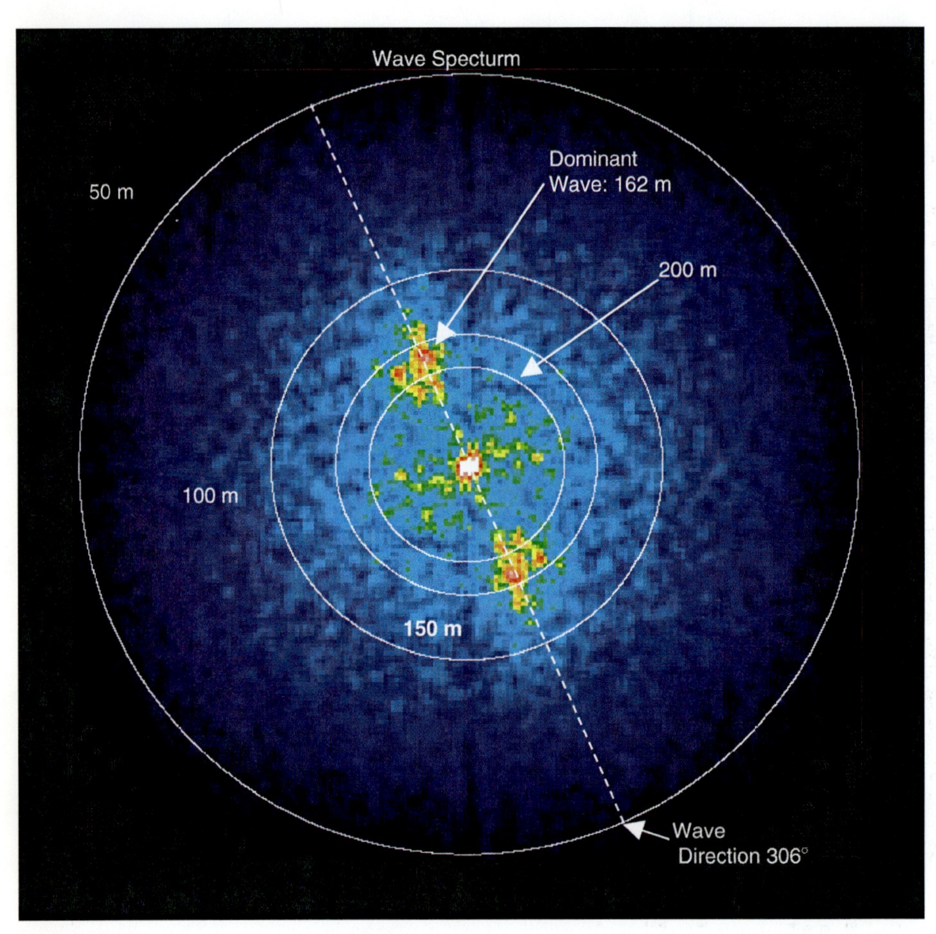

FIGURE 1.9
Spectrum of waves in the range direction using the alpha parameter from the Cloude–Pottier decomposition method. Wave direction is 306° and dominant wavelength is 162 m.

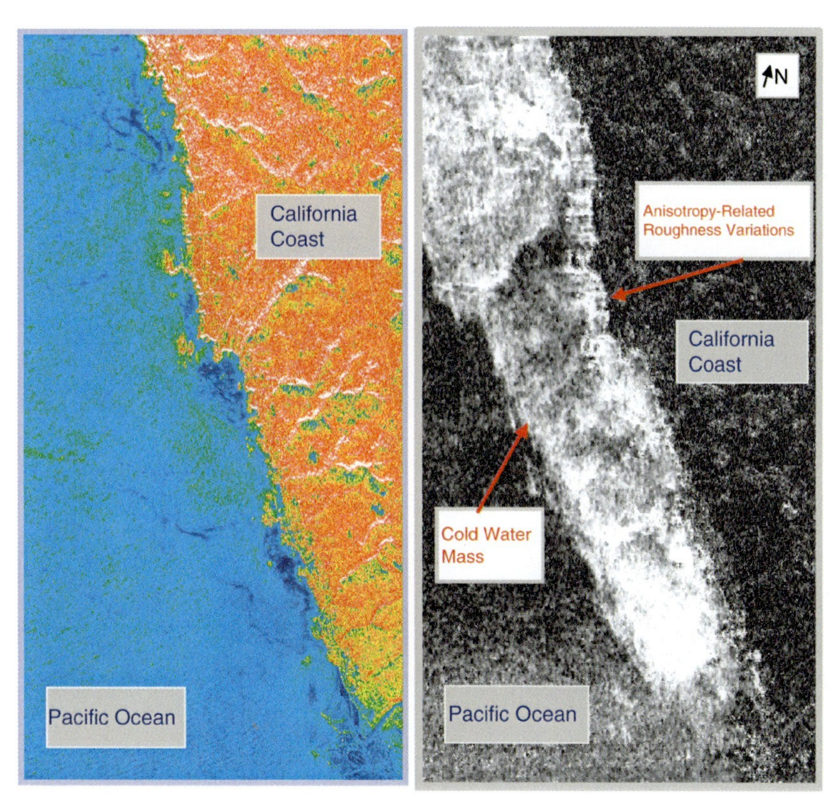

FIGURE 1.19
(a and b) (a) Variations in anisotropy at low wind speeds for a filament of colder, trapped water along the northern California coast. The roughness changes are not seen in the conventional VV-pol image, but are clearly visible in (b) an anisotropy image. The data is from coastal waters near the Mendocino Co. town of Gualala.

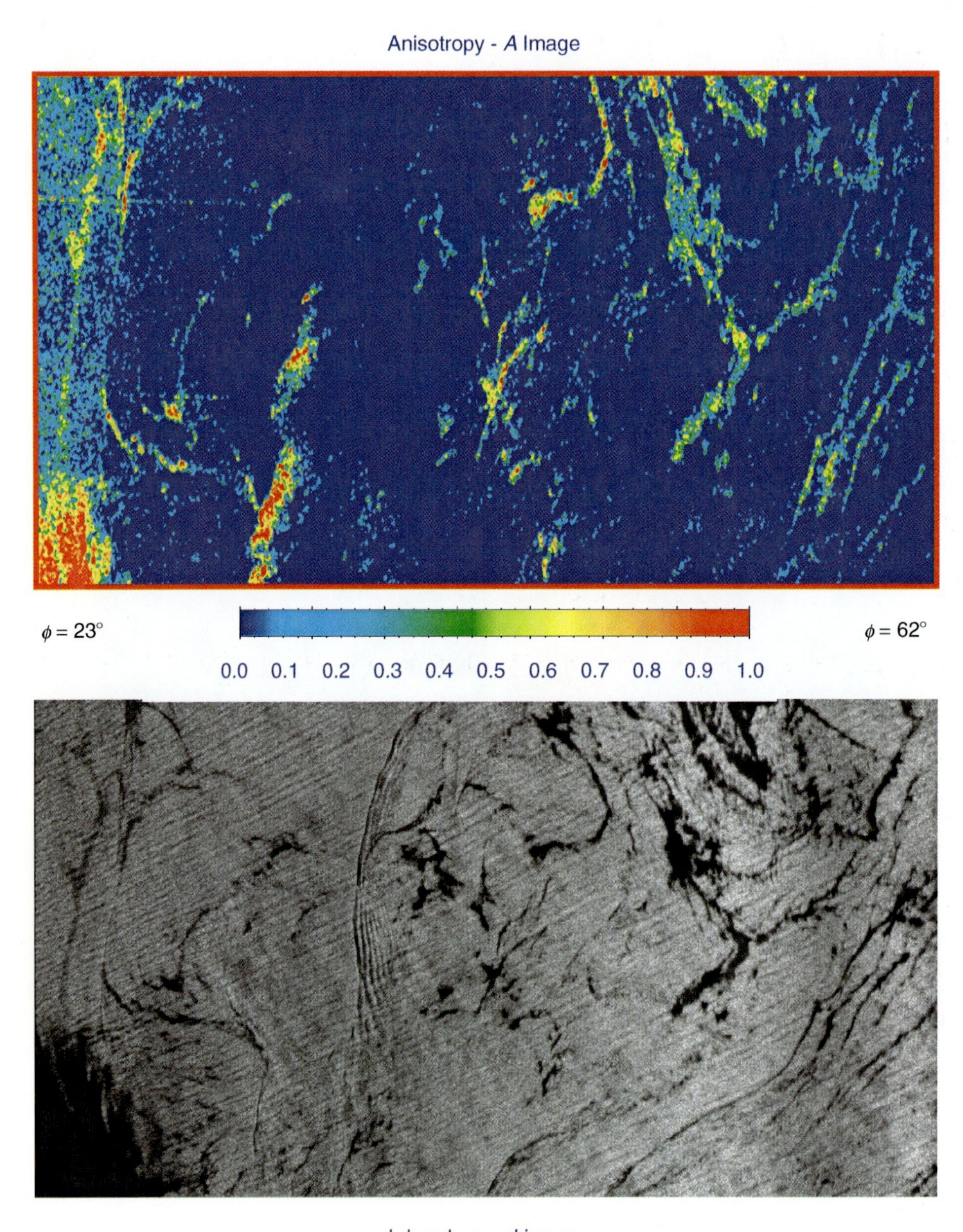

FIGURE 1.20
(a) Image of anisotropy values. The quantity, 1-A, is proportional to small-scale surface roughness and (b) a conventional L-band VV-pol image of the study area.

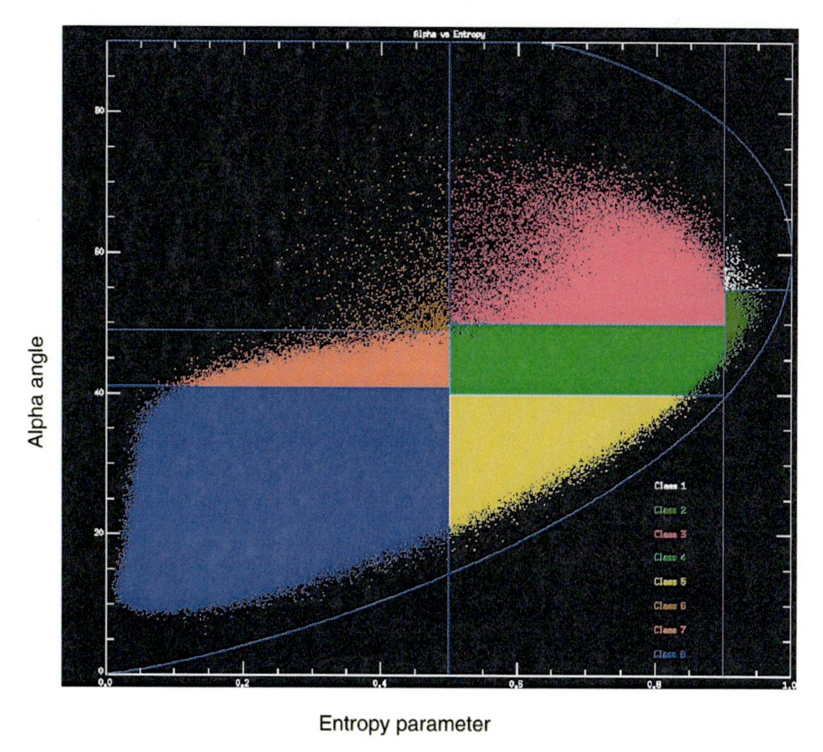

FIGURE 1.21
Alpha-entropy scatter plot for the image study area. The plot is divided into eight color-coded scattering classes for the Cloude–Pottier decomposition described in Ref. [6].

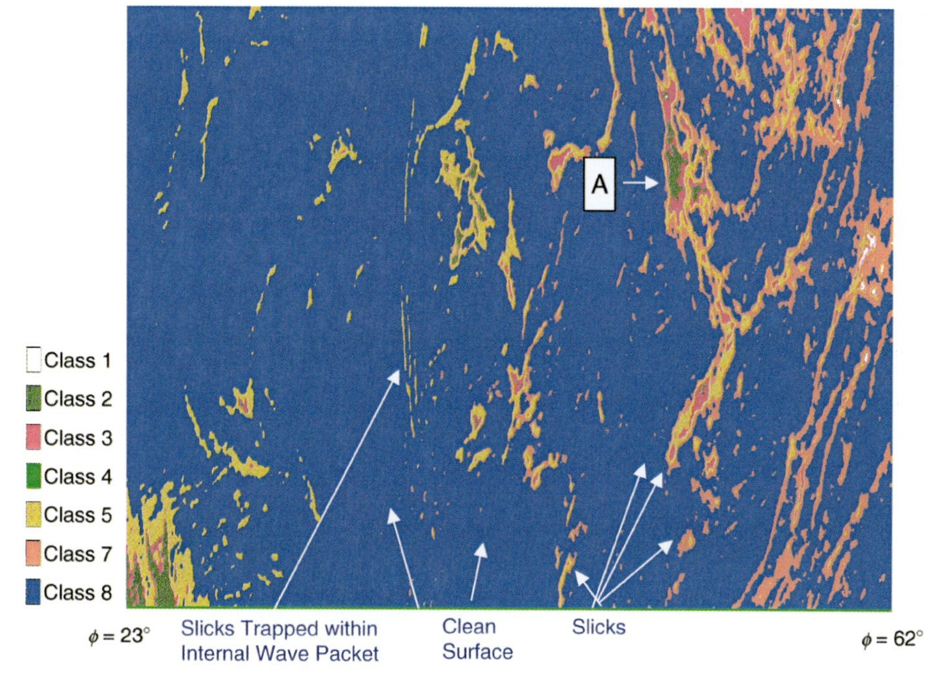

FIGURE 1.22
Classification of the slick-field image into $H/\bar{\alpha}$ scattering classes.

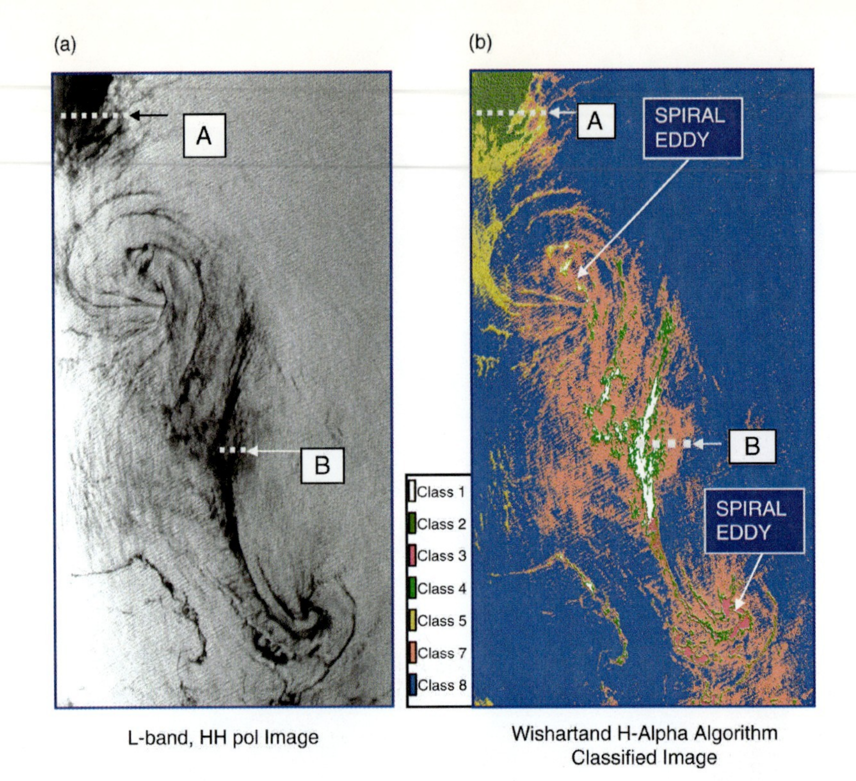

FIGURE 1.23
(a) L-band, HH-pol image of a second study image (CM6744) containing two strong spiral eddies marked by natural biogenic slicks and (b) classification of the slicks marking the spiral eddies. The image features were classified into eight classes using the H–$\bar{\alpha}$ values combined with the Wishart classifier.

FIGURE 1.24
Classification of the slick-field image into H/A/$\bar{\alpha}$ 14 scattering classes. The Classes 1–7 correspond to anisotropy A values 0.5 to 1.0 and the classes 8–14 correspond to anisotropy A values 0.0 to 0.49. The two lighter blue vertical features at the lower right of the image appear in all images involving anisotropy and are thought to be smooth slicks of lower concentration.

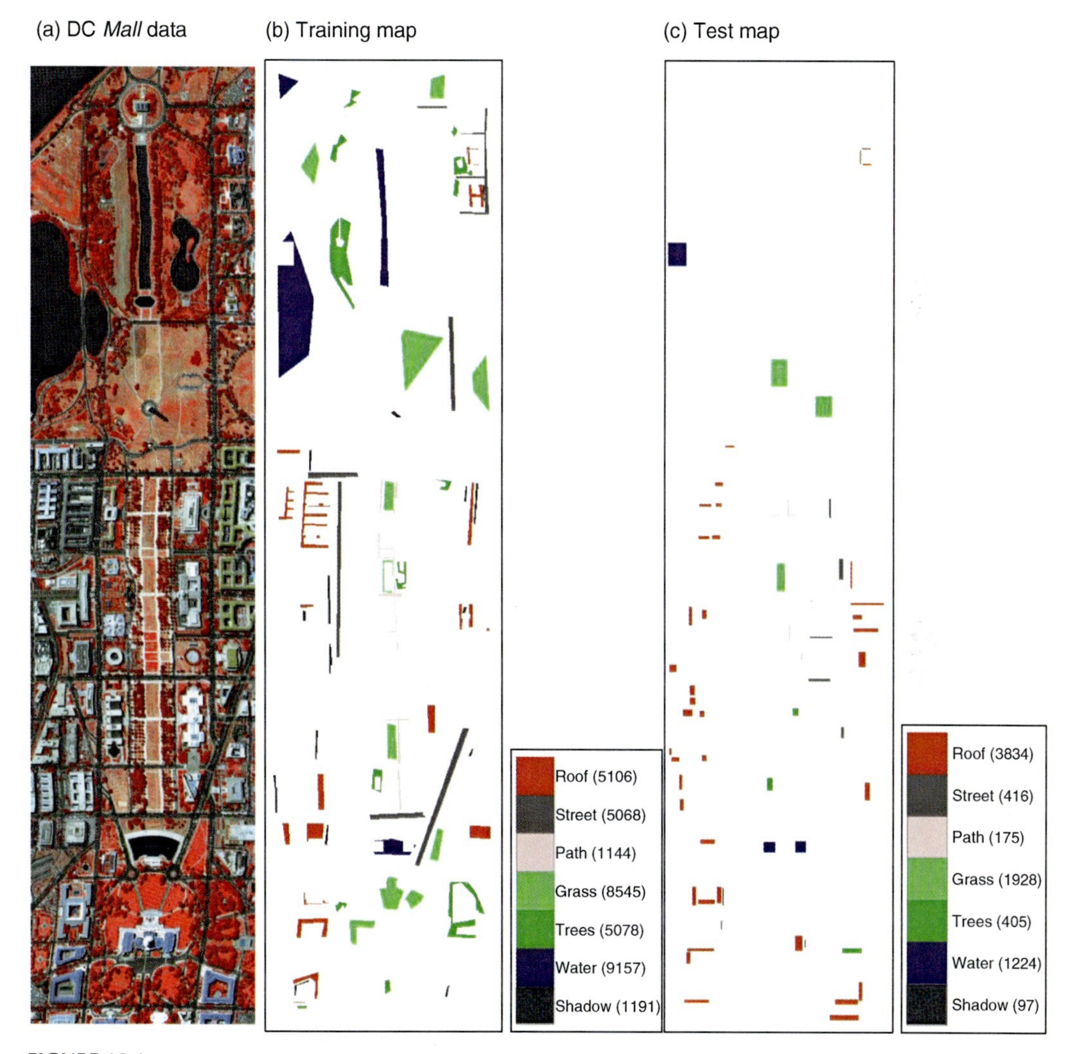

FIGURE 10.1
False color image of the *DC Mall* data set (generated using the bands 63, 52, and 36) and the corresponding ground-truth maps for training and testing. The number of pixels for each class are shown in parenthesis in the legend.

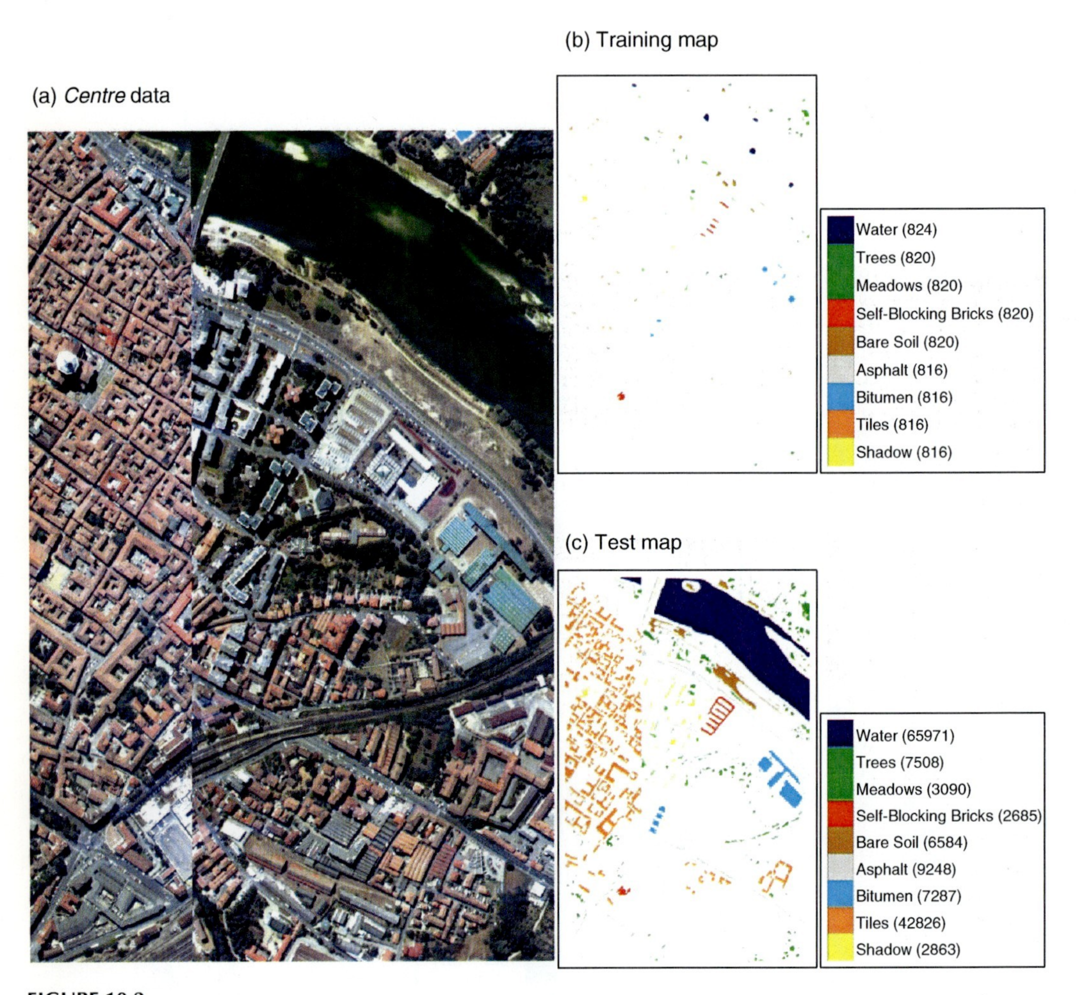

FIGURE 10.2
False color image of the *Centre* data set (generated using the bands 68, 30, and 2) and the corresponding ground-truth maps for training and testing. The number of pixels for each class are shown in parenthesis in the legend. (A missing vertical section in the middle was removed.)

FIGURE 10.3
False color image of the *University* data set (generated using the bands 68, 30, and 2) and the corresponding ground-truth maps for training and testing. The number of pixels for each class are shown in parenthesis in the legend.

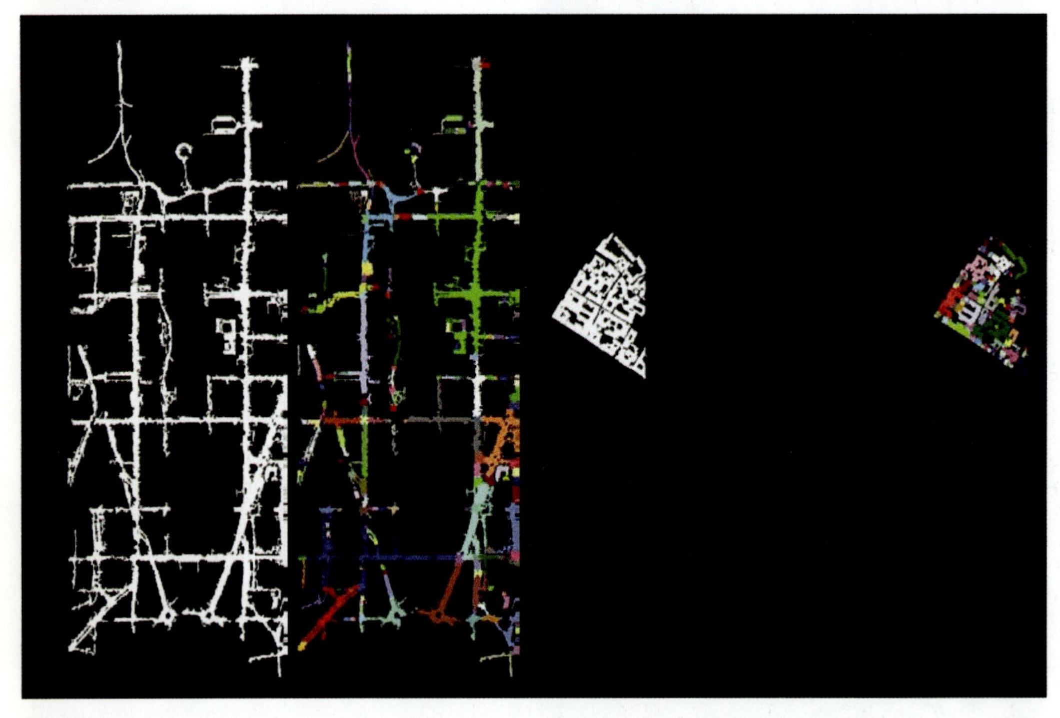

FIGURE 10.11
Examples for the region segmentation process. The iterative algorithm that uses mathematical morphology operators is used to split a large connected region into more compact subregions.

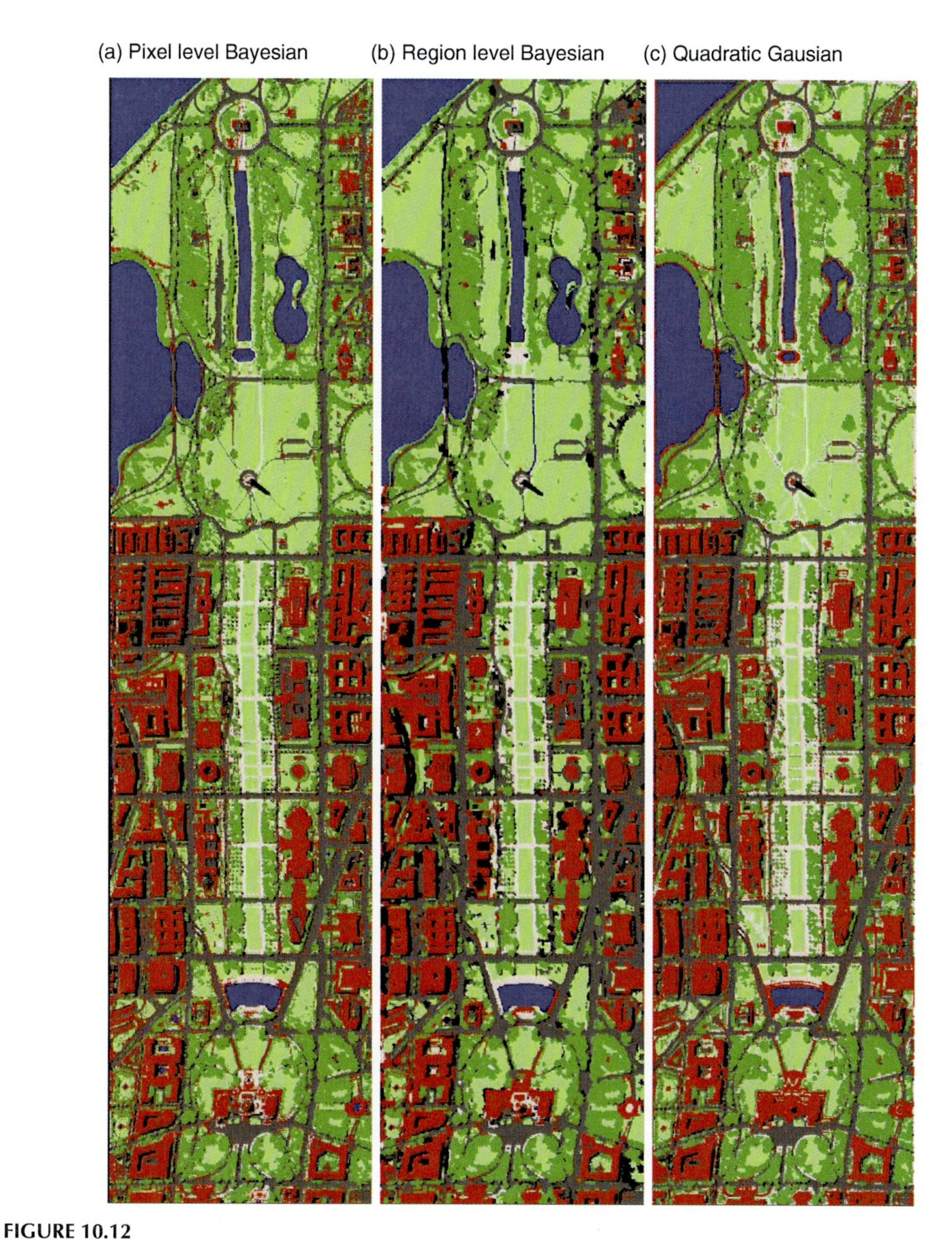

FIGURE 10.12
Final classification maps with the Bayesian pixel and region-level classifiers and the quadratic Gaussian classifier for the *DC Mall* data set. Class color codes were listed in Figure 10.1.

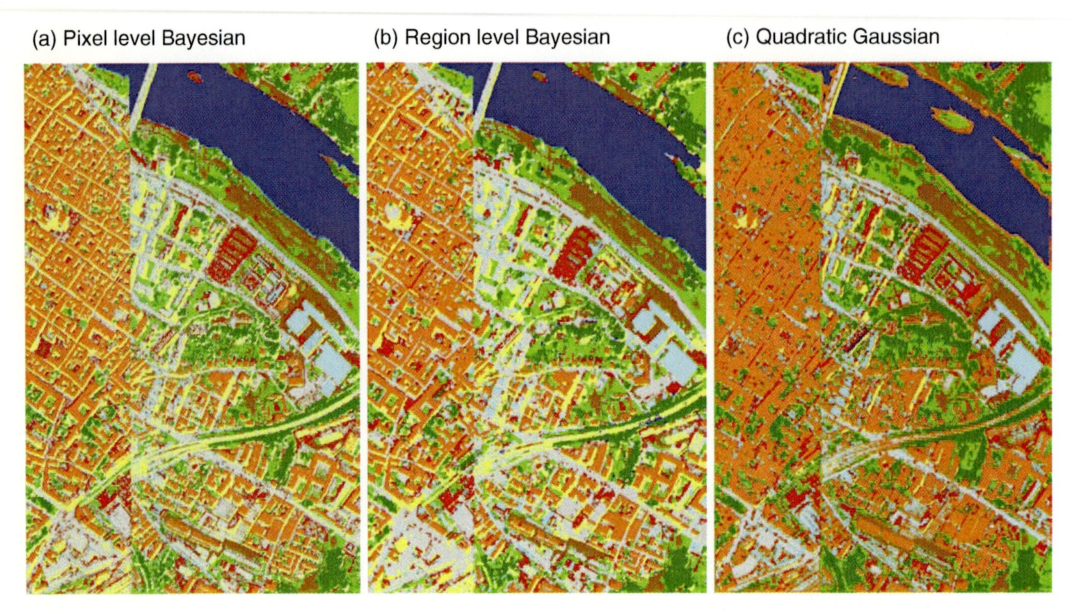

FIGURE 10.13
Final classification maps with the Bayesian pixel and region-level classifiers and the quadratic Gaussian classifier for the *Centre* data set. Class color codes were listed in Figure 10.2.

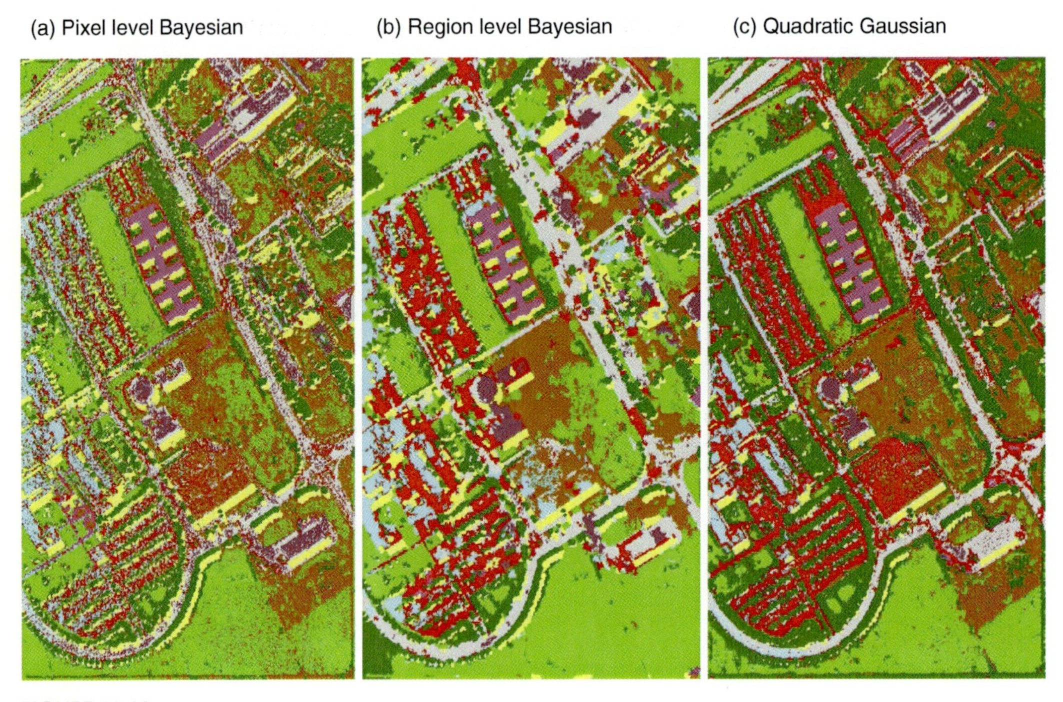

FIGURE 10.14
Final classification maps with the Bayesian pixel and region-level classifiers and the quadratic Gaussian classifier for the *University* data set. Class color codes were listed in Figure 10.3.

FIGURE 11.1
Example of multi-sensor visualization of an oil spill in the Baltic Sea created by combining an ENVISAT ASAR image with a Radarsat SAR image taken a few hours later.

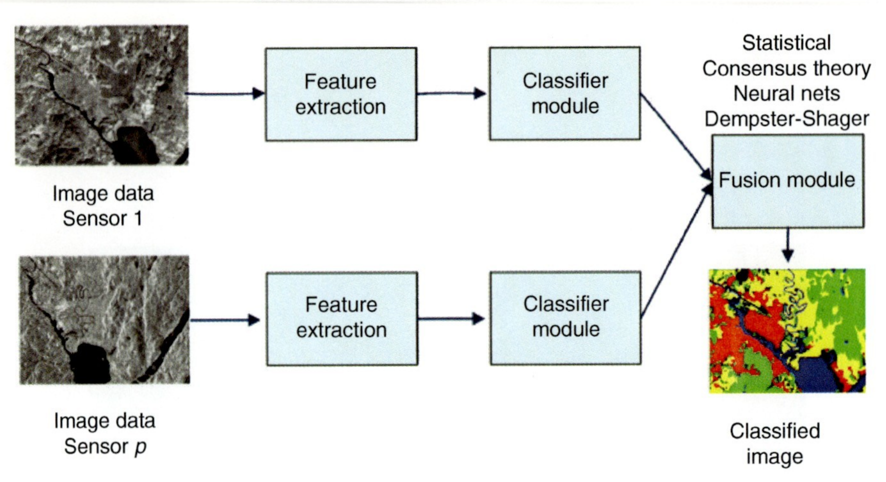

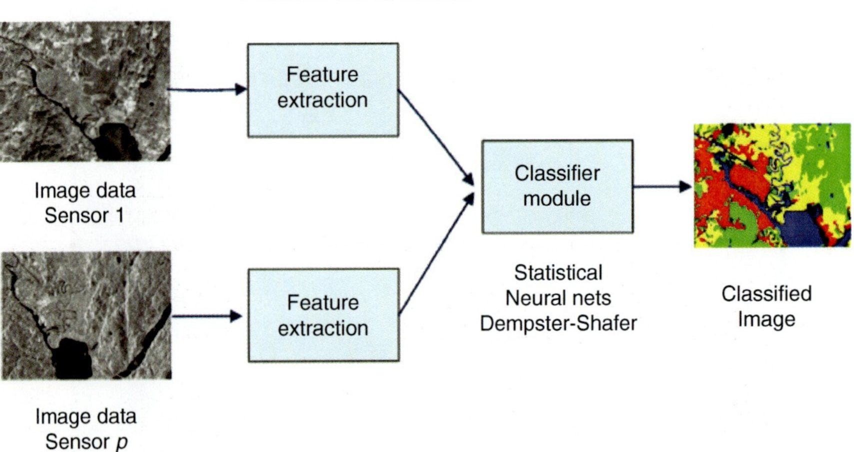

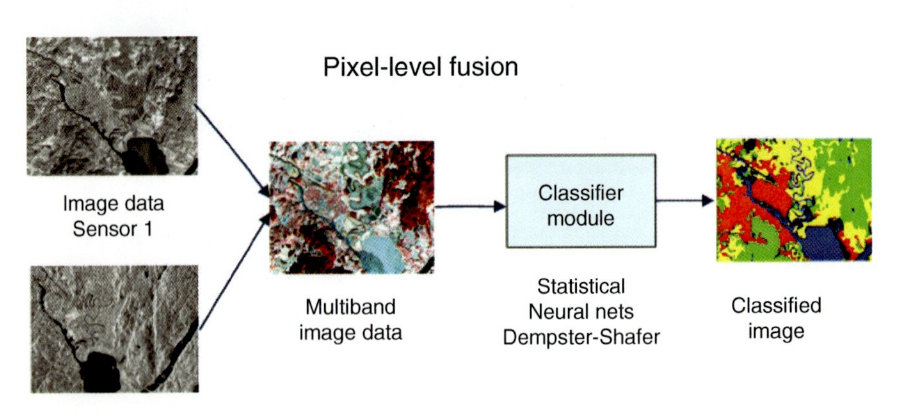

FIGURE 11.2
An illustration of data fusion on different levels.

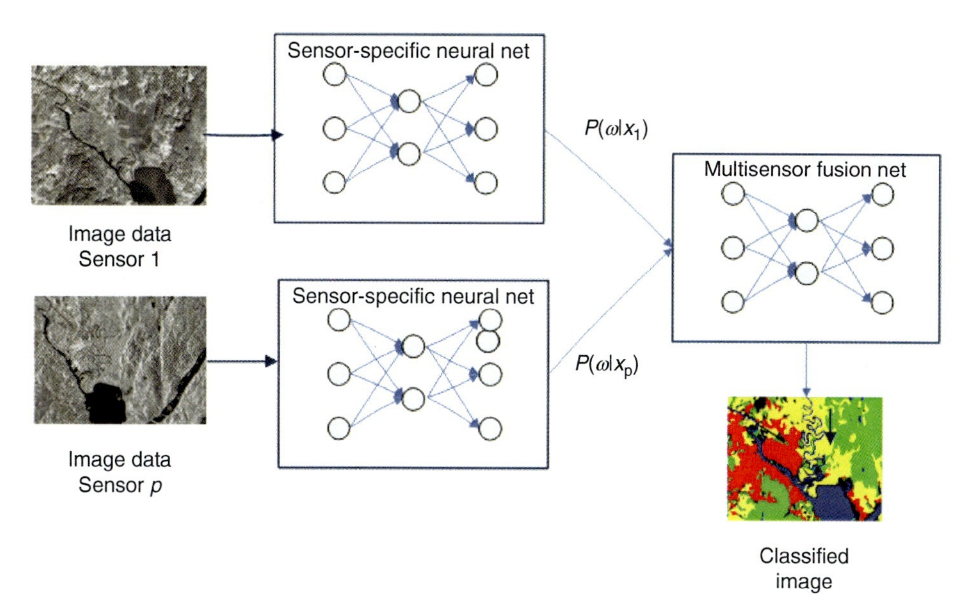

FIGURE 11.4
Network architecture for decision-level fusion using neural networks.

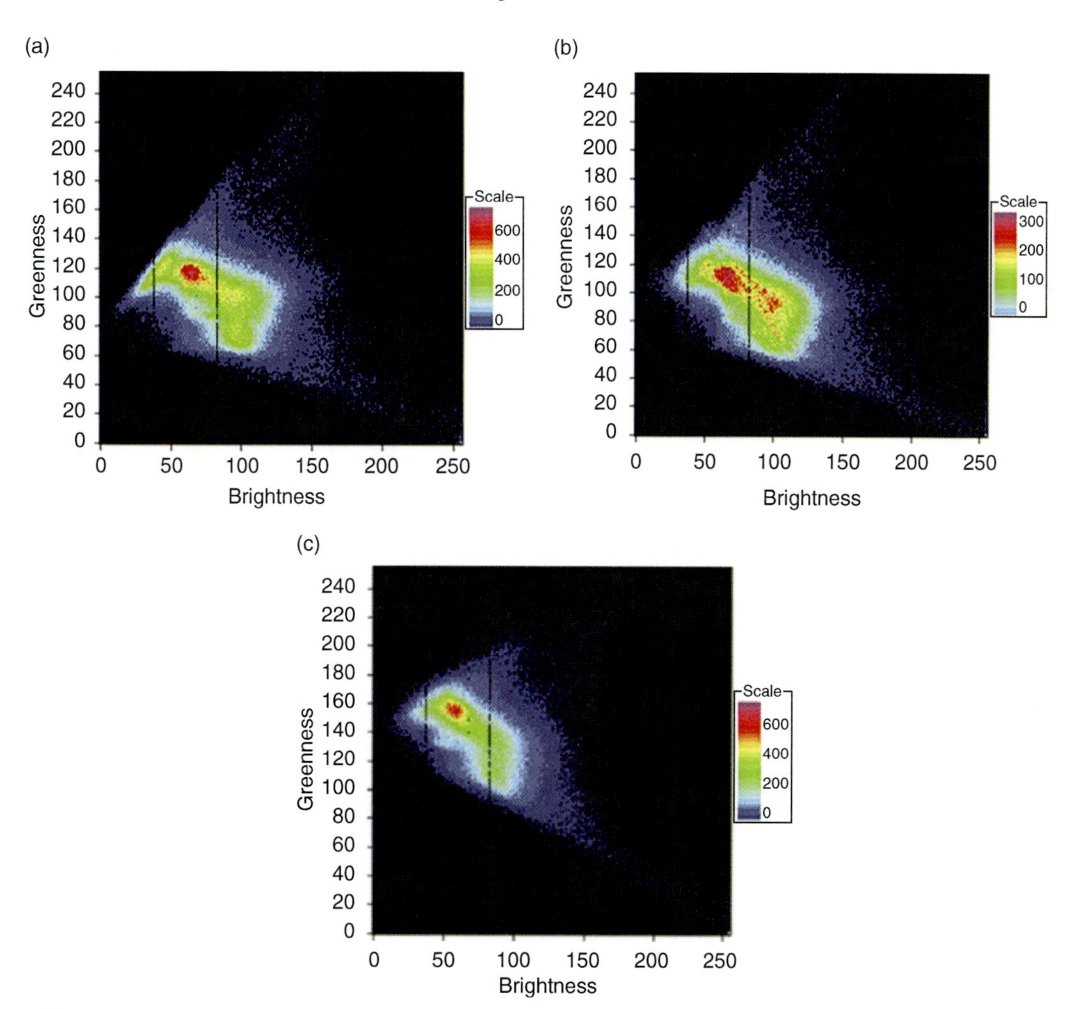

FIGURE 12.13
Greenness versus brightness, (a) original multi-spectral, (b) HT fusion, (c) PCA fusion.

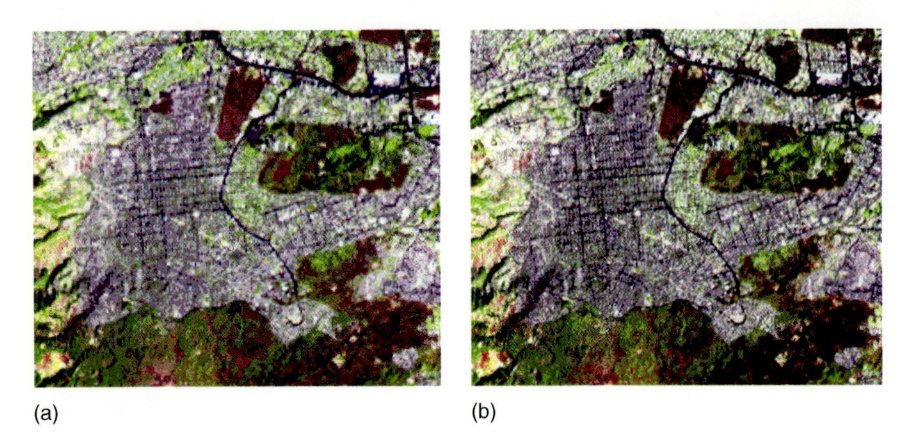

FIGURE 12.16
(a) Original multi-spectral, (b)Result of ETM+ and Radarsat image fusion with HT (Gaussian window with spread $\sigma = \sqrt{2}$ and window spacing $d = 4$) (RGB composition 5–4–3).

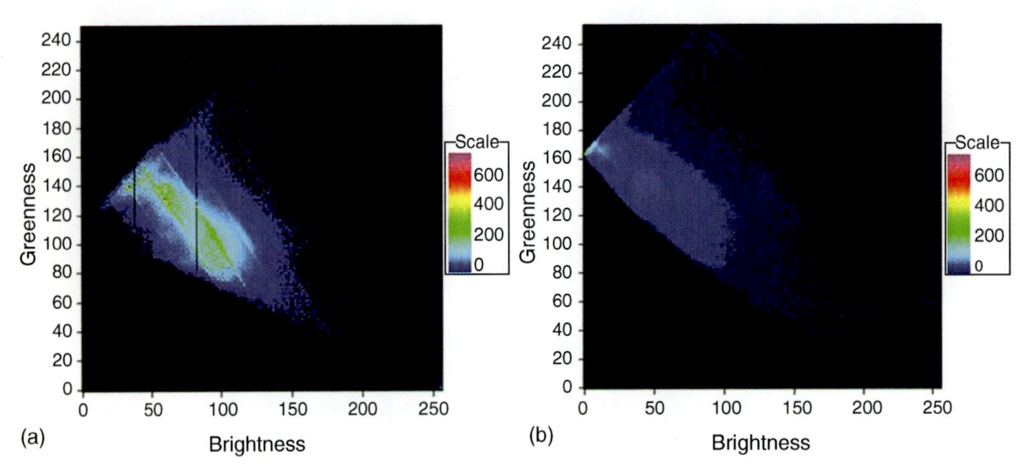

FIGURE 12.17
Greenness versus brightness, (a) original multi-spectral, (b) LANDSAT–SAR fusion with HT.

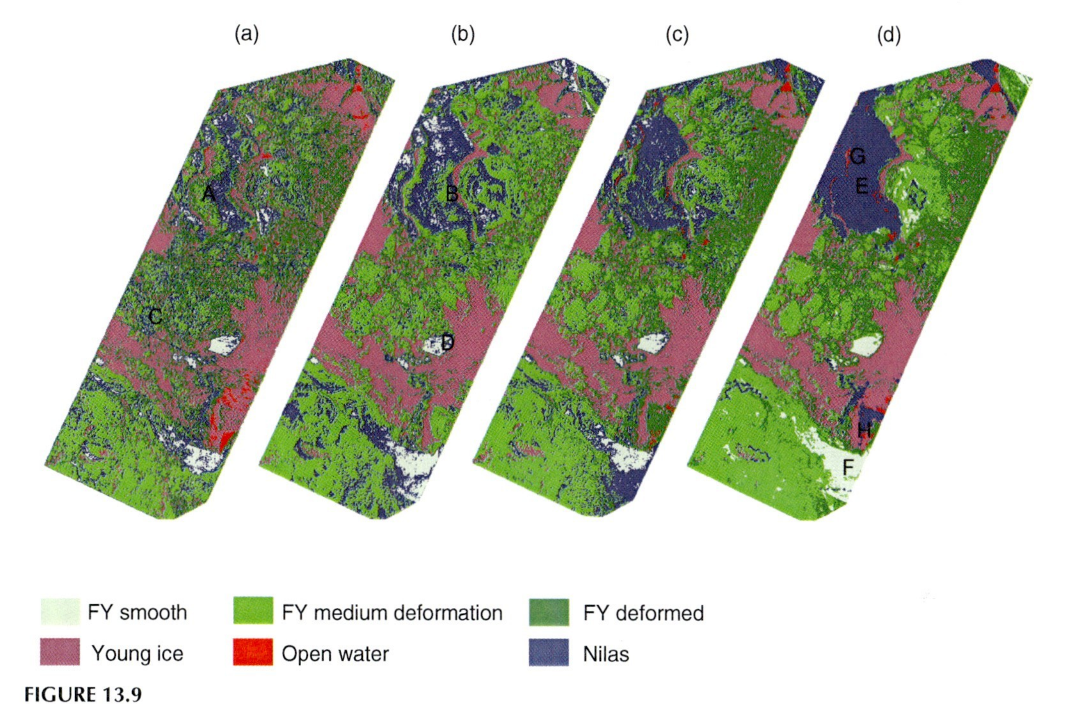

FIGURE 13.9
MLP sea ice classification maps obtained using (a) ERS, (b) RADARSAT, (c) ERS and RADARSAT, (d) ERS, RADARSAT, and Meteor images. The classifiers' parameters are given in Table 13.4.

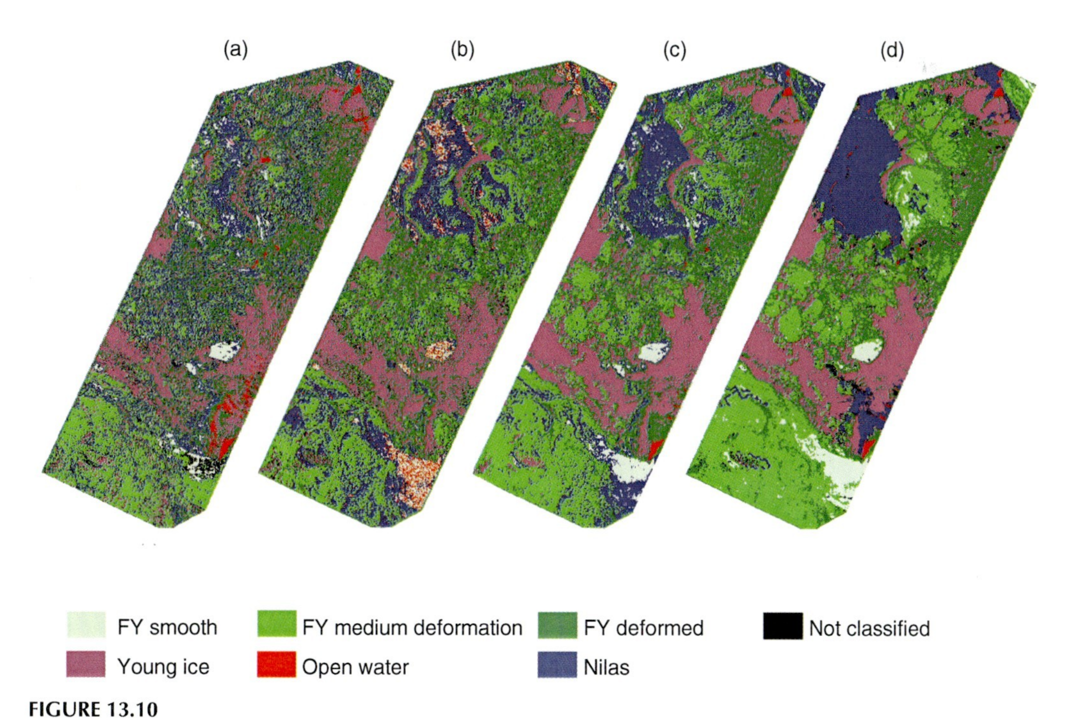

FIGURE 13.10
LDA sea ice classification maps obtained using: (a) ERS; (b) RADARSAT; (c) ERS and RADARSAT; and (d) ERS, RADARSAT, and Meteor images.

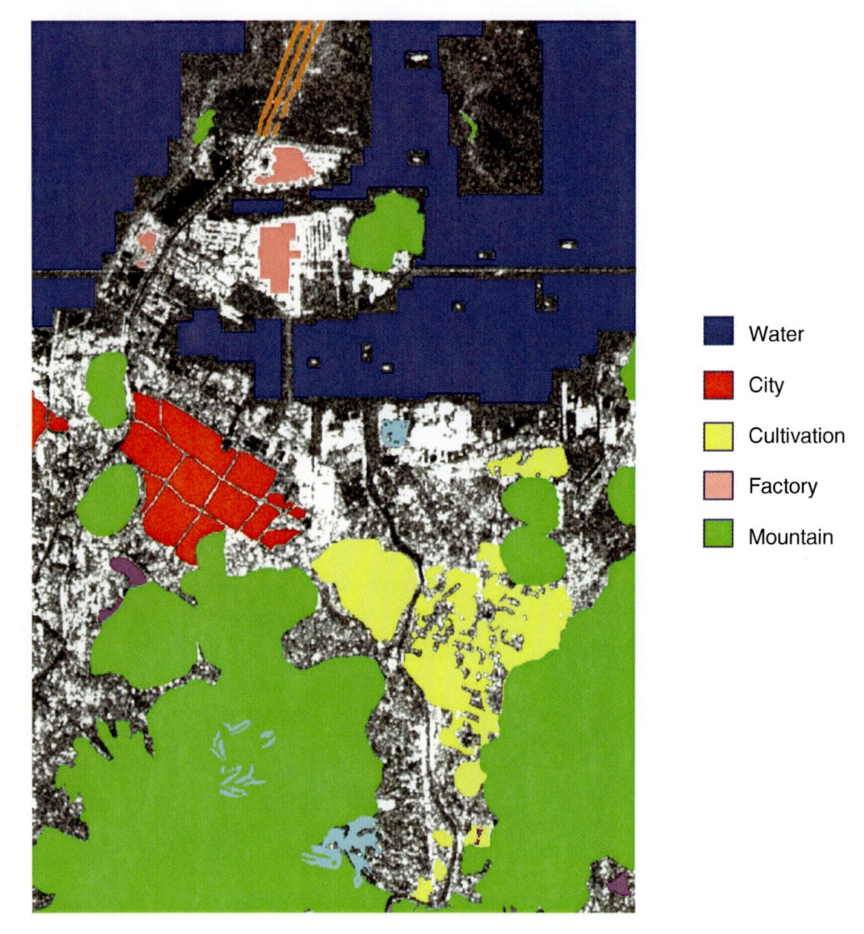

FIGURE 15.12
Ground-truth data for training.

TABLE 10.1

Confusion Matrix for Pixel-Level Classification of the *DC Mall* Data Set (Testing Subset) Using LDA, PCA, and Gabor Features

	Assigned								
True	**Roof**	**Street**	**Path**	**Grass**	**Trees**	**Water**	**Shadow**	**Total**	**% Agree**
Roof	3771	49	12	0	1	0	1	3834	98.3568
Street	0	412	0	0	0	0	4	416	99.0385
Path	0	0	175	0	0	0	0	175	100.0000
Grass	0	0	0	1926	2	0	0	1928	99.8963
Trees	0	0	0	0	405	0	0	405	100.0000
Water	0	0	0	0	0	1223	1	1224	99.9183
Shadow	0	4	0	0	0	0	93	97	95.8763
Total	3771	465	187	1926	408	1223	99	8079	99.0840

TABLE 10.2

Confusion Matrix for Pixel-Level Classification of the *Centre* Data Set (Testing Subset) Using LDA, PCA, and Gabor Features

	Assigned										
True	**Water**	**Trees**	**Meadows**	**Bricks**	**Bare Soil**	**Asphalt**	**Bitumen**	**Tiles**	**Shadow**	**Total**	**% Agree**
Water	65,877	0	1	0	1	7	0	0	85	65,971	99.8575
Trees	1	6420	1094	5	0	45	4	0	29	7598	84.4959
Meadows	0	349	2718	0	22	1	0	0	0	3090	87.9612
Bricks	0	0	0	2238	221	139	87	0	0	2685	83.3520
Bare soil	0	9	110	1026	5186	191	59	3	0	6584	78.7667
Asphalt	4	0	0	317	30	7897	239	5	756	9248	85.3914
Bitumen	4	0	1	253	22	884	6061	9	53	7287	83.1755
Tiles	0	1	0	150	85	437	116	41,826	211	42,826	97.6650
Shadow	12	0	0	3	0	477	0	0	2371	2863	82.8152
Total	65,898	6779	3924	3992	5567	10,078	6566	41,843	3505	148,152	94.8985

TABLE 10.3

Confusion Matrix for Pixel-Level Classification of the *University* Data Set (Testing Subset) Using LDA, PCA, and Gabor Features

	Assigned										
True	**Asphalt**	**Meadows**	**Gravel**	**Trees**	**Metal Sheets**	**Bare Soil**	**Bitumen**	**Bricks**	**Shadow**	**Total**	**% Agree**
Asphalt	4045	38	391	39	1	105	1050	875	87	6631	61.0014
Meadows	21	14,708	14	691	0	3132	11	71	1	18,649	78.8675
Gravel	91	14	1466	0	0	3	19	506	0	2099	69.8428
Trees	5	76	1	2927	0	40	1	2	12	3064	95.5287
Metal sheets	0	2	0	1	1341	0	0	1	0	1345	99.7026
Bare soil	34	1032	7	38	20	3745	32	119	2	5029	74.4681
Bitumen	424	1	7	1	0	1	829	67	0	1330	62.3308
Bricks	382	45	959	2	1	87	141	2064	1	3682	56.0565
Shadow	22	0	0	0	0	0	0	2	923	947	97.4657
Total	5024	15,916	2845	3699	1363	7113	2083	3707	1026	42,776	74.9205

TABLE 10.4

Confusion Matrix for Region-Level Classification of the *DC Mall* Data Set (Testing Subset) Using LDA, PCA, and Gabor Statistics, and Shape Features

	Assigned								
True	**Roof**	**Street**	**Path**	**Grass**	**Trees**	**Water**	**Shadow**	**Total**	**% Agree**
Roof	3814	11	5	0	0	1	3	3834	99.4784
Street	0	414	0	0	0	0	2	416	99.5192
Path	0	0	175	0	0	0	0	175	100.0000
Grass	0	0	0	1928	0	0	0	1928	100.0000
Trees	0	0	0	0	405	0	0	405	100.0000
Water	0	1	0	0	0	1223	0	1224	99.9183
Shadow	1	2	0	0	0	0	94	97	96.9072
Total	3815	428	180	1928	405	1224	99	8079	99.678

TABLE 10.5

Confusion Matrix for Region-Level Classification of the *Centre* Data Set (Testing Subset) Using LDA, PCA, and Gabor Statistics, and Shape Features

	Assigned										
True	**Water**	**Trees**	**Meadows**	**Bricks**	**Bare Soil**	**Asphalt**	**Bitumen**	**Tiles**	**Shadow**	**Total**	**% Agree**
Water	65,803	0	0	0	0	0	0	0	168	65,971	99.7453
Trees	0	6209	1282	28	22	11	5	0	41	7598	81.7189
Meadows	0	138	2942	0	10	0	0	0	0	3090	95.2104
Bricks	0	0	1	2247	173	31	233	0	0	2685	83.6872
Bare soil	1	4	59	257	6139	11	102	0	11	6584	93.2412
Asphalt	0	1	2	37	4	8669	163	0	372	9248	93.7392
Bitumen	0	0	0	24	3	726	6506	0	28	7287	89.2823
Tiles	0	0	0	39	13	220	2	42,380	172	42,826	98.9586
Shadow	38	0	2	2	0	341	12	0	2468	2863	86.2033
Total	65,842	6352	4288	2634	6364	10,009	7023	42,380	3260	148,152	96.7675

TABLE 10.6

Confusion Matrix for Region-Level Classification of the *University* Data Set (Testing Subset) Using LDA, PCA, and Gabor Statistics, and Shape Features

	Assigned										
True	**Asphalt**	**Meadows**	**Gravel**	**Trees**	**Metal Sheets**	**Bare Soil**	**Bitumen**	**Bricks**	**Shadow**	**Total**	**% Agree**
Asphalt	4620	7	281	4	0	52	344	1171	152	6631	69.6727
Meadows	8	17,246	0	1242	0	19	6	7	121	18,649	92.4768
Gravel	9	5	1360	2	0	0	0	723	0	2099	64.7928
Trees	39	37	0	2941	0	4	13	14	16	3064	95.9856
Metal sheets	0	0	0	0	1344	0	0	1	0	1345	99.9257
Bare soil	0	991	0	5	0	4014	0	19	0	5029	79.8171
Bitumen	162	0	0	0	0	0	1033	135	0	1330	77.6692
Bricks	248	13	596	33	5	21	125	2635	6	3682	71.5644
Shadow	16	0	0	0	1	0	0	1	929	947	98.0993
Total	5102	18,299	2237	4227	1350	4110	1521	4706	1224	42,776	84.4445

TABLE 10.7

Summary of Classification Accuracies Using the Pixel-Level and Region-Level Bayesian Classifiers and the Quadratic Gaussian Classifier

	DC Mall	*Centre*	*University*
Pixel-level Bayesian	99.0840	94.8985	74.9205
Region-level Bayesian	99.6782	96.7675	84.4445
Quadratic Gaussian	99.3811	93.9677	81.2792

Finally, comparative experiments were done by training and evaluating traditional maximum likelihood classifiers with the multi-variate Gaussian with full covariance matrix assumption for each class (quadratic Gaussian classifier) using the same training and test ground-truth data. The classification performances of all three classifiers (pixel-level Bayesian, region-level Bayesian, quadratic Gaussian) are summarized in Table 10.7. For qualitative comparison, the classification maps for all classifiers for all data sets were computed as shown in Figure 10.12 through Figure 10.14.

The results show that the proposed region-level features and Bayesian classifiers performed better than the traditional maximum likelihood classifier with the Gaussian density assumption for all data sets with respect to the ground-truth maps available. Using texture features, which model spatial neighborhoods of pixels, in addition to the spectral-based ones improved the performances of all classifiers. Using the Gabor filters at the third and fourth scales (corresponding to eight features) improved the results the most. (The confusion matrices presented show the performances of using these features instead of the original 16.) The reason for this is the high spatial image resolution where filters with a larger coverage include mixed effects from multiple structures within a pixel's neighborhood.

Using region-level information gave the most significant improvement for the *University* data set. The performances of pixel-level classifiers for *DC Mall* and *Centre* data sets using LDA- and PCA-based spectral and Gabor-based textural features were already quite high. In all cases, region-level classification performed better than pixel-level classifiers.

One important observation to note is that even though the accuracies of all classifiers seem quite high, some misclassified areas can still be found in the classification maps for all images. This is especially apparent in the results of pixel-level classifiers where many isolated pixels that are not covered by test ground-truth maps (e.g., the upper part of the *DC Mall* data, tiles on the left of the *Centre* data, many areas in the *University* data) were assigned wrong class labels because of the lack of spatial information and, hence, the context. The same phenomenon can be observed in many other results published in the literature. A more detailed ground truth is necessary for a more reliable evaluation of classifiers for high-resolution imagery. We believe that there is still a large margin for improvement in the performance of classification techniques for data received from state-of-the-art satellites.

10.8 Conclusions

We have presented an approach for classification of remotely sensed imagery using spatial techniques. First, pixel-level spectral and textural features were extracted and used for classification with nonparametric Bayesian classifiers. Next, an iterative

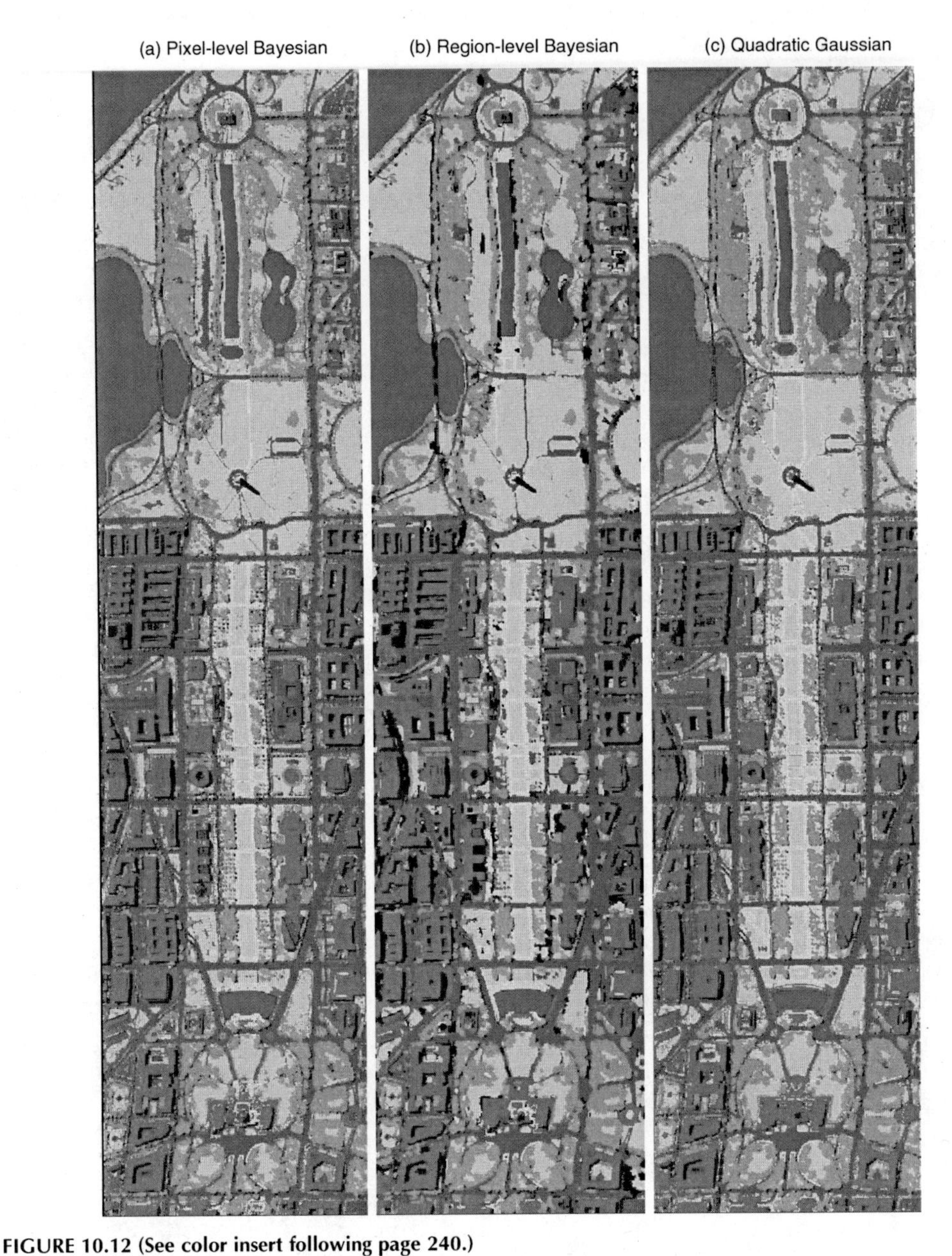

FIGURE 10.12 (See color insert following page 240.)
Final classification maps with the Bayesian pixel-and region-level classifiers and the quadratic Gaussian classifier for the *DC Mall* data set. Class color codes were listed in Figure 10.1.

split-and-merge algorithm was used to convert the pixel-level classification maps into contiguous regions. Then, spectral and textural statistics and shape features extracted from these regions were used with similar Bayesian classifiers to compute the final classification maps.

Comparative quantitative and qualitative evaluation using traditional maximum likelihood Gaussian classifiers in experiments with three different data sets with ground truth showed that the proposed region-level features and Bayesian classifiers performed better than the traditional pixel-level classification techniques. Even though the numerical results already look quite impressive, we believe that selection of the most discriminative

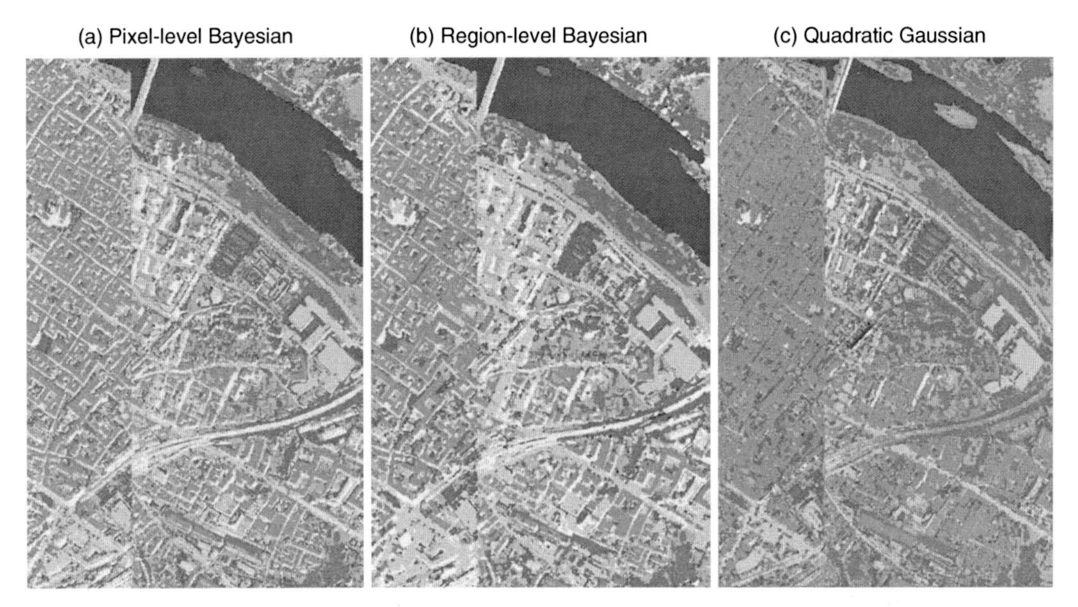

FIGURE 10.13 (See color insert following page 240.)
Final classification maps with the Bayesian pixel-and region-level classifiers and the quadratic Gaussian classifier for the *Centre* data set. Class color codes were listed in Figure 10.2.

subset of features and better segmentation of regions will bring further improvements in classification accuracy. We are also in the process of gathering ground-truth data with a larger coverage for better evaluation of classification techniques for images from high-resolution satellites.

FIGURE 10.14 (See color insert following page 240.)
Final classification maps with the Bayesian pixel- and region-level classifiers and the quadratic Gaussian classifier for the *University* data set. Class color codes were listed in Figure 10.3.

Acknowledgments

The author would like to thank Dr. David A. Landgrebe and Mr. Larry L. Biehl from Purdue University, Indiana, U.S.A., for the *DC Mall* data set, and Dr. Paolo Gamba from the University of Pavia, Italy, for the *Centre* and *University* data sets.

References

1. S.S. Durbha and R.L. King, Knowledge mining in earth observation data archives: a domain ontology perspective, in *Proceedings of IEEE International Geoscience and Remote Sensing Symposium*, September 1, 2004.
2. D.A. Landgrebe, *Signal Theory Methods in Multispectral Remote Sensing*, John Wiley & Sons, Inc., New York, 2003.
3. G.G. Wilkinson, Results and implications of a study of fifteen years of satellite image classification experiments, *IEEE Transactions on Geoscience and Remote Sensing*, 43(3), 433–440, 2005.
4. M. Datcu, H. Daschiel, A. Pelizzari, M. Quartulli, A. Galoppo, A. Colapicchioni, M. Pastori, K. Seidel, P.G. Marchetti, and S. D'Elia, Information mining in remote sensing image archives: system concepts, *IEEE Transactions on Geoscience and Remote Sensing*, 41(12), 2923–2936, 2003.
5. R.M. Haralick and L.G. Shapiro, *Computer and Robot Vision*, Addison-Wesley, Reading, MA, 1992.
6. R.L. Kettig and D.A. Landgrebe, Classification of multispectral image data by extraction and classification of homogeneous objects, *IEEE Transactions on Geoscience Electronics*, GE-14(1), 19–26, 1976.
7. C. Evans, R. Jones, I. Svalbe, and M. Berman, Segmenting multispectral Landsat TM images into field units, *IEEE Transactions on Geoscience and Remote Sensing*, 40(5), 1054–1064, 2002.
8. A. Sarkar, M.K. Biswas, B. Kartikeyan, V. Kumar, K.L. Majumder, and D.K. Pal, A MRF model-based segmentation approach to classification for multispectral imagery, *IEEE Transactions on Geoscience and Remote Sensing*, 40(5), 1102–1113, 2002.
9. J.C. Tilton, G. Marchisio, K. Koperski, and M. Datcu, Image information mining utilizing hierarchical segmentation, in *Proceedings of IEEE International Geoscience and Remote Sensing Symposium*, 2, 1029–1031, Toronto, Canada, June 2002.
10. G.G. Hazel, Object-level change detection in spectral imagery, *IEEE Transactions on Geoscience and Remote Sensing*, 39(3), 553–561, 2001.
11. T. Blaschke, Object-based contextual image classification built on image segmentation, in *Proceedings of IEEE GRSS Workshop on Advances in Techniques for Analysis of Remotely Sensed Data*, 113–119, Washington, DC, October 2003.
12. A. Rydberg and G. Borgefors, Integrated method for boundary delineation of agricultural fields in multispectral satellite images, *IEEE Transactions on Geoscience and Remote Sensing*, 39(11), 2514–2520, 2001.
13. S. Aksoy, K. Koperski, C. Tusk, G. Marchisio, and J.C. Tilton, Learning Bayesian classifiers for scene classification with a visual grammar, *IEEE Transactions on Geoscience and Remote Sensing*, 43(3), 581–589, 2005.
14. S. Aksoy and H.G. Akcay, Multi-resolution segmentation and shape analysis for remote sensing image classification, in *Proceedings of 2nd International Conference on Recent Advances in Space Technologies*, Istanbul, Turkey, June 9–11, 599–604, 2005.
15. S. Aksoy, K. Koperski, C. Tusk, and G. Marchisio, Interactive training of advanced classifiers for mining remote sensing image archives, in *Proceedings of ACM SIGKDD International Conference on Knowledge Discovery and Data Mining*, 773–782, Seattle, WA, August 22–25, 2004.
16. R.O. Duda, P.E. Hart, and D.G. Stork, *Pattern Classification*, John Wiley & Sons, Inc., New York, 2000.

17. B.S. Manjunath and W.Y. Ma, Texture features for browsing and retrieval of image data, *IEEE Transactions on Pattern Analysis and Machine Intelligence*, 18(8), 837–842, 1996.
18. S. Aksoy and R.M. Haralick, Feature normalization and likelihood-based similarity measures for image retrieval, *Pattern Recognition Letters*, 22(5), 563–582, 2001.
19. M. Schroder, H. Rehrauer, K. Siedel, and M. Datcu, Interactive learning and probabilistic retrieval in remote sensing image archives, *IEEE Transactions on Geoscience and Remote Sensing*, 38(5), 2288–2298, 2000.
20. C.M. Bishop, *Neural Networks for Pattern Recognition*, Oxford University Press, Oxford, 1995.
21. M.H. DeGroot, *Optimal Statistical Decisions*, McGraw-Hill, New York, 1970.
22. D. Geiger and D. Heckerman, A characterization of the Dirichlet distribution through global and local parameter independence, *The Annals of Statistics*, 25(3), 1344–1369, 1997, MSRTR-94-16.
23. T.M. Mitchell, *Machine Learning*, McGraw-Hill, New York, 1997.
24. R.F. Krichevskiy, Laplace's law of succession and universal encoding, *IEEE Transactions on Information Theory*, 44(1), 296–303, 1998.
25. S. Aksoy, C. Tusk, K. Koperski, and G. Marchisio, Scene modeling and image mining with a visual grammar, in C.H. Chen, ed., *Frontiers of Remote Sensing Information Processing*, World Scientific, 2003, 35–62.
26. J. Shi and J. Malik, Normalized cuts and image segmentation, *IEEE Transactions on Pattern Analysis and Machine Intelligence*, 22(8), 888–905, 2000.
27. D. Comaniciu and P. Meer, Mean shift: a robust approach toward feature space analysis, *IEEE Transactions on Pattern Analysis and Machine Intelligence*, 24(5), 603–619, 2002.
28. P. Paclik, R.P.W. Duin, G.M.P. van Kempen, and R. Kohlus, Segmentation of multi-spectral images using the combined classifier approach, *Image and Vision Computing*, 21(6), 473–482, 2003.

11

Data Fusion for Remote-Sensing Applications

Anne H.S. Solberg

CONTENTS

11.1 Introduction

Earth observation is currently developing more rapidly than ever before. During the last decade the number of satellites has been growing steadily, and the coverage of the Earth in space, time, and the electromagnetic spectrum is increasing correspondingly fast.

The accuracy in classifying a scene can be increased by using images from several sensors operating at different wavelengths of the electromagnetic spectrum. The interaction between the electromagnetic radiation and the earth's surface is characterized by certain properties at different frequencies of electromagnetic energy. Sensors with different wavelengths provide complementary information about the surface. In addition to image data, prior information about the scene might be available in the form of map data from geographic information systems (GIS). The merging of multi-source data can create a more consistent interpretation of the scene compared to an interpretation based on data from a single sensor.

This development opens up for a potential significant change in the approach of analysis of earth observation data. Traditionally, analysis of such data has been by means of analysis of a single satellite image. The emerging exceptionally good coverage in space, time, and the spectrum opens for analysis of time series of data, combining different sensor types, combining imagery of different scales, and better integration with ancillary data and models. Thus, data fusion to combine data from several sources is becoming increasingly more important in many remote-sensing applications.

This paper provides a tutorial on data fusion for remote-sensing applications. The main focus is on methods for multi-source image classification, but separate sections on multi-sensor image registration, multi-scale classification, and multi-temporal image classification are also included. The remainder of this chapter is organized in the following manner: in Section 11.2 the "multi" concept in remote sensing is presented. Multi-sensor data registration is treated in Section 11.3. Classification strategies for multi-sensor applications are discussed in Section 11.4. Multi-temporal image classification is discussed in Section 11.5, while multi-scale approaches are discussed in Section 11.6. Concluding remarks are given in Section 11.7.

11.2 The "Multi" Concept in Remote Sensing

The variety of different sensors already available or being planned creates a number of possibilities for data fusion to provide better capabilities for scene interpretation. This is referred to as the "multi" concept in remote sensing. The "multi" concept includes multi-temporal, multi-spectral or multi-frequency, multi-polarization, multi-scale, and multi-sensor image analysis. In addition to the concepts discussed here, imaging using multiple incidence angles can also provide additional information [1,2].

11.2.1 The Multi-Spectral or Multi-Frequency Aspect

The measured backscatter values for an area vary with the wavelength band. A land-use category will give different image signals depending on the frequency used, and by using different frequencies, a spectral signature that characterizes the land-use category can be found. A description of the scattering mechanisms for optical sensors can be found in

Ref. [3], while Ref. [4] contains a thorough discussion of the backscattering mechanisms in the microwave region. Multi-spectral optical sensors have demonstrated this effect for a substantial number of applications for several decades; they are now followed by high-spatial-resolution multi-spectral sensors such as Ikonos and Quickbird, and by hyperspectral sensors from satellite platforms (e.g., Hyperion).

11.2.2 The Multi-Temporal Aspect

The term multi-temporal refers to the repeated imaging of an area over a period. By analyzing an area through time, it is possible to develop interpretation techniques based on an object's temporal variations and to discriminate different pattern classes accordingly. Multi-temporal imagery allows, the study of the variation of backscatter of different areas with time, weather conditions, and seasons. It also allows monitoring of processes that change over time.

The principal advantage of multi-temporal analysis is the increased amount of information for the study area. The information provided for a single image is, for certain applications, not sufficient to properly distinguish between the desired pattern classes. This limitation can sometimes be resolved by examining the pattern of temporal changes in the spectral signature of an object. This is particularly important for vegetation applications. Multi-temporal image analysis is discussed in more detail in Section 11.5.

11.2.3 The Multi-Polarization Aspect

The multi-polarization aspect is related to microwave image data. The polarization of an electromagnetic wave refers to the orientation of the electric field during propagation. A review of the theory and features of polarization is given in Refs. [5,6].

11.2.4 The Multi-Sensor Aspect

With an increasing number of operational and experimental satellites, information about a phenomenon can be captured using different types of sensors.

Fusion of images from different sensors requires some additional preprocessing and poses certain difficulties that are not solved in traditional image classifiers. Each sensor has its own characteristics, and the image captured usually contains various artifacts that should be corrected or removed. The images also need to be geometrically corrected and co-registered. Because the multi-sensor images often are not acquired on the same data, the multi-temporal nature of the data must also often be explained.

Figure 11.1 shows a simple visualization of two synthetic aperture radar (SAR) images from an oil spill in the Baltic sea, imaged by the ENVISAT ASAR sensor and the Radarsat SAR sensor. The images were taken a few hours apart. During this time, the oil slick has drifted to some extent, and it has become more irregular in shape.

11.2.5 Other Sources of Spatial Data

The preceding sections have addressed spatial data in the form of digital images obtained from remote-sensing satellites. For most regions, additional information is available in the form of various kinds of maps, for example, topography, ground cover, elevation, and so on. Frequently, maps contain spatial information not obtainable from a single remotely sensed image. Such maps represent a valuable information resource in addition to the

FIGURE 11.1 (See color insert following page 240.)
Example of multi-sensor visualization of an oil spill in the Baltic sea created by combining an ENVISAT ASAR image with a Radarsat SAR image taken a few hours later.

satellite images. To integrate map information with a remotely sensed image, the map must be available in digital form, for example, in a GIS system.

11.3 Multi-Sensor Data Registration

A prerequisite for data fusion is that the data are co-registered, and geometrically and radiometrically corrected. Data co-registration can be simple if the data are georeferenced. In that case, the co-registration consists merely of resampling the images to a common map projection. However, an image-matching step is often necessary to obtain subpixel accuracy in matching. Complicating factors for multi-sensor data are the different appearances of the same object imaged by different sensors, and nonrigid changes in object position between multi-temporal images.

The image resampling can be done at .various stages of the image interpretation process. Resampling an image affects the spatial statistics of the neighboring pixel, which is of importance for many radar image feature extraction methods that might use speckle statistics or texture. When fusing a radar image with other data sources, a solution might be to transform the other data sources to the geometry of the radar image. When fusing a multi-temporal radar image, an alternative might be to use images from the same image mode of the sensor, for example, only ascending scenes with a given incidence angle range. If this is not possible and the spatial information from the original geometry is important, the data can be fused and resampling done after classification by the sensor-specific classifiers.

An image-matching step may be necessary to achieve subpixel accuracy in the co-registration even if the data are georeferenced. A survey of image registration methods is given by Zitova and Flusser [7]. A full image registration process generally consists of four steps:

- *Feature extraction*. This is the step where regions, edges, and contours can be used to represent tie-points in the set of images to be matched are extracted. This is a crucial step, as the registration accuracy can be no better than what is achieved for the tie-points.

Feature extraction can be grouped into area-based methods [8,9], feature-based methods [10–12], and hybrid approaches [7]. In area-based methods, the gray levels of the images are used directly for matching, often by statistical comparison of pixel values in small windows, and they are best suited for images from the same or highly similar sensors. Feature-based methods will be application-dependent, as the type of features to use as tie points needs to be tailored to the application. Features can be extracted either from the spatial domain (edges, lines, regions, intersections, and so on) or from the frequency domain (e.g., wavelet features). Spatial features can perform well for matching data from heterogeneous sensors, for example, optical and radar images. Hybrid approaches use both area-based and feature-based techniques by combining both a correlation-based matching with an edge-based approach, and they are useful in matching data from heterogeneous sensors.

- *Feature matching*. In this step, the correspondence between the tie-points or features in the sensed image and the reference image is found. Area-based methods for feature extraction use correlation, Fourier-transform methods, or optical flow [13]. Feature-based methods use the equivalence between correlation in the spatial domain and multiplication in the Fourier domain to perform matching in the Fourier domain [10,11]. Correlation-based methods are best suited for data from similar sensors. The optical flow approach involves estimation of the relative motion between two images and is a broad approach. It is commonly used in video analysis, but only a few studies have used it in remote-sensing applications [29,30].
- *Transformation selection* concerns the choice of mapping function and estimation of its parameters based on the established feature correspondence. The affine transform model is commonly used for remote-sensing applications, where the images normally are preprocessed for geometrical correction—a step that justifies the use of affine transforms.
- *Image resampling*. In this step, the image is transformed by means of the mapping function. Image values in no-integer coordinates are computed by the

appropriate interpolation technique. Normally, either a nearest neighbor or a bilinear interpolation is used. Nearest neighbor interpolation is applicable when no new pixel values should be introduced. Bilinear interpolation is often a good trade-off between accuracy and computational complexity compared to cubic or higher order interpolation.

11.4 Multi-Sensor Image Classification

The literature on data fusion in the computer vision and machine intelligence domains is substantial. For an extensive review of data fusion, we recommend the book by Abidi and Gonzalez [16]. Multi-sensor architectures, sensor management, and designing sensor setup are also thoroughly discussed in Ref. [17].

11.4.1 A General Introduction to Multi-Sensor Data Fusion for Remote-Sensing Applications

Fusion can be performed at the *signal, pixel, feature,* or *decision* level of representation (see Figure 11.2). In signal-based fusion, signals from different sensors are combined to create a new signal with a better signal-to-noise ratio than the original signals [18]. Techniques for signal-level data fusion typically involve classic detection and estimation methods [19]. If the data are noncommensurate, they must be fused at a higher level.

Pixel-based fusion consists of merging information from different images on a pixel-by-pixel basis to improve the performance of image processing tasks such as segmentation [20]. Feature-based fusion consists of merging features extracted from different signals or images [21]. In feature-level fusion, features are extracted from multiple sensor observations, then combined into a concatenated feature vector, and classified using a standard classifier. Symbol-level or decision-level fusion consists of merging information at a higher level of abstraction. Based on the data from each single sensor, a preliminary classification is performed. Fusion then consists of combining the outputs from the preliminary classifications.

The main approaches to data fusion in the remote-sensing literature are statistical methods [22–25], Dempster–Shafer theory [26–28], and neural networks [22,29]. We will discuss each of these approaches in the following sections. The best level and methodology for a given remote-sensing application depends on several factors: the complexity of the classification problem, the available data set, and the goal of the analysis.

11.4.2 Decision-Level Data Fusion for Remote-Sensing Applications

In the general multi-sensor fusion case, we have a set of images $X^1 \cdots X^P$ from P sensors. The class labels of the scene are denoted C. The Bayesian approach is to assign each pixel to the class that maximizes the posterior probabilities $P(C \mid X_1, \ldots, X_P)$

$$P(C|X^1, \ldots, X^P) = \frac{P(X^1, \ldots, X^P|C)P(C)}{P(X^1, \ldots, X^P)} \tag{11.1}$$

where $P(C)$ is the prior model for the class labels.

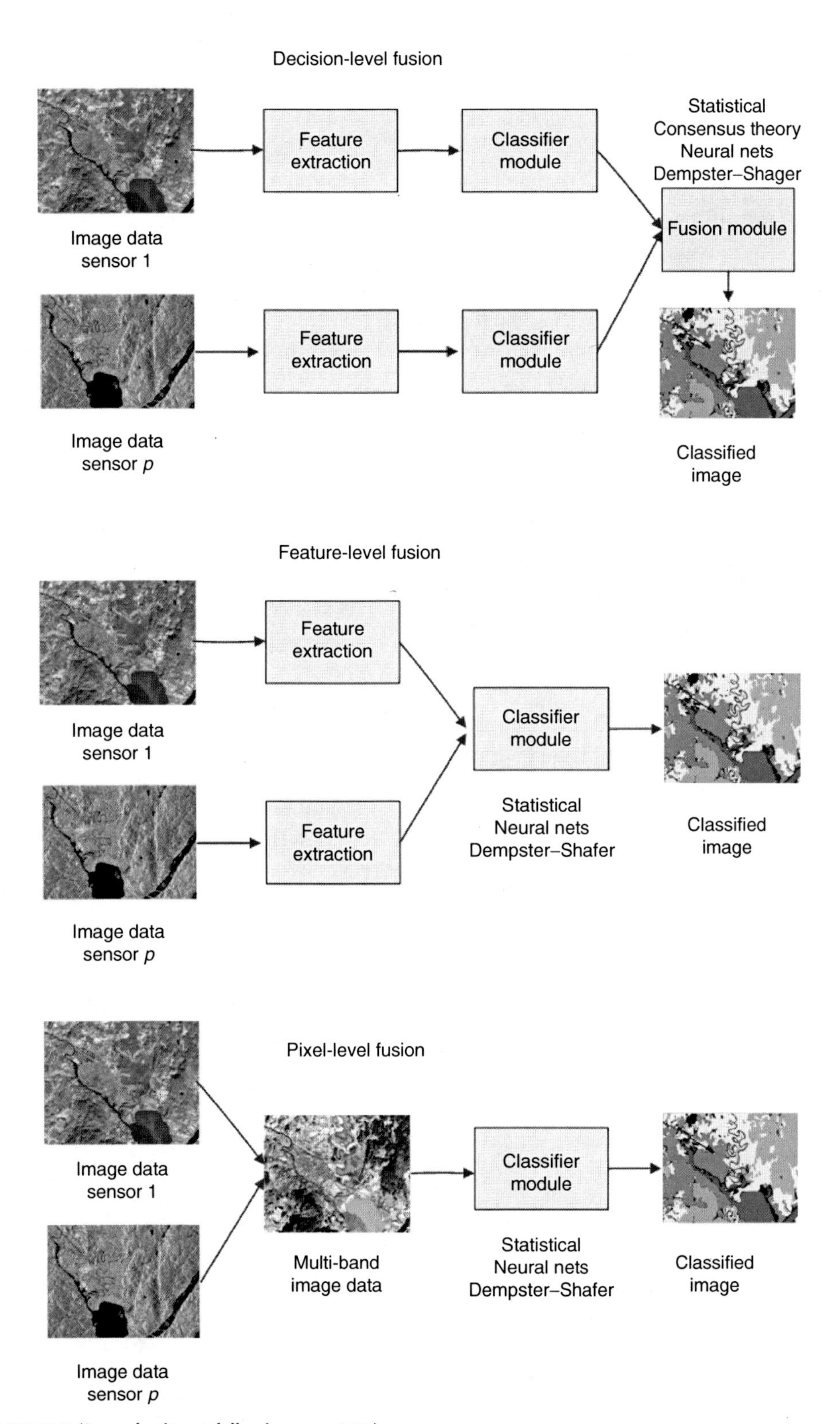

FIGURE 11.2 (See color insert following page 240.)
An illustration of data fusion on different levels.

For decision-level fusion, the following conditional independence assumption is used:

$$P(X^1, \ldots, X^P|C) \equiv P(X^1|C) \cdots P(X^P|C)$$

This assumption means that the measurements from the different sensors are considered to be conditionally independent.

11.4.3 Combination Schemes for Combining Classifier Outputs

In the data fusion literature [30], various alternative methods have been proposed for combining the outputs from the sensor-specific classifiers by weighting the influence of each sensor. This is termed consensus theory. The weighting schemes can be linear, logarithmic, or of a more general form (see Figure 11.3).

The simplest choice, the linear opinion pool (LOP), is given by

$$\text{LOP}(X^1, \ldots, X^P) = \sum_{p=1}^{P} P(X^p|C)^{\lambda_p} \tag{11.2}$$

The logarithmic opinion pool (LOGP) is given by

$$\text{LOGP}(X^1, \ldots, X^P) = \prod_{p=1}^{P} P(X^p|C)^{\lambda_p} \tag{11.3}$$

which is equivalent to the Bayesian combination if the weights λ_p are equal. This weighting scheme contradicts the statistical formulation in which the sensor's uncertainty is supposed to be modeled by the variance of the probability density function.

The weights are supposed to represent the sensor's reliability. The weights can be selected by heuristic methods based on their goodness [3] by weighting a sensor's influence by a factor proportional to its overall classification accuracy on the training data set. An alternative approach for a linear combination pool is to use a genetic algorithm [32].

An approach using a neural net to optimize the weights is presented in Ref. [30]. Yet another possibility is to choose the weights in such a way that they not only weigh the individual data sources but also the classes within the data sources [33].

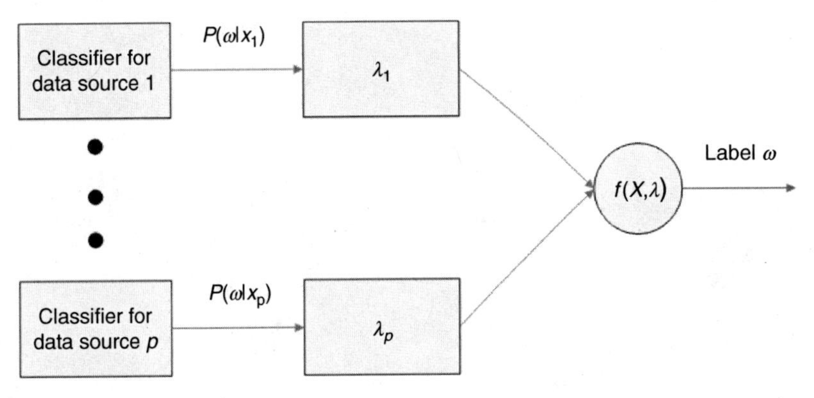

FIGURE 11.3
Schematic view of weighting the outputs from sensor-specific classifiers in decision-level fusion.

Benediktsson et al. [30,31] use a multi-layer perceptron (MLP) neural network to combine the class-conditional probability densities $P(X^p \mid C)$. This allows a more flexible, nonlinear combination scheme. They compare the classification accuracy using MLPs to LOPs and LOGPs, and find that the neural net combination performs best.

Benediktsson and Sveinsson [34] provide a comparison of different weighting schemes for an LOP and LOGP, genetic algorithm with and without pruning, parallel consensus neural nets, and conjugate gradient backpropagation (CGBP) nets on a single multi-source data set. The best results were achieved by using a CGBP net to optimize the weights in an LOGP.

A study that contradicts the weighting of different sources is found in Ref. [35]. In this study, three different data sets (optical and radar) were merged using the LOGP, and the weights were varied between 0 and 1. Best results for all three data sets were found by using equal weights.

11.4.4 Statistical Multi-Source Classification

Statistical methods for fusion of remotely sensed data can be divided into four categories: the augmented vector approach, stratification, probabilistic relaxation, and extended statistical fusion. In the augmented vector approach, data from different sources are concatenated as if they were measurements from a single sensor. This is the most common approach for many application-oriented applications of multi-source classification, because no special software is needed. This is an example of pixel-level fusion.

Such a classifier is difficult to use when the data cannot be modeled with a common probability density function, or when the data set includes ancillary data (e.g., from a GIS system). The fused data vector is then classified using ordinary single-source classifiers [36]. Stratification has been used to incorporate ancillary GIS data in the classification process. The GIS data are stratified into categories and then a spectral model for each of these categories is used [37].

Richards et al. [38] extended the methods used for spatially contextual classification based on probabilistic relaxation to incorporate ancillary data. The methods based on extended statistical fusion [10,43] were derived by extending the concepts used for classification of single-sensor data. Each data source is considered independently and the classification results are fused using weighted linear combinations.

By using a statistical classifier one often assumes that the data have a multi-variate Gaussian distribution. Recent developments in statistical classifiers based on regression theory include choices of nonlinear classifiers [11–13,18–20,26,28,33,38,39–56]. For a comparison of neural nets and regression-based nonlinear classifiers, see Ref. [57].

11.4.5 Neural Nets for Multi-Source Classification

Many multi-sensor studies have used neural nets because no specific assumptions about the underlying probability densities are needed [40,58]. A drawback of neural nets in this respect is that they act like a black box in that the user cannot control the usage of different data sources. It is also difficult to explicitly use a spatial model for neighboring pixels (but one can extend the input vector from measurements from a single pixel to measurements from neighboring pixels). Guan et al. [41] utilized contextual information by using a network of neural networks with which they built a quadratic regularizer. Another drawback is that specifying a neural network architecture involves specifying a large number of parameters. A classification experiment should take care in choosing them and apply different configurations, making the complete training process very time

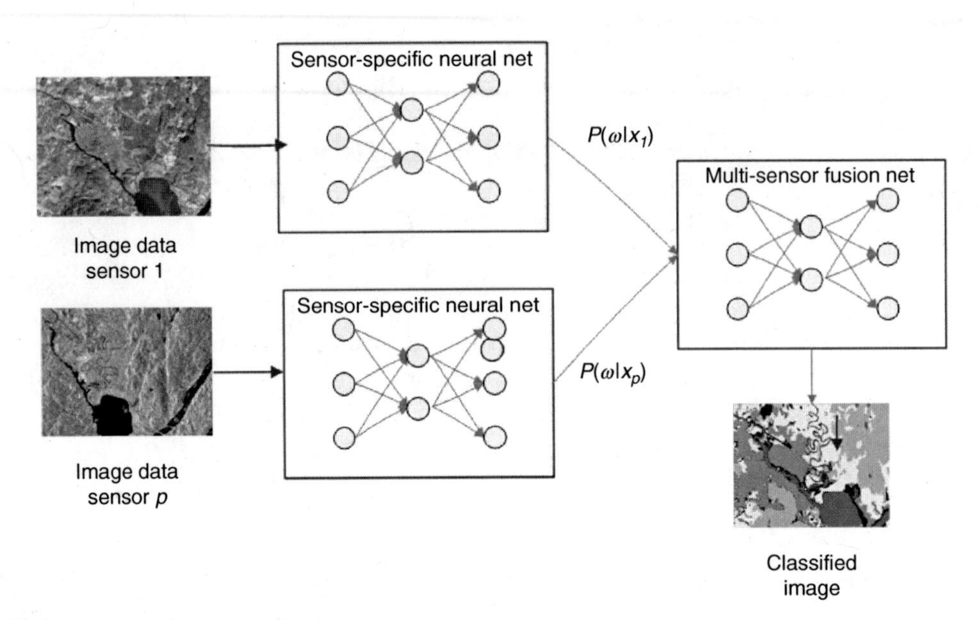

FIGURE 11.4 (See color insert following page 240.)
Network architecture for decision-level fusion using neural networks.

consuming [52,58]. Hybrid approaches combining statistical methods and neural networks for data fusion have also been proposed [30]. Benediktsson et al. [30] apply a statistical model to each individual source and use neural nets to reach a consensus decision. Most applications involving a neural net use an MLP or radial basis function network, but other neural network architectures can be used [59–61].

Neural nets for data fusion can be applied both at the pixel, feature, and decision level. For pixel- and feature-level fusion a single neural net is used to classify the joint feature vector or pixel measurement vector. For decision-level fusion, a network combination like the one outlined in Figure 11.4 is often used [29]. An MLP neural net is first used to classify the images from each source separately. Then, the outputs from the sensor-specific nets are fused and weighted in a fusion network.

11.4.6 A Closer Look at Dempster–Shafer Evidence Theory for Data Fusion

Dempster–Shafer theory of evidence provides a representation of multi-source data using two central concepts: plausibility and belief. Mathematical evidence theory was first introduced by Dempster in the 1960s, and later extended by Shafer [62].

A good introduction to Dempster–Shafer evidence theory for remote sensing data fusion is given in Ref. [28].

Plausibility (Pls) and belief (Bel) are derived from a mass function m, which is defined on the [0,1] interval. The belief and plausibility functions for an element A are defined as

$$\text{Bel}(A) = \sum_{B \subseteq A} m(B) \tag{11.4}$$

$$\text{Pls}(A) = \sum_{B \cap A \neq \emptyset} m(B). \tag{11.5}$$

They are sometimes referred to as lower and upper probability functions. The belief value of hypothesis A can be interpreted as the minimum uncertainty value about A, and its plausibility as the maximum uncertainty [28].

Evidence from p different sources is combined by combining the mass functions $m_1 \cdots m_p$ by

$$\text{If } K \neq 1,\ m(A) = \frac{\sum_{B_1 \cap \cdots \cap B_p = A}^{m(\emptyset)=0} \prod_{1 \leq i \leq p}}{1 - K} m_i(B_i)$$

where $K = \sum_{B_1 \cap \cdots \cap B_p = o} \prod_{1<i<p} m_i(B_i)$ is interpreted as a measure of conflict between the different sources.

The decision rule used to combine the evidence from each sensor varies from different applications, either maximum of plausibility or maximum of belief (with variations). The performance of Dempster–Shafer theory for data fusion does however depend on the methods used to compute the mass functions. Lee et al. [20] assign nonzero mass function values only to the single classes, whereas Hegarat-Mascle et al. [28] propose two strategies for assigning mass function values to sets of classes according to the membership for a pixel for these classes.

The concepts of evidence theory belong to a different school than Bayesian multi-sensor models. Researchers coming from one school often have a tendency to dislike modeling used in the alternative theory. Not many neutral comparisons of these two approaches exist. The main advantage of this approach is its robustness in the method by which information from several heterogeneous sources is combined. A disadvantage is the underlying basic assumption that the evidence from different sources is independent. According to Ref. [43], Bayesian theory assumes that imprecision about uncertainty in the measurements is assumed to be zero and uncertainty about an event is only measured by the probability. The author disagrees with this by pointing out that in Bayesian modeling, uncertainty about the measurements can be modeled in the priors. Priors of this kind are not always used, however. Priors in a Bayesian model can also be used to model spatial context and temporal class development. It might be argued that the Dempster–Shafer theory can be more appropriate for a high number of heterogeneous sources. However, most papers on data fusion for remote sensing consider two or maximum three different sources.

11.4.7 Contextual Methods for Data Fusion

Remote-sensing data have an inherent spatial nature. To account for this, contextual information can be incorporated in the interpretation process. Basically, the effect of context in an image-labeling problem is that when a pixel is considered in isolation, it may provide incomplete information about the desired characteristics. By considering the pixel in context with other measurements, more complete information might be derived.

Only a limited set of studies have involved spatially contextual multi-source classification. Richards et al. [38] extended the methods used for spatial contextual classification based on probabilistic relaxation to incorporate ancillary data. Binaghi et al. [63] presented a knowledge-based framework for contextual classification based on fuzzy set theory. Wan and Fraser [61] used multiple self-organizing maps for contextual classification. Le Hégarat-Mascle et al. [28] combined the use of a Markov random field model with the Dempster–Shafer theory. Smits and Dellepiane [64] used a multi-channel image segmentation method based on Markov random fields with adaptive neighborhoods. Markov random fields have also been used for data fusion in other application domains [65,66].

11.4.8 Using Markov Random Fields to Incorporate Ancillary Data

Schistad Solberg et al. [67,68] used a Markov random field model to include map data into the fusion. In this framework, the task is to estimate the class labels of the scene C given the image data X and the map data M (from a previous survey):

$$P(C|X, M) = P(X|C, M)P(C)$$

with respect to C.

The spatial context between neighboring pixels in the scene is modeled in $P(C)$ using the common Ising model. By using the equivalence between Markov random fields and Gibbs distribution

$$P(\cdot) = \frac{1}{Z} \exp -U(\cdot)$$

where U is called the energy function and Z is a constant; the task of maximizing $P(C|X,M)$ is equivalent to minimizing the sum

$$U = \sum_{i=1}^{P} U_{\text{data}(i)} + U_{\text{spatial, map}}$$

U_{spatial} is the common Ising model:

$$U_{\text{spatial}} = \beta_s \sum_{k \in N} I(c_i, c_k)$$

and

$$U_{\text{map}} = \beta_m \sum_{k \in M} t(c_i|m_k)$$

m_k is the class assigned to the pixel in the map, and $t(c_i|m_k)$ is the probability of a class transition from class m_k to class c_i. This kind of model can also be used for multi-temporal classification [67].

11.4.9 A Summary of Data Fusion Architectures

Table 11.1 gives a schematic view on different fusion architectures applied to remote-sensing data.

11.5 Multi-Temporal Image Classification

For most applications where multi-source data are involved, it is not likely that all the images are acquired at the same time. When the temporal aspect is involved, the classification methodology must handle changes in pattern classes between the image acquisitions, and possibly also use different classes.

TABLE 11.1

A Summary of Data Fusion Architectures

Pixel-level fusion	
Advantages	Simple. No special classifier software needed.
	Correlation between sources utilized.
	Well suited for change detection.
Limitations	Assumes that the data can be modeled using a common probability density function.
	Source reliability cannot be modeled.
Feature-level fusion	
Advantages	Simple. No special classifier software needed.
	Sensor-specific features give advantage over pixel-based fusion.
	Well suited for change detection.
Limitations	Assumes that the data can be modeled using a common probability density function.
	Source reliability cannot be modeled.
Decision-level fusion	
Advantages	Suited for data with different probability densities.
	Source-specific reliabilities can be modeled.
	Prior information about the source combination can be modeled.
Limitations	Special software often needed.

To find the best classification strategy for a multi-temporal data set, it is useful to consider the goal of the analysis and the complexity of the multi-temporal image data to be used. Multi-temporal image classification can be applied for different purposes:

- *Monitor and identify specific changes.* If the goal is to monitor changes, multi-temporal data are required either in the form of a combination of existing maps and new satellite imagery or as a set of satellite images. For identifying changes, different fusion levels can be considered. Numerous methods for

TABLE 11.2

A Summary of Decision-Level Fusion Strategies

Statistical multi-sensor classifiers	
Advantages	Good control over the process.
	Prior knowledge can be included if the model is adapted to the application.
	Inclusion of ancillary data simple using a Markov random field approach.
Limitations	Assumes a particular probability density function.
Dempster–Shafer multi-sensor classifiers	
Advantages	Useful for representation of heterogeneous sources.
	Inclusion of ancillary data simple.
	Well suited to model a high number of sources.
Limitations	Performance depends on selected mass functions.
	Not many comparisons with other approaches.
Neural net multi-sensor classifiers	
Advantages	No assumption about probability densities needed.
	Sensor-specific weights can easily be estimated.
	Suited for heterogeneous sources.
Limitations	The user has little control over the fusion process and how different sources are used.
	Involves a large number of parameters and a risk of overfitting.
Hybrid multi-sensor classifiers	
Advantages	Can combine the best of statistical and neural net or Dempster–Shafer approaches.
Limitations	More complex to use.

change detection exist, ranging from pixel-level to decision-level fusion. Examples of pixel-level change detection are classical unsupervised approaches like image math, image regression, and principal component analysis of a multi-temporal vector of spectral measurements or derived feature vectors like normalized vegetation indexes. In this paper, we will not discuss in detail these well-established unsupervised methods. Decision-level change detection includes postclassification comparisons, direct multi-date classification, and more sophisticated classifiers.

- *Improved quality in discriminating between a set of classes*. Sometimes, parts of an area might be covered by clouds, and a multi-temporal image set is needed to map all areas. For microwave images, the signature depends on temperature and soil moisture content, and several images might be necessary to obtain good coverage of all regions in an area as two classes can have different mechanisms affecting their signature. For this kind of application, a data fusion model that takes source reliability weighting into account should be considered. An example concerning vegetation classification in a series of SAR images is shown in Figure 11.5.
- *Discriminate between classes based on their temporal signature development*. By analyzing an area through time and studying how the spectral signature changes, it is possible to discriminate between classes that are not separable on a single

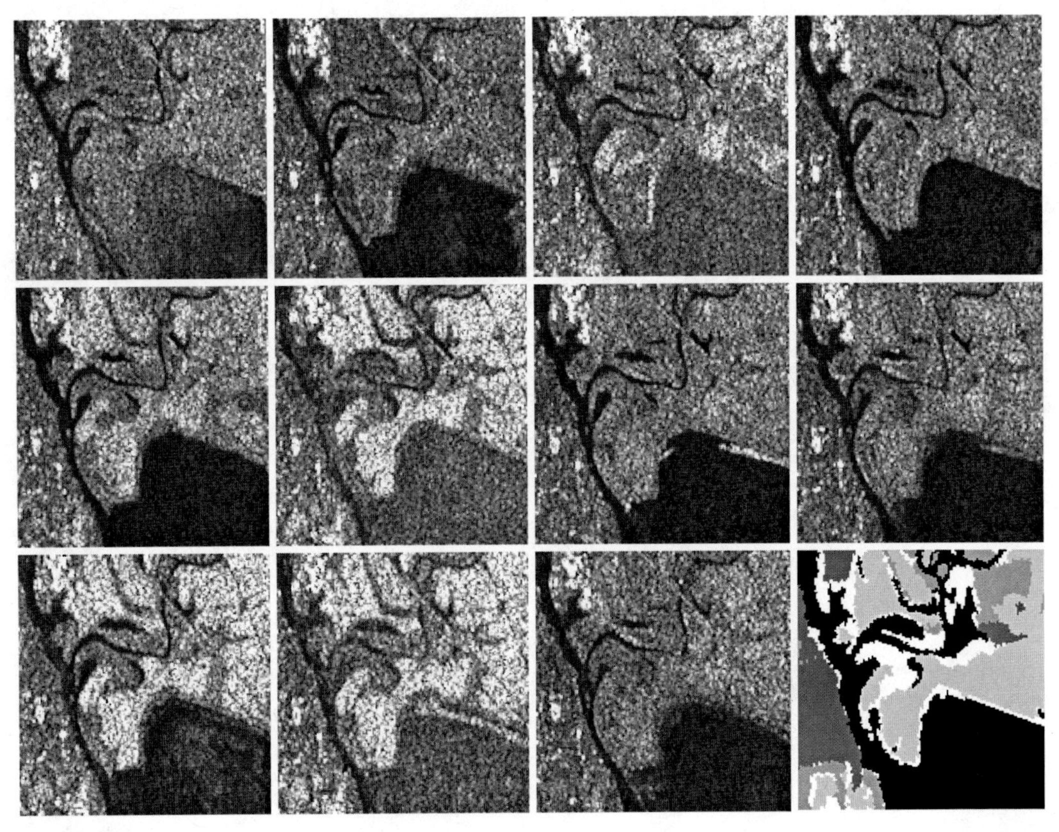

FIGURE 11.5
Multi-temporal image from 13 different dates during August–December 1991 for agricultural sites in Norway. The ability to identify ploughing activity in a SAR image depends on the soil moisture content at the given date.

image. Consider for example vegetation mapping. Based on a single image, we might be able to discriminate between deciduous and conifer trees, but not between different kinds of conifer or deciduous. By studying how the spectral signature varies during the growth season, we might also be able to discriminate between different vegetation species.

It is also relevant to consider the available data set. How many images can be included in the analysis? Most studies use bi-temporal data sets, which are easy to obtain. Obtaining longer time series of images can sometimes be difficult due to sensor repeat cycles and weather limitations. In Northern Europe, cloud coverage is a serious limitation for many applications of temporal trajectory analysis. Obtaining long time series tends to be easier for low- and medium-resolution images from satellites with frequent passes.

A principal decision in multi-temporal image analysis is whether the images are to be combined on the pixel level or the decision level. Pixel-level fusion consists of combining the multi-temporal images into a joint data set and performing the classification based on all data at the same time. In decision-level fusion, a classification is first performed for each time, and then the individual decisions are combined to reach a consensus decision. If no changes in the spectral signatures of the objects to be studied have occurred between the image acquisitions, then this is very similar to classifier combination [31].

11.5.1 Multi-Temporal Classifiers

In the following, we describe the main approaches for multi-temporal classification. The methods utilize temporal correlation in different ways. *Temporal feature correlation* means that the correlation between the pixel measurements or feature vectors at different times is modeled. *Temporal class correlation* means that the correlation between the class labels of a given pixel at different times is modeled.

11.5.1.1 Direct Multi-Date Classification

In direct compound or stacked vector classification, the multi-temporal data set is merged at the pixel level into one vector of measurements, followed by classification using a traditional classifier. This is a simple approach that utilizes temporal feature correlation. However, the approach might not be suited when some of the images are of lower quality due to noise. An example of this classification strategy is to use multiple self-organizing map (MSOM) [69] as a classifier for compound bi-temporal images.

11.5.1.2 Cascade Classifiers

Swain [70] presented the initial work on using cascade classifiers. In a cascade-classifier approach the temporal class correlation between multi-temporal images is utilized in a recursive manner. To find a class label for a pixel at time t_2, the conditional probability for observing class ω given the images x_1 and x_2 is modeled as

$$P(\omega|x_1, x_2)$$

Classification was performed using a maximum likelihood classifier. In several papers by Bruzzone and co-authors [71,72] the use of cascade classifiers has been extended to unsupervised classification using multiple classifiers (combining both maximum likelihood classifiers and radial basis function neural nets).

11.5.1.3 Markov Chain and Markov Random Field Classifiers

Schistad Solberg et al. [67] describe a method for classification of multi-source and multi-temporal images where the temporal changes of classes are modeled using Markov chains with transition probabilities. This approach utilizes temporal class correlation. In the Markov random field model presented in Ref. [25], class transitions are modeled in terms of Markov chains of possible class changes and specific energy functions are used to combine temporal information with multi-source measurements, and ancillary data. Bruzzone and Prieto [73] use a similar framework for unsupervised multi-temporal classification.

11.5.1.4 Approaches Based on Characterizing the Temporal Signature

Several papers have studied changes in vegetation parameters (for a review see Ref. [74]). In Refs. [50,75] the temporal signatures of classes are modeled using Fourier series (using temporal feature correlation). Not many approaches have integrated phenological models for the expected development of vegetation parameters during the growth season. Aurdal et al. [76] model the phenological evolution of mountain vegetation using hidden Markov models. The different vegetation classes can be in one of a predefined set of states related to their phenological development, and classifying a pixel consists of selecting the class that has the highest probability of producing a given series of observations. The performance of this model is compared to a compound maximum likelihood approach and found to give comparable results for a single scene, but more robust when testing and training on different images.

11.5.1.5 Other Decision-Level Approaches to Multi-Temporal Classification

Jeon and Landgrebe [46] developed a spatio-temporal classifier utilizing both the temporal and the spatial context of the image data. Khazenie and Crawford [47] proposed a method for contextual classification using both spatial and temporal correlation of data. In this approach, the feature vectors are modeled as resulting from a class-dependent process and a contaminating noise process, and the noise is correlated in both space and time. Middelkoop and Janssen [49] presented a knowledge-based classifier, which used land-cover data from preceding years. An approach to decision-level change detection using evidence theory is given in Ref. [43].

A summary of approaches for multi-temporal image classifiers is given in Table 11.3.

11.6 Multi-Scale Image Classification

Most of the approaches to multi-sensor image classification do not treat the multi-scale aspect of the input data. The most common approach is to resample all the images to be fused to a common pixel resolution.

In other domains of science, much work on combining data sources at different resolutions exists, for example, in epidemiology [77], in the estimation of hydraulic conductivity for characterizing groundwater flow [78], and in the estimation of environmental components [44]. These approaches are mainly for situations where the aim is to estimate an underlying *continuous* variable.

The remote-sensing literature contains many examples of multi-scale and multi-sensor data visualization. Many multi-spectral sensors, such as SPOT XS or Ikonos, provide a combination of multi-spectral band and a panchromatic band of a higher resolution. Several methods for visualizing such multi-scale data sets have been proposed, and

TABLE 11.3

A Discussion of Multi-Temporal Classifiers

Direct multi-date classifier	
Advantages	Simple. Temporal feature correlation between image measurements utilized.
Limitations	Is restricted to pixel-level fusion.
	Not suited for data sets containing noisy images.
Cascade classifiers	
Advantages	Temporal correlation of class labels considered.
	Information about special class transitions can be modeled.
Limitations	Special software needed.
Markov chain and MRF classifiers	
Advantages	Spatial and temporal correlation of class labels considered.
	Information about special class transitions can be modeled.
Limitations	Special software needed.
Temporal signature trajectory approaches	
Advantages	Can discriminate between classes not separable at a single point in time.
	Can be used either at feature level or at decision level.
	Decision-level approaches allow flexible modeling.
Limitations	Feature-level approaches can be sensitive to noise.
	A time series of images needed (can be difficult to get more than bi-temporal).

they are often based on overlaying a multi-spectral image on the panchromatic image using different colors. We will not describe such techniques in detail, but refer the reader to surveys like [51,55,79]. Van der Meer [80] studied the effect of multi-sensor image fusion in terms of information content for visual interpretation, and concluded that image fusion aiming at improving the visual content and interpretability was more successful for homogeneous data than for heteorogeneous data.

For classification problems, Puyou-Lascassies [54] and Zhukov et al. [81] considered unmixing of low-resolution data by using class label information obtained from classification of high-resolution data. The unmixing is performed through several sequential steps, but no formal model for the complete data set is derived. Price [53] proposed unmixing by relating the correlation between low-resolution data and high-resolution data resampled to low resolution, to correlation between high-resolution data and low-resolution data resampled to high resolution. The possibility of mixed pixels was not taken into account.

In Ref. [82], separate classifications were performed based on data from each resolution. The resulting resolution-dependent probabilities were averaged over the resolutions.

Multi-resolution tree models are sometimes used for multi-scale analysis (see, e.g., Ref. [48]). Such models yield a multi-scale representation through a quad tree, in which each pixel at a given resolution is decomposed into four child pixels at higher resolution, which are correlated. This gives a model where the correlation between neighbor pixels depends on the pixel locations in an arbitrary (i.e., not problem-related) manner.

The multi-scale model presented in Ref. [83] is based on the concept of a reference resolution and is developed in a Bayesian framework [84]. The reference resolution corresponds to the highest resolution present in the data set. For each pixel of the input image at the reference resolution it is assumed that there is an underlying discrete class. The observed pixel values are modeled conditionally on the classes. The properties of the class label image are described through an *a priori* model. Markov random fields have been selected for this purpose. Data at coarser resolutions are modeled as mixed pixels,

TABLE 11.4

A Discussion of Multi-Scale Classifiers

Resampling combined with single-scale classifier	
Advantages	Simple. Works well enough for homogeneous regions.
Limitations	Can fail in identifying small or detailed structures.
Classifier with explicit multi-scale model	
Advantages	Can give increased performance for small or detailed structures.
Limitations	More complex software needed. Not necessary for homogeneous regions.

that is, the observations are allowed to include contributions from several distinct classes. In this way it is possible to exploit spectrally richer images at lower resolutions to obtain more accurate classification results at the reference level, without smoothing the results as much as if we simply oversample the low-resolution data to the reference resolution prior to the analysis.

Methods that use a model for the relationship between the multi-scale data might offer advantages compared to simple resampling both in terms of increased classification accuracy and being able to describe relationships between variables measured at different scales. This can provide tools to predict high-resolution properties from coarser resolution properties. Of particular concern in the establishment of statistical relationships is the quantification of what is lost in precision at various resolutions and the associated uncertainty.

The potential of using multi-scale classifiers will also depend on the level of detail needed for the application, and might be related to the typical size of the structures one wants to identify in the images. Even simple resampling of the coarsest resolution to the finest resolution, followed by classification using a multi-sensor classifier, can help improve the classification result. The gain obtained by using a classifier that explicitly models the data at different scales depends not only on the set of classes used but also on the regions used to train and test the classifier. For scenes with a high level of detail, for example, in urban scenes, the performance gain might be large. However, it depends also on how the classifier performance is evaluated. If the regions used for testing the classifier are well inside homogeneous regions and not close to other classes, the difference in performance in terms of overall classification accuracy might not be large, but visual inspection of the level of detail in the classified images can reveal the higher level of detail.

A summary of multi-scale classification approaches is given in Table 11.4.

11.7 Concluding Remarks

A number of different approaches for data fusion in remote-sensing applications have been presented in the literature. A prerequisite for data fusion is that the data are co-registered and geometrically and radiometrically corrected.

In general, there is no consensus on which multi-source or multi-temporal classification approach works best. Different studies and comparisons report different results. There is still a need for a better understanding on which methods are most suited to different applications types, and also broader comparison studies. The best level and methodology for a given remote-sensing application depends on several factors: the complexity of the classification problem, the available data set, the number of sensors involved, and the goal of the analysis.

Some guidelines for selecting the methodology and architecture for a given fusion task are given below.

11.7.1 Fusion Level

Decision-level fusion gives best control and allows weighting the influence of each sensor. Pixel-level fusion can be suited for simple analysis, for example, fast unsupervised change detection.

11.7.2 Selecting a Multi-Sensor Classifier

If decision-level fusion is selected, three main approaches for fusion should be considered: the statistical approach, neural networks, or evidence theory. A hybrid approach can also be used to combine these approaches. If the sources are believed to provide data of different quality, weighting schemes for consensus combination of the sensor-specific classifiers should be considered.

11.7.3 Selecting a Multi-Temporal Classifier

To find the best classification strategy for a multi-temporal data set, the complexity of the class separation problem must be considered in light of the available data set. If the classes are difficult to separate, it might be necessary to use methods for characterizing the temporal trajectory of signatures. For pixel-level classification of multi-temporal imagery, the direct multi-date classification approach can be used. If specific knowledge about certain types of changes needs to be modeled, Markov chain and Markov random field approaches or cascade classifiers should be used.

11.7.4 Approaches for Multi-Scale Data

Multi-scale images can either be resampled to a common resolution or a classifier with implicit modeling of the relationship between the different scales can be used. For classification problems involving small or detailed structures (e.g., urban areas) or heteorogeneous sources, the latter is recommended.

Acknowledgment

The author would like to thank Line Eikvil for valuable input, in particular, regarding-multi-sensor image registration.

References

1. C. Elachi, J. Cimino, and M. Settle, Overview of the shuttle imaging radar-B preliminary scientific results, *Science*, 232, 1511–1516, 1986.

2. J. Cimino, A. Brandani, D. Casey, J. Rabassa, and S.D. Wall, Multiple incidence angle SIR-B experiment over Argentina: Mapping of forest units, *IEEE Trans. Geosc. Rem. Sens.*, 24, 498–509, 1986.
3. G. Asrar, *Theory and Applications of Optical Remote Sensing*, Wiley, New York, 1989.
4. F.T. Ulaby, R.K. Moore, and A.K. Fung, *Microwave Remote Sensing, Active and Passive, Vols. I–III*, Artech House Inc., 1981, 1982, 1986.
5. F.T. Ulaby and C. Elachi, *Radar Polarimetry for Geoscience Applications*, Artec House Inc., 1990.
6. H.A. Zebker and J.J. Van Zyl, Imaging radar polarimetry: a review, *Proc. IEEE*, 79, 1583–1606, 1991.
7. B. Zitova and J. Flusser, Image registration methods: a survey, *Image and Vision Computing*, 21, 977–1000, 2003.
8. P. Chalermwat and T. El-Chazawi, Multi-resolution image registration using genetics, in *Proc. ICIP*, 452–456, 1999.
9. H.M. Chen, M.K. Arora, and P.K. Varshney, Mutual information-based image registration for remote sensing data, *Int. J. Rem. Sens.*, 24, 3701–3706, 2003.
10. X. Dai and S. Khorram, A feature-based image registration algorithm using improved chain-code representation combined with invariant moments, *IEEE Trans. Geosc. Rem. Sens.*, 37, 17–38, 1999.
11. D.M. Mount, N.S. Netanyahu, and L. Le Moigne, Efficient algorithms for robust feature matching, *Pattern Recognition*, 32, 17–38, 1999.
12. E. Rignot, R. Kwok, J.C. Curlander, J. Homer, and I. Longstaff, Automated multisensor registration: Requirements and techniques, *Photogramm. Eng. Rem. Sens.*, 57, 1029–1038, 1991.
13. Z.-D. Lan, R. Mohr, and P. Remagnino, Robust matching by partial correlation, in *British Machine Vision Conference*, 651–660, 1996.
14. D. Fedorov, L.M.G. Fonseca, C. Kennedy, and B.S. Manjunath, Automatic registration and mosaicking system for remotely sensed imagery, in *Proc. 9th Int. Symp. Rem. Sens.*, 22–27, Crete, Greece, 2002.
15. L. Fonseca, G. Hewer, C. Kenney, and B. Manjunath, Registration and fusion of multispectral images using a new control point assessment method derived from optical flow ideas, in *Proc. Algorithms for Multispectral and Hyperspectral Imagery V*, 104–111, SPIE, Orlando, USA, 1999.
16. M.A. Abidi and R.C. Gonzalez, *Data Fusion in Robotics and Machine Intelligence*, Academic Press, Inc., New York, 1992.
17. N. Xiong and P. Svensson, Multi-sensor management for information fusion: issues and approaches, *Information Fusion*, 3, 163–180, 2002.
18. J.M. Richardson and K.A. Marsh, Fusion of multisensor data, *Int. J. Robot. Res.* 7, 78–96, 1988.
19. D.L. Hall and J. Llinas, An introduction to multisensor data fusion, *Proc. IEEE*, 85(1), 6–23, 1997.
20. T. Lee, J.A. Richards, and P.H. Swain, Probabilistic and evidential approaches for multisource data analysis, *IEEE Trans. Geosc. Rem. Sens.*, 25, 283–293, 1987.
21. N. Ayache and O. Faugeras, Building, registrating, and fusing noisy visual maps, *Int. J. Robot. Res.*, 7, 45–64, 1988.
22. J.A. Benediktsson and P.H. Swain, A method of statistical multisource classification with a mechanism to weight the influence of the data sources, in *IEEE Symp. Geosc. Rem. Sens. (IGARSS)*, 517–520, Vancouver, Canada, July 1989.
23. S. Wu, Analysis of data acquired by shuttle imaging radar SIR-A and Landsat Thematic Mapper over Baldwin county, Alabama, in *Proc. Mach. Process. Remotely Sensed Data Symp.*, 173–182, West Lafayette, Indiana, June 1985.
24. A.H. Schistad Solberg, A.K. Jain, and T. Taxt, Multisource classification of remotely sensed data: Fusion of Landsat TM and SAR images, *IEEE Trans. Geosc. Rem. Sens.*, 32, 768–778, 1994.
25. A. Schistad Solberg, Texture fusion and classification based on flexible discriminant analysis, in *Int. Conf. Pattern Recogn. (ICPR)*, 596–600, Vienna, Austria, August 1996.
26. H. Kim and P.H. Swain, A method for classification of multisource data using interval-valued probabilities and its application to HIRIS data, in *Proc. Workshop Multisource Data Integration Rem. Sens.*, 75–82, NASA Conference Publication 3099, Maryland, June 1990.
27. J. Desachy, L. Roux, and E-H. Zahzah, Numeric and symbolic data fusion: a soft computing approach to remote sensing image analysis, *Pattern Recognition Letters*, 17, 1361–1378, 1996.

28. S.L. Hégarat-Mascle, I. Bloch, and D. Vidal-Madjar, Application of Dempster–Shafer evidence theory to unsupervised classification in multisource remote sensing, *IEEE Trans. Geosc. Rem. Sens.*, 35, 1018–1031, 1997.
29. S.B. Serpico and F. Roli, Classification of multisensor remote-sensing images by structured neural networks, *IEEE Trans. Geosc. Rem. Sens.*, 33, 562–578, 1995.
30. J.A. Benediktsson, J.R. Sveinsson, and P.H. Swain, Hybrid consensys theoretic classification, *IEEE Trans. Geosc. Rem. Sens.*, 35, 833–843, 1997.
31. J.A. Benediktsson and I. Kanellopoulos, Classification of multisource and hyperspectral data based on decision fusion, *IEEE Trans. Geosc. Rem. Sens.*, 37, 1367–1377, 1999.
32. B.C.K. Tso and P.M. Mather, Classification of multisource remote sensing imagery using a genetic algorithm and Markov random fields, *IEEE Trans. Geosc. Rem. Sens.*, 37, 1255–1260, 1999.
33. M. Petrakos, J.A. Benediktsson, and I. Kannelopoulos, The effect of classifier agreement on the accuracy of the combined classifier in decision level fusion, *IEEE Trans. Geosc. Rem. Sens.*, 39, 2539–2546, 2001.
34. J.A. Benediktsson and J. Sveinsson, Multisource remote sensing data classification based on consensus and pruning, *IEEE Trans. Geosc. Rem. Sens.*, 41, 932–936, 2003.
35. A. Solberg, G. Storvik, and R. Fjørtoft, A comparison of criteria for decision fusion and parameter estimation in statistical multisensor image classification, in *IEEE Symp. Geosc. Rem. Sens. (IGARSS'02)*, July 2002.
36. D.G. Leckie, Synergism of synthetic aperture radar and visible/infrared data for forest type discrimination, *Photogramm. Eng. Rem. Sens.*, 56, 1237–1246, 1990.
37. S.E. Franklin, Ancillary data input to satellite remote sensing of complex terrain phenomena, *Comput. Geosci.*, 15, 799–808, 1989.
38. J.A. Richards, D.A. Landgrebe, and P.H. Swain, A means for utilizing ancillary information in multispectral classification, *Rem. Sens. Environ.*, 12, 463–477, 1982.
39. J. Friedman, Multivariate adaptive regression splines (with discussion), *Ann. Stat.*, 19, 1–141, 1991.
40. P. Gong, R. Pu, and J. Chen, Mapping ecological land systems and classification uncertainties from digital elevation and forest-cover data using neural networks, *Photogramm. Eng. Rem. Sens.*, 62, 1249–1260, 1996.
41. L. Guan, J.A. Anderson, and J.P. Sutton, A network of networks processing model for image regularization, *IEEE Trans. Neural Networks*, 8, 169–174, 1997.
42. T. Hastie, R. Tibshirani, and A. Buja, Flexible discriminant analysis by optimal scoring, *J. Am. Stat. Assoc.*, 89, 1255–1270, 1994.
43. S. Le Hégarat-Mascle and R. Seltz, Automatic change detection by evidential fusion of change indices, *Rem. Sens. Environ.*, 91, 390–404, 2004.
44. D. Hirst, G. Storvik, and A.R. Syversveen, A hierarchical modelling approach to combining environmental data at different scales, *J. Royal Stat. Soc., Series C*, 52, 377–390, 2003.
45. J.-N. Hwang, D. Li, M. Maechelr, D. Martin, and J. Schimert, Projection pursuit learning networks for regression, *Eng. Appl. Artif. Intell.*, 5, 193–204, 1992.
46. B. Jeon and D.A. Landgrebe, Classification with spatio-temporal interpixel class dependency contexts, *IEEE Trans. Geosc. Rem. Sens.*, 30, 663–672, 1992.
47. N. Khazenie and M.M. Crawford, Spatio-temporal autocorrelated model for contextual classification, *IEEE Trans. Geosc. Rem. Sens.*, 28, 529–539, 1990.
48. M.R. Luettgen, W. Clem Karl, and A.S. Willsky, Efficient multiscale regularization with applications to the computation of optical flow, *IEEE Trans. Image Process.*, 3(1), 41–63, 1994.
49. J. Middelkoop and L.L.F. Janssen, Implementation of temporal relationships in knowledge based classification of satellite images, *Photogramm. Eng. Rem. Sens.*, 57, 937–945, 1991.
50. L. Olsson and L. Eklundh, Fourier series for analysis of temporal sequences of satellite sensor imagery, *Int. J. Rem. Sens.*, 15, 3735–3741, 1994.
51. G. Pajares and J.M. de la Cruz, A wavelet-based image fusion tutorial, *Pattern Recognition*, 37, 1855–1871, 2004.
52. J.D. Paola and R.A. Schowengerdt, The effect of neural-network structure on a multispectral land-use/land-cover classification, *Photogramm. Eng. Rem. Sens.*, 63, 535–544, 1997.
53. J.C. Price, Combining multispectral data of differing spatial resolution, *IEEE Trans. Geosci. Rem. Sens.*, 37(3), 1199–1203, 1999.

54. P. Puyou-Lascassies, A. Podaire, and M. Gay, Extracting crop radiometric responses from simulated low and high spatial resolution satellite data using a linear mixing model, *Int. J. Rem. Sens.*, 15(18), 3767–3784, 1994.
55. T. Ranchin, B. Aiazzi, L. Alparone, S. Baronti, and L. Wald, Image fusion—the ARSIS concept and some successful implementations, *ISPRS J. Photogramm. Rem. Sens.*, 58, 4–18, 2003.
56. B.D. Ripley, Flexible non-linear approaches to classification, in *From Statistics to Neural Networks. Theory and Pattern Recognition Applications*, V. Cherkassky, J.H. Friedman, and H. Wechsler, eds., 105–126, NATO ASI series F: Computer and systems sciences, springer-Verlag, Heidelberg, 1994.
57. A.H. Solberg, Flexible nonlinear contextual classification, *Pattern Recognition Letters*, 25, 1501–1508, 2004.
58. A.K. Skidmore, B.J. Turner, W. Brinkhof, and E. Knowles, Performance of a neural network: mapping forests using GIS and remotely sensed data, *Photogramm. Eng. Rem. Sens.*, 63, 501–514, 1997.
59. J.A. Benediktsson, J.R. Sveinsson, and O.K. Ersoy, Optimized combination of neural networks, in *IEEE Int. Symp. Circuits and Sys. (ISCAS'96)*, 535–538, Atlanta, Georgia, May 1996.
60. G.A. Carpenter, M.N. Gjaja, S. Gopal, and C.E. Woodcock, ART neural networks for remote sensing: Vegetation classification from Landsat TM and terrain data, in *IEEE Symp. Geosc. Rem. Sens. (IGARSS)*, 529–531, Lincoln, Nebraska, May 1996.
61. W. Wan and D. Fraser, A self-organizing map model for spatial and temporal contextual classification, in *IEEE Symp. Geosc. Rem. Sens. (IGARSS)*, 1867–1869, Pasadena, California, August 1994.
62. G. Shafer, *A Mathematical Theory of Evidence*, Princeton University Press, 1976.
63. E. Binaghi, P. Madella, M.G. Montesano, and A. Rampini, Fuzzy contextual classification of multisource remote sensing images, *IEEE Trans. Geosc. Rem. Sens.*, 35, 326–340, 1997.
64. P.C. Smits and S.G. Dellepiane, Synthetic aperture radar image segmentation by a detail preserving Markov random field approach, *IEEE Trans. Geosc. Rem. Sens.*, 35, 844–857, 1997.
65. P.B. Chou and C.M. Brown, Multimodal reconstruction and segmentation with Markov random fields and HCF optimization, in *Proc. 1988 DARPA Image Understanding Workshop*, 214–221, 1988.
66. W.A. Wright, A Markov random field approach to data fusion and colour segmentation, *Image Vision Comp.*, 7, 144–150, 1989.
67. A.H. Schistad Solberg, T. Taxt, and Anil K. Jain, A Markov random field model for classification of multisource satellite imagery, *IEEE Trans. Geosc. Rem. Sens.*, 34, 100–113, 1996.
68. A.H. Schistad Solberg, Contextual data fusion applied to forest map revision, *IEEE Trans. Geosc. Rem. Sens.*, 37, 1234–1243, 1999.
69. W. Wan and D. Fraser, Multisource data fusion with multiple self-organizing maps, *IEEE Trans. Geosc. Rem. Sens.*, 37, 1344–1349, 1999.
70. P.H. Swain, Bayesian classification in a time-varying environment, *IEEE Trans. Sys. Man Cyber.*, 8, 879–883, 1978.
71. L. Bruzzone and R. Cossu, A multiple-cascade-classifier system for a robust and partially unsupervised updating of land-cover maps, *IEEE Trans. Geosc. Rem. Sens.*, 40, 1984–1996, 2002.
72. L. Bruzzone and D.F. Prieto, Unsupervised retraining of a maximum-likelihood classifier for the analysis of multitemporal remote-sensing images, *IEEE Trans. Geosc. Rem. Sens.*, 39, 456–460, 2001.
73. L. Bruzzone and D.F. Prieto, An adaptive semiparametric and context-based approach to unsupervised change detection in multitemporal remote-sensing images, *IEEE Trans. Image Proc.*, 11, 452–466, 2002.
74. P. Coppin, K. Jonkheere, B. Nackaerts, and B. Muys, Digital change detection methods in ecosystem monitoring: A review, *Int. J. Rem. Sens.*, 25, 1565–1596, 2004.
75. L. Andres, W.A. Salas, and D. Skole, Fourier analysis of multi-temporal AVHRR data applied to a land cover classification, *Int. J. Rem. Sens.*, 15, 1115–1121, 1994.
76. L. Aurdal, R. B. Huseby, L. Eikvil, R. Solberg, D. Vikhamar, and A. Solberg, Use of hiddel Markov models and phenology for multitemporal satellite image classification: applications to mountain vegetation classification, in *MULTITEMP 2005*, 220–224, May 2005.
77. N.G. Besag, K. Ickstadt, and R.L. Wolpert, Spatial poisson regression for health and exposure data measured at disparate resolutions, *J. Am. Stat. Assoc.*, 452, 1076–1088, 2000.

78. M.M. Daniel and A.S. Willsky, A multiresolution methodology for signal-level fusion and data assimilation with applications to remote sensing, *Proc. IEEE*, 85(1), 164–180, 1997.
79. L. Wald, *Data Fusion: Definitions and Achitectures—Fusion of Images of Different Spatial Resolutions*, Ecole des Mines Press, 2002.
80. F. Van der Meer, What does multisensor image fusion add in terms of information content for visual interpretation? *Int. J. Rem. Sens.*, 18, 445–452, 1997.
81. B. Zhukov, D. Oertel, F. Lanzl, and G. Reinhäckel, Unmixing-based multisensor multiresolution image fusion, *IEEE Trans. Geosci. Rem. Sens.*, 37(3), 1212–1226, 1999.
82. M.M. Crawford, S. Kumar, M.R. Ricard, J.C. Gibeaut, and A. Neuenshwander, Fusion of airborne polarimetric and interferometric SAR for classification of coastal environments, *IEEE Trans. Geosci. Rem. Sens.*, 37(3), 1306–1315, 1999.
83. G. Storvik, R. Fjørtoft, and A. Solberg, A Bayesian approach to classification in multiscale remote sensing data, *IEEE Trans. Geosc. Rem. Sens.*, 43, 539–547, 2005.
84. J. Besag, Towards Bayesian image analysis, *J. Appl. Stat.*, 16(3), 395–407, 1989.

12

The Hermite Transform: An Efficient Tool for Noise Reduction and Image Fusion in Remote-Sensing

Boris Escalante-Ramírez and Alejandra A. López-Caloca

CONTENTS

12.1 Introduction

In this chapter, we introduce the Hermite transform (HT) as an efficient tool for remote-sensing image processing applications. The HT is an image representation model that mimics some of the more important properties of human visual perception, namely the local orientation analysis and the Gaussian derivative model of early vision. We limit our discussion to the cases of noise reduction and image fusion. However many different applications can be tackled within the scheme of direct-inverse HT.

It is generally acknowledged that visual perception models must involve two major processing stages: (1) initial measurements and (2) high-level interpretation. Fleet and Jepson [1] pointed out that the early measurement is a rich encoding of image structure in terms of generic properties from which structures that are more complex are easily detected and analyzed. Such measurement processes should be image-independent and require no previous or concurrent interpretation. Unfortunately, it is not known what primitives are necessary and sufficient for interpretation or even identification of meaningful features. However, we know that, for image processing purposes, linear operators that exhibit special kind of symmetries related to translation, rotation, and magnification are of particular interest. A family of generic neighborhood operators fulfilling these requirements is that formed by the so-called Gaussian derivatives [2]. These operators have long been used in computer vision for feature extraction [3,4], and are relevant in visual system modeling [5]. Formal integration of these operators is achieved in the HT introduced first by Martens [6,7], and recently reformulated as a multi-scale image representation model for local orientation analysis [8,9]. This transform can take many alternative forms corresponding to different ways of coding local orientations in the image.

Young showed that Gaussian derivatives model the measured receptive field data more accurately than the Gabor functions do [10]. Like the receptive fields, both Gabor functions and Gaussian derivatives are spatially local and consist of alternating excitatory and inhibitory regions within a decaying envelope. However, the Gaussian derivative analysis is found to be more efficient because it takes advantage of the fact that Gaussian derivatives comprise an orthogonal basis if they belong to the same point of analysis. Gaussian derivatives can be interpreted as local generic operators in a scale-space representation described by the isotropic diffusion equation [2]. In a related work, the Gaussian derivatives have been interpreted as the product of Hermite polynomials and a Gaussian window [6], where windowed images are decomposed into a set of Hermite polynomials. Some mathematical models based on these operators at a single spatial scale have been described elsewhere [6,11]. In the case of the HT, it has been extended to the multi-scale case [7–9], and has been successfully used in different applications such as noise reduction [12], coding [13], and motion estimation for the case of image sequences [14].

Applications to local orientation analysis are a major concern in this chapter. It is well known that local orientation estimation can be achieved by combining the outputs from polar separable quadrature filters [15]. Freeman and Adelson developed a technique to steer filters by linearly combining basis filters oriented at a number of specific directions [16]. The possibilities are, in fact, infinite because the set of basis functions required to steer a function is not unique [17]. The Gaussian derivative family is perhaps the most common example of such functions.

In the first part of this chapter we introduce the HT as an image representation model, and show how local analysis can be achieved from a steered HT.

In the second part we build a noise-reduction algorithm for synthetic aperture radar (SAR) images based on the steered HT that adapts to the local image content and to the multiplicative nature of speckle.

In the third section, we fuse multi-spectral and panchromatic images from the same satellite (Landsat ETM+) with different spatial resolutions. In this case we show how the proposed method improves spatial resolution and preserves the spectral characteristics, that is, the biophysical variable interpretation of the original images remains intact.

Finally, we fuse SAR and multi-spectral Landsat ETM+ images, and show that in this case spatial resolution is also improved while spectral resolution is preserved. Speckle reduction in the SAR image is achieved, along with image fusion, within the analysis–synthesis process of the fusion scheme.

Both fusion and speckle-reduction algorithms are based on the detection of relevant image structures (primitives) during the analysis stage. For this purpose, Gaussian-derivative filters at different scales can be used. Local orientation is estimated so that the transform can be rotated at every position of the analysis window. In the case of noise reduction, transform coefficients are classified based on structure dimensionality and energy content so that those belonging to speckle are discarded. With a similar criterion, transform coefficients from different image sources are classified to select coefficients from each image that contribute to synthesize the fused image.

12.2 The Hermite Transform

12.2.1 The Hermite Transform as an Image Representation Model

The HT [6,7] is a special case of polynomial transform. It can be regarded as an image description model. Firstly, windowing with a local function $\omega(x, y)$ takes place at several positions over the input image. Next, local information at every analysis window is expanded in terms of a family of orthogonal polynomials. The polynomials $G_{m,n-m}(x, y)$ used to approximate the windowed information are determined by the analysis window function and satisfy the orthogonal condition:

$$\int_{-\infty}^{+\infty}\int_{-\infty}^{+\infty} \omega^2(x, y)G_{m,n-m}(x, y)G_{l,k-l}(x, y)\,dx\,dy = \delta_{nk}\delta_{ml} \tag{12.1}$$

for $n, k = 0, \ldots, \infty$; $m = 0, \ldots, n$; $l = 0, \ldots, k$; where δ_{nk} denotes the Kronecker function.

Psychophysical insights suggest using a Gaussian window function, which resembles the receptive field profiles of human vision, that is,

$$\omega(x, y) = \frac{1}{2\pi\sigma^2}\exp\left(-\frac{(x^2 + y^2)}{2\sigma^2}\right) \tag{12.2}$$

The Gaussian window is separable into Cartesian coordinates; it is isotropic, thus it is rotationally invariant and its derivatives are good models of some of the more important retinal and cortical cells of the human visual system [5,10].

In the case of a Gaussian window function, the associated orthogonal polynomials are the Hermite polynomials [18]:

$$G_{n-m,m}(x, y) = \frac{1}{\sqrt{2^n(n-m)!m!}}H_{n-m}\left(\frac{x}{\sigma}\right)H_m\left(\frac{y}{\sigma}\right) \tag{12.3}$$

where $H_n(x)$ denotes the nth Hermite polynomial.

The original signal $L(x, y)$, where (x, y) are the pixel coordinates, is multiplied by the window function $\omega(x - p, y - q)$, at positions (p, q) that conform the sampling lattice S.

Through replication of the window function over the sampling lattice, a periodic weighting function is defined as $W(x,y) = \sum_{(p,q)\in S} \omega(x - p, y - q)$. This weighting function must be different from zero for all coordinates (x, y), then:

$$L(x, y) = \frac{1}{W(x, y)} \sum_{(p,q)\in S} L(x, y)\omega(x - p, y - q) \tag{12.4}$$

The signal content within every window function is described as a weighted sum of polynomials $G_{m,n-m}(x, y)$ of m degree in x and $n - m$ in y. In a discrete implementation, the Gaussian window function may be approximated by the binomial window function, and in this case, its orthogonal polynomials $G_{m,n-m}(x, y)$ are known as the Krawtchouck polynomials.

In either case, the polynomial coefficients $L_{m,n-m}(p, q)$ are calculated by convolution of the original image $L(x, y)$ with the function filter $D_{m,n-m}(x, y) = G_{m,n-m}(-x, -y)$ $\omega^2(-x,-y)$ followed by subsampling at positions (p, q) of the sampling lattice S, i.e.,

$$L_{m,n-m}(p, q) = \int_{-\infty}^{+\infty} \int_{-\infty}^{+\infty} L(x, y) D_{m,n-m}(x-p, y-q)\, dx\, dy \tag{12.5}$$

For the case of the HT, it can be shown [18] that the filter functions $D_{m,n-m}(x, y)$ correspond to Gaussian derivatives of order m in x and $n - m$ in y, in agreement with the Gaussian derivative model of early vision [5,10].

The process of recovering the original image consists of interpolating the transform coefficients with the proper synthesis filters. This process is called an inverse polynomial transform and is defined by

$$\hat{L}(x, y) = \sum_{n=0}^{\infty} \sum_{m=0}^{n} \sum_{(p,q)\in S} L_{m,n-m}(p, q) P_{m,n-m}(x-p, y-q) \tag{12.6}$$

The synthesis filters $P_{m,n-m}(x, y)$ of order m and $n - m$ are defined by

$$P_{m,n-m}(x, y) = \frac{G_{m,n-m}(x, y)\omega(x, y)}{W(x, y)} \quad \text{for } m = 0, \ldots, n \quad \text{and} \quad n = 0, \ldots, \infty.$$

Figure 12.1 shows the analysis and synthesis stages of a polynomial transform. Figure 12.2 shows a HT calculated on a satellite image.

To define a polynomial transform, some parameters have to be chosen. First, we have to define the characteristics of the window function. The Gaussian window is the best option from a perceptual point of view and from the scale-space theory. Other free parameters are the size of the Gaussian window spread (σ) and the distance between adjacent window positions (sampling lattice). The size of the window functions must be related to the spatial scale of the image structures that are to be analyzed. Fine local changes are better detected with small windows, but on the contrary, representation of low-resolution objects needs large windows. To overcome this compromise, multi-resolution representations are a good alternative. For the case of the HT, a multi-resolution extension has recently been proposed [8,9].

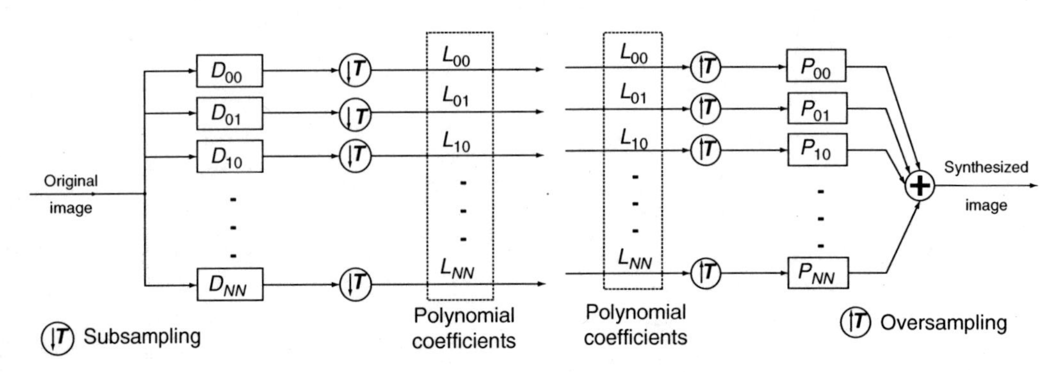

FIGURE 12.1
Analysis and synthesis with the polynomial transform.

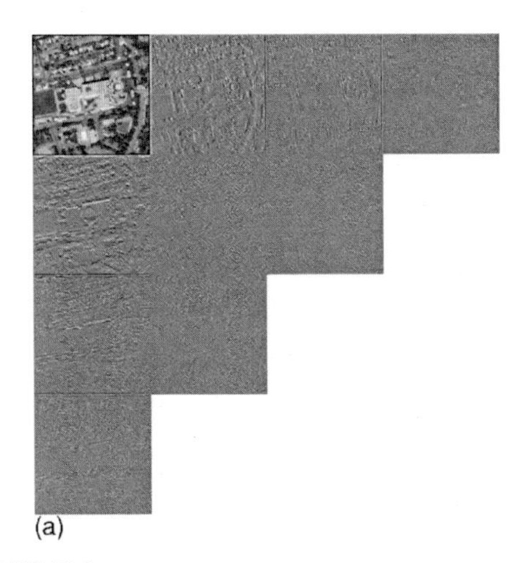

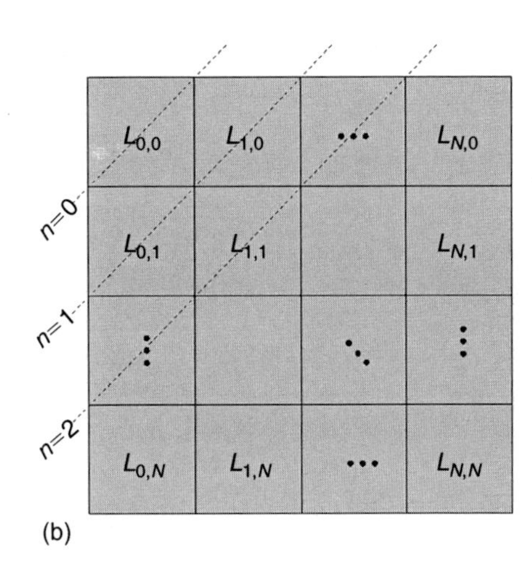

FIGURE 12.2
(a) Hermite transform calculated on a satellite image. (b) Diagram showing the coefficient orders. Diagonals depict zero-order coefficients ($n = 0$), first-order coefficients ($n = 1$), etc. Gaussian window with spread $\sigma = \sqrt{2}$ and subsampling $d = 4$ was used.

12.2.2 The Steered Hermite Transform

The HT has the advantage that high-energy compaction can be obtained through adaptively steering the transform [19]. The term steerable filters describes a set of filters that are rotated copies of each other, and a copy of the filter in any orientation which is then constructed as a linear combination of a set of basis filters. The steering property of the Hermite filters can be considered because the filters are products of polynomials with a radially symmetric window function. The $N + 1$ Hermite filters of Nth-order form a steerable basis for each individual filter of order N. Based on the steering property, the Hermite filters at each position in the image adapt to the local orientation content. This adaptability results in significant compaction.

For orientation analysis purposes, it is convenient to work with a rotational version of the HT. The polynomial coefficients can be computed through a convolution of the image with the filter functions $D_m(x)D_{n-m}(y)$; the properties of the filter functions are separable in spatial and polar domains and the Fourier transform of the filter functions are expressed in polar coordinates considering $\omega_x = \omega \cos\theta$ and $\omega_y = \omega \sin\theta$,

$$d_m(\omega_x)d_{n-m}(\omega_y) = g_{m,n-m}(\theta) \cdot d_n(\omega) \tag{12.7}$$

where $d_n(\omega)$ is the Fourier transform for each filter function, and the radial frequency of the filter function of the nth order Gaussian derivative is given by

$$d_n(\omega) = \frac{1}{\sqrt{2^n n!}}(-\mathrm{j}\omega\sigma)^n \exp\left(-(\omega\sigma)^2/4\right) \tag{12.8}$$

and the orientation selectivity of the filter is expressed by

$$g_{m,n-m}(\theta) = \sqrt{\binom{n}{m}} \cos^m\theta \cdot \sin^{n-m}\theta \tag{12.9}$$

In terms of orientation frequency functions, this property of the Hermite filters can be expressed by

$$g_{m,n-m}(\theta - \theta_0) = \sum_{k=0}^{n} c_{m,k}^{n}(\theta_0) g_{n-k,k}(\theta) \tag{12.10}$$

where $c_{m,k}^{n}(\theta_0)$ is the steering coefficient. The Hermite filter rotation at each position over the image is an adaptation to local orientation content. Figure 12.3 shows the directional Hermite decomposition over an image. First a HT was applied and then the coefficients of this transform were rotated towards the local estimated orientation, according to a maximum oriented energy criterion at each window position. For local 1D patterns, the steered HT provides a very efficient representation. This representation consists of a parameter θ, indicating the orientation of the pattern, and a small number of coefficients, representing the profile of the pattern perpendicular to its orientation. For a 1D pattern with orientation θ, the following relation holds:

$$L_{n-m,m}^{\theta} = \begin{cases} \sum_{k=0}^{n} g_{n-k,k}(\theta) L_{n-k,k}, & \mathrm{m} = 0 \\ 0, & \mathrm{m} > 0 \end{cases} \tag{12.11}$$

For such a pattern, steering over θ results in a compaction of energy into the coefficients $L_{n,0}^{\theta}$, while all other coefficients are set to zero.

The energy content can be expressed through the Hermite coefficients (Parseval Theorem) as

$$E_{\infty} = \sum_{n=0}^{\infty} \sum_{m=0}^{n} [L_{n-m,m}]^2 \tag{12.12}$$

The energy up to order N, E_N is defined as the addition of all squared coefficients up to N order.

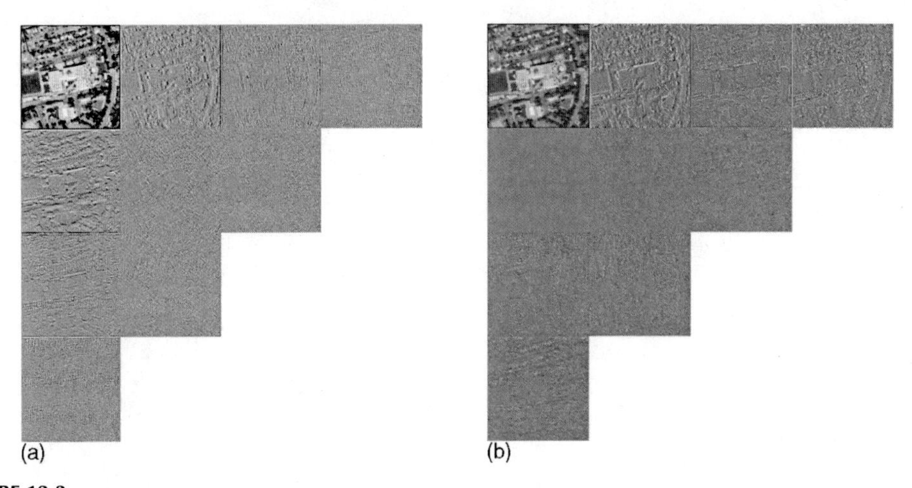

FIGURE 12.3
Steered Hermite transform. (a) Original coefficients. (b) Steered coefficients. It can be noted that most coefficient energy is concentrated on the upper row.

The steered Hemite transform offers a way to describe 1D patterns on the basis of their orientation and profile. We can differentiate 1D energy terms and 2D energy terms. That is, for each local signal we have

$$E_N^{1D}(\theta) = \sum_{n=1}^{N} \left[L_{n,0}^{\theta}\right]^2, \tag{12.13}$$

$$E_N^{2D}(\theta) = \sum_{n=1}^{N} \sum_{m=1}^{n} \left[L_{n-m,m}^{\theta}\right]^2 \tag{12.14}$$

12.3 Noise Reduction in SAR Images

The use of SAR images instead of visible and multi-spectral images is becoming increasingly popular, because of their capability of imaging even in the case of cloud-covered remote areas In addition to the all-weather capacity, there are several well-known advantages of SAR data over other imaging systems [20]. Unfortunately, the poor quality of SAR images makes it very difficult to perform direct information extraction tasks. Even more, the incorporation of external reference data (in-situ measurements) is frequently needed to guaranty a good positioning of the results. Numerous filters have been proposed to remove speckle in SAR imagery; however, in most cases and even in the most elegant approaches, filtering algorithms have a tendency to smooth speckle as well as information. For numerous applications, low-level processing of SAR images remains a partially unsolved problem. In this context, we propose a restoration algorithm that adaptively smoothes images. Its main advantage is that it retains subtle details.

The HT coefficients are used to discriminate noise from relevant information such as borders and lines in a SAR image. Then an energy mask containing relevant image locations is built by thresholding the first-order transform coefficient energy E_1: $E_1 = L_{0,1}^2 + L_{1,0}^2$ where $L_{0,1}$ and $L_{1,0}$ are the first-order coefficients of the HT. These coefficients are obtained by convolving the original image with the first-order derivatives of a Gaussian function, which are known to be quasi-optimal edge detectors [21]; therefore, the first-order energy can be used to discriminate edges from noise by means of a threshold scheme.

The optimal threshold is set considering two important characteristics of SAR images. First, one-look amplitude SAR images have a Rayleigh distribution and the signal-to-noise ratio (SNR) is approximately 1.9131. Second, in general, the SNR of multi-look SAR images does not change over the whole image; furthermore, $\text{SNR}_{N\text{looks}} = 1.9131\sqrt{N}$, which yields for a homogeneous region l:

$$\sigma_l = \frac{\mu_1}{1.9131\sqrt{N}} \tag{12.15}$$

where σ_l is the standard deviation of the region l, μ_l is its mean value, and N is the number of looks of the image.

The first-order coefficient noise variance in homogeneous regions is given by

$$\sigma^2 = \alpha\sigma_l^2, \tag{12.16}$$

where

$$\alpha = |R_L(x, y) * D_{1,0}(x, y) * D_{1,0}(-x, -y)|_{x=y=0}$$

R_L is the normalized autocorrelation function of the input noise, and $D_{1,0}$ is the filter used to calculate the first-order coefficient. Moreover, the probability density function (PDF) of $L_{1,0}$ and $L_{0,1}$ in uniform regions can be considered Gaussian, according to the Central Limit Theorem, then, the energy PDF is exponential:

$$P(E_1) = \frac{1}{2\sigma^2} \exp\left(-\frac{E_1}{2\sigma^2}\right) \tag{12.17}$$

Finally, the threshold is fixed:

$$T = 2 \ln\left(\frac{1}{P_R}\right)\sigma^2 \tag{12.18}$$

where P_R is the probability (percentage) of noise left on the image and will be set by the user. A careful analysis of this expression reveals that this threshold adapts to the local content of the image since Equation 12.15 and Equation 12.16 show the dependence of σ on the local mean value μ_l, the latter being approximated by the Hermite coefficient L_{00}.

With the locations of relevant edges detected, the next step is to represent these locations as one-dimensional patterns. This can be achieved by steering the HT as described in the previous section so that the steering angle θ is determined by the local edge orientation. Next, only coefficients $L^{\theta}_{n,0}$ are preserved, all others are set to zero.

In summary, the noise reduction strategy consists of classifying the image in either zero-dimensional patterns consisting of homogeneous noisy regions, or one-dimensional patterns containing noisy edges. The former are represented by the zeroth order coefficient, that is, the local mean value, and the latter by oriented 1D Hermite coefficients. When an inverse HT is performed over these selected coefficients, the resulting synthesized image consists of noise-free sharp edges and smoothed homogeneous regions. Therefore the denoised image preserves sharpness and thus, image quality. Some speckle remains in the image because there is always a compromise between the degree of noise reduction and the preservation of low-contrast edges. The user controls the balance of this compromise by changing the percentage of noise left P_R on the image according to Equation 12.18.

Figure 12.4 shows the algorithm for noise reduction, and Figure 12.5 through Figure 12.8 show different results of the algorithm.

12.4 Fusion Based on the Hermite Transform

Image fusion has become a useful tool to enhance information provided by two or more sensors by combining the most relevant features of each image. A wide range of disciplines including remote sensing and medicine have taken advantage of fusion techniques, which in recent years have evolved from simple linear combinations to sophisticated methods based on principal components, color models, and signal transformations

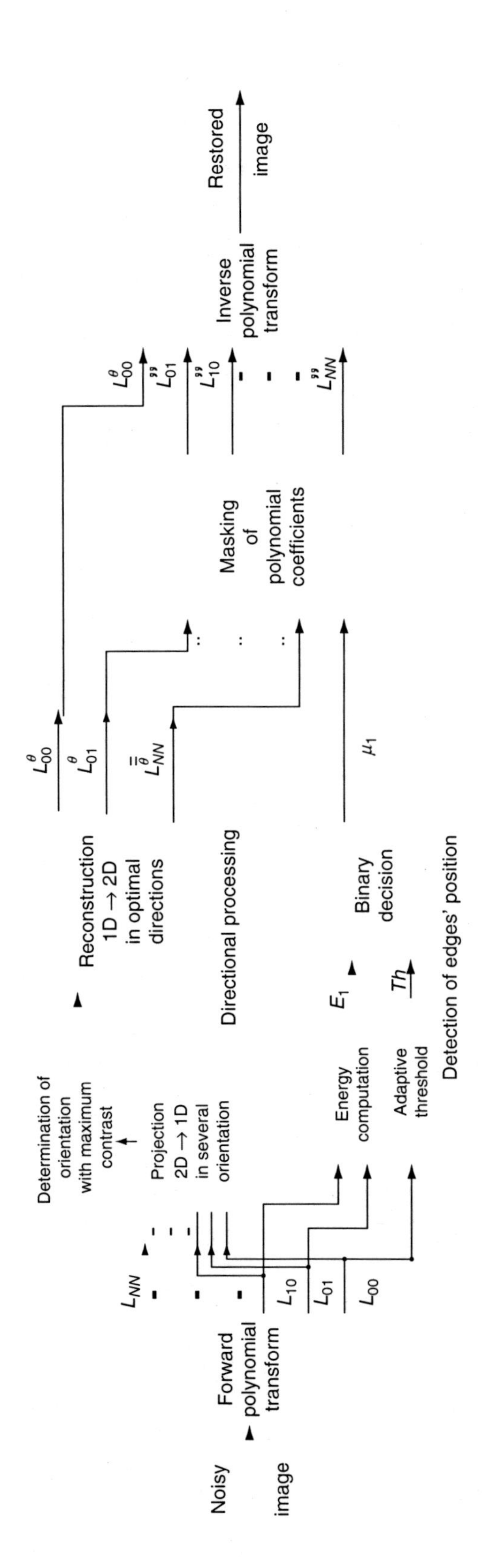

FIGURE 12.4
Noise-reduction algorithm.

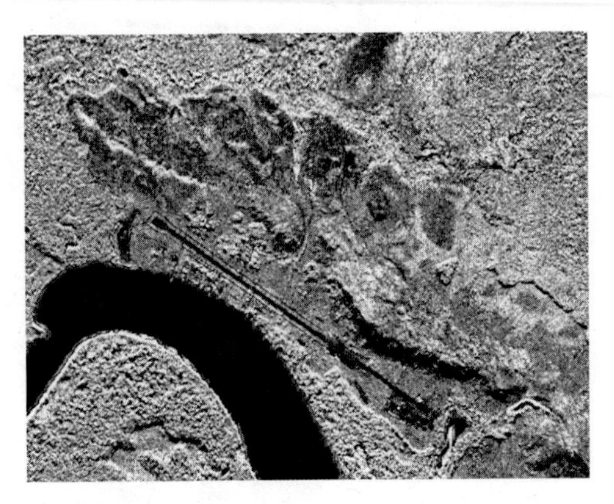
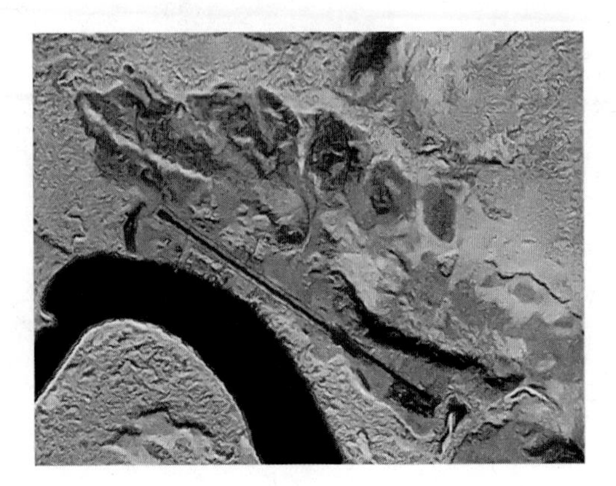

FIGURE 12.5
Left: Original SAR AeS-1 image. Right: Image after noise reduction.

among others [22–25]. Recently, multi-resolution techniques such as image pyramids and wavelet transforms have been successfully used [25–27]. Several authors have shown that, for image fusion, the wavelet transform approach offers good results [1,25,27]. Comparisons of Mallat's and "à trous" methodologies have been studied [28]. Furthermore, multi-sensor image fusion algorithms based on intensity modulation have been proposed for SAR and multi-band optical data fusion [29].

Information in the fused image must lead to improved accuracy (from redundant information) and improved capacity (from complementary information). Moreover, from a visual perception point of view, patterns included in the fused image must be perceptually relevant and must not include distracting artifacts. Our approach aims at analyzing images by means of the HT, which allows us to identify perceptually relevant patterns to be included in the fusion process while discriminating spurious artifacts.

The steered HT has the advantage of energy compaction. Transform coefficients are selected with an energy compaction criterion from the steered Hermite transform;

FIGURE 12.6
Left: Original SEASAT image. Right: Image after noise reduction.

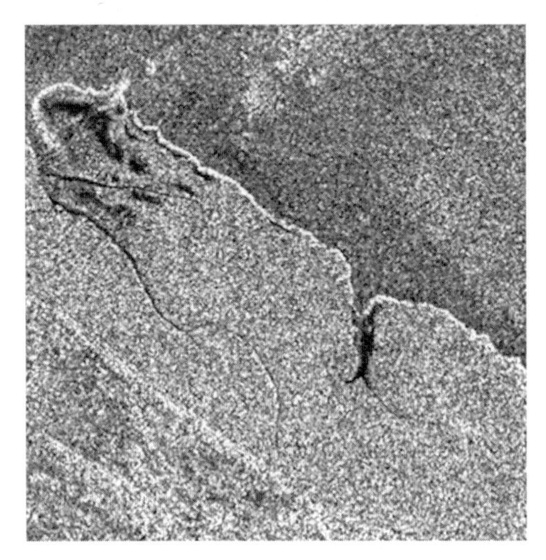

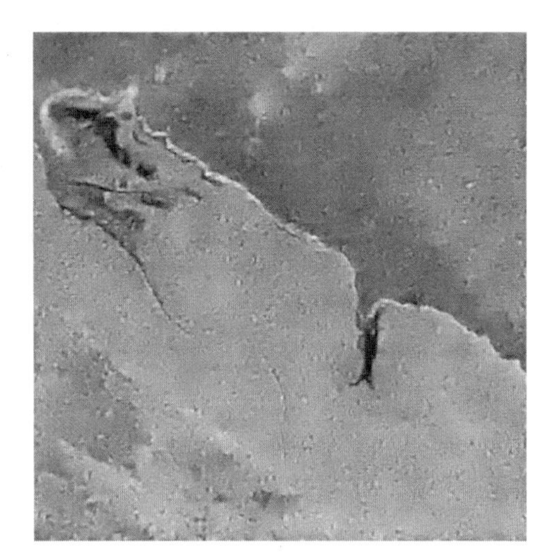

FIGURE 12.7
Left: Original ERS1 image. Right: Image after noise reduction.

therefore, it is possible to reconstruct an image with few coefficients and still preserve details such as edges and textures.

The general framework for fusion through HT includes five steps: (1) HT of the image. (2) Detection of maximum energy orientation with the energy measure $E_N^{1D}(\theta)$ at each window position. In practice, one estimator of the optimal orientation θ can be obtained through $\tan(\theta) = L_{0,1}/L_{1,0}$, where $L_{0,1}$ and $L_{1,0}$ are the first-order HT coefficients. (3) Adaptive steering of the transform coefficients, as described in previous sections. (4) Coefficient selection based on the method of verification of consistency [27]. This selection rule uses the maximum absolute value within a 5×5 window over the image (area of activity). The window variance is computed and used as a measurement of the activity associated with the central pixel of the window. In this way, a significant value indicates the presence of a dominant pattern in the local area. A map of binary decision is then

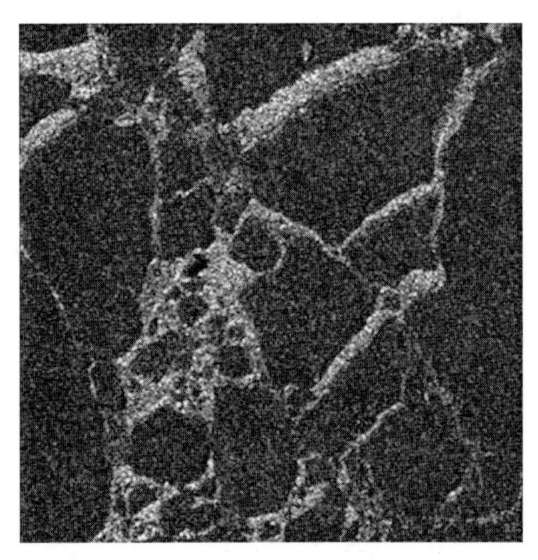

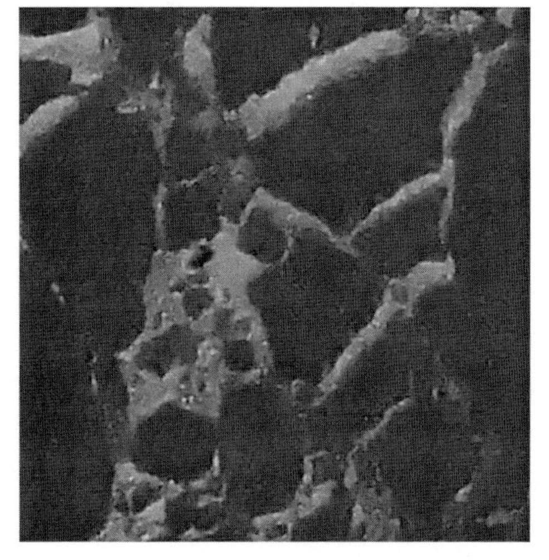

FIGURE 12.8
Left: Original ERS1 image. Right: Image after noise reduction.

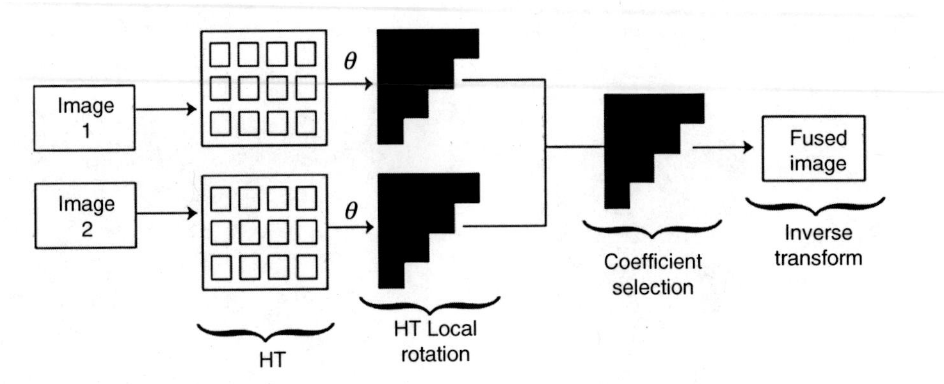

FIGURE 12.9
Fusion scheme with the Hermite transform.

created for the registry of the results. This binary map is subjected to consistency verification. (5) The final step of the fusion is the inverse transformation from the selected coefficients and their corresponding optimal θ. Figure 12.9 shows a simplified diagram of this method.

12.4.1 Fusion Scheme with Multi-Spectral and Panchromatic Images

Our objective of image fusion is to generate synthetic images with a higher resolution that attempts to preserve the radiometric characteristics of the original multi-spectral data. It is desirable that any procedure that fuses high-resolution panchromatic data with low-resolution multi-spectral data preserves, as much as possible, the original spectral characteristics.

To apply this image fusion method, it is necessary to resample the multi-spectral images so that their pixel size is the same as that of the panchromatic image's. The steps for fusing multi-spectral and panchromatic images are as follows: (1) Generate new panchromatic images, whose histograms match those of each band of the multi-spectral image. (2) Apply the HT with local orientation extraction and detection of maximum energy orientation. (3) Select the coefficients based on the method of verification of consistency. (4) Inverse transformation with the optimal θ resulting from the selected coefficient set. This process of fusion is depicted in Figure 12.10.

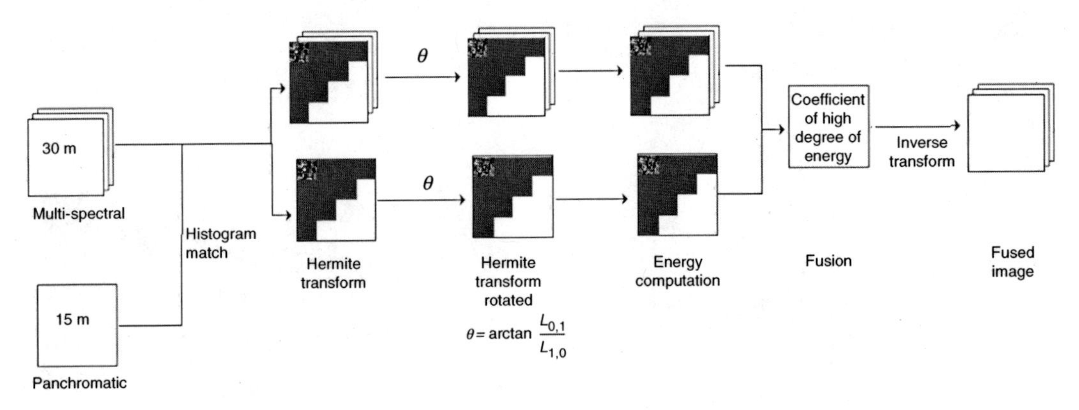

FIGURE 12.10
Hermite transform fusion for multi-spectral and panchromatic images.

12.4.2 Experimental Results with Multi-Spectral and Panchromatic Images

The proposed fusion scheme with multi-spectral images has been tested on optical data. We fused multi-spectral images from Landsat ETM+ (30 m) with its panchromatic band (15 m). We show in Figure 12.11 how the proposed method can help improve spatial resolution.

To evaluate the efficiency of the proposed method we calibrate the images so that digital values are transformed to reflectance values. Calibrated images were compared before and after fusion, by means of the Tasselep cap transformation (TCT) [30–32]. The TCT method is reported in [33]. The TCT transforms multi-spectral spatial values to a new domain based on biophysical variables, namely brightness, greenness, and a third component of the scene under study. It is deduced that the brightness component is a weighted sum of all the bands, based on the reflectance variation of the ground. The greenness component describes the contrast between near-infrared and visible bands with the mid-infrared bands. It is strongly related to the amount of green vegetation in the scene. Finally, the third component gives a measurement of the humidity content of the ground. Figure 12.12 shows the brightness, greenness, and third components obtained from HT fusion results.

The TCT was applied to the original multi-spectral image, the HT fusion result, and principal component analysis (PCA) fusion method.

To understand the variability of TCT results on the original, HT fusion and PCA fusion images, the greenness, and brightness components were compared. The greenness and brightness components define the plane of vegetation in ETM+ data. These results are displayed in Figure 12.13. It can be noticed that, in the case of PCA, the brightness and greenness content differs considerably from the original image, while in the case of HT they are very similar to the original ones. A linear regression analysis of the TCT components (Table 12.1) shows that the brightness and greenness components of the HT-fused image present a high linear correlation with the original image values. In other words, the biophysical properties of multi-spectral images are preserved when using the HT for image fusion, in contrast to the case of PCA fusion.

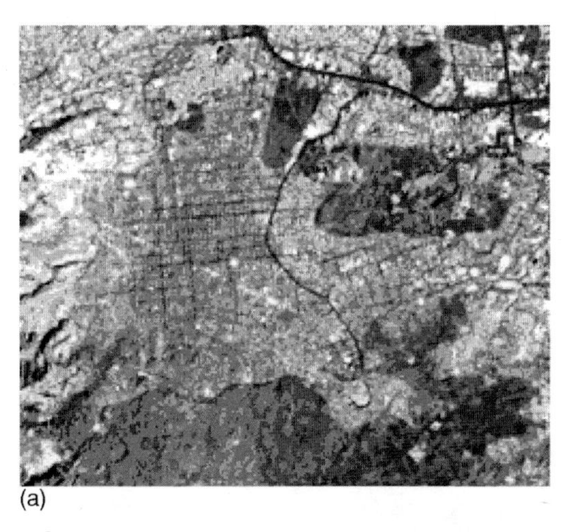
(a)

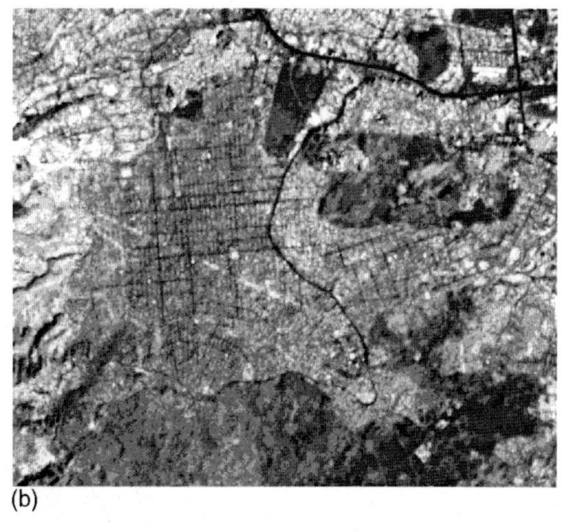
(b)

FIGURE 12.11
(a) Original Landsat 7 ETM+ image of Mexico city (resampled to 15 m to match geocoded panchromatic), (b) Resulting image of ETM+ and panchromatic band fusion with Hermite transform (Gaussian window with spread $\sigma = \sqrt{2}$ and window spacing T = 4) (RGB composition 5–4–3).

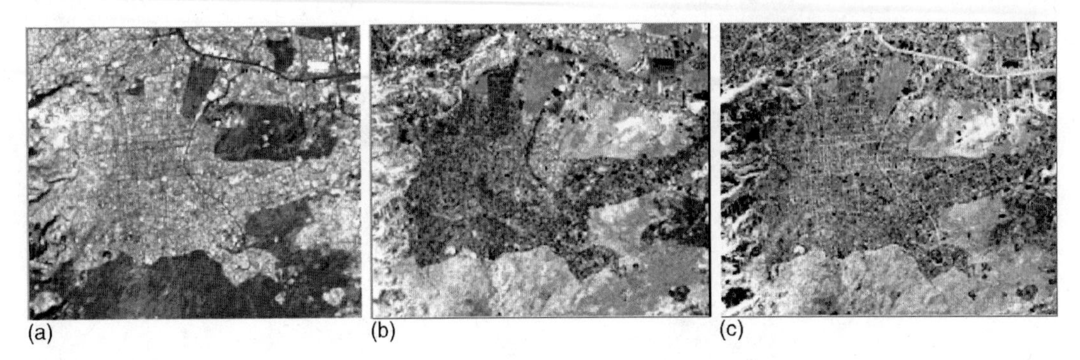

FIGURE 12.12
(a) Brightness, (b) greenness, and (c) third component in Hermite transform in a fused image.

12.4.3 Fusion Scheme with Multi-Spectral and SAR Images

In the case of SAR images, the characteristic noise, also known as speckle, imposes additional difficulties to the problem of image fusion. In spite of this limitation, the use of SAR images is becoming more popular due to their immunity to cloud coverage. Speckle removal, as described in previous section, is therefore a mandatory task in fusion

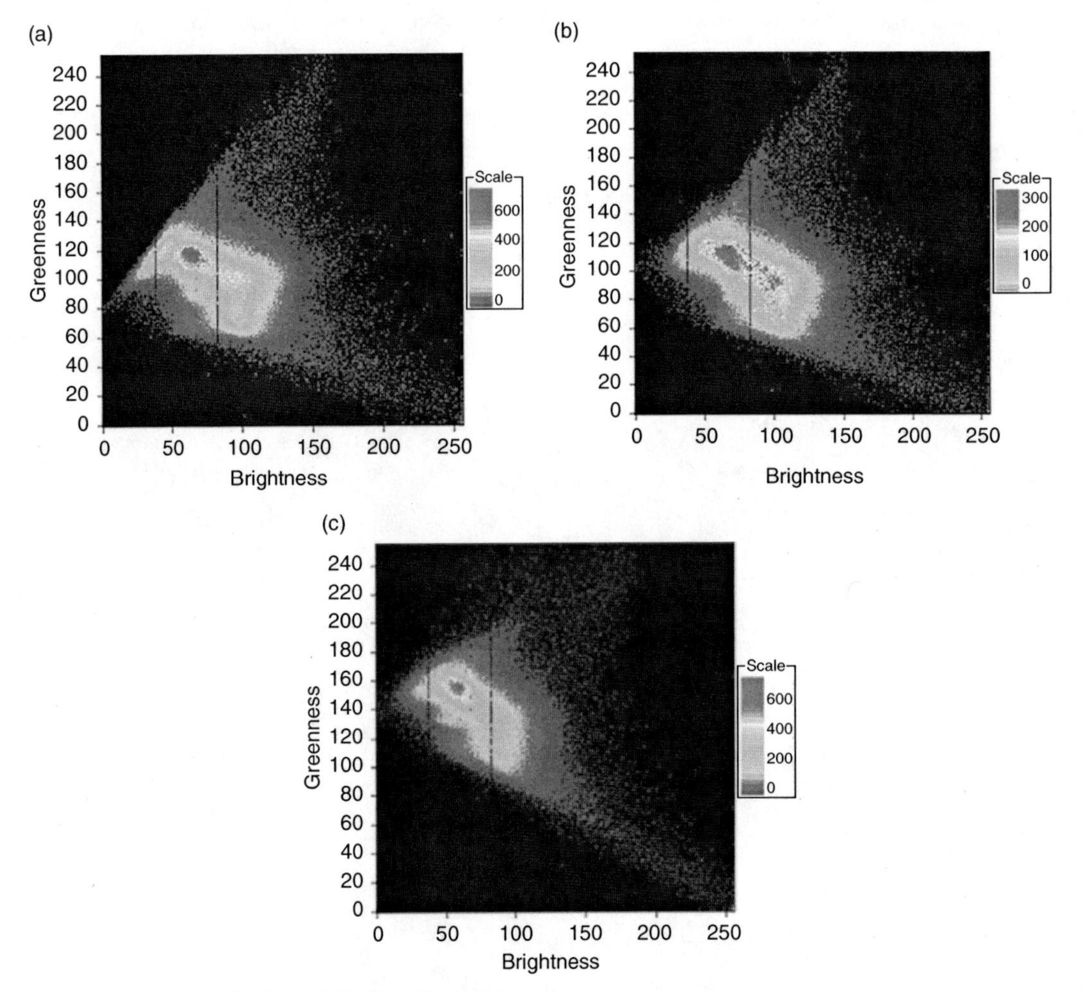

FIGURE 12.13 **(See color insert following page 240.)**
Greenness versus brightness, (a) original multi-spectral, (b) HT fusion, (c) PCA fusion.

TABLE 12.1

Linear Regression Analysis of TCT Components: Correlation Factors of Original Image with HT Fusion and PCA Fusion Images

	Brightness (Original/HT)	Brightness (Original/PCA)	Greenness (Original/HT)	Greenness (Original/PCA)	Third Component (Original/HT)	Third Component (Original/PCA)
Correlation factor	1.00	0.93	0.99	0.98	0.97	0.94

applications involving SAR imagery. The HT allows us to achieve both noise reduction and image fusion.

It is easy to figure out that local orientation analysis for the purpose of noise reduction can be combined with image fusion in a single direct-inverse HT scheme. Figure 12.14 shows the complete methodology to reduce noise and fuse Landsat ETM+ with SAR images.

12.4.4 Experimental Results with Multi-Spectral and SAR Images

We fused multi-sensor images, namely SAR Radarsat (8 m) and multi-spectral Landsat ETM+ (30 m), with the HT and showed that in this case too spatial resolution was improved while spectral resolution was preserved. Speckle reduction in the SAR image was achieved, along with image fusion, within the analysis–synthesis process of the fusion scheme proposed. Figure 12.15 shows the result of panchromatic and SAR image HT fusion including speckle reduction. Figure 12.16 illustrates the result of multi-spectral and SAR image HT fusion. No significant distortion in the spectral and radiometric information is detected.

A comparison of the TCT of the original multi-spectral image and the fused image can be seen in Figure 12.17. There is a variation between both plots; however, the vegetation

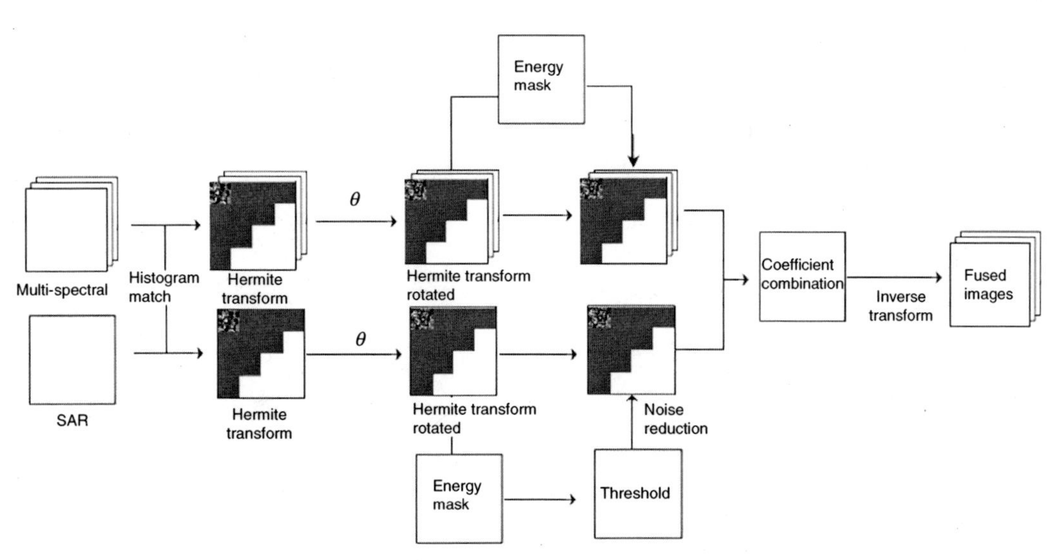

FIGURE 12.14
Noise reduction and fusion for multi-spectral and SAR images.

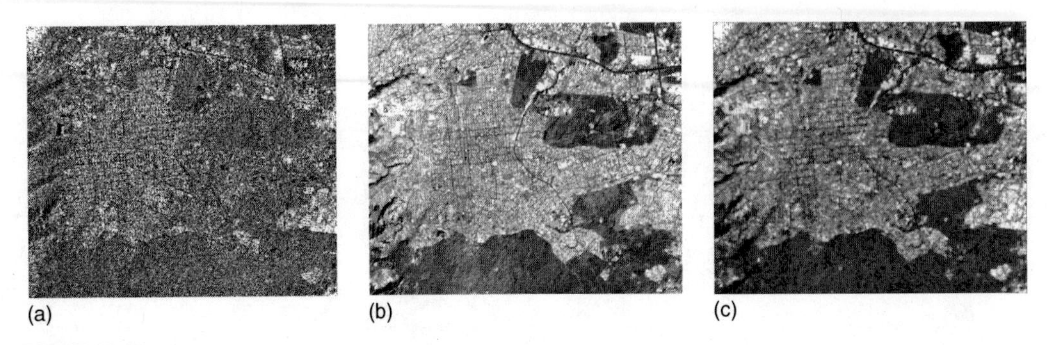

FIGURE 12.15
(a) Radarsat image with speckle (1998). (b) Panchromatic Landsat-7 ETM+ (1998). (c) Resulting image fusion with noise reduction.

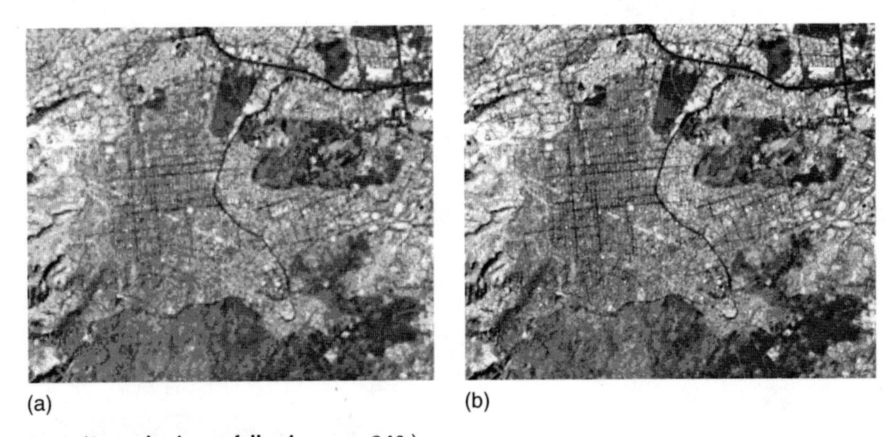

FIGURE 12.16 (See color insert following page 240.)
(a) Original multi-spectral. (b) Result of ETM+ and Radarsat image fusion with HT (Gaussian window with spread $\sigma = \sqrt{2}$ and window spacing $d = 4$) (RGB composition 5–4–3).

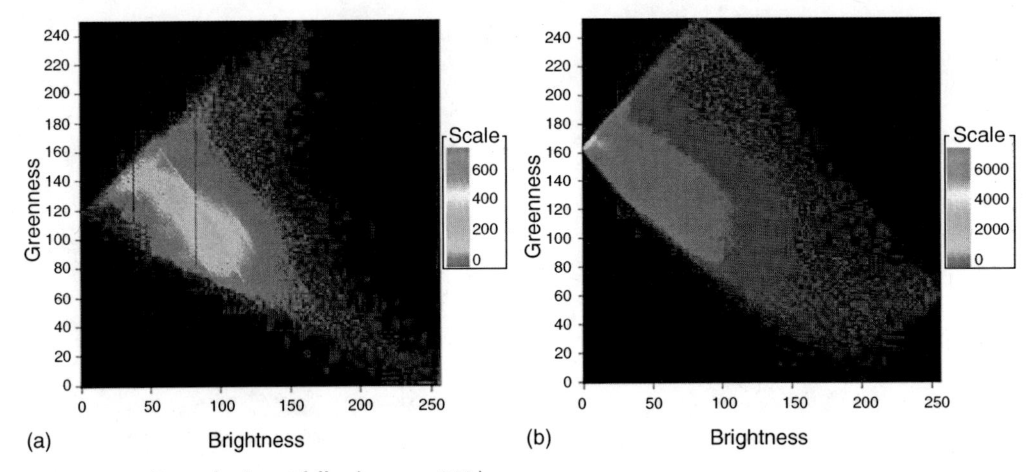

FIGURE 12.17 (See color insert following page 240.)
Greenness versus brightness: (a) original multi-spectral, (b) LANDSAT–SAR fusion with HT.

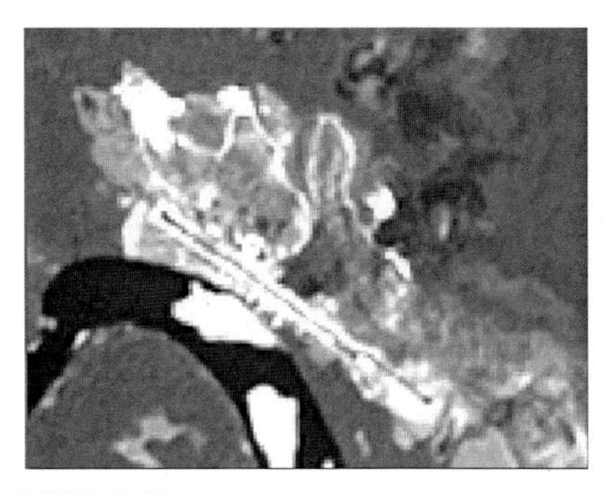

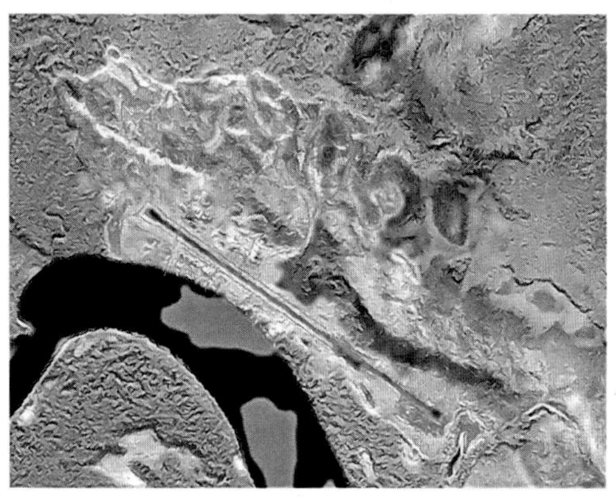

FIGURE 12.18
Left: Original first principal component of a 25 m resolution LANDSAT TM5 image. Right: Result of fusion with SAR AeS-1 denoised image of Figure 12.5.

plane remains similar, meaning that the fused image still can be used to interpret biophysical properties.

Another fusion result is displayed in Figrue 12.18. In this case, the 5 m resolution SAR AeS-1 denoised image displayed on right side of Figure 12.5 is fused with its corresponding 25 m resolution LANDSAT TM5 image. Multi-spectral bands were analyzed with principal components. The first component is shown on the left in Figure 12.18. Fusion of the latter with the SAR AeS-1 image is shown on the right. Note the resolution improvement of the fused image in comparison with the LANDSAT image.

12.5 Conclusions

In this chapter the HT was introduced as an efficient image representation model that can be used for noise reduction and fusion in remote perception imagery. Other applications such as coding and motion estimation have been demonstrated in related works [13,14].

In the case of noise reduction in SAR images, the adaptive algorithm presented here allows us to preserve image sharpness while smoothing homogeneous regions.

The proposed fusion algorithm based on the HT integrates images with different spatial and spectral resolutions, either from the same or different image sensors. The algorithm is intended to preserve both the highest spatial and spectral resolutions of the original data.

In the case of ETM+ multi-spectral and panchromatic image fusion, we demonstrated that the HT fusion method did not lose the radiometric properties of the original multi-spectral image; thus, the fused image preserved biophysical variable interpretation. Furthermore, the spatial resolution of the fused images was considerably improved.

In the case of SAR and ETM+ image fusion, spatial resolution of the fused image was also improved, and we showed for this case how noise reduction could be incorporated within the fusion scheme.

These algorithms present several common features, namely, detection of relevant image primitives, local orientation analysis, and Gaussian derivative operators, which are common to some of the more important characteristics of the early stages of human vision.

The algorithms presented here are formulated in a single spatial scale scheme, that is, the Gaussian window of analysis is fixed; however, multi-resolution is also an important characteristic of human vision and has also proved to be an efficient way to construct image processing solutions. Multi-resolution image processing algorithms are straightforward to build from the HT by means of hierarchical pyramidal structures that replicate, at each resolution level, the analysis–synthesis image processing schemes proposed here. Moreover, a formal approach to the multi-resolution HT for local orientation analysis has been recently developed, clearing the way to propose new multi-resolution image processing tasks [8,9].

Acknowledgments

This work was sponsored by UNAM grant PAPIIT IN105505 and by the Center for Geography and Geomatics Research "Ing. Jorge L. Tamayo".

References

1. D.J. Fleet and A.D. Jepson, Hierarchical construction of orientation and velocity selective filters, *IEEE Transactions on Pattern Analysis and Machine Intelligence*, 11(3), 315–325, 1989.
2. J. Koenderink and A.J. Van Doorn, Generic neighborhood operators, *IEEE Transactions on Pattern Analysis and Machine Intelligence*, 14, 597–605, 1992.
3. J. Bevington and R. Mersereau, Differential operator based edge and line detection, *Proceedings ICASSP*, 249–252, 1987.
4. V. Torre and T. Poggio, On edge detection, *IEEE Transactions on Pattern Analysis and Machine Intelligence*, 8, 147–163, 1986.
5. R. Young, The Gaussian derivative theory of spatial vision: analysis of cortical cell receptive field line-weighting profiles, *General Motors Research Laboratory, Report* 4920, 1986.
6. J.B. Martens, The Hermite transform—theory, *IEEE Transactions on Acoustics, Speech and Signal Processing*, 38(9), 1607–1618, 1990.
7. J.B. Martens, The Hermite transform—applications, *IEEE Transactions on Acoustics, Speech and Signal Processing*, 38(9), 1595–1606, 1990.
8. B. Escalante-Ramírez and J.L. Silvan-Cardenas, Advanced modeling of visual information processing: a multiresolution directional-oriented image transform based on Gaussian derivatives, *Signal Processing: Image Communication*, 20, 801–812, 2005.
9. J.L. Silván-Cárdenas and B. Escalante-Ramírez, The multiscale hermite transform for local orientation analysis, *IEEE Transactions on Image Processing*, 15(5), 1236–1253, 2006.
10. R. Young, Oh say, can you see? The physiology of vision, *Proceedings of SPIE*, 1453, 92–723, 1991.
11. Z.-Q. Liu, R.M. Rangayyan, and C.B. Frank, Directional analysis of images in scale space, [On line] *IEEE Transactions on Pattern Analysis and Machine Intelligence*, 13(11), 1185–1192, 1991; http://iel.ihs.com:80/cgi-bin/.
12. B. Escalante-Ramírez and J.-B. Martens, Noise reduction in computed tomography images by means of polynomial transforms, *Journal of Visual Communication and Image Representation*, 3(3), 272–285, 1992
13. J.L. Silván-Cárdenas and B. Escalante-Ramírez, Image coding with a directional-oriented discrete Hermite transform on a hexagonal sampling lattice, *Applications of Digital Image Processing XXIV* (A.G. Tescher, Ed.), *Proceedings of SPIE*, 4472, 528–536, 2001.

14. B. Escalante-Ramírez, J.L. Silván-Cárdenas, and H. Yuen-Zhou, Optic flow estimation using the Hermite transform, *Applications of Digital Image Processing XXVII* (A.G. Tescher, Ed.), *Proceedings of SPIE*, 5558, 632–643, 2004.
15. G. Granlund and H. Knutsson, *Signal Processing for Computer Vision*, Kluwer, Dordrecht, The Netherlands, 1995.
16. W.T. Freeman and E.H. Adelson, The design and use of steerable filters, *IEEE Transactions on Pattern Analysis and Machine Intelligence*, 13(9), 891–906, 1991.
17. M. Michaelis and G. Sommer, A lie group-approach to steerable filters, *Pattern Recognition Letters*, 16(11), 1165–1174, 1995.
18. G. Szegö, *Orthogonal Polynomials*, American Mathematical Society, Colloquium Publications, 1959.
19. A.M. Van Dijk and J.B. Martens, Image representation and compression with steered Hermite transform, *Signal Processing*, 56, 1–16, 1997.
20. F. Leberl, *Radargrammetric Image Processing*, Artech House, Inc, 1990.
21. J.F. Canny, Finding edges and lines in images, MIT Technical Report 720, 1983.
22. C. Pohl and J.L. van Genderen, Multisensor image fusion in remote sensing: concepts, methods and applications, *International Journal of Remote Sensing*, 19(5), 823–854, 1998.
23. Y. Du, P.W. Vachon, and J.J. van der Sanden, Satellite image fusion with multiscale wavelet analysis for marine applications: preserving spatial information and minimizing artifacts (PSIMA), *Canadian Journal of Remote Sensing*, 29, 14–23, 2003.
24. T. Feingersh, B.G.H. Gorte, and H.J.C. van Leeuwen, Fusion of SAR and SPOT image data for crop mapping, *Proceedings of the International Geoscience and Remote Sensing Symposium, IGARSS*, pp. 873–875, 2001.
25. J. Núñez, X. Otazu, O. Fors, A. Prades, and R. Arbiol, Multiresolution-based image fusion with additive wavelet decomposition, *IEEE Transactions on Geoscience and Remote Sensing*, 37(3), 1204–1211, 1999.
26. T. Ranchin and L. Wald, Fusion of high spatial and spectral resolution images: the ARSIS concepts and its implementation, *Photogrammetric Engineering and Remote Sensing*, 66(1), 49–61, 2000.
27. H. Li, B.S. Manjunath, and S.K. Mitra, Multisensor image fusion using the wavelet transform, *Graphical Models and Image Processing*, 57(3), 235–245, 1995.
28. M. González-Audícana, X. Otazu, O. Fors, and A. Seco, Comparison between Mallat's and the 'à trous' discrete wavelet transform based algorithms for the fusion of multispectyral and panchromatic images, *International Journal of Remote Sensing*, 26(3,10), 595–614, 2005.
29. L. Alparone, S. Baronti, A. Garzelli, and F. Nencini, Landsat ETM+ and SAR image fusion based on generalized intensity modulation, *IEEE Transactions on Geoscience and Remote Sensing*, 42(12), 2832–2839, 2004.
30. E.P. Crist and R.C. Cicone, A physically based transformation of thematic mapper data—the TM Tasselep Cap, *IEEE Transactions on Geoscience and Remote Sensing*, 22(3), 256–263, 1984.
31. E.P. Crist and R.J. Kauth, The tasseled cap de-mystified, *Photogrammetric Engineering and Remote Sensing*, 52(1), 81–86, 1986.
32. E.P. Crist and R.C. Cicone, Application of the Tasseled Cap concept to simulated thematic mapper data, *Photogrammetric Engineering and Remote Sensing*, 50(3), 343–352, 1984.
33. C. Huang, B. Wylie, L. Yang, C. Homer, and. G. Zylstra, Derivation of a tasseled cap transform. Based on Landsat 7 at satellite reflectance, Raytheon ITSS, USGS EROS Data Center, Sioux Falls, SD 57198, USA www.nr.usu.edu/~regap/download/documents/t-cap/usgs-tcap.pdf.

13

Multi-Sensor Approach to Automated Classification of Sea Ice Image Data

A.V. Bogdanov, S. Sandven, O.M. Johannessen, V.Yu. Alexandrov, and L.P. Bobylev

CONTENTS

13.1 Introduction

Satellite radar systems have an important ability to observe the Earth's surface, independent of cloud and light conditions. This property of satellite radars is particularly useful in high latitude regions, where harsh weather conditions and the polar night restrict the use of optical sensors. Regular observations of sea ice using space-borne radars started in 1983 when the Russian OKEAN side looking radar (SLR) system became operational. The wide swath (450 km) SLR images of 0.7–2.8 km spatial resolution were used to support ship transportation along the Northern Sea Route and to provide ice information to facilitate other polar activities. Sea ice observation using high-resolution synthetic aperture radar (SAR) from satellites began with the launch of Seasat in 1978, which operated for only three months, and continued with ERS from 1991, RADARSAT from 1996, and ENVISAT from 2002. Satellite SAR images with a typical pixel size of 30 m to 100 m, allow observation of a number of sea ice parameters such as floe parameters [1], concentration [2], drift [3], ice type classification [4,5,6], leads [7], and ice edge processes. RADARSAT wide swath SAR images, providing enlarged spatial coverage, are now in operation at several sea ice centers [8]. The ENVISAT advanced SAR (ASAR) operating at several imaging modes, including single polarization wide swath (400 km) and alternating polarization narrow swath (100 km) modes can improve the classification of several ice types and open water (OW) using dual polarization images.

Several methods for sea ice classification have been developed and tested [9–11]. The straightforward and physically plausible approach is based on the application of sea ice microwave scattering models for the inverse problem solution [12]. This is, however, a difficult task because the SAR signature depends on many sea ice characteristics [13]. A common approach in classification is to use empirically determined sea ice backscatter coefficients obtained from field campaigns [14,15]. Classical statistical methods based on Bayesian theory [16] are known to be optimal if the form of the probability density function (PDF) is known and can be parameterized in the algorithm. A Bayesian classifier, developed at the Alaska SAR Facility (ASF) [4], assumes a Gaussian distribution of sea ice backscatter coefficients [17]. Utilization of backscatter coefficients only, limits the number of ice classes that can be distinguished and decreases the accuracy of classification because backscatter coefficients of several sea ice types and OW overlap significantly [18]. Incorporation of other image features with a non-Gaussian distribution requires modeling of the joint PDF of features from different sensors, which is difficult to achieve. The classification errors can be grouped into two categories: (1) labeling inconsistencies and (2) classification-induced errors [19]. The errors in the first group are due to mixed pixels, transition zones between different ice regimes, temporal change of physical properties, sea ice drift, within-class variability, and limited training and test data sets. The errors in the second group are errors induced by the classifier. These errors can be due to the selection of an improper classifier for the given problem, its parameters, learning algorithms, input features, etc.—the problems traditionally considered within pattern recognition and classification domains.

Fusion of data from several observation systems can greatly reduce errors in labeling inconsistency. These can be satellite and aircraft images obtained at different wavelengths and polarizations, data in cartographic format represented by vectors and polygons (i.e., bathymetry profiles, currents, meteorological information), and expert knowledge. Data fusion can improve the classification and extend the use of the algorithms to larger geographical areas and several seasons. Data fusion can be done using statistical methods, the theory of belief functions, fuzzy logic and fuzzy set theory, neural networks, and expert systems [20]. Some of these methods have been successfully applied to sea ice classification [21]. Haverkamp et al. [11] combined a number of SAR-derived sea ice parameters and expert geophysical knowledge in the rule-based expert system. Beaven [22] used a combination of ERS-1 SAR and special sensor microwave/imager (SSM/I) data to improve estimates of ice concentration after the onset of freeze-up. Soh and Tsatsoulis [23] used information from various data sources in a new fusion process based on Dempster–Shafer belief theory. Steffen and Heinrichs [24] merged ERS SAR and Landsat thematic mapper data using a maximum likelihood classifier. These studies demonstrated the advantages that can be gained by fusing different types of data. However, there is still forthcoming work to compare different sea sensor data fusion algorithms and assess their performances using ground-truth data. In this study we investigate and analyze the performance of an artificial neural network model applied for sea ice classification and compare its performance with the performance of the linear discriminant analysis (LDA) based algorithm.

Artificial neural network models received high attention during the last decades due to their ability to approximate complex input–output relationships using a training data set, perform without any prior assumptions on the statistical model of the data, generalize well on the new, previously unseen data (see Ref. [25] and therein), and be less affected by noise. These properties make neural networks especially attractive for the sensor data fusion and classification. Empirical comparisons of neural network–based algorithms with the standard parametric statistical classifiers [26,27] showed that the neural network model, being distribution free, can outperform the statistical methods on the condition that a sufficient number of representative training samples is presented to the neural network. It also avoids the problem of determining the amount of influence a source should have in the classification [26]. Standard statistical parametric classifiers require a statistical model and thus work well when the used statistical model (usually multivariate normal) is in good correspondence with the observed data. There are not many comparisons of neural network models with nonparametric statistical algorithms. However, there are some indications that these algorithms can work at least as well as neural network approaches [28].

Several researchers proposed neural network models for sea ice classification. Key et al. [29] applied a backpropagation neural network to fuse the data of two satellite radiometers. Sea ice was among 12 surface and cloud classes identified on the images. Hara et al. [30] developed an unsupervised algorithm that combines learning vector quantization and iterative maximum likelihood algorithms for the classification of polarimetric SAR images. The total classification accuracy, estimated using three ice classes, comprised 77.8% in the best case (P-band). Karvonen [31] used a pulse-coupled neural network for unsupervised sea ice classification in RADARSAT SAR images. Although these studies demonstrated the usefulness of neural network models when applied to sea ice classification, the algorithms still need to be extensively tested under different environmental conditions using ground-truth data. It is unclear whether neural network models outperform traditional statistical classifiers and generalize well on the test data set. It is also unclear which input features and neural network structure should be used in classification.

This study analyzes the performance of a multi-sensor data fusion algorithm based on a multi-layer neural network also known as multi-layer *perceptron* (MLP) applied for sea ice classification. The algorithm fuses three different types of satellite images: ERS, RADARSAT SAR, and low-resolution visible images; each type of data carries unique information on sea ice properties. The structure of the neural network is optimized for the sea ice classification using a pruning method that removes redundant connections between neurons. The analysis presented in this study consists of the following steps: Firstly, we use a set of *in situ* sea ice observations to estimate the contribution of different sensor combinations to the total classification accuracy. Secondly, we evaluate the positive effect of SAR image texture features included in the ice classification algorithm, utilizing only tonal image information. Thirdly, we verify the performance of the classifier by comparing it with the performance of the standard statistical approach. As a benchmark, and for comparison, we use an LDA-based algorithm [6] that resides in an intermediate position between parametric and nonparametric algorithms such as the *K*-nearest-neighbor classifier. Finally, the whole image area is classified and analyzed to give additional evidence of the generalization properties of the classifier; and the results of automatic classification are compared with manually prepared classification maps.

In the following sections we describe the multi-sensor image sets used and the *in situ* data (Section 13.2), the MLP and LDA-based classification algorithms (Section 13.3), and finally discuss the results of our experiments in Section 13.4.

13.2 Data Sets and Image Interpretation

13.2.1 Acquisition and Processing of Satellite Images

In our experiments, we used a set of spatially overlapping ERS-2 SAR low-resolution images (LRI), RADARSAT Scan SAR Wide beam mode image, and Meteor 3/5 TV optical image, acquired on April 30, 1998. The characteristics of the satellite data are summarized in Table 13.1. The RADARSAT Scan SAR scene and the corresponding fragment of the Meteor image, covering a part of the coastal Kara Sea with the Ob and Yenisey estuaries, are shown in Figure 13.1. ERS SAR image has the narrowest swath width (100 km) among the three sensors. Thus the size of the image fragments (Figure 13.2) used for fusion is limited by the spatial coverage of the two ERS SAR images available for the study shown in Figure 13.2a. The images contain various stages and forms of first year, young, and new ice. The selection of test and training regions in the images is done using *in situ* observations made onboard the Russian nuclear icebreaker "Sovetsky Soyuz," which sailed through the area as shown in Figure 13.1a by a white line. Compressed SAR images were transmitted to the icebreaker via INMARSAT in near real-time and were available onboard for ice navigation. The satellite images onboard enabled direct identification of various sea ice types observed in SAR images and verification of their radar signatures.

The SAR data were received and processed into images at Kongsberg Satellite Services in Tromsø, Norway. The ScanSAR image is 500 km wide and has 100 m spatial resolution (Table 13.1), which corresponds to a pixel spacing of 50 m. The image was filtered and down-sampled to have the same pixel size (100 m) as the ERS SAR LRI with 200 m spatial resolution (Table 13.1). Further processing includes antenna pattern correction, range spreading loss compensation, and a correction for incidence angle. The resulting pixel

TABLE 13.1

The Main Parameters of the Satellite Systems and Images Used in the Study

Sensor	Wavelength and Band	Polarization	Swath Width	Spatial Resolution/ Number of Looks	Range of Incidence Angles
RADARSAT Scan SAR Wide beam mode	5.66 cm C-band	HH	500 km	100 m 4×2	20°–49°
ERS SAR low resolution image (LRI)	5.66 cm C-band	VV	100 km	200 m ~30	20°–26°
Meteor-3/5 MR-900 TV camera system	0.5–0.7 μm VIS	Nonpolarized, panchromatic	2600 km	~2 km	~46.6°(left)–46.6° (right-looking)

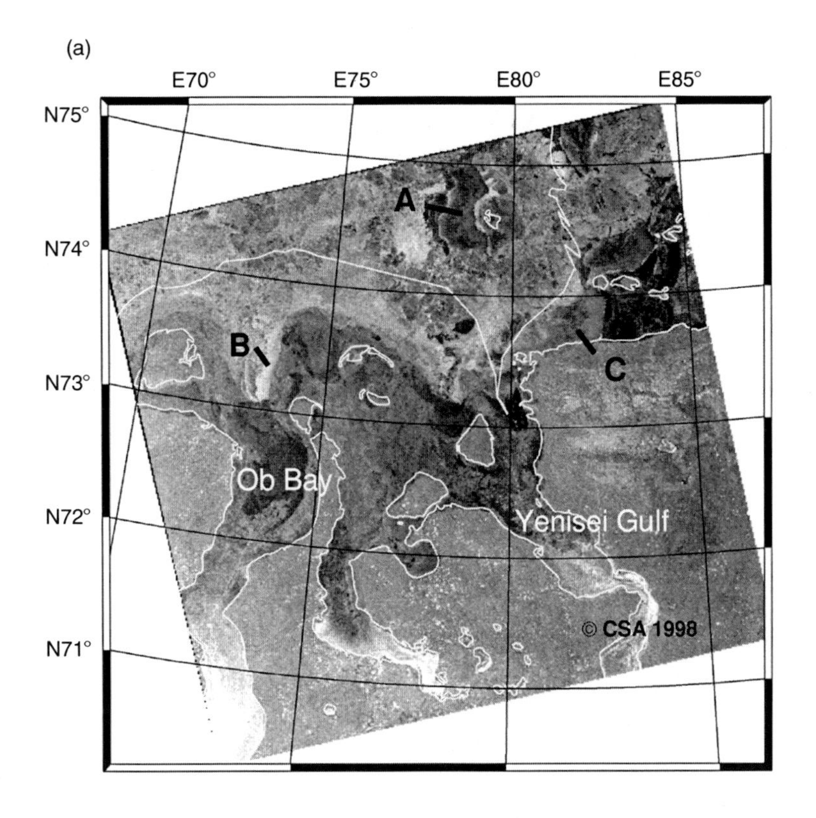

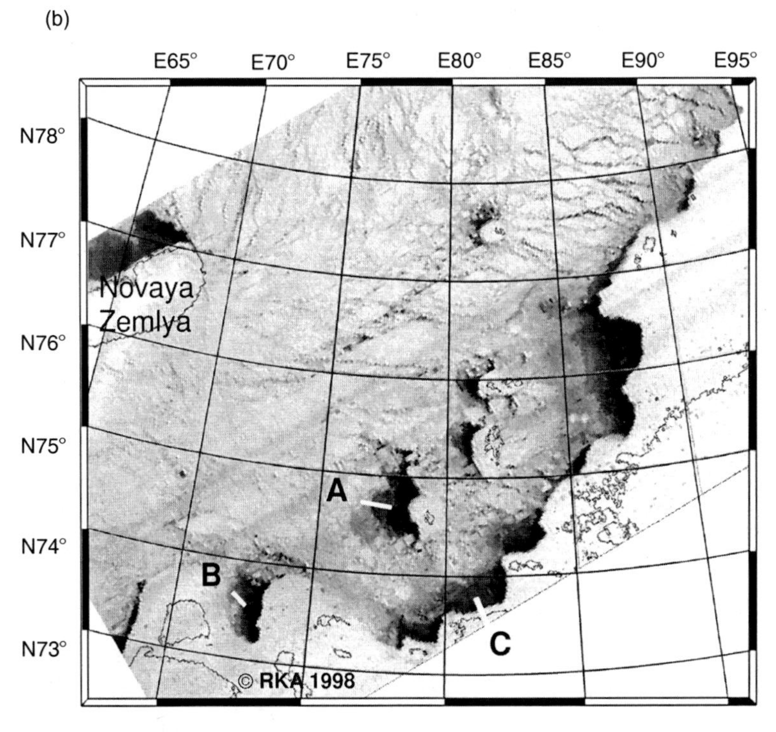

FIGURE 13.1
RADARSAT Scan SAR (a) and Meteor 3/5 TV (b) images acquired on April 30, 1998. The icebreaker route and coastal line are shown. Flaw polynyas are marked by letters A, B, and C.

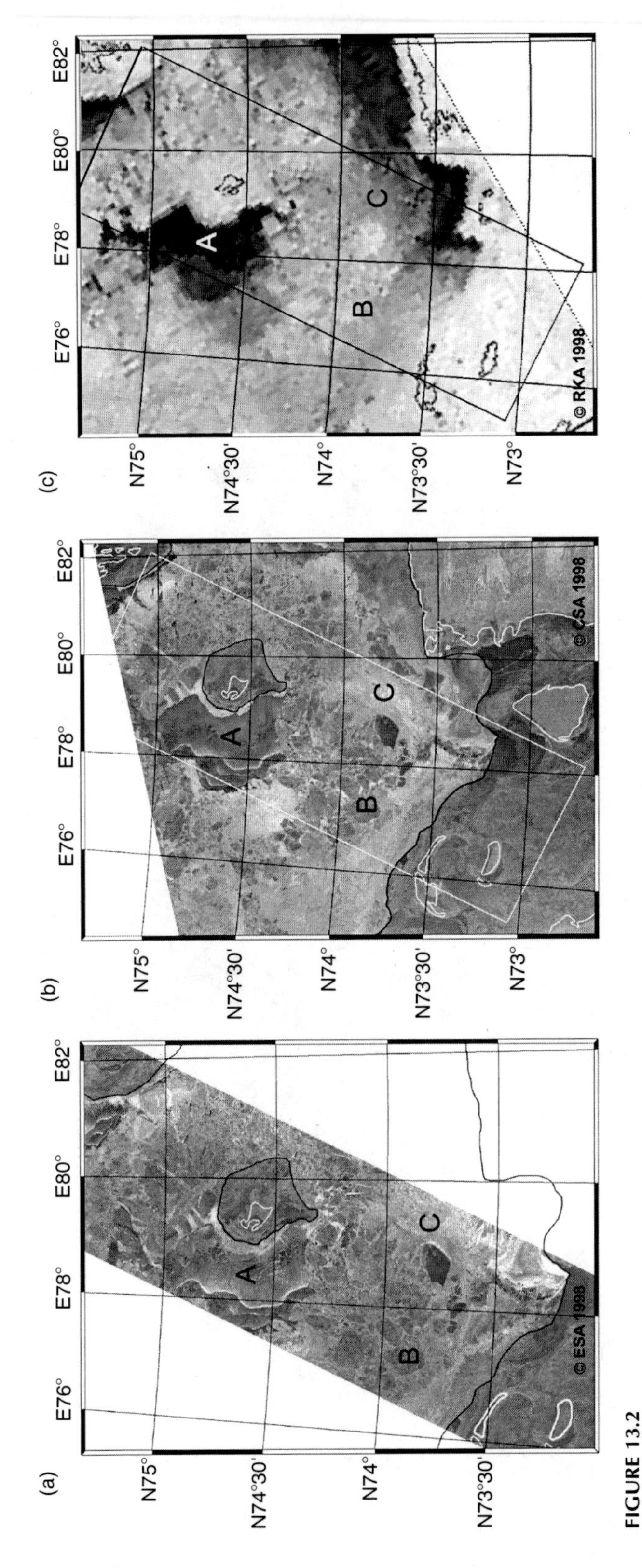

FIGURE 13.2
Satellite images used for data fusion: (a) mosaic of ERS-2 SAR images, (b) a part of the RADARSAT Scan SAR image, and (c) a part of Meteor-3/5 TV image (April 30, 1998). Coastal line, fast ice edge [dark lines in (a) and (b)], and ERS image border are overlaid. The letters mark: (A) nilas, new ice, and open water, (B) first-year ice, and (C) young ice.

value is proportional to the logarithm of the backscatter coefficient. The scaling factor and a fixed offset, normally provided in CEOS radiometric data record, are used to obtain absolute values of the backscatter coefficients (sigma-zero) in decibels [32]. These parameters are not available for the relevant operational quantized 8-bit product, making retrieval of absolute values of sigma-zero difficult. However, in a supervised classification procedure it is important that only relative values of image brightness within a single image and across different images used in classification are preserved. Variations of backscatter coefficients of sea ice in the range direction are relatively large, varying from 4 dB (for dry multi-year ice) to 8 dB (for wet ice) [33], due to the large range of incidence angles from 20° to 49°. The range-varying normalization, using empirical dependencies for the first-year (FY) ice dominant in the images, was applied to reduce this effect [33]. The uncompensated radiometric residuals for the other ice types presented in the images increase classification error. The latter effect may be reduced by application of texture and other statistical local parameters, or by restricting the range of incidence angles and training classification algorithms separately within each range. In this study we apply texture features, which depend on relative image values and thus should be less sensitive to the variations of image brightness in range direction.

The two ERS SAR LRI (200 m spatial resolution) were processed in a similar way to the RADARSAT image. The image pixel value is proportional to the square root of backscatter coefficients [34], which is different from the RADARSAT pixel value representation where a logarithm function is used. The absolute values of the backscatter coefficients can be obtained using calibration constants provided by the European Space Agency (ESA) [34], but for this study we used only the pixel values derived from the processing described above.

The visual image was obtained in the visible spectrum (0.5–0.7 μm) by the MR-900 camera system used onboard the Meteor-3/5 satellite. The swath width of the sensor is 2600 km and the spatial resolution is ~2 km. For fusion purposes the coarse image is resampled to the same pixel size as RADARSAT and ERS images. Even though no clouds are observed in the image, small or minor clouds might be present but not visible in the image due to the ice-dominated background.

For spatial alignment, the images were georeferenced using corner coordinates and ground control points and then transformed to the universal transverse mercator (UTM) geographical projection. The corresponding pixels of the spatially aligned and resampled images cover approximately the same ice on the ground. Because the images are acquired with time delay reaching 8 h 42 min for RADARSAT—Meteor images (Table 13.2) and several kilometers ice drift occur during this period, a certain mismatch of the ice features in the images are present. This is corrected for as much as possible, but there are still minor errors in the co-location of ice features due to rotation and local convergence or divergence of the drifting ice pack. The fast ice does not introduce this error due to its stationarity.

TABLE 13.2

Satellite Images and *In Situ* Data

Sensor	Date/Time (GMT)	No. of Images	No. of *In Situ* Observations
RADARSAT Scan SAR	30 April 1998/11:58	1	56
ERS-2 SAR	30 April 1998/06:39	3	25
Meteor-3 TV camera MR 900	30 April 1998/03:16	1	>56

13.2.2 Visual Analysis of the Images

The ice in the area covered by the visual image shown in Figure 13.1b mostly consists of thick and medium FY ice of different deformations, identified by a bright signature in the visible image and various grayish signatures in the ScanSAR image in Figure 13.1a. Due to dominant easterly and southeasterly winds in the region before and during the image acquisition, the ice drifted westwards, creating the coastal polynyas with OW and very thin ice, characterized by the dark signatures in the optical image. Over new and young ice types, the brightness of ice in the visual image increases as the ice becomes thicker. Over FY ice types, increases in ice thickness are masked by high albedo snow cover. The coarse spatial resolution of the TV image reduces the discrimination ability of the classifier, which is especially noticeable in regions of mixed sea ice. However, two considerations need to be taken into account: firstly, the texture features computed over the relatively large SAR image regions are themselves characterized by the lower spatial resolution, and secondly, the neural network–based classifier providing nonlinear input–output mapping can theoretically mediate the later affects by combining low and high spatial resolution data.

The physical processes of scattering, reflectance, and attenuation of microwaves determine sea ice radar signatures [35]. The scattered signal received depends on the surface and volume properties of ice. For thin, high salinity ice types the attenuation of microwaves in the ice volume is high, and the backscatter signal is mostly due to surface scattering. Multi-year ice characterized by strong volume scattering is usually not observed in the studied region. During the initial stages of sea ice growth, the sea ice exhibits a strong change in its physical and chemical properties [36]. Radar signatures of thin sea ice starting to form at different periods of time and growing under different ambient conditions are very diverse. Polynyas appearing dark in the visual image (Figure 13.1b, regions A, B, and C) are depicted by various levels of brightness in the SAR image in Figure 13.1a. The dark signature of the SAR image in region A corresponds mostly to grease ice formed on the water surface. The low image brightness of grease ice is primarily due to its high salinity and smooth surface, which results in a strong specular reflection of the incident electromagnetic waves. At C-band, this ice is often detectable due to the brighter scattering of adjacent, rougher, OW. Smooth nilas also appears dark in the SAR image but the formation of salt flowers or its rafting strongly increases the backscatter. The bright signature of the polynya in region B could be due to the formation of pancake ice, brash ice, or salt flowers on the surface of the nilas.

A common problem of sea ice classification of SAR images acquired at single frequency and polarization is the separation of OW and sea ice, since backscatter of OW changes as a function of wind speed. An example of ice-free polynya can be found in region C in Figure 13.1a where OW and thin ice have practically the same backscatter. This vast polynya (Taimyrskaya), expanding far northeast, can be easily identified in the Meteor TV image in Figure 13.1b, region C, due to its dark signature (RADARSAT SAR image covers only a part of it). As mentioned before, dark signature in the visual image mainly corresponds to the thin ice and OW. Therefore, to some extent, the visual image is complementary to SAR data, enabling separation of FY ice with different surface roughness from thinner sea ice and OW. SAR images, on the other hand, can be used for classification of FY ice of different surface roughness, and separation of thin ice types.

13.2.3 Selection of Training Regions

For supervised classification it is necessary to define sea ice classes and to select in the image training regions for each class. The classes should generally correspond to the

TABLE 13.3

Description of the Ice Classes and the Number of Training and Test Feature Vectors for Each Class

Sea Ice Class	Description	No. of Training Vectors	No. of Test Vectors
1. Smooth first-year ice	Very smooth first-year ice of medium thickness (70–120 cm)	150	130
2. Medium deformation first-year ice	Deformed medium and thick (>120 cm) first-year ice, deformation is 2–3 using the 5-grade scale	1400	1400
3. Deformed first-year ice	The same as above, but with deformation 3–5	1400	1400
4. Young ice	Gray (10–15 cm) and gray–white (15–30 cm) ice, small floes (20–100 m) and ice cake, contains new ice in between floes space	1400	1400
5. Nilas	Nilas (5–10 cm), grease ice, areas of open water	1400	1400
6. Open water	Mostly open water, at some places formation of new ice on water surface	30	26

World Meteorological Organization (WMO) terminology [37], so that the produced sea ice maps can be used in practical applications. WMO defines a number of sea ice types and parameters, but the WMO classification is not necessarily in agreement with the classification that can be retrieved from satellite images. In defining the sea ice classes, we combine some of the ice types into the larger classes based on *a priori* knowledge of their separation in the images and some practical considerations. For navigation in sea ice it is more important to identify thicker ice types, their deformations, and OW regions. Since microwave backscatter from active radars such as SAR is sensitive to various stages of new and young ice, multi-year FY ice and surface roughness, we have selected the following six sea ice classes for use in the classification: smooth, medium deformation, deformed FY ice, young ice, nilas, and open water OW. From their description given in Table 13.3 it is seen that the defined ice classes contain inclusions of other ice types because it is usually difficult to find "pure" ice types extended over large areas in the studied region. The selected training and test regions for different sea ice classes, overlaid on the RADARSAT SAR image, are shown by the rectangles in Figure 13.3. These are homogeneous areas that represent "typical" ice signatures as known *a priori* based on the combined analysis of the multi-sensor data set, image archive, bathymetry, meteorological data, and *in situ* observations. The *in situ* ice observations from the icebreaker were done along the whole sailing route between Murmansk and the Yenisei estuary. In this study we have mainly used the observations falling into the image fragment used for fusion, or located nearby, as shown in Figure 13.3a. Examples of photographs of various ice types are shown in Figure 13.4.

13.3 Algorithms Used for Sea Ice Classification

13.3.1 General Methodology

To assess the improvement of classification accuracy that can be achieved by combining data from the three sensors we trained and tested several classifiers using different combinations of image features stacked in feature vectors. A set of the feature vectors

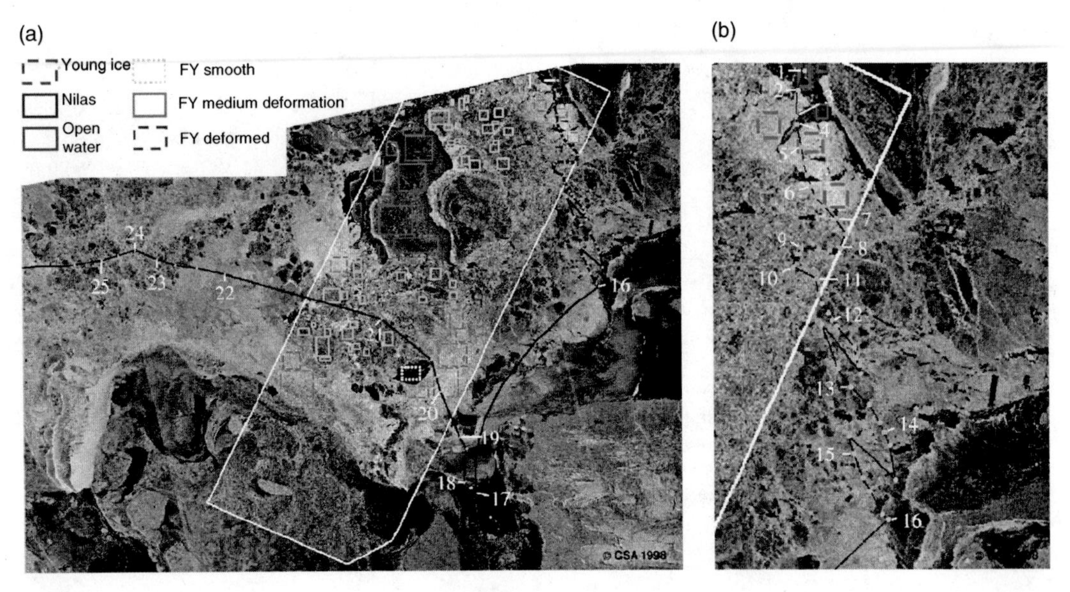

FIGURE 13.3
Selection of the training and test regions: (a) fragment of the RADARSAT Scan SAR image (April 30, 1998) with the ship route and the image regions for different ice classes overlaid and (b) enlarged part of the same fragment.

computed for different ice classes is randomly separated into training and test data sets. The smaller dimensionality subsets have been produced from the original data sets containing all features and were used for training and validation of both algorithms in the experiments described below.

13.3.2 Image Features

The SAR image features used for the feature vectors are in three main groups: image moments, gray level co-occurrence matrix (GLCM) texture, and autocorrelation function–based features. These features describe local statistical image properties within a small region of an image. They have been investigated in several studies [6,9,10,38–41] and are in general found to be useful for sea ice classification. A set of the most informative features differs from study to study, and it may depend on several factors including geographical region, ambient conditions, etc. Application of texture usually increases classification accuracy; however, it cannot fully resolve ambiguities between different sea ice types, so that incorporation of other information is required.

The texture features are often understood as a description of spatial variations of image brightness in a small image region. Some texture features can be used to describe regular patterns in the region, while others depend on the overall distribution of brightness. Texture has been used for a long time by sea ice image interpreters for visual classification of different sea ice types in radar images. For example, multi-year ice is characterized by a patchy image structure explained by the formation of numerous melting ponds on its surface during summer and then freezing in winter. Another example is the network of bright linear segments corresponding to ridges in the deformed FY ice. The texture depends on the spatial resolution of the radar, the spatial scale of sea ice surface, and volume inhomogeneity. There is currently a lack of information on large-scale sea ice properties, and as a consequence, on mechanisms of texture formation.

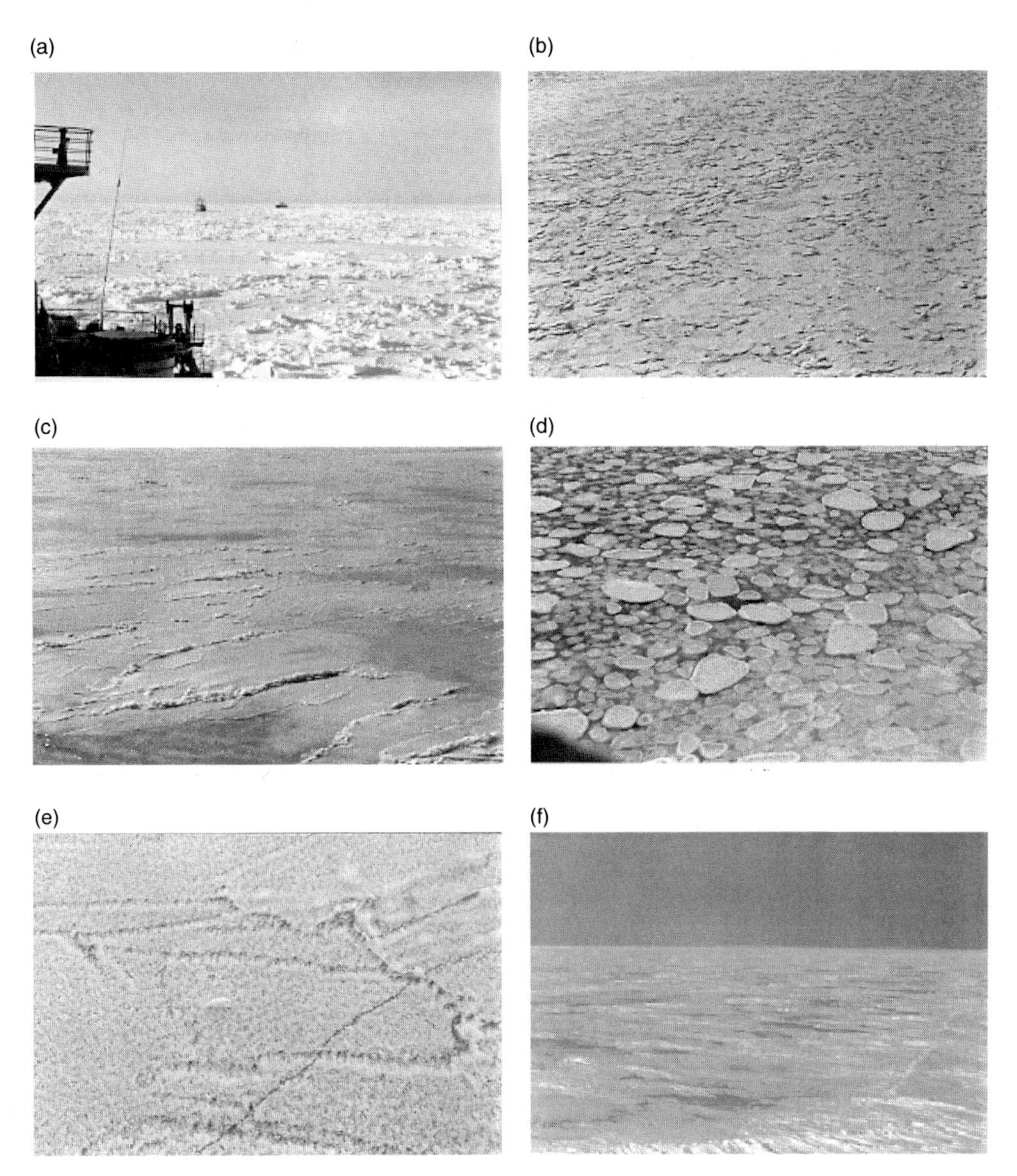

FIGURE 13.4
Photographs of different sea ice types outlining different mechanisms of ice surface roughness formation. (a) Deformed first-year ice. (b) Gray–white ice, presumably formed from congealed pancake ice. (c) Rafted nilas. (d) Pancake ice formed in the marginal ice zone from newly formed ice (grease ice, frazil ice) influenced by surface waves. (e) Frost flowers on top of the first-year ice. (f) Level fast ice (smooth first-year ice).

In supervised classification the texture features are computed over the defined training regions and the classifier is trained to recognize similar patterns in the newly acquired images. Several texture patterns can correspond to one ice class, which implies the existence of several disjointed regions in feature space for the given class. The latter, however, is not observed in our data. The structure of data in the input space is affected by several factors including definition of ice classes, selection of the training regions, and existence of smooth transitions between different textures. In this study the training and test data have been collected over a relatively small geographic area where image and *in situ* data are overlapped. In contrast to this local approach, the ice texture investigation can be carried out using training regions selected over a relatively large geographic area and across different seasons based on visual analysis of images [38]. Selection of ice types that may have several visually distinct textures can facilitate formation of disjointed or complex form clusters in the feature space pertinent for one ice type. Note that in this case MLP should show better results than the LDA-based algorithm.

The approach to texture computation is closely related to the classification approach adopted to design multi-season, large geographic area classification system using (1) a single classifier with additional inputs indicating area and season (month number), (2) a set (ensemble) of local classifiers designed to classify ice within a particular region and season, and (3) a multiple classifier system (MCS). The trained classifier presented in this paper can be considered as a member of a set of classifiers, each of which performs a simpler job than a single multi-season, multi-region classifier.

The image moments used in this study are mean value, second-, third-, and fourth-order moments, and central moments computed over the distribution of pixel values within a small computation window. The GLCM-based texture features include homogeneity, contrast, entropy, inverse difference moment [42], cluster prominence, and cluster shade. The autocorrelation function–based features are decorrelation lengths computed along 0°, 45°, and 90° directions. In total, 16 features are used for SAR image classification. Only the mean value was used for the visual image because of its lower spatial resolution. The texture computation parameters are selected experimentally, taking into account the results of previous investigations [6,9,10,38–41].

There are several important parameters that need to be defined for GLCM: (1) the computation window size; (2) the displacement value, also called interpixel distance; (3) the number of quantization levels; and (4) orientation. We took into account that the studied region contains mixed sea ice types, while defining these parameters. With increasing window size and interpixel distance (which is related to the spatial scale of inhomogeneities "captured" by the algorithm), computed texture would be more affected by the composition of ice types within the computational window rather than the properties of ice. Therefore in the hard classification approach adopted here, we selected the smaller window size equal to 5×5 pixels and interpixel distance equal to 2. This implies that we explore moderate scale ice texture. The use of macro texture information (larger displacement values) or multi-scale information (a range of different displacement values), recommended in the latest and comprehensive ice texture study [38], would require a soft classification approach in our case. To reduce the computational time, the range of image gray levels is usually quantized into a number of separate bins. The image quantization, generally leading to the loss of image information, does not strongly influence the computation of texture parameters on the condition that a sufficient number of bins are used (>16–32) [38]. In our experiments the range of image gray levels is quantified to the 20 equally spaced bins (see Ref. [38] for the discussion on different quantization schemes); the GLCM is averaged for the three different directions 0°, 45°, and 90° to account for possible rotation of ice. The training data set is prepared by moving the computational window within the defined training regions. For each nonoverlapping placement of the window, the image features are computed in three images and stacked in a vector. The number of feature vectors computed for different ice classes is given in Table 13.3.

13.3.3 Backpropagation Neural Network

In our experiments we used a multi-layer feedforward neural network trained by a standard backpropagation algorithm [43,44]. Backpropagation neural networks also known as MLP [45] are structures of highly interconnected processing units, which are usually organized in layers. MLP can be considered as a universal approximator of functions that learns or approximates the nonlinear input–output mapping function using a training data set. During training the weights between processing units are iteratively adjusted to minimize an error function, usually the root-mean-square (RMS) error function. The simple method for finding the weight updates is the *steepest descent* algorithm in which the weights are changed in the direction of the largest reduction of the error, that is, in the direction where the gradient of the error function with respect to the weights is negative. This method has some limitations [25], including slow convergence in the areas characterized by substantially different curvatures along different directions in the error surface as, for example, in the long, steep-sided valley. To speed up the convergence, we used a modification of the method that adds a momentum term [44] to the equation:

$$\Delta w_{\tau} = -\eta \nabla E_{\tau} + \mu \Delta w_{\tau-1} \tag{13.1}$$

where Δw_{τ} is the weight change at iteration τ, ∇E_{τ} is the gradient of the error function with respect to the weights evaluated at the current iteration, η is the learning rate parameter, and μ is the *momentum constant*, $0 < |\mu| < 1$. Due to the inclusion of the second term, the changes of weights, having the same sign in steady downhill regions of the error surface, are accumulated during successive iterations, which increases the step size of the algorithm. In the regions where oscillations take place, contributions from the momentum terms change sign and thus tend to cancel each other, reducing the step size of the algorithm. The gradients ∇E_{τ} are computed using the known backpropagation algorithm [43,44].

13.3.4 Linear Discriminant Analysis Based Algorithm

An LDA-based algorithm is proposed by Wackerman and Miller [6] for sea ice classification in the marginal ice zone (MIZ) using single channel SAR data. In this study it is applied for data fusion of different sensors. LDA is a known method for the reduction of dimensionality of the input space, which can be used at the preprocessing stage of the classification algorithm, to reduce the number of input features. This method is used to project the original, usually high-dimensional input space onto a lower dimensional one. The projection of n-dimensional data vector $\vec{x}$ is done using the linear transformation $\vec{y} = V^{\mathrm{T}}\vec{x}$, where $\vec{y}$ is the vector of dimension m ($m<n$) and V is the $n \times m$ transformation matrix. Elements of the transformation matrix are found by maximizing Fisher's criteria, which is a measure of separability between classes. For a two-class problem it is defined as [46]:

$$\frac{\vec{v}_{ij}^{\mathrm{T}} B_{ij} \vec{v}_{ij}}{\vec{v}_{ij}^{\mathrm{T}} W_{ij} \vec{v}_{ij}} \tag{13.2}$$

where $W_{ij} = \sum(\vec{x}_i - \vec{m}_i)(\vec{x}_i - \vec{m}_i)^{\mathrm{T}} + \sum(\vec{x}_j - \vec{m}_j)(\vec{x}_j - \vec{m}_j)^{\mathrm{T}}$ is the total *within-class* covariance matrix, given as a sum of the two covariance matrices of ith and jth ice classes, $\vec{m}_i$ and $\vec{m}_j$ are the mean feature vectors of classes i and j, respectively; B_{ij} is the *between-class* matrix, given by $B_{ij} = (\vec{m}_i - \vec{m}_j)(\vec{m}_i - \vec{m}_j)^{\mathrm{T}}$, and $\vec{v}_{ij}$ is the transformation (projection) vector to which matrix V reduces in the two-class case. Vector $\vec{v}_{ij}$ defines a new direction in feature space, along which separation of classes i and j is maximal. It can be shown that vector $\vec{v}_{ij}$

maximizing the clustering metric in Equation 13.2 is the eigenvector with the maximum eigenvalue λ that satisfies the equation [47]:

$$W_{ij}^{-1} B_{ij} \vec{v}_{ij} = \lambda \vec{v}_{ij} \quad (13.3)$$

In general, case classification of vectors $\vec{y}$ can be performed using traditional statistical classifiers. The central limit theorem is applied since $\vec{y}$ represents a weighted sum of random variables and conditional PDFs of $\vec{y}$ are assumed to be multi-variate normal. The method is distribution-free in the sense that "it is a reasonable criterion for constructing a linear combination" [47]. It is shown to be statistically optimal if the input features are multi-variate normal [25]. Another assumption that needs to be satisfied when applying LDA is the equivalence of the class conditional covariance matrices for each class. These assumptions are difficult to satisfy in practice. However, the slight violation of these criteria does not strongly degrade the performance of the classifier [47].

In the limiting case LDA can be used to project the input space in one dimension only. By projecting feature vectors of pairs of classes, the multi-class classification problem can be decomposed into two-class problems. The constructed classifier is a piecewise linear classifier. For training of the classification algorithm and finding parameters of the classifier the following steps are performed [6]:

1. The mean vectors $\vec{m}_i$ $(i = 1, \ldots, c)$, the between-class B_{ij}, and within-class covariance matrices W_{ij} $(i = 1, \ldots, c; i = 1, \ldots, c; i \neq j)$ are estimated using the training data set;
2. The transformation vectors $\vec{v}_{ij}$ $(i = 1, \ldots, c;\ j = 1, \ldots, c;\ i \neq j)$ are found as eigenvectors of the matrix $W_{ij}^{-1} B_{ij}$ solving Equation 13.3 (since $B_{ij}\vec{v}_{ij}$ has the same direction as $\vec{m}_i - \vec{m}_j$ and the scaling factor is not important, the $\vec{v}_{ij}$ can also be found as $\vec{v}_{ij} = W_{ij}^{-1}(\vec{m}_i - \vec{m}_j)$ [48]).
3. The feature training vectors for ice classes i and j are projected on lines defined by $\vec{v}_{ij}$, computed in the previous step. The threshold t_{ij} between two classes is found as an intersection of two histograms. In total there are $c^2 - c$ thresholds for c ice classes.

During the classification stage, to see if a new vector $\vec{x}$ belongs to class i, $c-1$ projections $p_j = (\vec{x} - \vec{m}_i)^T \vec{v}_{ij}$, $(j = 1, \ldots, c-1; i \neq j)$ are computed and $\vec{x}$ assigned to the class i if $p_j < t_{ij}$ for all j. If the latter condition is not satisfied vector $\vec{x}$ is left unclassified.

13.4 Results

13.4.1 Analysis of Sensor Brightness Scatterplots

The image brightness[1] distribution for the six classes in the ERS and RADARSAT image is presented in the scatter diagram in Figure 13.5a. As mentioned before, the absolute values of backscatter coefficients were not available for the RADARSAT image and therefore only relative values between ice classes can be analyzed from the scatter diagram. It shows the variability of image brightness for most of the ice classes, exclusive

[1]We use this term as equivalent to image value or image digital number (DN).

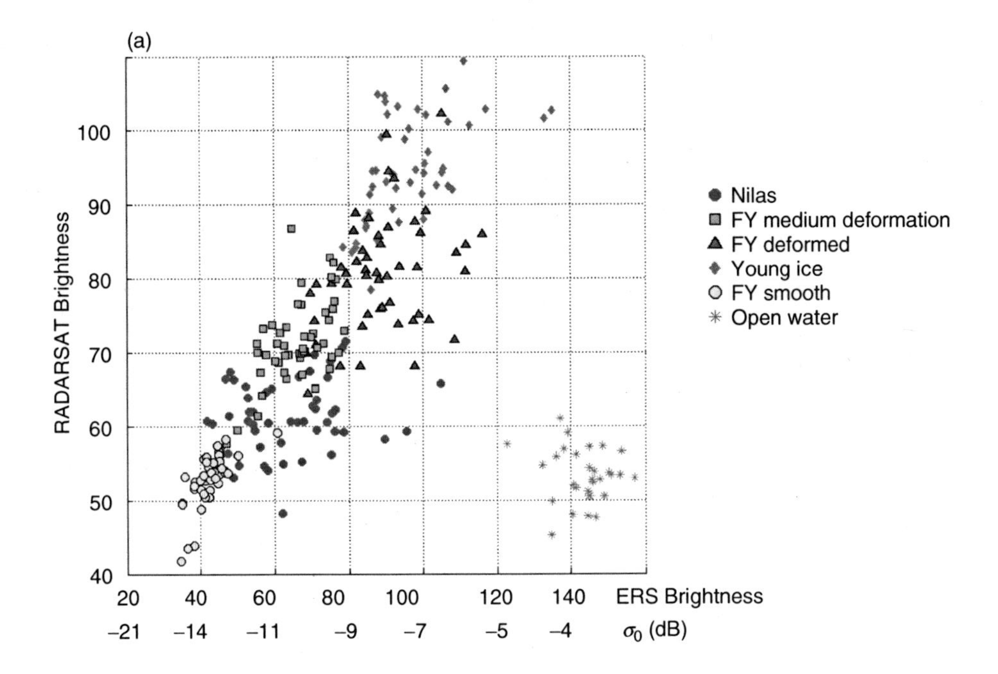

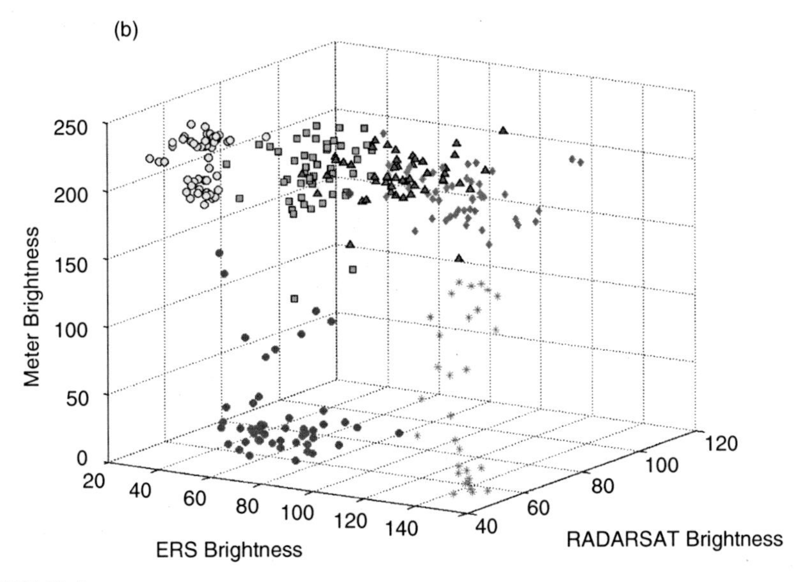

FIGURE 13.5
Image brightness scattergrams for (a) ERS, RADARSAT and (b) ERS, RADARSAT SAR, and Meteor television camera images plotted using a subset of training feature vectors.

FY smooth ice is relatively high, and the clusters in two-dimensional subspace are overlapped. The exception is OW for which polarization ratio is as high as 16 dB [49] (VV/HH), estimated using the CMOD4 model (a semiempirical C-band model that describes the dependency of backscatter signal on wind speed and image geometry [50]), with wind speed values of 3–4 m/s measured onboard the ship. The corresponding cluster is located away from those of sea ice, thus separation of OW from ice can be achieved using both HH and VV SAR sensors if the wind speed is in a certain range. The changes of wind speed and, consequently, changes of OW radar signatures from those used for training, can decrease classification accuracy. In the latter case incorporation of visual data (see Figure 13.5b), where OW and nilas can be well separated from the ice independently on wind speed, is quite important. The yearly averaged wind speeds at the Kara Sea are in the range of 5–8 m/s for different coastal stations [51], which suggests that the calm water conditions (wind speed below 3 m/s) are less common than wind roughed OW conditions.

The variation in backscatter from FY ice mostly depends on surface roughness, thus FY smooth ice appears dark in the images. Nilas has higher backscatter than FY smooth ice due to rafting and formation of frost flowers on its surface. An example of frost flowers is shown in Figure 13.4e. The relatively high variation of backscatter can be partially explained by the spatial variation of surface roughness, and the existence of OW areas between the ice. These factors may also influence the other ice type signatures (FY deformed and young ice) often containing OW areas in small leads between ice floes. Formation of new ice in leads and existence of brash ice between larger ice floes can also modify ice radar signatures. The backscatter from young ice in the diagram is high due to the small size of the ice floes. The raised, highly saline ice edges are perfect scatterers and the backscatter signal integrated over large areas of young ice is often typically high.

13.4.2 MLP Training

In our study we used a standard software package—Stuttgart Neural Network Simulator (SNNS) [52]—developed at the University of Stuttgart in collaboration with other universities and research institutes. In addition to a variety of implemented neural network architectures it provides a convenient graphical user interface.

The proper representation of the neural network inputs and outputs is important for its performance. There are several choices of target coding schemes. The "one from n" or "winner-takes-all" coding scheme is often used for classification. According to this scheme a desired target value of an output unit corresponding to true class of the input vector is unity, and the target values of all other units are zero. In cases where significantly different numbers of training vectors are used for each class, the neural network biases strongly in favour of the classes with the largest membership, as shown in Ref. [53]. Neural network training using this technique implicitly encodes proportions of samples within classes as *prior* probabilities. In our training data set the number of training vectors for each class does not correspond to their prior probabilities. These probabilities depend on many factors and are generally unknown. We assumed them to be equal and adopted the following modification of the coding scheme [53]:

$$T_i = \begin{cases} 1/\sqrt{N_i} & \text{if } \vec{x}_i \in \text{ class } i \\ 0 & \text{otherwise,} \end{cases}$$

where T_i is the target value of class i, and N_i is the number of patterns in that class. Since the number of vectors in several classes are large, their target values do not differ much from zero. Therefore we linearly scaled the target values to span the range [0,1].

Different parameters of the MLP used in our experiments are presented in Table 13.4 and Table 13.5 along with classification results. In notation nu_1–nu_2–nu_3, nu_i ($i = 1-3$) is the number of units in the input, hidden, and output layers of the neural network, respectively. The number of input units of the network is equivalent to the number of

TABLE 13.4

MLP Percentage of Correct Classification on the Test Data Set

Image Features/BPNN Parameters	FY Smooth	FY Medium	FY Def.	Young Ice	Nilas	OW	Total
ERS mean and texture/16–20–6, 100 cycles	82.3	55.1	71.9	82.9	52.4	76.9	66.0
RADARSAT mean and texture/16–20–6, 100 cycles	68.5	70.9	51.2	89.5	72.6	40.0	70.7
ERS, RADARSAT mean and texture/32–40–6, 300 cycles	88.5	78.3	80.5	93.8	82.4	96.2	83.9
ERS, RADARSAT mean and texture, Meteor mean/33–40–6, 300 cycles	99.2	87.1	84.6	94.1	98.2	100.0	91.2

input image features and nu_3 corresponds to the number of ice classes. We found sufficient to have one layer of hidden neurons. The number of units in this layer is determined empirically. In Section 13.4.6 we use a pruning method to evaluate the number of hidden neurons more precisely. This number should generally correspond to the complexity of the task. The more neurons are in the hidden layer, the more precise the achieved approximation of the input–output mapping function. It is known that the fine approximation of the training data set does not necessarily lead to the improved generalization of the neural network, i.e., the ability to classify previously unseen data. Therefore, nu_2 is determined as a compromise between the two factors and it usually increases with increasing the nu_1.

The following parameters are used for the MLP training: the learning rate η and the momentum parameter α are set to 0.5 and 0.2, respectively. The weights of the MLP are randomly initialized before training. They are updated after each presentation of the training vector to the neural network, that is, *online learning* is used. At each training cycle all vectors of the training data set are selected at random and sequentially provided to the classifier. The activation function is the standard *logistic sigmoid* function. In the experiments below we present accuracies estimated using the test data set. The absolute differences between the total classification accuracies of the test data set (Table 13.4 and Table 13.6) and those of the training data set (not shown) are less than 1.1 percent point, which indicates good generalization of the trained algorithms.

TABLE 13.5

MLP Percentage of Correct Classification on the Test Data Set Using Only Mean Values of Sensor Brightness and a Reduced Set of Texture Features (Last Row)

Image Features/MLP Parameters	FY Smooth	FY Medium	FY Def.	Young Ice	Nilas	OW	Total
ERS and RADARSAT mean values/2–10–6, 80 cycles	91.5	68.3	71.1	82.7	78.1	100.0	75.5
ERS, RADARSAT, and Meteor mean values/3–10–6, 80 cycles	99.2	87.0	69.4	84.0	97.0	100.0	84.8
ERS and RADARSAT mean values, fourth-order central moment, cluster shade, Meteor mean value/7–6–6, 200 cycles	99.2	84.6	81.8	93.1	97.3	100.0	89.5

TABLE 13.6

LDA Percentage of Correct Classification on the Test Data Set

Image Features	FY Smooth	FY Medium	FY Def.	Young Ice	Nilas	OW	Total
ERS mean and texture	93.1	50.6	67.9	81.4	55.2	92.3	64.6
RADARSAT mean and texture	43.1	60.6	65.9	89.1	60.9	46.2	68.5
ERS and RADARSAT mean and texture	91.5	74.4	80.6	93.2	81.3	100.0	82.6
ERS, RADARSAT mean and texture, Meteor mean	99.2	86.6	81.7	93.4	97.4	100.0	90.1

13.4.3 MLP Performance for Sensor Data Fusion

13.4.3.1 ERS and RADARSAT SAR Image Classification

The estimated accuracy for separate classification of ERS and RADARSAT images by the MLP-based algorithm are shown in the first two rows of Table 13.4. The mean value of image brightness and image texture are used in both cases. As evident from the table, the single source classification provides poor accuracy for several ice classes, such as FY ice, nilas, and OW. This is expected since corresponding clusters are difficult to separate along axes of the scatterplot in Figure 13.5a. OW, which is visually dark in RADARSAT and bright in ERS images, is largely misclassified as FY smooth ice in RADARSAT image and classified with higher accuracy in ERS images. The evaluated total classification accuracies are 66.0% and 70.7% for ERS and RADARSAT, respectively.

13.4.3.2 Fusion of ERS and RADARSAT SAR Images

The joint use of ERS and RADARSAT image features increases the number of correctly classified test vectors to the value of about 83.9% which is 17.9 and 13.2 percent points higher than those values obtained using ERS and RADARSAT SAR separately. It is known that the radar signatures of sea ice vary across different Arctic regions (see for example Ref. [54] for multi-year ice) and are subject to the seasonal changes [35] that should influence the classifier performance. The number of ice types discriminated is maximal in winter and less in other seasons. From late spring to late summer, only two classes, namely, ice and water are most likely to be separated [4]. This limits the classifier applicability to the particular season that the classifier is trained for and to a certain geographical region.

As seen from the table, the improvements are observed for all ice classes. The classification accuracy of OW is much higher, due to the incorporation of polarimetric information. This increase is as much as 56.2 and 19.3 percent points compared with RADARSAT and ERS images classified separately. Classification of the dual polarization data not only facilitates separation of classes with large polarization ratios, but also those with relatively small ratios, which have similar tonal signatures on ERS and RADARSAT images. The improvement in classification of young ice, medium, and rough FY ice with a pronounced texture, can be due to incorporation of texture features in the classification discussed in the next sections. The latter factor can also explain the higher increase in accuracy observed in our experiment as compared with previous studies [30], which demonstrated 10 percent points increase in the total classification accuracy (63.6% vs. 52.4%) for the three ice types using fully polarimetric data obtained at C-band, although the direct comparison of the results is rather difficult.

13.4.3.3 Fusion of ERS, RADARSAT SAR, and Meteor Visible Images

Combination of visible and SAR images improves estimated accuracy up to 91.2% (the last row in Table 13.4). The increase in classification accuracy gained by fusion of a single polarization SAR (ERS-1) with Landsat TM images was demonstrated earlier by using maximum likelihood classifier [24]. Our results show that visual data are also useful in combination with the multi-polarization SAR data. A significant improvement of about 16 percent points is observed for nilas, since the optical sensor has a large capacity to discriminate between nilas and OW (see Figure 13.5b). An increase in classification accuracy is also observed for the smooth and deformed FY ice because less of their test vectors are misclassified as nilas. The improvement is insignificant for OW and young ice because these classes have already been well separated using the polarimetric data set. It should be mentioned that the importance of optical data for classification of OW using C-band should increase in the near range because polarization ratio (VV/HH) decreases with decreasing incident angle [55] and is small in the directions close to nadir. Another important factor favoring incorporation of optical clear sky data is the generalization of the classification algorithm over the images containing OW areas, acquired under different wind conditions (not available in this study). Since backscattering from OW, among other factors, depends on wind speed and direction [55], it would be more difficult for a classifier utilizing only polarimetric information to generalize well over OW regions, unless wind speed and direction, obtained from the other data sources (scatterometer derived wind fields, ground-truth data) are presented to the classifier.

13.4.4 LDA Algorithm Performance for Sensor Data Fusion

The classification accuracy of the LDA algorithm, estimated using the same combinations of image features as for MLP are presented in Table 13.6. The MLP slightly outperforms the LDA algorithm with an accuracy increase of 1.2–2.2 percent points for different feature combinations. In interpreting this result we should mention the close relationship between neural networks and traditional statistical classifiers. It is shown that a feed forward neural network without hidden layers approximates an LDA [25], and multilayer neural network with at least one hidden layer of neurons provides nonlinear discrimination between classes. For the performance and selection of the classifier, a structure of the input space, the form of clusters, and their closeness in the feature space, are of primary importance. If the clusters have a complex form and are close to each other, construction of nonlinear decision boundaries may improve discrimination. By experimentally comparing LDA- and MLP-based algorithms, we compare linear and nonlinear discriminant methods applied for the sea ice classification problem. Our results suggest that the used feature space is fairly simple, and that the ice classes can be linearly separated piecewise. Construction of nonlinear decision boundaries can only slightly improve classification results. The assessed classification accuracies for individual classes are different. Some ice classes are better classified by the LDA algorithm, but there is a reduction in classification accuracy for the other classes. Taking into account the longer training time for MLP, compared with the LDA algorithm (5 min 59 sec vs. only 12 sec respectively[2]) the latter may be preferable when the time factor is important. However, a greater reduction of training and classification time can be gained by using less input features or even just using image brightness values, since computation of texture is time consuming (31 min 39 sec, 16 parameters over the 2163×2763 single image in nonoverlapping window[3]). Therefore, in the next section we will empirically evaluate the usefulness of texture computation for a multi-sensor data set.

13.4.5 Texture Features for Multi-Sensor Data Set

Assuming that texture is useful for classification of single polarization SAR data [6,9,10,38–41], the question that we would like to address is "are the texture features still useful for classification of the multi-polarization data set?" For this purpose the MLP is trained using only mean values of image brightness, and the test results obtained (Table 13.5, the first two rows) are compared with those presented earlier (Table 13.4, the last two rows), where texture is used. By comparing them, one notices the increase of total classification accuracy from 75.5% (no texture) to 83.9% when texture is computed for a combination of ERS and RADARSAT. The improvement gained by the incorporation of texture is largest for the FY rough ice, young ice, and nilas.

It is interesting to mention that computation of texture features for the dual polarization data set (VV and HH) provides almost the same accuracy (83.9%) as reached by the fusion of data from all three sensors without texture computation (84.8%), as shown in the second row of Table 13.5. This is useful because visual images may be unavailable due to the weather or low light conditions so that texture computation for the polarimetric data set can, at least partially, compensate for the absence of optical data. As expected, sea ice classes with a developed texture, such as young and deformed first ice, are better classified when texture features are computed for the dual polarization data set, while the estimated classification accuracy of other classes are higher using brightness from all three sensors without texture computation. The texture features are still useful when data from all three sensors are available, gaining an improvement of 6.4 percent points (91.2% vs. 84.8%).

13.4.6 Neural Network Optimization and Reduction of the Number of Input Features

In the previous sections we considered texture features as comprising one group. Nevertheless, some of the selected features may carry similar information on sea ice classes, thus increasing the dimensionality of the input feature space and necessitating the use of larger training data sets. To exclude the redundant features and simultaneously optimize the structure of the neural network we apply pruning methods [56] that trim or remove units or links of the neural network (depending on the method), and evaluate some parameters of neural network performance. The stepwise LDA is one of the methods that can be used for the analysis and selection of the input features for the linear classifier; however, application of this method may not be optimal for the nonlinear MLP.

---

[2]33 input features, 300 training cycles for MLP, Sun Blade 100 workstation with 500 MHz UltraSPARC II CPU are used.

[3]The computation, perhaps, can be made faster by the code optimization.

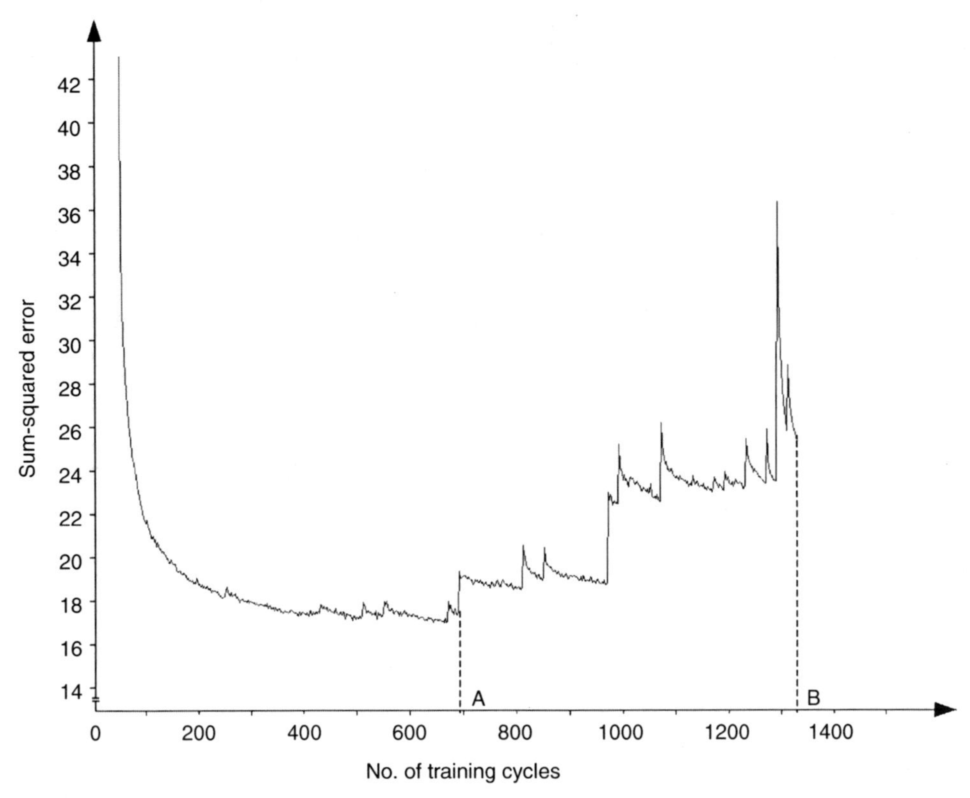

FIGURE 13.6
Change of the sum-squared error during initial training and pruning of the MLP.

The largest trained neural network (33–40–6) is selected for the experiments. During pruning the redundant units or links are sequentially removed. After this the neural network is additionally trained to recover from the change in its structure. Figure 13.6 shows the change of sum-squared error during initial training and pruning using the *skeletonization* algorithm [57]. This is a sensitivity type algorithm that estimates the relevance of each unit by computing derivative of a linear error function with respect to the switching off and on of the unit using *attenuation strength* parameter. These errors propagate backward using a similar algorithm to that used for training of the MLP. The picks in the error plot correspond to removal of the input and hidden units. Approximately half of the input and several hidden units are removed at the middle of the pruning (letter A). As shown, their removal at this stage does not strongly effect classification. Increasing the number of removed units causes the MLP performance on the training data set to decrease. This process continues until only three input units, corresponding to the three sensor image brightnesses and six hidden units, are left (letter B in Figure 13.6). In the next section the two neural networks similar to those obtained at the end and at the middle of pruning will be analyzed.

13.4.6.1 Neural Network with No Hidden Layers

We start with the description of the neural network consisting of three input and no hidden units. Its Hinton diagram is shown in Figure 13.7. Connections between the input and output units are represented by squares in the diagram. The size and color of the

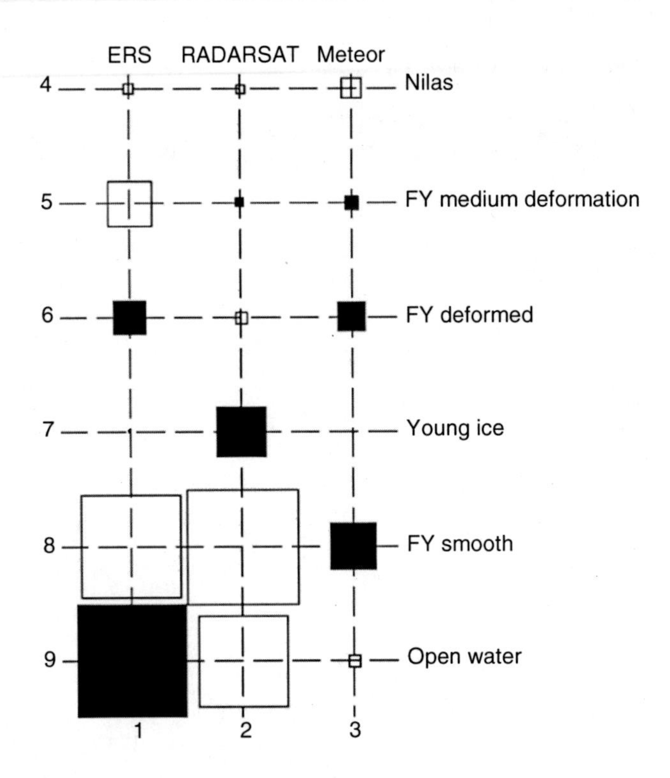

FIGURE 13.7
The Hinton diagram of the MLP with no hidden layers (3–6). The squares at the diagram represent weights between input (no. 1–3) and output units (no. 4–9) of the neural network. The size of the squares lineally scales with the absolute value of weights from 20.0 to 0, dark and white colors mark positive and negative weights, respectively.

squares correspond to the strength of the connections, and their sign, respectively. In a neural network without hidden layers these weights affect the incoming signals, directly defining the contribution of each sensor (i.e., input unit) to the class outputs. As known, this neural network provides a linear discrimination between classes.

Analyzing the diagram, we find the large Meteor image values, which were changed by positive weight connections, (last column in the diagram), increase the FY ice output signal, and correspondingly, a posterior probability of the input vector to belong to the FY ice. Simultaneously, the probability for nilas class is reduced due to the negative weight between the respective input and output units. The high ERS image values increase FY deformed ice output through the positive weight connection (first column in the diagram) and decrease it for smooth and medium deformation FY ice. The test vector most probably belongs to OW when the ERS image value is high and the RADARSAT image value is low. There is no redundancy in using these sensors: if FY deformed ice can be separated using a combination of ERS and Meteor sensors, RADARSAT SAR data need not be involved. The latter is also true for the young ice. The other sensors can indirectly influence the assignment of the vector to a particular class by decreasing or increasing outputs for the other classes.

13.4.6.2 Neural Network with One Hidden Layer

Neural networks obtained in the middle stages of the pruning (Figure 13.6, letter A) correspond to some compromise between neural network complexity (i.e., number of input and hidden units) and the RMS error for the training data set. An increase in this error does not necessarily imply a decrease in the classification rate for the test data set, since generalization of the neural network may improve. This neural network still has a number of hidden units and can thus provide nonlinear discrimination between ice classes. The outputs of the neural network on presenting the feature vector to the classifier are proportional to the class conditional probabilities [25] for the type of neural network and target coding schema used here. During several pruning trials, neural networks of different structure and with different combinations of input features are obtained, which implies the existence of several minima in the multi-dimensional error surface. The application of pruning methods does not guarantee finding a global minimum, thus several MLPs with similar performances on the test data set are obtained.

The Hinton diagram of the MLP with the structure 7–6–6 is shown in Figure 13.8. In addition to the brightness for each of the three sensors, it uses two texture features: a fourth-order central statistical moment and a cluster shade for ERS and RADARSAT images. For its analysis it is convenient to represent the performance of each hidden neuron as a linear combination of several input features, weighted by a sigmoid activation function. There are six such combinations (see first six rows of the diagram), corresponding to the six hidden neurons. In the first two of these, the individual sensor brightness values play a dominant role through the large positive weights. The contribution of each combination to the formation of different class outputs can be analyzed by looking at weights between the hidden and output units (shown in the last six columns of the diagram and marked by arrows). The role of the first two combinations is similar to

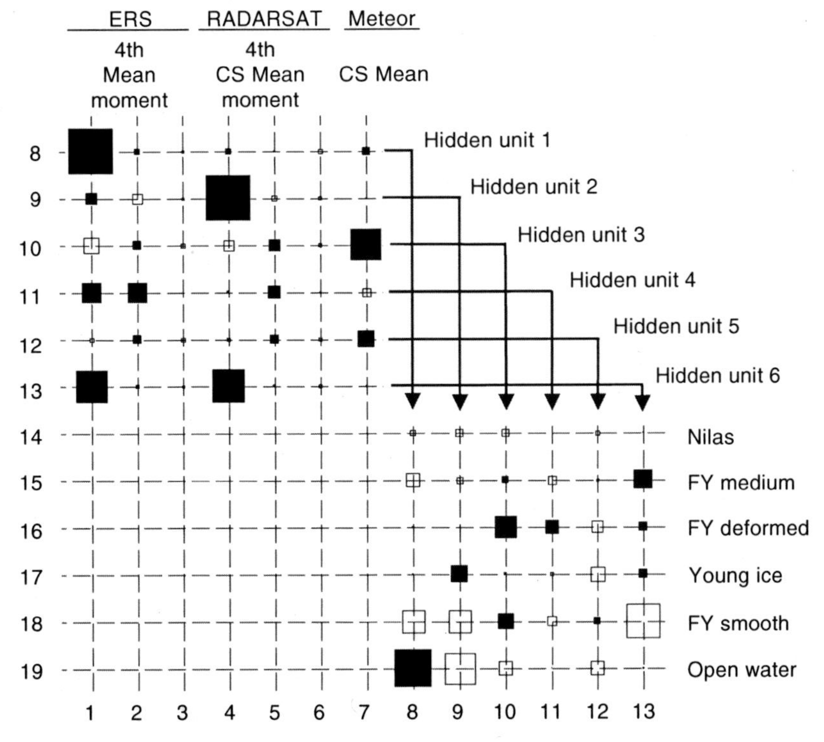

FIGURE 13.8
The Hinton diagram of the MLP with one hidden layer of neurons (7–6–6). The squares at the upper left-hand corner of the diagram represent weights between input (no. 1–7) and hidden units (no. 8–13), the squares at the lower right-hand corner of the diagram represent weights between hidden and outputs units (no. 14–19) of the neural network. The size of the squares lineally scales with the absolute value of weights from 15.0 to 0, dark and white colors mark positive and negative weights, respectively. The used abbreviation: CS, cluster shade.

the individual sensor contributions in the network without hidden neurons described earlier. However, the availability of the sixth combination, where the ERS and RADARSAT image brightness are combined together, modifies the weight structure. For example, a negative contribution of ERS brightness (1st hidden unit) to the FY medium ice output is partially compensated by positively weighted ERS brightness of the 6th combination. Eventually, only RADARSAT brightness plays a dominant role in output of FY medium ice.

From the diagram it is seen that a cluster shade feature contributes little to the sea ice separation due to the small weights connecting the corresponding input units with hidden units (2nd and 6th columns). Perhaps this feature can be removed when data from all three sensors are used in classification since inclusion of cluster shade provides only about a 0.5 percent point increase in classification accuracy (5–6–6, 200 cycles). The fourth-order statistical moment mostly takes part in formation of FY deformed and young ice outputs through the 3rd, 4th, and 5th hidden units. This observation is supported by the increase in classification accuracy for these classes when the texture features are used (last row of Table 13.5). It is evident from the classification results that the role of texture increases with decreasing number of sensors used in classification (8.4 vs. 6.4 percent points for 2 and 3 sensors, respectively). More texture features in addition to those shown in the diagram may be required for the two-sensor image classification (ERS and RADARSAT). The suitable candidates are variation and entropy. Analysis of the pruning sequences suggests that texture features computed over VV and HH polarization images are not fully redundant (see also Ref. [58]); therefore, it seems reasonable to use the same feature sets for ERS and RADARSAT images.

13.4.7 Classification of the Whole Image Scene

Sea ice maps produced by different classifiers (Table 13.4 and Table 13.6) are presented in Figure 13.9 and Figure 13.10 for the MLP and LDA-based algorithms, respectively.

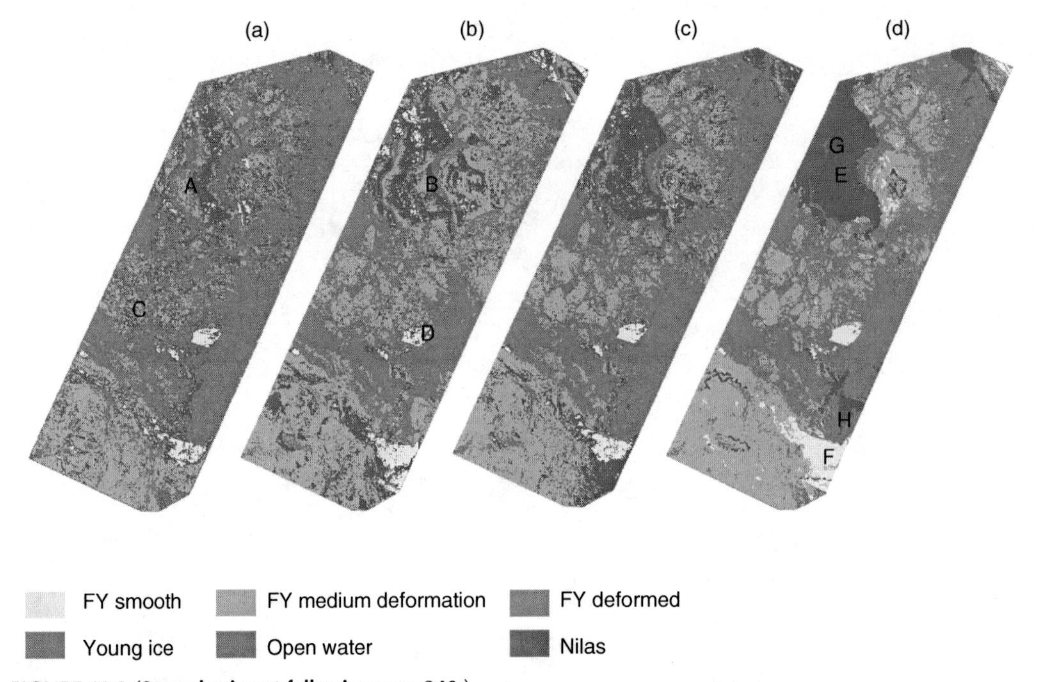

FIGURE 13.9 (See color insert following page 240.)
MLP sea ice classification maps obtained using: (a) ERS; (b) RADARSAT; (c) ERS and RADARSAT; and (d) ERS, RADARSAT, and Meteor images. The classifiers' parameters are given in Table 13.4.

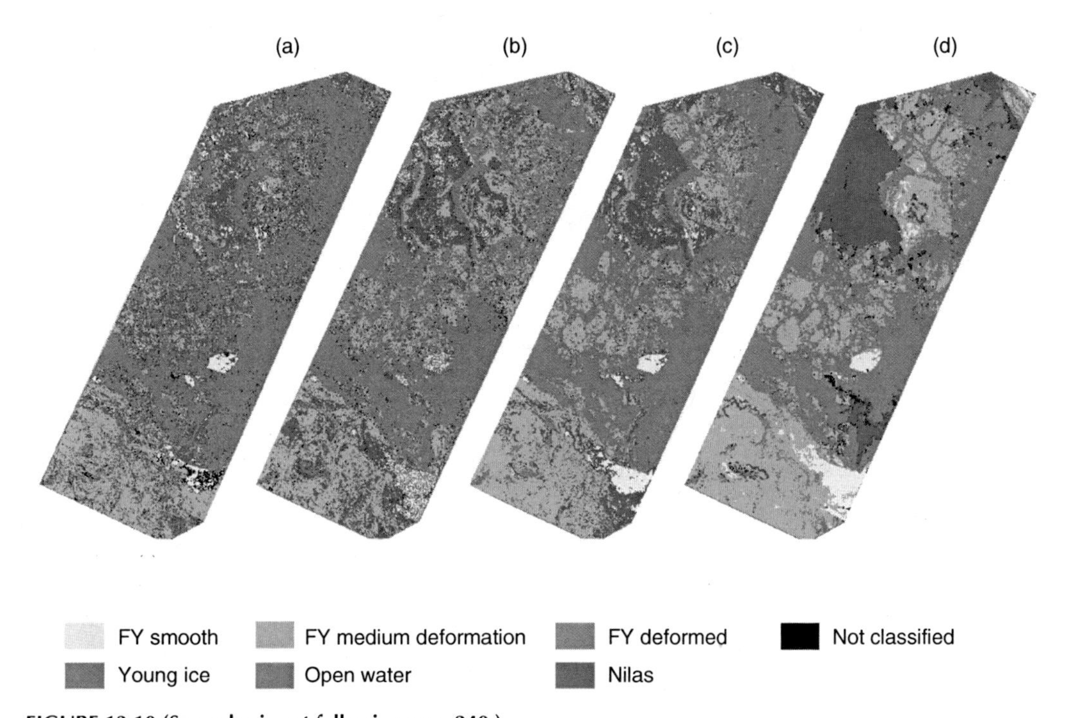

FIGURE 13.10 (See color insert following page 240.)
LDA sea ice classification maps obtained using: (a) ERS; (b) RADARSAT; (c) ERS and RADARSAT; and (d) ERS, RADARSAT, and Meteor images.

No postclassification filtering or smoothing is performed. The color legend is based on the WMO-defined color code with the following discrepancies: firstly, different gradations of green reflect ice surface roughness and, secondly, the red color marks OW, to separate it from nilas. The ERS and RADARSAT image classification maps are visually rather poor, and this is independent of the type of algorithm used. Some parts of the polynya (regions A and B in Figure 13.9) are misclassified as FY ice and, conversely, some areas of the FY ice (regions C and D) are misclassified as nilas. This also applies to the LDA classification maps (Figure 13.10), which appear a bit more "noisy" because some image areas are left unclassified by the LDA algorithm (dark pixels) and some areas of smooth FY ice are largely misclassified as OW in the case of the RADARSAT image in Figure 13.10b. Due to classification errors, some of the objects cannot be visually identified in the images in the case of the single sensor classification.

Sea ice classification is improved by combining the two SAR images (Figure 13.9c and Figure 13.10c). OW, young, and FY ice are much better delineated using dual polarization information. Classification of sea ice using three data sources (Figure 13.9d and Figure 13.10d) in general corresponds well to the visual expert classification shown in Figure 13.11, but with more details in comparison to the usual manually prepared sea ice maps. To generalize ice zones well (some applications do not require detailed ice information) and reduce classification time, larger ice zones are usually manually outlined, partial concentrations of ice types within these zones are (often subjectively) evaluated and presented in an ice chart using the WMO egg code (not used in Figure 13.11). As can be seen from the examples the detailed presentation of information in automatically produced sea ice maps can be useful in tactical ice navigation, where small-scale ice information is needed. For applications that do not require a high level of detail, the partial ice concentrations can be calculated more precisely based on the results of

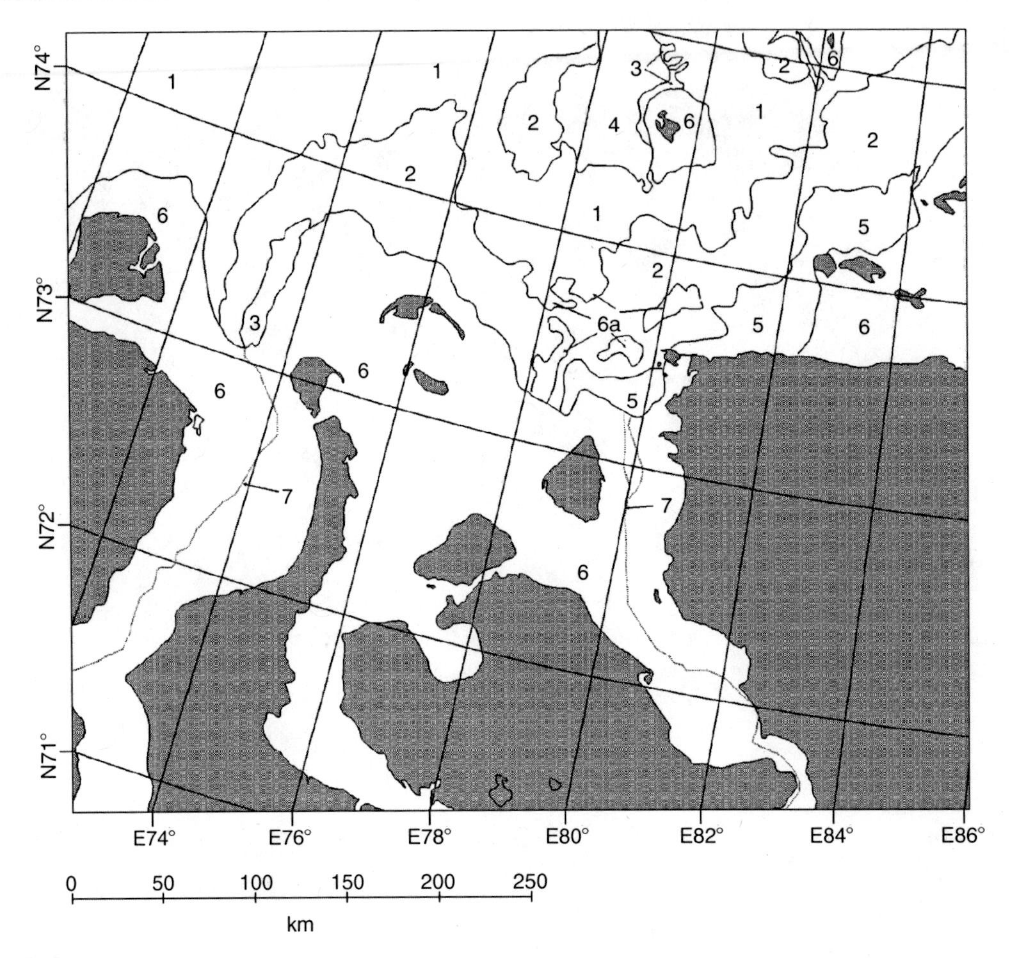

FIGURE 13.11
Expert interpretation of RADARSAT Scan SAR image (30 April 1998) made by N. Babitch (Murmansk Shipping Company). (1) The darker areas are mainly large thick FY ice while the brighter areas are young ice. The darkest areas between the FY and young ice is nilas. (2) Heavily deformed thin and medium thick FY ice with some inclusions of young ice. (3) Nilas (dark) and young ice (bright). (4) A difficult area for interpretation. Probably it is a polynya with a mixture of water and thin ice types such as grease ice and nilas. (5) Mixture of open water and very thin ice. (6) Fast ice of various age and deformation. The darker areas are undeformed ice while brighter areas are more deformed ice. The brightest signature in the river estuaries can be due to freshwater ice from the rivers. (6a) Larger floes of fast ice drifting out from the Yenisei estuary. (7) The bright lines in the fast ice are ship tracks. (From O.M. Johannessen et al., SAR sea ice interpretation guide. NERSC Tech. Rep. No. 227, Bergen, Norway, 2002. With Permission.)

automatic classification. The major difference between expert and automatic classifications is in the assignment of bright segment no. 2 in the upper right hand corner of Figure 13.11, which has been assigned to heavily deformed FY ice by ice expert and to young ice in our case. Some discrepancies in classification results can arise because *in situ* and optical data were not used for the expert classification.

In many cases, the incorporation of low-resolution visible information reduces the noise in the classification, while preserving small-scale ice features. The nilas and fast ice (regions E and F in Figure 13.9d, respectively) are correctly classified, while some relatively thin strips of rafted ice in polynya (region G) are still retained in the classified images. Classification maps produced by the algorithms look rather similar except for some minor details: the strips of rafted ice (region G) are better delineated by the

MLP-based algorithm and the region of OW in the low right hand corner of the images (marked by letter H in Figure 13.9d) is better classified by LDA-based algorithm. It is interesting to mention that in the latter case only, the transition from OW to nilas and then to young ice is correctly outlined in the rather complex region (H). Such transition zones are observed many times during the expedition and this classification result seems reasonable. It is also interesting to see how fast ice in region F in Figure 13.9d is delineated by different classifiers (Figure 13.9 and Figure 13.10a–d).

13.5 Conclusions

In this paper we have presented a sea ice classification algorithm that fuses data of different satellite sensors. The algorithm is based on a multi-layer backpropagation neural network and utilizes ERS, RADARSAT SAR and low-resolution optical images. Radiometric correction, spatial alignment, and co-registration of images from these three sensors have been done before their fusion. The selection of image test and training regions is based on *in situ* observations made onboard the Russian nuclear icebreaker "Sovetsky Soyuz" in April, 1998. The image features used include image moments, gray-level co-occurrence matrix texture, and autocorrelation function–based features for the SAR image and mean brightness for visible image.

The empirical assessment of the algorithms via the independent test data set showed that the performance of MLP with only one data source (SAR) is rather poor (66.0% and 70.7% for the ERS and RADARSAT SAR, respectively), even if texture features are used in the classification. A substantial improvement in sea ice classification is achieved by combining ERS and RADARSAT SAR images obtained at two different polarizations (VV and HH). Fusion of these two SAR data types increases the classification accuracy up to 83.9%. This noticeable improvement is observed for all ice classes. These results suggest that an improvement in ice classification can be also expected for ENVISAT ASAR operating at alternating polarization mode and for future RADARSAT-2 SAR with its several multi-polarization modes. Incorporation of visual data provides additional information, especially on the nilas and OW classes. The estimated total classification accuracy reaches 91.2% when the low-resolution visual data are combined with the dual polarization data set. In cases where visual data are unavailable, computation of texture features is found to be particularly useful, since it can, at least partially, compensate for the reduction in accuracy due to the absence of low-resolution visual data.

Both of the considered MLP and LDA algorithms show similar results when applied for sea ice classification. This implies that the ice classes can be linearly separated piecewise in multi-dimensional feature space. Incorporation of texture features into the classification improves the separation of various ice classes, although its contribution decreases with an increasing number of sensors used in classification. Computation of texture is time consuming. Application of a pruning method allowed us to reduce the number of input features of the neural network, to adjust the number of hidden neurons, and, as a result, to decrease classification time. These methods are not free from entrapment in the local minima of the error surface, thus the application of evolutionary optimization methods for the neural network structure optimization is a possible future development. The other interesting direction of research can be in combining the outputs of several classifiers in an MCS [60]. The differences in classifier performances for different ice classes (especially in single polarization image classification) can be accommodated in MCS to improve overall classification accuracy.

In this study we used an extensive set of *in situ* observations collected on board the icebreaker sailing in ice; however, the studied area and the number of observations are still quite small to describe the variety of possible ice conditions. The presented results should therefore be taken with care, because they are obtained for a relatively small geographical region and during the short time period in cold conditions. With new data acquisition it is necessary to assess the generalization of the algorithm to a larger geographical region and to a larger number of seasons.

Acknowledgments

This work was fulfilled during the first author's Ph.D. study at the Nansen Environmental and Remote Sensing Centres in St. Petersburg (Russia) and Bergen (Norway). The study was sponsored by the Norwegian Research Council. The first author would like to thank his scientific supervisor Dr. V.V. Melentyev and Prof. V.D. Pozdniakov for their help and valuable advice. The ice data sets were obtained through "Ice Routes" (WA-96-AM-1136) and "ARCDEV" (WA-97-SC2191) EU Projects. The authors are also grateful to Dr. C.C. Wackerman (Veridian Systems Division, Ann Arbor, Michigan), SNNS developers for providing the software packages, and Dr. G. Schöner and Dr. C. Igel (INI, Bochum) for their suggestions for improving the manuscript.

References

1. L.K. Soh, C. Tsatsoulis, and B. Holt, Identifying ice floes and computing ice floe distributions in SAR images, in *Analysis of SAR Data of the Polar Oceans*, C. Tsatsoulis and R. Kwok, Eds., New York: Springer-Verlag, pp. 9–34, 1998.
2. D. Haverkamp and C. Tsatsoulis, Information fusion for estimation of summer MIZ ice concentration from SAR imagery, *IEEE Trans. Geosci. Rem. Sens.*, 37(3), 1278–1291, 1999.
3. J. Banfield, Automated tracking of ice floes: a stochastic approach, *IEEE Trans. Geosci. Rem. Sens.*, 29(6), 905–911, 1991.
4. R. Kwok, E. Rignot, B. Holt, and R. Onstott, Identification of sea ice types in spaceborne synthetic aperture radar data, *J. Geophys. Res.*, 97(C2), 2391–2402, 1992.
5. Y. Sun, A. Calrlström, and J. Askne, SAR image classification of ice in the Gulf of Bothnia, *Int. J. Rem. Sens.*, 13(13), 2489–2514, 1992.
6. C.C. Wackerman and D.L. Miller, An automated algorithm for sea ice classification in the marginal ice zone using ERS-1 synthetic aperture radar imagery, Tech. Rep., ERIM, Ann Arbor, MI, 1996.
7. M.M. Van Dyne, C. Tsatsoulis, and F. Fetterer, Analyzing lead information from SAR images, *IEEE Trans. Geosci. Rem. Sens.*, 36(2), 647–660, 1998.
8. C. Bertoia, J. Falkingham, and F. Fetterer, Polar SAR data for operational sea ice mapping, in *Analysis of SAR Data of the Polar Oceans*, C. Tsatsoulis and R. Kwok, Eds., New York: Springer-Verlag, pp. 201–234, 1998.
9. M.E. Shokr, Evaluation of second-order texture parameters for sea ice classification from radar images, *J. Geophys. Res.*, 96(C6), 10625–10640, 1991.
10. D.G. Barber, M.E. Shokr, R.A. Fernandes, E.D. Soulis, D.G. Flett, and E.F. LeDrew, A comparison of second-order classifiers for SAR sea ice discrimination, *Photogram. Eng. Remote Sens.*, 59(9), 1397–1408, 1993.

11. D. Haverkamp, L.K. Soh, and C. Tsatsoulis, A comprehensive, automated approach to determining sea ice thickness from SAR data, *IEEE Trans. Geosci. Rem. Sens.*, 33(1), 46–57, 1995.
12. A.K. Fang, *Microwave Scattering and Emission Models and Their Applications*, Norwood, MA: Artech House, 1994.
13. D.P. Winebrenner, J. Bredow, M.R. Drinkwater, A.K. Fung, S.P. Gogineni, A.J. Gow, T.C. Grenfell, H.C. Han, J.K. Lee, J.A. Kong, S. Mudaliar, S. Nghiem, R.G. Onstott, D. Perovich, L. Tsang, and R.D. West., Microwave sea ice signature modelling, in *Geophysical Monograph 68: Microwave Remote Sensing of Sea Ice*, F.D. Carsey, Ed., Washington, DC: American Geophysical Union, pp. 137–175, 1992.
14. R.G. Onstott, R.K. Moore, and W.F. Weeks, Surface-based scatterometer results of Arctic sea ice, *IEEE Trans. Geosci. Rem. Sens.*, GE-17, 78–85, 1979.
15. O.M. Johannessen, W.J. Campbell, R. Shuchman, S. Sandven, P. Gloersen, E.G. Jospberger, J.A. Johannessen, and P.M. Haugan, Microwave study programs of air–ice–ocean interactive processes in the Seasonal Ice Zone of the Greenland and Barents seas, in *Geophysical Monograph 68: Microwave Remote Sensing of Sea Ice*, F.D. Carsey, Ed., Washington, DC: American Geophysical Union, pp. 261–289, 1992.
16. K. Fukunaga, *Introduction to Statistical Pattern Recognition*, 2nd ed., New York: Academic Press, 1990.
17. F.M. Fetterer, D. Gineris, and R. Kwok, Sea ice type maps from Alaska Synthetic Aperture Radar Facility imagery: an assessment, *J. Geophys. Res.*, 99(C11), 22443–22458, 1994.
18. S. Sandven, O.M. Johannessen, M.W. Miles, L.H. Pettersson, and K. Kloster, Barents Sea seasonal ice zone features and processes from ERS 1 synthetic aperture radar: seasonal Ice Zone Experiment 1992, *J. Geophys. Res.*, 104(C7), 15843–15857, 1999.
19. R.A. Schowengerdt, *Remote Sensing—Models and Methods for Image Processing*, New York: Academic Press, 1997.
20. D.L. Hall, *Mathematical Techniques in Multisensor Data Fusion*, Norwood, MA: Artech House, 1992.
21. M.J. Collins, Information fusion in sea ice remote sensing, in *Geophysical Monograph 68: Microwave Remote Sensing of Sea Ice*, F.D. Carsey, Ed., Washington, DC: American Geophysical Union, pp. 431–441, 1992.
22. S.G. Beaven and S.P. Gogineni, Fusion of satellite SAR with passive microwave data for sea remote sensing, in *Analysis of SAR Data of the Polar Oceans*, C. Tsatsoulis and R. Kwok, Eds., New York: Springer-Verlag, pp. 91–109, 1998.
23. L.K. Soh, C. Tsatsoulis, D. Gineris, and C. Bertoia, ARKTOS: an intelligent system for SAR sea ice image classification, *IEEE Trans. Geosci. Rem. Sens.*, 42(1), 229–248, 2004.
24. K. Steffen and J. Heinrichs, Feasibility of sea ice typing with synthetic aperture radar (SAR): Merging of Landsat thematic mapper and ERS 1 SAR satellite imagery, *J. Geophys. Res.*, 99(C11), 22413–22424, 1994.
25. C.M. Bishop, *Neural Networks for Pattern Recognition*, Oxford: Clarendon Press, 1995.
26. J.A. Benediktsson, P.H. Swain, and O.K. Ersoy, Neural network approaches versus statistical methods in classification of multisource remote sensing data, *IEEE Trans. Geosci. Rem. Sens.*, 28(4), 540–552, 1990.
27. J.D. Paola and R.A. Schowengerdt, A detailed comparison of backpropagation neural network and maximum-likelihood classifiers for urban land use classification, *IEEE Trans. Geosci. Rem. Sens.*, 33(4), 981–996, 1995.
28. C. Eckes and B. Fritzke, Classification of sea-ice with neural networks—results of the EU research project ICE ROUTS, Int. Rep. 2001–02, Institut für Neuroinformatik, Ruhr-Universität Bochum, Bochum, Germany, March 2001.
29. J. Key, J.A. Maslanik, and A.J. Schweiger, Classification of merged AVHRR and SMMR Arctic data with neural networks, *Photogram. Eng. Rem. Sens.*, 55(9), 1331–1338, 1989.
30. Y. Hara, R.G. Atkins, R.T. Shin, J.A. Kong, S.H. Yueh, and R. Kwok, Application of neural networks for sea ice classification in polarimetric SAR images, *IEEE Trans. Geosci. Rem. Sens.*, 33(3), 740–748, 1995.
31. J.A. Karvonen, Baltic sea ice SAR segmentation and classification using modified pulse-coupled neural networks, *IEEE Trans. Geosci. Rem. Sens.*, 42(7), 1566–1574, 2004.

32. RADARSAT data products specifications, Doc. RSI-GS-026, ver. 3.0, RADARSAT International, May 2000.
33. K. Kloster, A report on backscatter normalisation and calibration of C-band SAR, NERSC Tech. Notes, Bergen, Norway, 1997.
34. H. Laur, P. Bally, P. Meadows, J. Sanchez, B. Schaettler, and E. Lopinto, ERS SAR calibration. Derivation of the backscattering coefficient σ^0 in ESA ERS SAR PRI products, Tech. Rep. ES-TN-RS-PM-HL09, issue 2, ESA/ESRIN, Frascati, Italy, 1996.
35. R.G. Onstott, SAR and scatterometer signatures of sea ice, in *Geophysical Monograph 68: Microwave Remote Sensing of Sea Ice*, F.D. Carsey, Ed., Washington, DC: American Geophysical Union, pp. 73–104, 1992.
36. T.C. Grenfell, D. Cavalieri, D. Comiso., and K. Steffen, Considerations for microwave remote sensing of thin sea ice, in *Geophysical Monograph 68: Microwave Remote Sensing of Sea Ice*, F.D. Carsey, Ed., Washington, DC: American Geophysical Union, pp. 291–301, 1992.
37. WMO, *Sea Ice Nomenclature*, World Meteorological Organization, Geneva, Switzerland, 1970.
38. L.K. Soh and C. Tsatsoulis, Texture analysis of SAR sea ice imagery using gray level co-occurrence matrices, *IEEE Trans. Geosci. Rem. Sens.*, 37(2), 780–795, 1999.
39. Q.A. Holmes, D.R. Nüesch, and R.A. Shuchman, Textural analysis and real-time classification of sea ice types using digital SAR data, *IEEE Trans. Geosci. Rem. Sens.*, GE 22(2), 113–120, 1984.
40. J.A. Nystuen and F.W. Garcia, Jr., Sea ice classification using SAR backscatter statistics, *IEEE Trans. Geosci. Rem. Sens.*, 30(3), 502–509, 1992.
41. M.J. Collins, C.E. Livingstone, and R.K. Raney, Discrimination of sea ice in the Labrador marginal ice zone from synthetic aperture radar image texture, *Int. J. Rem. Sens.*, 18(3), 535–571, 1997.
42. R.M. Haralic, K. Shanmugam, and I. Dinstein, Textural features for image classification, *IEEE Trans. Syst. Man Cybern.*, SMS-3(6), 610–621, 1973.
43. P.J. Werbos, Beyond regression: New tools for prediction and analysis in the behavioural sciences, *Ph.D. Thesis*, Harvard University, Cambridge, 1974.
44. D.E. Rumelhart, G.E. Hinton, and R.J. Williams, Learning representations by back-propagating errors, *Nature*, 323, 533–536, 1986.
45. F. Rosenblatt, The Perceptron: a probabilistic model for information storage and organization in the brain, *Psychol. Rev.*, 65, 386–408, 1958.
46. R.A. Fisher, The use of multiple measurement in taxonomic problems, *Ann. Eugen.*, 7, 179–188, 1936.
47. P.A. Lachenbruch, *Discriminant Analysis*, New York: Hafner Press, 1975.
48. R.O. Duda and P.E. Hart, *Pattern Classification and Scene Analysis*, New York: John Wiley & Sons, 1973.
49. K. Kloster, Personal communication, May 1999.
50. A.C.M. Stoffelen and D.L.T. Andreson, Scatterometer data interpretation: estimation and validation of the transfer function CMOD4, *J. Geophys. Res.*, 102(C3), 5767–5780, 1997.
51. Atlas of the Arctic, Moscow: Glavnoye Upravlenie Geodezii i Kartografii, 1985.
52. A. Zell, G. Mamier, M. Vogt, N. Mache, R. Hübner, S. Döring, et al., SNNS Stuttgart neural network simulator user manual, ver. 4.1, Rep. no. 6/95, Inst. Parallel Distributed High Performance Syst., University of Stuttgart, Stuttgart, Germany, 1995.
53. A.R. Webb and D. Lowe, The optimised internal representation of multiplayer classifier networks performs nonlinear discriminant analysis, *Neural Netw.*, 3, 367–375, 1990.
54. R. Kwok and G.F. Cunningham, Backscatter characteristics of the winter ice cover in the Beaufort Sea, *J. Geophys. Res.*, 99(C4), 7787–7802, 1994.
55. R.H. Stewart, *Methods of Satellite Oceanography*, Berkeley: University of California Press, 1985.
56. J. Siestma and R.J.F. Dow, Creating artificial neural networks that generalize, *Neural Netw.*, 4, 67–79, 1991.
57. P. Smolensky and M. Mozer, Skeletonization: a technique for trimming the fat from a network via relevance assessment, in *Advances in Neural Information Processing Systems (NIPS)*, San Mateo: Morgan Kaufmann, pp. 107–115, 1989.

58. M.E. Shokr, L.J.Wilson, and D.L. Surdu-Miller, Effect of radar parameters on sea ice tonal and textural signatures using multi-frequency polarimetric SAR data, *Photogram. Eng. Rem. Sens.*, 61(12), 1463–1473, 1995.
59. O.M. Johannessen et al., SAR sea ice interpretation guide. NERSC Tech. Rep. No. 227, Bergen, Norway, 2002.
60. A.V. Bogdanov, G. Schöner, A. Steinhage, and S. Sandven, Multiple classifier system based on attractor dynamics, in *Proc. IEEE Symp. Geosci. Rem. Sens. (IGARSS)*, Toulouse, France, July 2003.

14

Use of the Bradley–Terry Model to Assess Uncertainty in an Error Matrix from a Hierarchical Segmentation of an ASTER Image

Alfred Stein, Gerrit Gort, and Arko Lucieer

CONTENTS

14.1 Introduction

Remotely sensed images are increasingly being used for collection of spatial information. A wide development in sensor systems has occurred during the last decades, resulting in improved spatial, temporal, and spectral resolution. The collection of data by remote sensing is generally more efficient and cheaper than by direct observation and measurement on the ground, although still of a varying quality. Data collected by sensors may be affected by atmospheric factors between sensors and the values reflected on the earth's surface, local impurities on the earth's surface, technical deficiencies of sensors and other factors. In addition, only the reflection of the sensor's signal or of the sunlight on the earth's surface is being measured, and no direct measurements are made. Consequently, the quality of maps produced by remote sensing needs to be assessed [1].

Another major issue of interest concerns the ontological interpretation of remote-sensing imagery. The truth as it supposedly exists at the earth surface is due to what one wishes to see. This can be formalized in terms of formal ontologies, although these may be subject to changes in time as well, and also a subjective interpretation is often being given.

Image classification and segmentation are important concepts in this respect. A segmentation of an image yields segments of more or less homogeneity supposed to be applicable on the earth's surface. In fact, a segmentation procedure can be done mathematically, without any more knowledge of the earth than is present in the data, hence governed by the conditions of the sensor on the earth's surface and in the atmosphere. Image classification leads to segmentation with meaningful contents. Meaningful is either what is described by the ontologies, or what a researcher aims to find. There still is a large discrepancy between the image representation on one hand and the earth's surface processes on the other hand.

After a classification is being carried out, its accuracy can be determined if ground truth is available. Classification accuracy refers to the extent to which the classified image or map corresponds with the description of a class on the earth's surface. This is commonly described by an error matrix, in which the overall accuracy and the accuracy of the individual classes are calculated. The κ-statistic is then used for testing on homogeneity [2]. Its use is somewhat controversial, because its value depends strongly on the marginal distributions [3]. Indeed, it measures the strength of agreement without taking into account the strength of disagreement. Another criticism, which we will not address in this chapter, is that it requires the presence of well-identifiable objects. In many ontologies, though, we realize that objects at the earth's surface are fuzzy or vague. In such cases, it appears to be more convenient to compare membership values.

This chapter aims to focus on an alternative to the κ-statistic. To do so, we consider on pairwise comparisons, dealing with the structure of disagreement between categories. A way to do so is by using the Bradley–Terry (BT) model [4]. Based on the logistic regression model, it may provide more details on the strength and direction of disagreement. The chapter builds on a previous study published recently [5]. It extends that paper by addressing another case study, providing more details at various places and by using a very large number of test points. We also provide additional discussion on the qualities of the BT model.

The aim of research described in this chapter is to study the use of the BT model as an alternative measure for association in the remotely sensed image for the κ-statistic. To do so, we implement and interpret a test of significance of the BT model for paired preferences in a multi-variate texture-based segmentation of geological classes from an ASTER image in Mongolia. An error matrix is obtained by validation with an existing geological map.

14.2 Concepts and Methods

A common aim in classifying a multi-spectral image is to automatically categorize all pixels in the image into classes [6]. Each pixel in the image is assigned in a Boolean fashion into one class. As a result, a thematic layer from the multi-spectral image emerges. Several classification algorithms applicable to thematic mapping from satellite images exist [7]. Supervised image classification depends on differences in the reflection on the earth's surface and hence, in bare areas, on the composition of the material on the earth's surface. Pixel-by-pixel spectral information is used as the basis for automated land-cover classification. A classification algorithm gives a classified thematic layer. Such a form of classification yields, therefore, both a segmentation of the image (i.e., different segments emerge) and a classification (i.e., each segment has a meaning within some ontology). This is particularly the case if a supervised classification is carried out. A feature of this method, often recognized as a drawback, is that it yields a large number of very small segments, hence a highly fragmented thematic layer.

Next, the thematic layer is assessed in terms of accuracy by comparing image class samples to reference samples [8]. Representative samples are selected for each class from different locations of the image. The amount of training data usually represents between 1% and 5% of the pixels [9].

Accuracy assessment of classification is usually carried out by evaluating error matrices. An error matrix is the square contingency table. The columns represent the reference data and the rows represent the classified data [2,9]. On the basis of the error matrix, overall classification accuracy, producer's accuracy, and the user's accuracy are calculated. In addition, the κ-statistic for each individual category within the matrix can be calculated. In all these cases, the accuracy of any individual pixel is associated in a strictly Boolean fashion, that is, the classification is either correct or incorrect.

14.2.1 The κ-Statistic

The κ-statistic derived from the error matrix is a measure of agreement [2]. It is based on the difference between the actual agreement in the error matrix and the chance agreement. The sample outcome is the $\hat{\kappa}$ statistic, an estimate of κ. The actual agreement refers to the correctly classified pixel indicated by the major diagonal of the error matrix. κ measures the accuracy of classification in terms of whether the classification is better than random or not. To describe κ we assume a multinomial sampling model: each class label is derived from a finite set of possible labels. To calculate the maximum likelihood estimate $\hat{\kappa}$ of κ and a test statistic z, the error matrix is represented in mathematical form. Then $\hat{\kappa}$ is given by

$$\hat{\kappa} = \frac{p_0 - p_c}{1 - p_c} \tag{14.1}$$

where p_0 and p_c are the actual agreement and the chance agreement. Let n_{ij} equal the number of samples classified into category i, as belonging to category j in the reference data. The $\hat{\kappa}$ value can be calculated using the following formula:

$$\hat{\kappa} = \frac{n\sum_{i=1}^{k} n_{ii} - \sum_{i=1}^{k} n_{i+}n_{+i}}{n^2 - \sum_{i=1}^{k} n_{i+}n_{+i}} \tag{14.2}$$

where k is the number of classes, n_{ii} is the number of correctly classified pixels of category i, n_{i+} is the total number of pixels classified as category i, n_{+i} is the total number of actual pixels in category i, and n is the total number of pixels.

The approximate large sample variance of κ equals:

$$\mathrm{var}(\hat{\kappa}) = \frac{1}{n}\left(\frac{\vartheta_1(1-\vartheta_1)}{(1-\vartheta_2)^2} + \frac{2(1-\vartheta_1)(2\vartheta_1\vartheta_2 - \vartheta_3)}{(1-\vartheta_2)^3} + \frac{2(1-\vartheta_2)^2(\vartheta_4 - 4\vartheta_2^2)}{(1-\vartheta_2)^4}\right) \tag{14.3}$$

where

$$\vartheta_1 = \frac{1}{n}\sum_{i=1}^{k} n_{ii}$$

$$\vartheta_2 = \frac{1}{n^2}\sum_{i=1}^{k} n_{i+}n_{+i}$$

$$\vartheta_3 = \frac{1}{n^2}\sum_{i=1}^{k} n_{ii}(n_{i+} + n_{+i})$$

$$\vartheta_4 = \frac{1}{n^3}\sum_{i=1}^{k} n_{ii}(n_{i+} + n_{+i})^2$$

Each of the terms ϑ_i can be easily calculated, and hence the κ-statistic has found its place in a wide range of (semi-)automatic classification algorithms.

Apart from being a measure of accuracy, the conditional κ coefficient κ_i is defined as the conditional agreement for the ith category. It can also be used to test the individual class (category) agreement. The maximum likelihood estimate of κ_i is given by

$$\hat{\kappa}_i = \frac{n \cdot n_{ii} - n_{i+}n_{+i}}{n \cdot n_{i+} - n_{i+}n_{+i}} \tag{14.4}$$

Its approximate large sample variance is given by

$$\text{var}(\hat{\kappa}_i) = \frac{n \cdot (n_{i+} - n_{ii})}{n_{i+}(n - n_{+i})^3}[n_{i+}n_{ii}(n_i + n_{+i} - n \cdot n_{ii}) + n \cdot n_{ii}(n_{ii} - n_{i+} - n_{+i} + n)] \tag{14.5}$$

A test of significance for a single error matrix is based on comparing the obtained classification with a random classification, that is, with assigning random labels to the individual classes at each of the test points. For this test the null hypothesis equals H_0: $\kappa = 0$, that is, the association does not significantly differ from that obtained from a random classification. The alternative hypothesis equals H_1: $\kappa \neq 0$, that is, the value of κ significantly differs from that obtained from a random classification. The test statistic equals

$$z = \frac{\hat{\kappa}}{\sqrt{\text{var}(\hat{\kappa})}} \tag{14.6}$$

A similar test of significance can be carried out for each individual category using the value of the conditional κ_i and its variance.

Different classifications may yield different error matrices. Possibly, one classification significantly improves on the other one. Comparing two error matrices, we might therefore be willing to test the null hypothesis H_0: $\kappa_1 = \kappa_2$, against the alternative hypothesis H_1: $\kappa_1 \neq \kappa_2$, that is, no significant difference occurs in association between the two error matrices. A test statistic for the significance of these two error matrices is given by

$$z_{12} = \frac{\hat{\kappa}_1 - \hat{\kappa}_2}{\sqrt{\text{var}(\hat{\kappa}_1) + \text{var}(\hat{\kappa}_2)}} \tag{14.7}$$

Computation of the z or the z_{12} statistic then allows us to evaluate H_0, as the distribution of z and that of z_{12} is approximately a normal distribution. H_0 is rejected if $z > z_{\alpha/2}$ (or equivalently $z_{12} > z_{\alpha/2}$), where $z_{\alpha/2}$ is the $1 - \alpha/2$ confidence level of the two-tailed standard normal distribution. A similar argument applies if a one-sided test is to be carried out, for example, to investigate a claim that some new classification procedure is better than another one.

14.2.2 The BT Model

The BT model makes a pairwise comparison among n individuals, classes, or categories. It focuses on misclassifications without considering the total number and the number of correctly classified pixels (or objects). First we follow the model in this original form. Below, we address the issue of including also the total number of classified pixels. In the BT model, classes are ordered according to magnitude on the basis of misclassifications. The ordering is estimated on the basis of pairwise comparisons. Pairwise comparisons model the preference of one individual (class, category) over another [3].

The BT model has found applications in various fields, notably in sports statistics. The original paper [4] considered basketball statistics, whereas a well worked-out example on tennis appears in Ref. [3]. Both examples include multiple confrontations of various teams against one another. The BT model can be seen as the logit model for paired preference data. A logit model is a generally applicable statistical model for the relation between the probability p that an effect occurs and an explanatory variable x by using two parameters α and β. It is modeled as $\ln(p/(1-p)) = \alpha + \beta\ x$, ensuring that probability values are between 0 and 1 [10]. To apply the BT model to the error matrix, we consider the pair of classes C_i and C_j and let Π_{ij} denote the probability that C_i is classified as C_j and let Π_{ji} denote that C_j is classified as C_i. Obviously, $\Pi_{ji} = 1-\Pi_{ij}$.

The BT model has parameters β_i such that

$$\text{logit}(\prod_{ij}) = \log(\prod_{ij}/\prod_{ji}) = \beta_i - \beta_j \tag{14.8}$$

To interpret this equation, we note that equal probabilities emerge for C_i being classified as C_j and C_j being classified as C_i if $\Pi_{ij} = \Pi_{ji} = \frac{1}{2}$, hence if $\text{logit}(\Pi_{ij}) = 0$, therefore $\beta_i = \beta_j$. If $\Pi_{ij} > \frac{1}{2}$, i.e., if C_i is more likely to be classified as C_j than C_j to be classified as C_i, then $\beta_i > \beta_j$. A value of β_i larger than that of β_j indicates a preference of misclassification of C_i to C_j above that of C_j to C_i.

By fitting this model to the error matrix, one obtains the estimates $\hat{\beta}_i$ of the β_i, and the estimated probabilities $\hat{\Pi}_{ij}$ that C_i is classified as C_j are given by

$$\hat{\Pi}_{ij} = \frac{\exp(\hat{\beta}_i - \hat{\beta}_j)}{1 + \exp(\hat{\beta}_i - \hat{\beta}_j)} \tag{14.9}$$

Similarly, from the fitted values of the BT model, we can derive fitted values for misclassification,

$$\hat{\Pi}_{ij} = \frac{\hat{\mu}_{ij}}{\hat{\mu}_{ij} - \hat{\mu}_{ji}} \tag{14.10}$$

where $\hat{\mu}_{ij}$ is the expected count value of C_i over C_j, that is, the fitted value for the model. The fitted value of the model can be derived from the output of SAS or SPSS.

14.2.3 Formulating and Testing a Hypothesis

In practice, we may be interested to formulate a test on the parameters. In particular we are interested in whether there is equality in performance of a classifier for the different classes that can be distinguished. Therefore, let H_0 be the null hypothesis: the class parameters are equal, that is, no significant difference exists between β_i and β_j. The alternative hypothesis equals H_1: $\beta_i \neq \beta_j$. To test H_0 against H_1, we compare $\hat{\beta}_i - \hat{\beta}_j$ to its asymptotic standard error (ASE). To find the ASE, we note that $\text{var}(\hat{\beta}_i - \hat{\beta}_j) = \text{var}(\hat{\beta}_i) + \text{var}(\hat{\beta}_j) - 2\,\text{cov}(\hat{\beta}_i, \hat{\beta}_j)$. Using the estimated variance–covariance matrix, each ASE is calculated as the square root of the sum of two diagonal values minus twice an off-diagonal value. An approximate 95% confidence interval for $\beta_i - \beta_j$ is then seen to be equal to $\hat{\beta}_i - \hat{\beta}_j \pm 1.96 \times \text{ASE}$. If the value 0 occurs within the confidence interval, then the difference between the two classes is not significantly different from zero, that is, no significant difference between the class parameters occurs, and H_0 is not rejected. On the other hand, if the value 0 does not occur within the confidence interval, then the difference between the two classes is significantly different from zero, indicating a significant difference between the class parameter and H_0 is rejected in favor of H_1.

14.3 Case Study

To illustrate the BT model in an actual remote-sensing study, we return to a case described in Ref. [11]. It was shown how spectral and spatial data collected by means of remote sensing can provide information about geological aspects of the earth surface. In particular in remote, barren areas, remote-sensing imagery can provide useful information on the geological constitution. If surface observations are combined with geologic knowledge and insights, geologists are able to make valid inferences about subsurface materials. The study area is located in the Dundgovi Aimag province in Southern Mongolia (longitude: 105°50′–106°26′ E and lattitude: 46°01′–46°18′ N). The total area is 1415.58 km^2. The area is characterized by an arid, mountainous-steppe zone with elevations between 1300 and 1700 m. Five geological units are distinguished: Cretaceous basalt (K1), Permian–Triassic sandstone (PT), Proterozoic granite (yPR), Triassic–Jurassic granite (yT3-J1) (an intrusive rock outcrop), and Triassic–Jurassic andesite (aT3-J1).

For the identification of general geological units we use images from the advanced spaceborne thermal emission and reflection radiometer (ASTER) satellite, acquired on 21 May, 2002. The multi-spectral ASTER data cover the visible, near infrared, shortwave and thermal infrared portions of the electromagnetic spectrum, in 14 discrete channels. Level 1B data as used in this study are radiometrically calibrated and geometrically co-registered for all ASTER bands. The combination of ASTER shortwave infrared (SWIR) bands is highly useful for the extraction of information on rock and soil types. Figure 14.1 shows a color composite of ASTER band combination 9, 6, and 4 in the SWIR range. An important aspect of this study is the validation of geological units derived with segmentation (Figure 14.2a). Reference data in the form of a geological map were obtained by expert field observation and image interpretation (Figure 14.2b).

Textural information derived from remotely sensed imagery can be helpful in the identification of geological units. These units are commonly mapped based on field observations or interpretation of aerial photographs. Geological units often show charac-

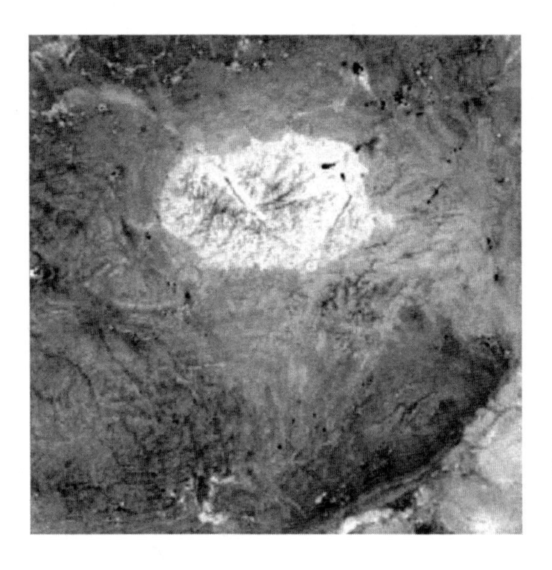

FIGURE 14.1
SWIR band combination of the ASTER image for the study area in Mongolia, showing different geological units.

teristic image texture features, for example, in the form of fracture patterns. Pixel-based classification methods might, therefore, fail to identify these units. A texture-based segmentation approach, taking into account the spatial relations between pixels, can be helpful in identifying geological units from an image scene.

For segmentation, we applied a hierarchical splitting algorithm to identify areas with homogeneous texture in the image. Similar to split-and-merge segmentation each square image block in the image is split into four sub-blocks forming a quadtree structure. The criterion used to determine if an image block is divided is based on a comparison between uncertainty of a block and uncertainty of its sub-blocks. Uncertainty is defined as the ratio between the similarity values (G-statistic), computed for an image block B, of the two most likely reference textures. This measure is also known as the confusion index (CI). The image is segmented such that uncertainty is minimized. Reference textures are defined by two-dimensional histograms of the local binary pattern and the variance texture measures. To test for similarity between an image block texture and a reference texture, the G-statistic is applied. Finally, a partition of the image with objects labeled with reference texture class labels is obtained [12].

14.3.1 The Error Matrix

Segmentation and classification resulted in a thematic layer with geological classes. Comparison of this layer with the geological map yielded the error matrix (Table 14.1). Accuracy of the overall classification equals 71.0%, and the κ-statistic equals 0.51. A major source for incorrect segmentation is caused by the differences in detail between the segmentation results and the geological map. In the map, only the main geological units are given, where the segmentation provides many more details. A majority filter of 15×15 pixels was applied to filter out the smallest objects from the ASTER segmentation map. Visually, the segmentation is similar to the geological map. However, the K1 unit is much more abundant in the segmentation map. The original image clearly shows a distinctly different texture from the surrounding area. Therefore, this area is segmented as a K1 instead of as a PT unit. The majority filtering did not provide higher accuracy values, as the total accuracy only increased by 0.5%.

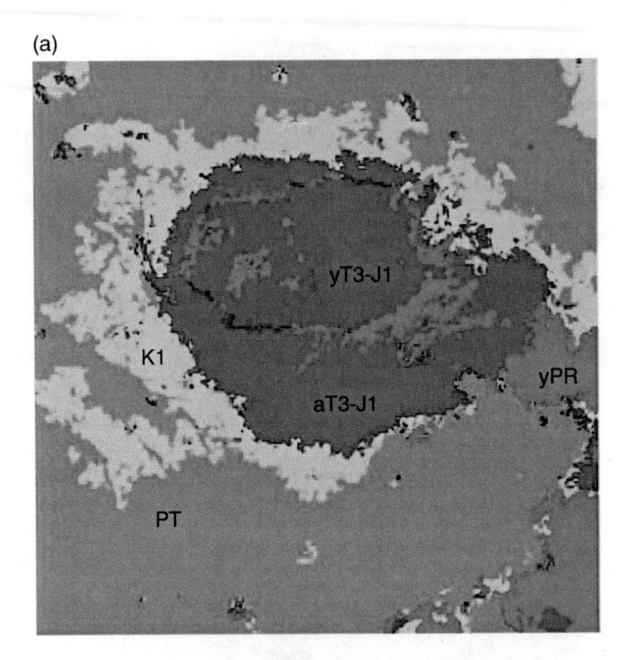

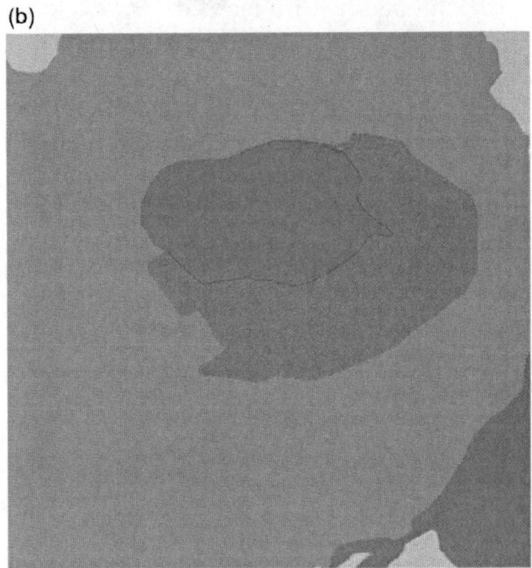

FIGURE 14.2
(a) Segmentation of the ASTER image. (b) The geological map used as a reference.

14.3.2 Implementation in SAS

The BT model was implemented as a logit model in the statistical package SAS (Table 14.2). We applied proc Genmod, using the built-in binomial probability distribution (DIST = BIN) and the logit link function. The covb option provides the estimated covariance matrix of the model parameter estimators, and estimated model parameters are obtained with the obstats option. In the data matrix, a dummy variable is set for each geological class. The variable for C_i is 1 and that for C_j is –1 if C_i is classified over C_j. The logit model has these variates as explanatory variables. Each line further lists the error value ('k') of C_i over C_j and the sum of the error values of C_i over C_j and C_j over C_i, ('n'). The intercept term is excluded.

14.3.3 The BT Model

The BT model was fitted to the error matrix as shown in Table 14.1. Table 14.3 shows the parameter estimates $\hat{\beta}_i$ for each class. These values give the ranking of the category in

TABLE 14.1

Error Matrix for Classification

Classified as	PT	yT3-J1	K1	Reality aT3-J1	yPR	Total
PT	518,326	304	8,772	3,256	13,716	544,374
yT3-J1	805	86,103	0	2,836	3,848	93,592
K1	140,464	95	8,123	13,645	3,799	166,126
aT3-J1	50,825	2,997	827	95,244	2,621	152,514
yPR	23,788	8,082	60	21,226	31,569	84,725
Total	734,208	97,581	17,782	136,207	55,553	1,041,331

Note: PT (Permian and Triassic formation), yT3-J1 (upper Triassic and lower Jurassic granite), K1 (lower Cretaceous basalt), aT3-J1 (upper Triassic and lower Jurassic andesite) and yPR (Penterozoic granite). The overall classification accuracy is 71.0%, the overall κ-statistic equals 0.51.

comparison with the reference category. Table 14.3 also shows standard errors for the $\hat{\beta}_i$s for each class. The β parameter for class yPR is not estimated, being the last in the input series, and is hence set equal to 0. The highest $\hat{\beta}$ coefficient equal to 1.545 is observed for class PT (Permian and Triassic formation) and the lowest value is equal to –1.484 for class K1 (lower Cretaceous basalt). Standard errors are relatively small (below 0.016), indicating that all coefficients significantly differ from 0. Hence the significantly highest erroneous classification occurs for the class PT and the lowest for class K1.

Estimated $\hat{\beta}_i$ values are used in turn to determine misclassification of one class over another (Table 14.4). This anti-symmetric matrix shows again relatively high values for differences of geological classes with PT and lower values for differences with class K1. Next, the probability of a misclassification is calculated using Equation 14.9 (Table 14.5). For example, the estimate of probability of misclassifying PT as upper Triassic and lower Jurassic granite (yT3-J1) is 0.28, whereas that of a misclassification of yT3-J1 as PT equals

TABLE 14.2

```
data matrix;
input PT yT3J1 K1 aT3J1 yPR k1 n1 k2 n2 w;
k = (k1/n1);
n = (k1/n1 + k2/n2);
cards;
```

Data Matrix	Input	PT	yT3-J1	K1	aT3-J1	yPR *k n*; Cards
−1	1	0	0	0	304	1,109
−1	0	1	0	0	8,772	149,236
−1	0	0	1	0	3,256	54,081
−1	0	0	0	1	13,716	37,504
0	−1	1	0	0	0	95
0	−1	0	1	0	2,836	5,833
0	−1	0	0	1	3,848	11,930
0	0	−1	1	0	13,645	14,472
0	0	−1	0	1	3,799	3,859
0	0	0	−1	1	2,621	23,847

```
proc genmod data = matrix;
model k/n = PT yT3J1 K1 aT3J1 yPR/NOINT DIST = BIN link = logit covb
obstats;
run;
```

TABLE 14.3

Estimated Parameters $\hat{\beta}_i$ and Standard Deviations se($\hat{\beta}_i$) for the BT Model

	PT	yT3-J1	K1	aT3-J1	yPR
$\hat{\beta}_i$	1.545	0.612	−1.484	0.317	0.000
se($\hat{\beta}_i$)	0.010	0.016	0.014	0.010	0.000

Note: See Table 14.1.

TABLE 14.4

Values for $\hat{\beta}_i - \hat{\beta}_j$ Comparing Differences of One Geological Class over Another

	PT	yT3-J1	K1	aT3-J1	yPR
PT	0.000	−0.932	−3.029	−1.228	−1.545
yT3-J1	0.932	0.000	−2.097	−0.295	−0.612
K1	3.029	2.097	0.000	1.801	1.484
aT3-J1	1.228	0.295	−1.801	0.000	−0.317
yPR	1.545	0.612	−1.484	0.317	0.000

Note: See Table 14.1.

0.72: it is therefore much more likely that PT is misclassified as yT3-J1, than yT3-J1 as PT. Table 14.6 shows the observed and fitted entries in the error matrix (in brackets) for the BT model. We notice that the estimated and observed differences are relatively close. An exception is the expected entry for the yPR–aT3-J1 combination, where fitting is clearly erroneous.

To test the significance of differences, ASEs are calculated, yielding a symmetric matrix (Table 14.7). Low values (less than 0.02) emerge for the different class combinations, mainly because of the large number of pixels. Next, we test for the significance of differences (Table 14.8), with H_0 being equal to the hypothesis that $\hat{\beta}_i = \hat{\beta}_j$, and the alternative hypothesis H_1 that $\hat{\beta}_i \neq \beta_j$. Because of the large number of pixels, H_0 is rejected for all class combinations. This means that there is a significant difference between the parameter values for each of these classes.

We now turn to the standardized data, where also the diagonal values of the error matrix are included in the calculations.

TABLE 14.5

Estimated Probabilities $\hat{\Pi}_{ij}$ Using Equation 14.9 of Misclassifying C_i over C_j Using Standardized Data

	PT	yT3-J1	K1	aT3-J1	yPR
PT		0.28	0.05	0.23	0.18
yT3-J1	0.72		0.11	0.43	0.35
K1	0.95	0.89		0.86	0.82
aT3-J1	0.77	0.57	0.14		0.42
yPR	0.82	0.65	0.18	0.58	

Note: See Table 14.1.

TABLE 14.6

Observed and Expected Entries in the Error Matrix

	PT	yT3-J1	K1	aT3-J1	yPR
PT		304	8,772	3,256	13,716
		(313)	(6,885)	(12,255)	(6,595)
yT3-J1	805		0	2,836	3,848
	(796)		(10)	(2,489)	(4,194)
K1	140,464	95		13,645	3,799
	(142,351)	(85)		(12,421)	(3,146)
aT3-J1	50,825	2,997	827		2,621
	(41,826)	(3,344)	(2,051)		(41,240)
yPR	23,788	8,082	60	21,226	
	(30,909)	(7,736)	(713)	(56,625)	

Note: See Table 14.1.

TABLE 14.7

Asymptotic Standard Errors for the Parameters in the BT Model, Using the Standard Deviations and the Covariances between the Parameters

	PT	yT3-J1	K1	aT3-J1	yPR
PT		0.017	0.011	0.008	0.010
yT3-J1	0.017		0.019	0.016	0.016
K1	0.011	0.019		0.012	0.014
aT3-J1	0.008	0.016	0.012		0.010
yPR	0.010	0.016	0.014	0.010	

Note: See Table 14.1.

TABLE 14.8

Test of Significances in Differences between Two Classes

		$\hat{\beta}_i - \hat{\beta}_j$	ASE	*t*-Ratio	H_0
PT	yT-J1	−0.932	0.017	−54.84	Reject
	K1	−3.029	0.011	−280.99	Reject
	aT-J1	−2.097	0.019	−108.02	Reject
	yPr	−1.228	0.008	−145.39	Reject
yT3-J1	K1	−0.295	0.016	−18.08	Reject
	aT-J1	1.801	0.012	145.01	Reject
	yPr	−1.545	0.010	−155.96	Reject
K1	aT-J1	−0.612	0.016	−39.47	Reject
	yPr	1.484	0.014	108.69	Reject
aT3-J1	yPr	−0.317	0.010	−32.77	Reject

Note: See Table 14.1.

14.3.4 The BT Model for Standardized Data

When applying the BT model, diagonal values are excluded. This may have the following effect. If C_1 contains k_1 pixels identified as C_2, and C_2 contains k_2 pixels identified as C_1, then the BT model analyzes the binomial fraction $k_1/(k_1+k_2)$. The denominator $k_1 + k_2$, however, depends also on the total number of incorrectly identified pixels in the two classes, say n_1 for C_1 and n_2 for C_2. Indeed, suppose that the number of classified pixels doubles for one particular class. One would have the assumption that this does not affect the misclassification probability. This does not apply to the BT model as presented above. As a solution, we standardized the counts per row by defining $k = (k_1/n_1)/(k_1/n_1 + k_2/n_2)$ and $n = 1$. This has as an effect that the diagonals in the error matrix are scaled to 1, and that matrix elements not on the diagonal contain the number of misclassified pixels for each class combination. Furthermore, a large number of classified pixels should lead to lower standard deviations than a low number of classified pixels. To take this into account, a weighting was done, with weights equal to $k_1 + k_2$ being equal to the number of incorrect classifications.

Again, a generalized model is defined, with as a dependent variable the ratio k/n and the same explanatory variables as in Table 14.2. The SAS input file is given in Table 14.9. The first five columns, describing particular class combinations, are similar to those in Table 14.2, and the final five columns are described above. The variables in the model have also been described in Table 14.2. Estimated parameters $\hat{\beta}_i$, together with their standard deviations are given in Table 14.10. We note that the largest $\hat{\beta}_i$ value again occurs for the class PT (Permian and Triassic formation) and the lowest value for class K1 (lower Cretaceous basalt). Class PT therefore has the highest probability of being misclassified and class K1 has the lowest probability. Being negative, this probability is smaller than that for Penterozoic granite (yPR). In contrast to the first analysis, none of the parameters significantly differs from 0, as is shown by the relatively large standard deviations.

Subsequent calculations are carried out to compare differences in classes, as was done earlier. For example, misclassification probabilities corresponding to Table 14.5 are now in Table 14.11 and observed and expected entries in the error matrix are in Table 14.12. We now observe first that some modeled values are extremely good (such as the combinations between PT and yT3-J1) and second, some modeled values are totally different

TABLE 14.9

```
data matrix;
input PT yT3J1 K1 aT3J1 yPR k1 n1 k2 n2 w;
k = (k1/n1);
n = (k1/n1 + k2/n2);
cards;
```

−1	1	0	0	0	304	518,326	805	86,103	1,109
−1	0	1	0	0	8,772	518,326	140,464	8,123	149,236
−1	0	0	1	0	3,256	518,326	50,825	95,244	54,081
−1	0	0	0	1	13,716	518,326	23,788	31,569	37,504
0	−1	1	0	0	0	86,103	95	8,123	95
0	−1	0	1	0	2,836	86,103	2,997	95,244	5,833
0	−1	0	0	1	3,848	86,103	8,082	31,569	11,930
0	0	−1	1	0	13,645	8,123	827	95,244	14,472
0	0	−1	0	1	3,799	8,123	60	31,569	3,859
0	0	0	−1	1	2,621	95,244	21,226	31,569	23,847

```
proc genmod data = matrix;
model k/n = PT yT3J1 K1 aT3J1 yPR/NOINT DIST = BIN covb obstats;
run;
```

TABLE 14.10

Estimated Parameters $\hat{\beta}_i$ and Standard Deviations se($\hat{\beta}_i$) Using the Weighted BT Model for Standardized Data

	PT	yT3-J1	K1	aT3-J1	yPR
$\hat{\beta}_i$	4.813	1.887	−3.337	2.245	0.000
se($\hat{\beta}_i$)	5.404	4.585	6.297	3.483	0.000

TABLE 14.11

Estimated Probabilities $\hat{\Pi}_{ij}$ Using Equation 14.9 of Misclassifying C_i over C_j Using Standardized Data

	PT	yT3-J1	K1	aT3-J1	yPR
PT		0.05	0.00	0.07	0.01
yT3-J1	0.95		0.01	0.59	0.13
K1	1.00	0.99		1.00	0.97
aT3-J1	0.93	0.41	0.00		0.10
yPR	0.99	0.87	0.03	0.90	

from the observed values, in particular, negative values still emerge. These differences occur in particular in the right corner of the matrix. Testing of differences between different classes (Table 14.13) can again be carried out in a similar way, with again significance occurring in all differences, due to the very large number of pixels.

14.4 Discussion

This chapter focuses on the BT model for summarizing the error matrix. The model is applied to the error matrix, derived from a segmentation of an image using a hierarchical

TABLE 14.12

Observed and Expected Entries in the Error Matrix

	PT	yT3-J1	K1	aT3-J1	yPR
PT		304	8772	3256	13716
		(304)	(117179)		(22340)
yT3-J1	805		0	2836	3848
	(804)		(0)	(2780)	(4005)
K1	140464	95		13645	3799
	(10515)	(0)		(9342)	*
aT3-J1	50825	2997	827		2621
	*	(3057)	(1208)		*
yPR	23788	8082	60	21226	
	(14605)	(7766)	*	*	

Note: See Table 14.1; entries marked * indicate an estimate of a negative value.

TABLE 14.13

Test of Significances in Differences between Two Classes Using Standardized Data

		$\hat{\beta}_i - \hat{\beta}_j$	ASE	*t*-Ratio	H_0
PT	yT3-J1	−2.926	0.017	−172.09	Reject
	K1	−8.150	0.011	−756.06	Reject
	aT3-J1	−5.224	0.019	−269.15	Reject
	yPR	−2.568	0.008	−304.09	Reject
yT3-J1	K1	0.358	0.016	21.94	Reject
	aT3-J1	5.582	0.012	449.40	Reject
	yPR	−4.813	0.010	−485.93	Reject
K1	aT3-J1	−1.887	0.016	−121.64	Reject
	yPR	3.337	0.014	244.36	Reject
aT3-J1	yPR	−2.245	0.010	−232.07	Reject

segmentation algorithm. The image is classified on geological units. The κ-statistic, measuring the accuracy of the whole error matrix, considers the actual agreement and chance agreement, but ignores asymmetry in the matrix. In addition, the conditional κ-statistic measures the accuracy of agreement within each category, but does not consider the preference of one category over another category. These measures of accuracy only consider the agreement of classified pixels and reference pixels. In this study we extended these measures with those from the BT model to include chance agreement and disagreement. The BT model in its original form studies preference of one category over another. A pairwise comparison between classes gives additional parameters as compared to other measures of accuracy. The model also yields expected against observed values, estimated parameters, and probabilities of misclassification of one category over another category.

Using the BT model, we can determine both the agreement within a category and disagreement in relation to another category. Parameters computed from this model can be tested for statistical significance. This analysis does not take into account the categories with zero values in combination with other categories. A formal testing procedure can be implemented, using ASEs. The class parameters β_i provide a ranking of the categories. The BT model shows that a class, which is easier to recognize, is less confused with other classes.

At this stage it is difficult to say which of the two implemented BT models is most useful and appropriate for application. It appears that on the one hand the standardized model allows us to give an interpretation that is fairer and more stable in the long run, but the sometimes highly erroneous estimates of misclassification are a major drawback for application. The original, unstandardized BT model may be applicable for an error matrix as demonstrated in this study, but a large part of the available information is ignored. One reason is that the error matrix is typically different from the error matrix as applied in Ref. [11].

Positional and thematic accuracy of the reference data is crucial for a successful accuracy assessment. Often the positional and thematic errors in the reference data are unknown or are not taken into account. Vagueness in the class definition and the spatial extent of objects is not included in most accuracy assessments. To take into account uncertainty in accuracy assessment, membership values or error index values could be used.

14.5 Conclusions

We conclude that the BT model can be used for a statistical analysis of an error matrix obtained by a hierarchical classification of a remotely sensed image. This model relies on the key assumption that misclassification of one class as another is one minus the probability of misclassifying the other class as the first class. The model provides parameters and estimates for differences between classes. As such, it may serve as an extension to the κ-statistic. As this study has shown, more directed information is obtained, including a statement whether these differences are significantly different from zero.

References

1. Foody, M.G., 2002, Status of land cover classification accuracy assessment, *Rem. Sens. Environ.*, 80, 185–201.
2. Congalton, R.G., 1994, A Review of assessing the accuracy of classifications of remotely sensed data, in *Remote Sensing Thematic Accuracy Assessment: A compendium*, Fenstermaker, K.L., ed., *Am. Soc. Photogramm. Rem. Sens.*, Bethseda, pp. 73–96.
3. Agresti, A., 1996, *An Introduction to Categorical Data Analysis*, John Wiley & Sons, Inc., New York.
4. Bradley, R.A. and Terry, M.E., 1952, Rank analysis of incomplete block designs I. The method of paired comparisons, *Biometrika*, 39, 324–345.
5. Stein, A., Aryal, J., and Gort, G., 2005, Use of the Bradley–Terry model to quantify association in remotely sensed images, *IEEE Trans. Geosci. Rem. Sens.*, 43, 852–856.
6. Lillesand, T.M. and Kiefer, R.W., 2000, *Remote sensing and Image Interpretation*, 4th edition, John Wiley & Sons, Inc., New York.
7. Gordon, A.D., 1980, *Classification*, Chapman & Hall, London.
8. Janssen, L.L.F. and Gorte, B.G.H., 2001, Digital image classification, in *Principles of Remote Sensing*, L.L.F. Janssen and G.C. Huurneman, eds., 2nd edition, ITC, Enschede, pp. 73–96.
9. Richards, J.A. and Jia, X., 1999, *Remote Sensing Digital Image Analysis*, 3rd edition, Springer-Verlag, Berlin.
10. Hosmer, D.W. and Lemeshow, S., 1989, *Applied Logistic Regression*, John Wiley & Sons, Inc., New York.
11. Lucieer, L., Tsolmongerel, O., and Stein, A., 2004, Texture-based segmentation for identification of geological units in remotely sensed imagery, in *Proc. ISSDQ '04*, A.U. Frank and E. Grum, eds., pp. 117–120.
12. Lucieer, A., Stein, A., and Fisher, P., 2005, Texture-based segmentation of high-resolution remotely sensed imagery for identification of fuzzy objects, *Int. J. Rem. Sens.*, 26, 2917–2936.

15

SAR Image Classification by Support Vector Machine

Michifumi Yoshioka, Toru Fujinaka, and Sigeru Omatu

CONTENTS

15.1 Introduction

Remote sensing is the term used for observing the strength of electromagnetic radiation that is radiated or reflected from various objects on the ground level with a sensor installed in a space satellite or in an aircraft. The analysis of acquired data is an effective means to survey vast areas periodically [1]. Land map classification is one of the analyses. The land map classification classifies the surface of the Earth into categories such as water area, forests, factories, or cities. In this study, we will discuss an effective method for land map classification by using synthetic aperture radar (SAR) and support vector machine (SVM). The sensor installed in the space satellite includes an optical and a microwave sensor. SAR as an active-type microwave sensor is used for land map classification in this study. A feature of SAR is that it is not influenced by weather conditions [2–9]. As a classifier, SVM is adopted, which is known as one of the most effective methods in pattern and texture classification; texture patterns are composed of many pixels and are used as input features for SVM [10–12]. Traditionally, the maximum likelihood method has been used as a general classification technique for land map classification. However, the categories to be classified might not achieve high accuracy because the method assumes normal distribution of the data of each category. Finally, the effectiveness of our proposed method is shown by simulations.

15.2 Proposed Method

The outline of the proposed method is described here. At first, the target images from SAR are divided into an area of 8×8 pixels for the calculation of texture features. The texture features that serve as input data to the SVM are calculated using gray level co-occurrence matrix (GLCM), C_{ij}, and gray level difference matrix (GLDM), D_k. The term GLCM means the co-occurrence probability that neighbor pixels i and j become the same gray level, and GLDM means the gray level difference of neighbor pixels whose distance is k. The definitions of texture features based on GLCM and GLDM are as follows:

Energy (GLCM)

$$E = \sum_{i,j} C_{ij}^2 \tag{15.1}$$

Entropy (GLCM)

$$H = -\sum_{i,j} C_{ij} \log C_{ij} \tag{15.2}$$

Local homogeneity

$$L = \sum_{i,j} \frac{1}{1+(i-j)^2} C_{ij} \tag{15.3}$$

Inertia

$$I = \sum_{i,j} (i-j)^2 C_{ij} \tag{15.4}$$

Correlation

$$C = \sum_{i,j} \frac{(i-\mu_i)(j-\mu_j)}{\sigma_i \sigma_j} C_{ij}$$

$$\mu_i = \sum_i i \sum_j C_{ij}, \quad \mu_j = \sum_j j \sum_i C_{ij}$$

$$\sigma_i^2 = \sum_i (i-\mu_i)^2 \sum_j C_{ij}, \quad \sigma_j^2 = \sum_j (j-\mu_j)^2 \sum_i C_{ij} \tag{15.5}$$

Variance

$$V = \sum_{i,j} (i-\mu_i) C_{ij} \tag{15.6}$$

Sum average

$$S = \sum_{i,j} (i+j) C_{ij} \tag{15.7}$$

Energy (GLDM)

$$E_d = \sum_k D_k^2 \tag{15.8}$$

Entropy (GLDM)

$$H_d = -\sum_k D_k \log \{D_k\} \tag{15.9}$$

Mean

$$M = \sum_k k\, D_k \tag{15.10}$$

Difference variance

$$V = \sum_k \{k - M\}^2 D_k \tag{15.11}$$

The next step is to select effective texture features as an input to SVM as there are too many texture features to feed SVM [(7 GLCMs + 4 GLDMs) × 8 bands = totally 88 features]. Kullback–Leibler distance is adopted as the selection method of features in this study. The definition of Kullback–Leibler distance between two probability density functions $p(x)$ and $q(x)$ is as follows:

$$L = \int p(x) \log \frac{p(x)}{q(x)} \mathrm{d}x \tag{15.12}$$

Using the above as the distance measure, the distance indicated in the selected features between two categories can be compared, and the feature combinations whose distance is large are selected as input to the SVM. However, it is difficult to calculate all combinations of 88 features for computational costs. Therefore, in this study, each 5-feature combination from 88 is tested for selection.

Then the selected features are fed to the SVM for classification. The SVM classifies the data into two categories at a time. Therefore, in this study, input data are classified into two sets, that is, a set of water and cultivation areas or a set of city and factory areas in the first stage. In the second stage, these two sets are classified into two categories, respectively. In this step, it is important to reduce the learning costs of SVM since the remote sensing data from SAR are too large for learning. In this study, we propose a reduction method of SVM learning costs using the extraction of surrounding part data based on the distance in the kernel space because the boundary data of categories determine the SVM learning efficiency. The distance $d(x)$ of an element x in the kernel space from the category to which the element belongs is defined as follows using the kernel function $\Phi(\mathbf{x})$:

$$\begin{aligned} d^2(\mathbf{x}) &= \left\| \Phi(\mathbf{x}) - \frac{1}{n}\sum_{k=1}^{n} \Phi(\mathbf{x}_k) \right\|^2 \\ &= \left(\Phi(\mathbf{x}) - \frac{1}{n}\sum_{k=1}^{n} \Phi(\mathbf{x}_k) \right)^t \left(\Phi(\mathbf{x}) - \frac{1}{n}\sum_{l=1}^{n} \Phi(\mathbf{x}_l) \right) \\ &= \Phi(\mathbf{x})^t\Phi(\mathbf{x}) - \frac{1}{n}\sum_{l=1}^{n} \Phi(\mathbf{x})^t\Phi(\mathbf{x}_l) - \frac{1}{n}\sum_{k=1}^{n} \Phi(\mathbf{x}_k)^t\Phi(\mathbf{x}) + \frac{1}{n^2}\sum_{k=1}^{n}\sum_{l=1}^{n} \Phi(\mathbf{x}_k)^t\Phi(\mathbf{x}_l) \end{aligned} \tag{15.13}$$

Here, x_k denotes the elements of category, and n is the total number of elements.

Using the above distance $d(x)$, the relative distance $r_1(x)$ and $r_2(x)$ can be defined as

$$r_1(x) = \frac{d_2(x) - d_1(x)}{d_1(x)} \tag{15.14}$$

$$r_2(x) = \frac{d_1(x) - d_2(x)}{d_2(x)} \tag{15.15}$$

In these equations, $d_1(x)$ and $d_2(x)$ indicate the distance of the element x from the category 1 or 2, respectively. A half of the total data that has small relative distance is extracted and fed to the SVM. To evaluate this extraction method by comparing with the traditional method based on Mahalanobis distance, the simulation is performed using sample data 1 and 2 illustrated in Figure 15.1 through Figure 15.4, respectively. The distribution of samples 1 and 2 is Gaussian. The centers of distributions are (−0.5,0), (0.5,0) in class 1 and 2 of sample 1, and (−0.6), (0.6) in class 1, and (0,0) in class 2 of sample 2, respectively. The variances of distributions are 0.03 and 0.015, respectively. The total number of data is 500 per class. The kernel function used in this simulation is as follows:

$$K(\mathbf{x},\mathbf{x}') = \Phi(\mathbf{x})^T\Phi(\mathbf{x}') = \exp\left(-\frac{\|\mathbf{x} - \mathbf{x}'\|^2}{2\sigma^2}\right) \tag{15.16}$$

$$2\sigma^2 = 0.1$$

As a result of the simulation illustrated in Figure 15.2, Figure 15.3, Figure 15.5, and Figure 15.6, in the case of sample 1, both the proposed and the Mahalanobis-based method classify

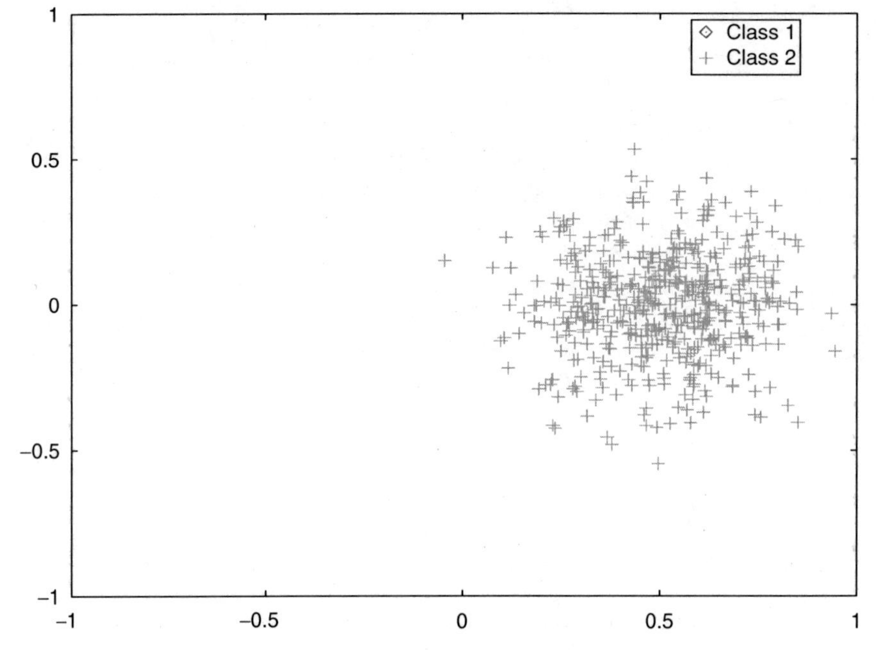

FIGURE 15.1
Sample data 1.

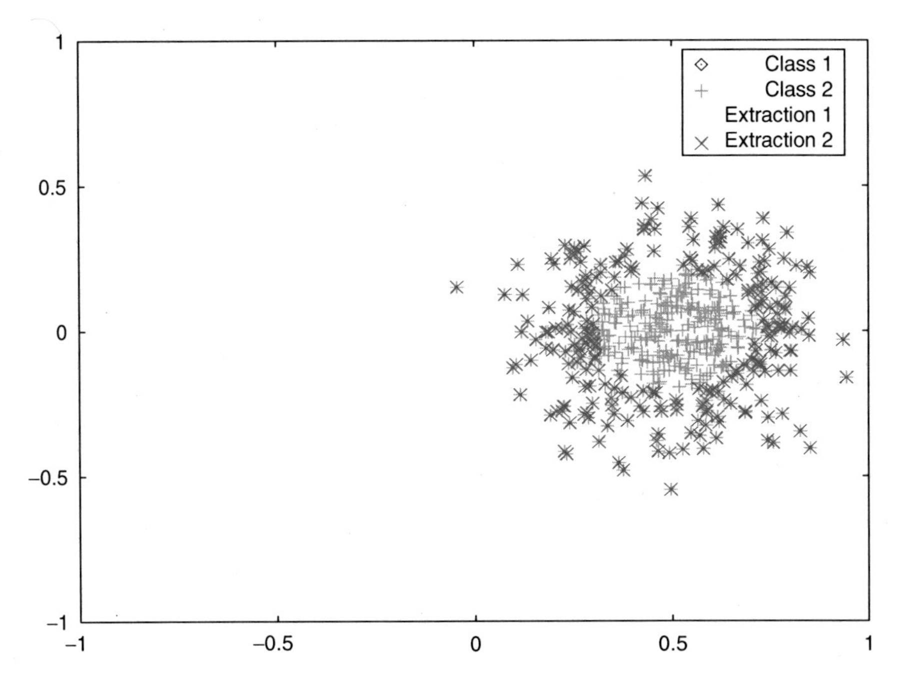

FIGURE 15.2
Extracted boundary elements by proposed method (sample 1).

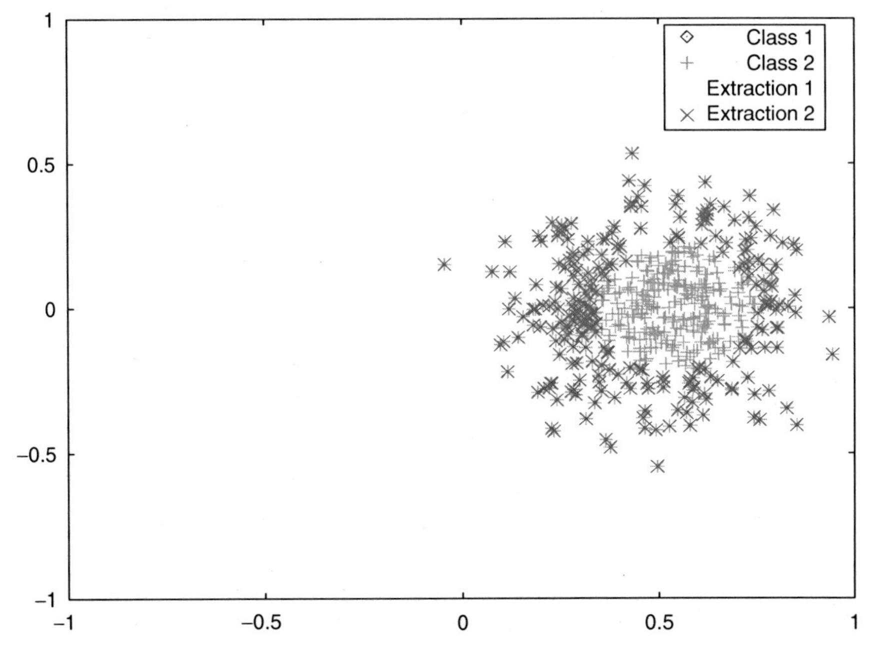

FIGURE 15.3
Extracted boundary elements by Mahalanobis distance (sample 1).

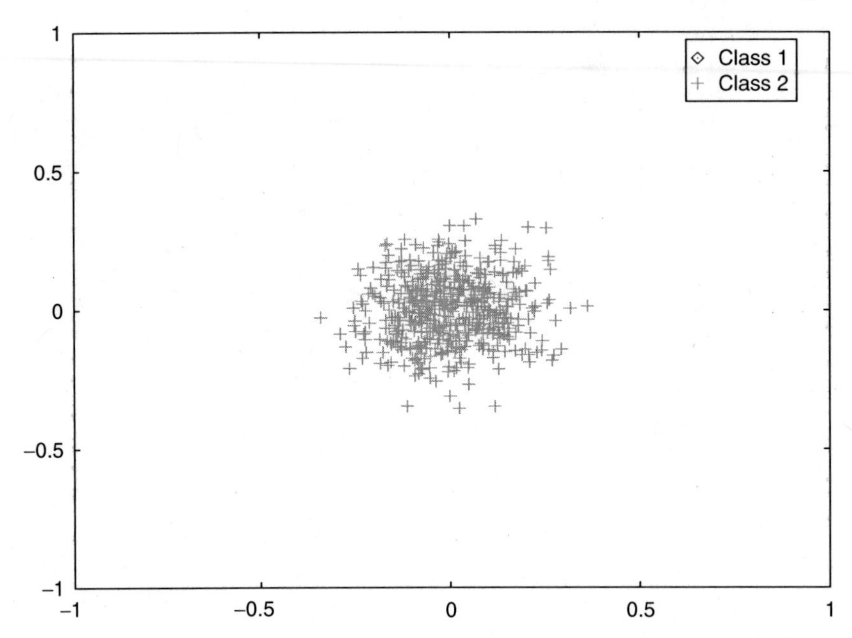

FIGURE 15.4
Sample data 2.

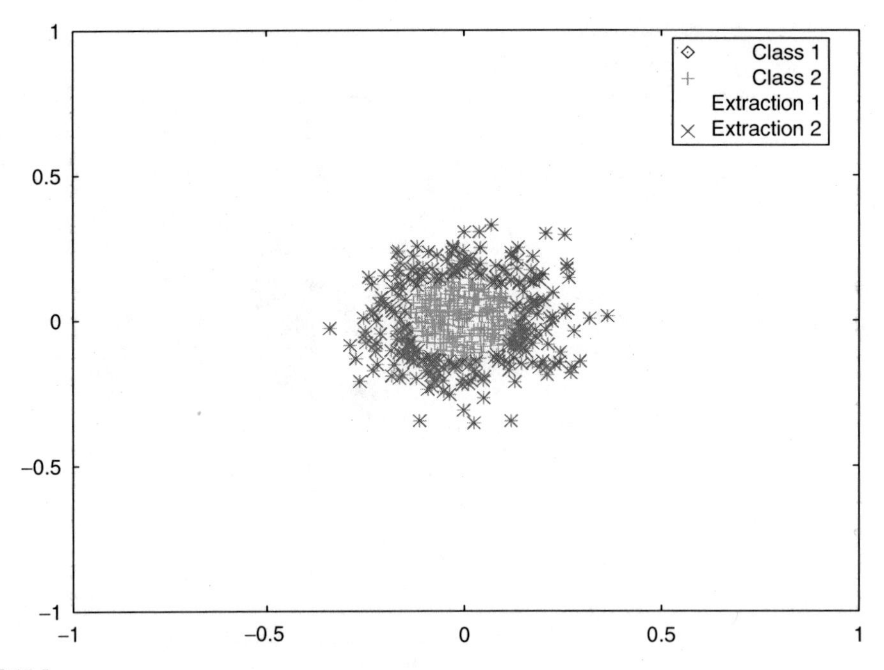

FIGURE 15.5
Extracted boundary elements by proposed method (sample 2).

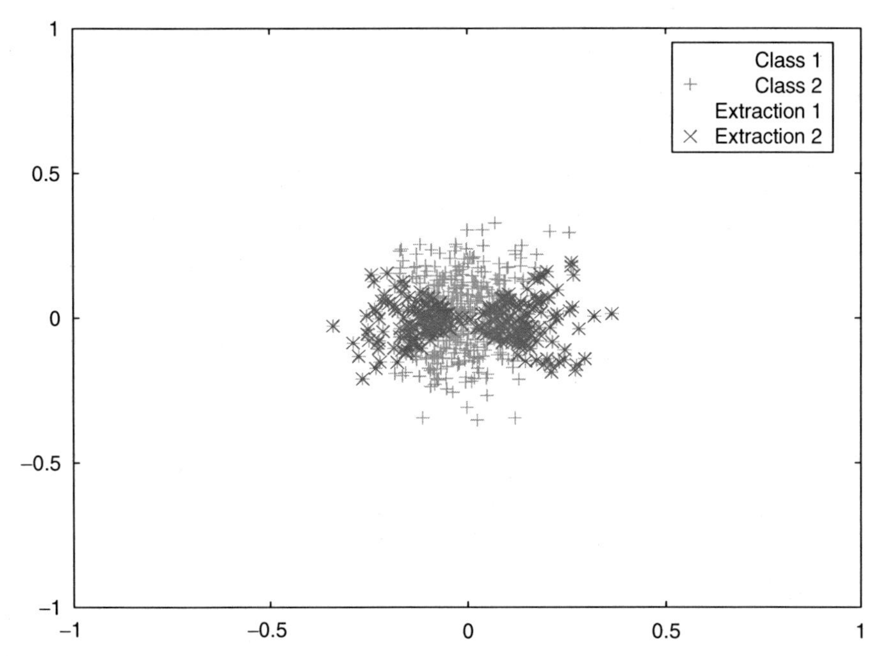

FIGURE 15.6
Extracted boundary elements by Mahalanobis distance (sample 2).

data successfully. However, in the case of sample 2, the Mahalanobis-based method fails to classify data though the proposed method succeeds. This is because Mahalanobis-based method assumes that the data distribution is spheroidal. The distance function in those methods illustrated in Figure 15.7 through Figure 15.10 clearly shows the reason for the classification ability difference of those methods.

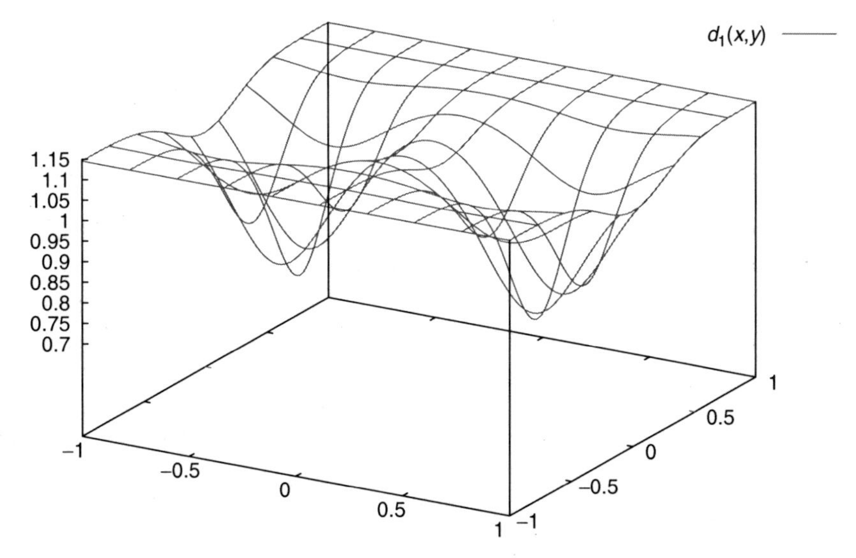

FIGURE 15.7
Distance $d_1(x)$ for class 1 in sample 2 (proposed method).

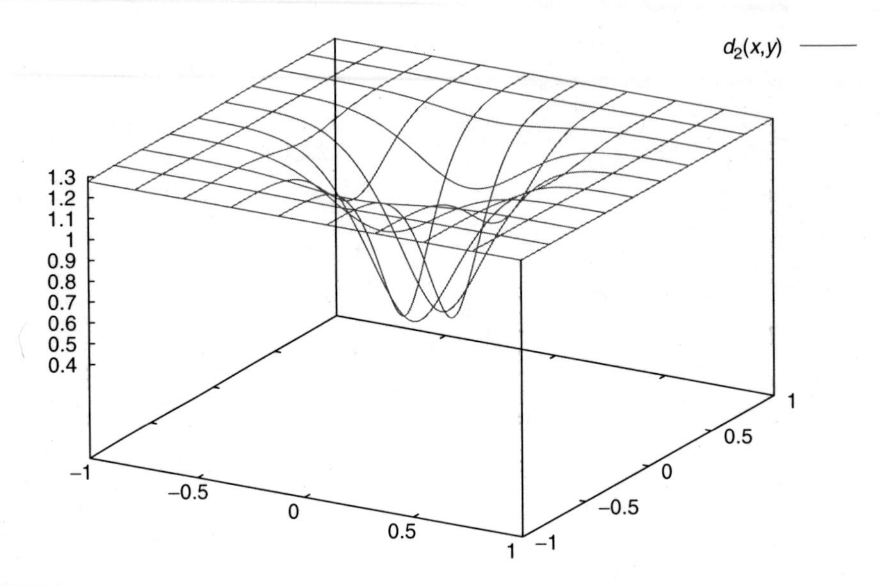

FIGURE 15.8
Distance $d_2(x)$ for class 2 in sample 2 (proposed method).

15.3 Simulation

15.3.1 Data Set and Condition for Simulations

The target data (Figure 15.11) used in this study for the classification are the observational data by SIR-C. The SIR-C device is a SAR system that consists of two wavelengths: L-band (wavelength 23 cm) and C-band (wavelength 6 cm) and four polarized electromagnetic radiations. The observed region is Kagawa Prefecture, Sakaide City, Japan (October 3,

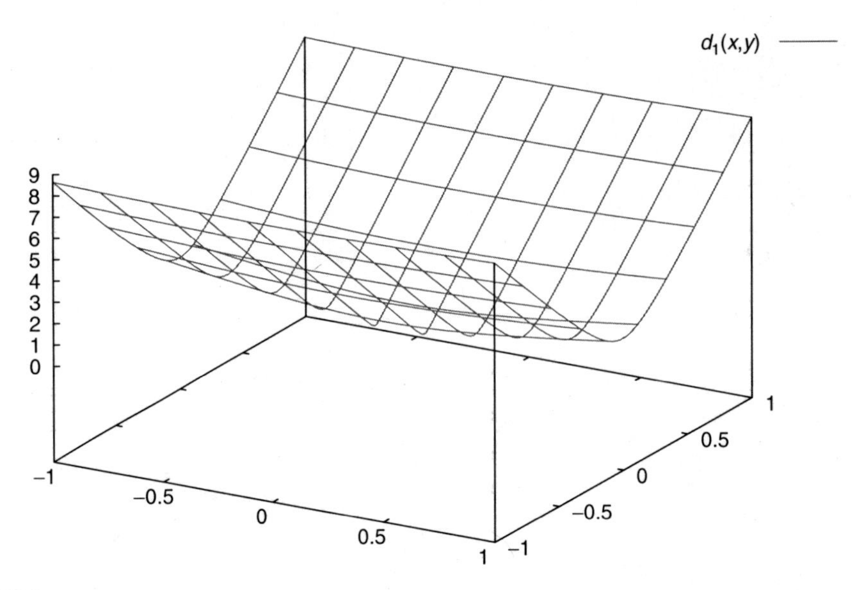

FIGURE 15.9
Distance $d_1(x)$ for class 2 in sample 2 (Mahalanobis).

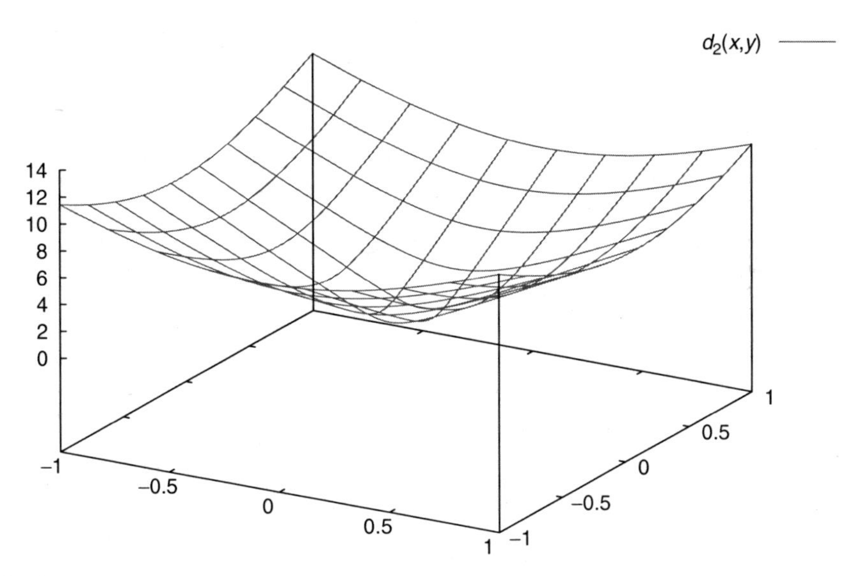

FIGURE 15.10
Distance $d_2(x)$ for class 2 in sample 2 (Mahalanobis).

1994). The image sizes are 1000 pixels in height, 696 pixels in width, and each pixel has 256 gray levels in eight bands.

To extract the texture features, the areas of 8×8 pixels on the target data are combined and classified into four categories "water area," "cultivation region," "city region," and "factory region." The mountain region is not classified because of the backscatter. The ground-truth data for training are shown in Figure 15.12 and the numbers of sample data in each category are shown in Table 15.1.

The selected texture features based on Kullback–Leibler distance mentioned in the previous section are shown in Table 15.2.

The kernel function of SVM is Gaussian kernel with the variance $\sigma^2 = 0.5$, and soft margin parameter C is 1000. The SVM training data are 100 samples randomly selected from ground-truth data for each category.

15.3.2 Simulation Results

The final result of simulation is shown in Table 15.3. In this table, "selected" implies the classification accuracy with selected texture features in Table 15.2, and "all" implies the accuracy with all 88 kinds of texture features. The final result shows the effectiveness of feature selection for improving classification accuracy.

15.3.3 Reduction of SVM Learning Cost

The learning time of SVM depends on the number of sample data. Therefore, the computational cost reduction method for SVM learning is important for complex data sets such as "city" and "factory" region in this study. Then, the reduction method proposed in the previous section is applied, and the effectiveness of this method is evaluated by comparing the learning time with traditional methods. The numbers of data are from 200 to 4000, and the learning times of the SVM classifier are measured in two cases. In the

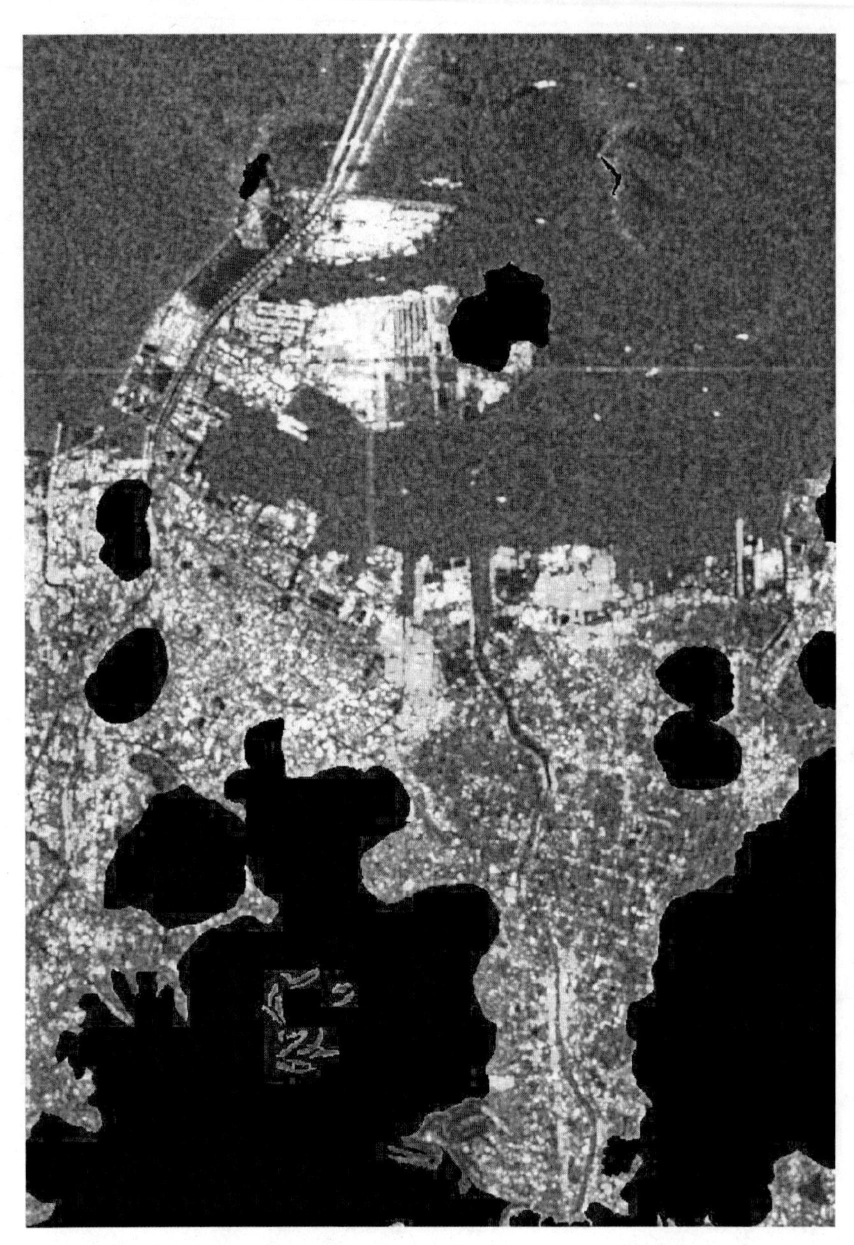

FIGURE 15.11
Target data.

first case, all data are used for learning and in the second case, by using the proposed method, data for learning are reduced by half. The selected texture features for classification are the energy (band 1), the entropy (band 6), and the local homogeneity (band 2), and the SVM kernel is Gaussian. Figure 15.13 shows the result of the simulation. The CPU of the computer used in this simulation is Pentium 4/2GHz. The result of the simulation clearly shows that the learning time is reduced by using the proposed method. The learning time is reduced to about 50% on average.

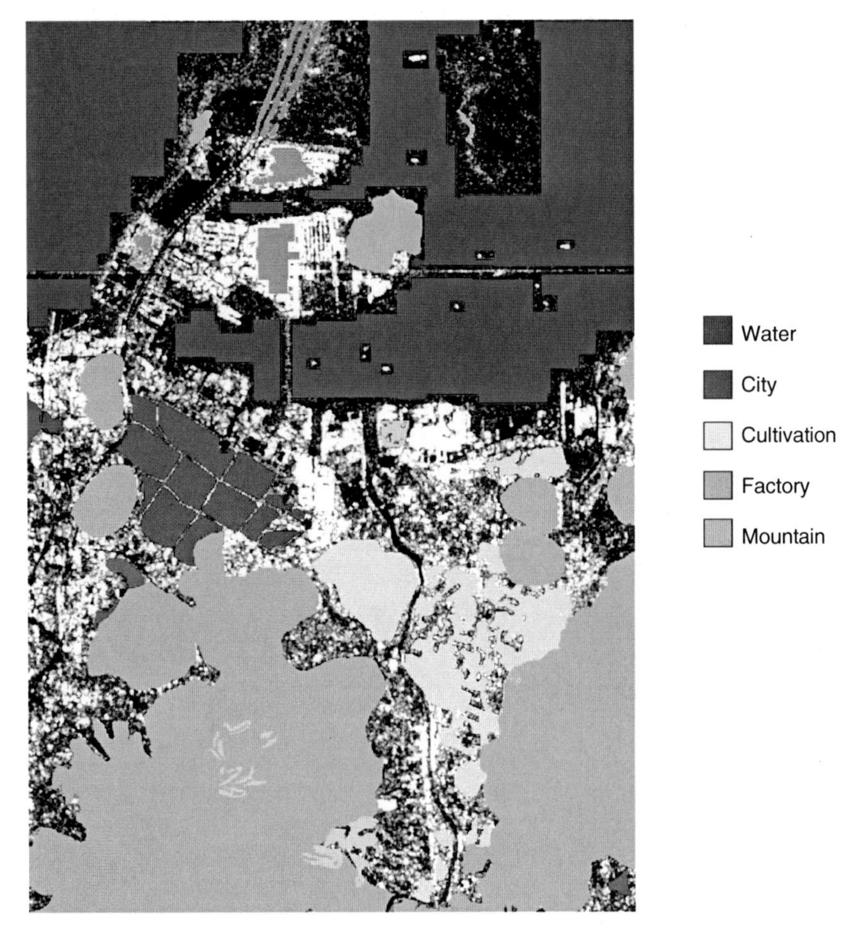

FIGURE 15.12 (See color insert following page 240.)
Ground-truth data for training.

TABLE 15.1

Number of Data

Category	Number of Data
Water	133201
Cultivation	16413
City	11378
Factory	2685

TABLE 15.2

Selected Features.

Category	Texture Features
Water, cultivation/city, factory	Correlation (1, 2), sum average (1)
Water/factory	Variance (1)
City/factory	Energy (1), entropy (6), local homogeneity (2)

Numbers in parentheses indicate SAR band

TABLE 15.3
Classification Accuracy (%)

	Water	Cultivation	City	Factory	Average
Selected	99.82	94.37	93.18	89.77	94.29
All	99.49	94.35	92.18	87.55	93.39

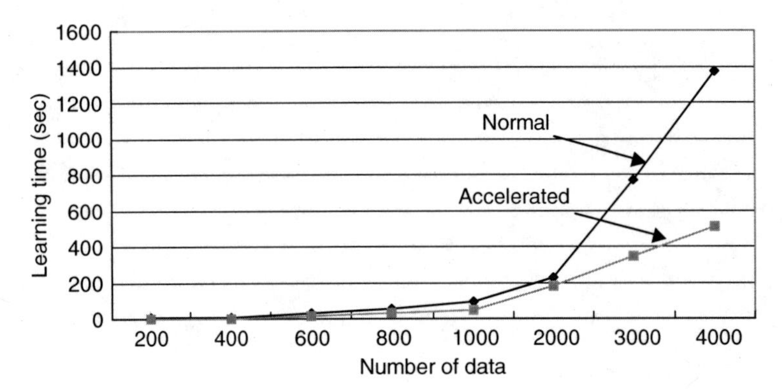

FIGURE 15.13
SVM learning Time.

15.4 Conclusions

In this chapter, we have proposed the automatic selection of texture feature combinations based on the Kullback–Leibler distance between category data distributions, and the computational cost reduction method for SVM classifier learning. As a result of simulations, by using our proposed texture feature selection method and the SVM classfier, higher classification accuracy is achieved when compared with traditional methods. In addition, it is shown that our proposed SVM learning method can be applied to more complex distributions than applying traditional methods.

References

1. Richards J.A., *Remote Sensing Digital Image Analysis*, 2nd ed., Springer-Verlag, Berlin, p. 246, 1993.
2. Hara Y., Atkins R.G., Shin R.T., Kong J.A., Yueh S.H., and Kwok R., Application of neural networks for sea ice classification in polarimetric SAR images, *IEEE Transactions on Geoscience and Remote Sensing*, 33, 740, 1995.
3. Heermann P.D. and Khazenie N., Classification of multispectral remote sensing data using a back-propagation neural network, *IEEE Transactions on Geoscience and Remote Sensing*, 30, 81, 1992.

4. Yoshida T., Omatu S., and Teranishi M., Pattern classification for remote sensing data using neural network, *Transactions of the Institute of Systems, Control and Information Engineers*, 4, 11, 1991.
5. Yoshida T. and Omatu S., Neural network approach to land cover mapping, *IEEE Transactions on Geoscience and Remote Sensing*, 32, 1103, 1994.
6. Hecht-Nielsen R., *Neurocomputing*, Addison-Wesley, New York, 1990.
7. Ulaby F.T. and Elachi C., *Radar Polarimetry for Geoscience Applications*, Artech House, Norwood, 1990.
8. Van Zyl J.J., Zebker H.A., and Elach C., Imaging radar polarization signatures: theory and observation, *Radio Science*, 22, 529, 1987.
9. Lim H.H., Swartz A.A., Yueh H.A., Kong J.A., Shin R.T., and Van Zyl J.J., Classification of earth terrain using polarimetric synthetic aperture radar images, *Journal of Geophysical Research*, 94, 7049, 1989.
10. Vapnik V.N., *The Nature of Statistical Learning Theory*, 2nd ed., Springer, New York, 1999.
11. Platt J., Sequential minimal optimization: A fast algorithm for training support vector machines, Technical Report MSR-TR-98-14, Microsoft Research, 1998.
12. Joachims T., Making large-scale SVM learning practical, In B. Schölkopf, C.J.C. Burges, and A.J. Smola, Eds., *Advanced in Kernel Method—Support Vector Learning*, MIT Press, Cambridge, MA, 1998.

16

Quality Assessment of Remote-Sensing Multi-Band Optical Images

Bruno Aiazzi, Luciano Alparone, Stefano Baronti, and Massimo Selva

CONTENTS

16.1 Introduction

Information theoretic assessment is a branch of image analysis aimed at defining and measuring the quality of digital images and is presently an open problem [1–3]. By resorting to Shannon's information theory [4], the concept of quality can be related to the information conveyed to a user by an image or, in general, by multi-band data, that is, to the mutual information between the unknown noise-free digitized signal (either radiance or reflectance in the visible-near infrared (VNIR) and short-wave infrared (SWIR) wavelengths, or irradiance in the middle infrared (MIR), thermal infrared (TIR), and far infrared (FIR) bands) and the corresponding noise-affected observed digital samples.

Accurate estimates of the *entropy* of an image source can only be obtained provided the data are uncorrelated. Hence, data decorrelation must be considered to suppress or largely reduce the correlation existing in natural images. Indeed, entropy is a measure of statistical information, that is, of uncertainty of symbols emitted by a source. Hence, any observation noise introduced by the imaging sensor results in an increment in entropy, which is accompanied by a decrement of the information content useful in application contexts.

Modeling and estimation of the noise must be preliminarily carried out [5] to quantify its contribution to the entropy of the *observed* source. Modeling of information sources is also important to assess the role played by the signal-to-noise ratio (SNR) in determining the extent to which an increment in radiometric resolution can increase the amount of information available to users.

The models that are exploited are simple, yet adequate, for describing first-order statistics of memoryless information sources and autocorrelation functions of noise processes, typically encountered in digitized raster data. The mathematical tractability of models is fundamental for deriving an information theoretic closed-form solution yielding the entropy of the noise-free signal from the entropy of the observed noisy signal and the estimated noise model parameters.

This work focuses on measuring the quality of multi-band remotely sensed digitized images. Lossless data compression is exploited to measure the information content of the data. To this purpose, extremely advanced lossless compression methods capable of attaining the ultimate compression ratio, regardless of any issues of computational complexity [6,7], are utilized. In fact, the bit rate achieved by a reversible compression process takes into account both the contribution of the "observation" noise (i.e., information regarded as statistical uncertainty, whose relevance is null to a user) and the intrinsic information of hypothetically noise-free samples.

Once the parametric model of the noise, assumed to be possibly non-Gaussian and both spatially and spectrally autocorrelated, has been preliminarily estimated, the mutual information between noise-free signal and recorded noisy signal is calculated as the difference between the entropy of the noisy signal and the entropy derived from the parametric model of the noise. Afterward, the amount of information that the digitized samples would convey if they were ideally recorded without observation noise is estimated. To this purpose, an entropy model of the source is defined. The inversion of the model yields an estimate of the information content of the noise-free source starting from the code rate and the noise model. Thus, it is possible to establish the extent to which an increment in the radiometric resolution, or equivalently in the SNR, obtained due to technological improvements of the imaging sensor can increase the amount of information that is available to the users' applications. This objective measurement of quality fits better the subjective concept of quality, that is, the capability of achieving a desired objective as the number of spectral bands increases. Practically, (mutual) information, or equivalently SNR, is the sole quality index for hyperspectral imagery, generally used for detection and identification of materials and spectral anomalies, rather than for conventional multi-spectral classification tasks.

The remainder of this chapter is organized as follows. The information theoretic fundamentals underlying the analysis procedure are reviewed in Section 16.2. Section 16.3 presents the information theoretic procedure step by step: noise model, estimation of noise parameters, source decorrelation by differential pulse code modulation (DPCM), and parametric entropy modeling of memoryless information sources via generalized Gaussian densities. Section 16.4 reports experimental results on a hyperspectral image acquired by the Airborne Visible InfraRed Imaging Spectrometer (AVIRIS) radiometer and on a test superspectral image acquired by the Advanced Spaceborne Thermal Emission and Reflection Radiometer (ASTER) imaging radiometer. Concluding remarks are drawn in Section 16.5.

16.2 Information Theoretic Problem Statement

If we consider a discrete multi-dimensional signal as an information source S, its average information content is given by its entropy $H(S)$[8]. An acquisition procedure originates an observed digitized source $\hat{S}$, whose information is the entropy $H(\hat{S})$. $H(\hat{S})$ is not adequate for measuring the amount of acquired information, since the observed source generally does not coincide with the digitized original source, mainly because of the observation noise. Furthermore, the source is not exactly band-limited by half of the sampling frequency; hence, the nonideal sampling is responsible for an additional amount of noise generated by the aliasing phenomenon. Therefore, only a fraction of the original source information is conveyed by the digitized noisy signal.

The amount of source information that is not contained in the digitized samples is measured by the conditional entropy $H(S\,|\hat{S})$ or equivocation, which is the residual uncertainty on the original source when the observed source is known. The contribution of the overall noise (i.e., *aliasing* and acquisition noise) to the entropy of the digitized source is measured by the conditional entropy $H(\hat{S}\mid S)$, which represents the uncertainty on the observed source $\hat{S}$ when the original source S is known. Therefore, the larger the acquisition noise, the larger $H(\hat{S})$, even if the amount of information of the original source that is available from the observed (noisy) source is not increased, but diminished by the presence of the noise. A suitable measure of the information content of a recorded source is instead represented by the mutual information:

$$\begin{aligned} I(S;\hat{S}) &= H(S) - H(S|\hat{S}) \\ &= H(\hat{S}) - H(\hat{S}|S) \\ &= H(S) + H(\hat{S}) - H(S,\hat{S}). \end{aligned} \tag{16.1}$$

Figure 16.1 describes the relationship existing between the entropy of the original and the recorded source and mutual information and joint entropy $H(S,\hat{S})$.

In the following sections, the procedure reported in Figure 16.2 for estimating the mutual information $I(S;\hat{S})$ and the entropy of the noise-free source $H(S)$ is described later. The estimation relies on parametric noise and source modeling that is also capable of describing non-Gaussian sources usually encountered in a number of application contexts.

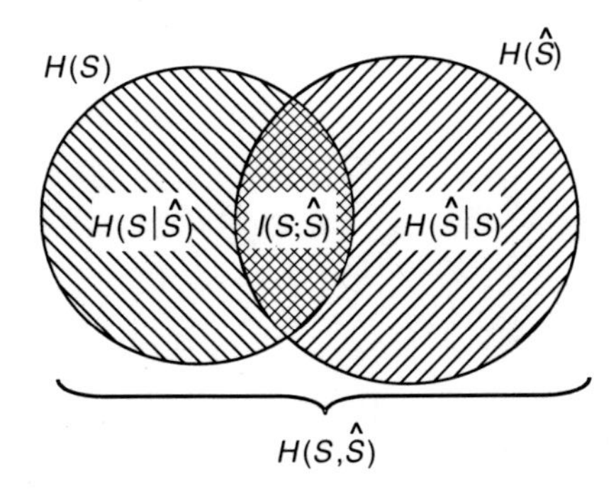

FIGURE 16.1
Relationship between entropies $H(S)$ and $H(\hat{S})$, equivocation $H(S\,|\hat{S})$, conditional entropy $H(\hat{S}|S)$, mutual information $I(S;\hat{S})$, and joint entropy $H(S,\hat{S})$.

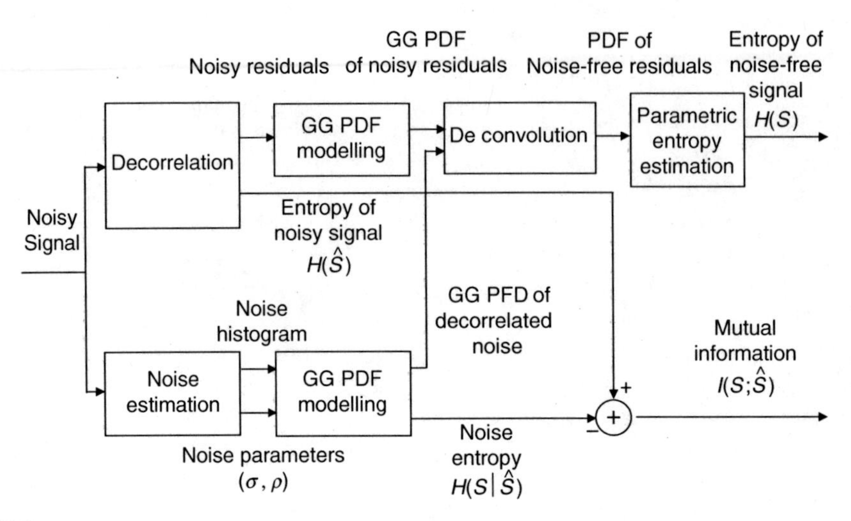

FIGURE 16.2
Flowchart of the information theoretic assessment procedure for a digital signal.

16.3 Information Assessment Procedure

16.3.1 Noise Modeling

This section focuses on modeling the noise affecting digitized observed signal samples. Unlike coherent or systematic disturbances, which may occur in some kind of data, the noise is assumed to be due to a fully stochastic process. Let us assume an additive signal-independent non-Gaussian model for the noise

$$g(i) = f(i) + n(i) \tag{16.2}$$

in which $g(i)$ is the recorded noisy signal level and $f(i)$ the noise-free signal at position (i). Both $g(i)$ and $f(i)$ are regarded as nonstationary non-Gaussian autocorrelated stochastic processes. The term $n(i)$ is a zero-mean process, independent of f, stationary and autocorrelated. Let its variance σ_n^2 and its correlation coefficient (CC) ρ be constant.

Let us assume for the stationary zero-mean noise a first-order Markov model, uniquely defined by the ρ and the σ_n^2

$$n(i) = \rho \cdot n(i-1) + \varepsilon_n(i) \tag{16.3}$$

in which $\varepsilon_n(i)$ is an uncorrelated random process having variance

$$\sigma_{\varepsilon_n}^2 = \sigma_n^2 \cdot (1 - \rho^2) \tag{16.4}$$

The variance of Equation 16.2 can be easily calculated as

$$\sigma_g^2(i) = \sigma_f^2(i) + \sigma_n^2 \tag{16.5}$$

due to the independence between signal and noise components and to the spatial stationarity of the latter. From Equation 16.3, it stems that the autocorrelation, $R_{nn}(m)$, of $n(i)$ is an exponentially decaying function of the correlation coefficient ρ:

$$R_{nn}(m) \underline{\Delta} E[n(i)n(i+m)] = \rho^{|m|}\sigma_n^2 \tag{16.6}$$

The zero-mean additive signal-independent correlated noise model (Equation 16.3) is relatively simple and mathematically tractable. Its accuracy has been validated for two-dimensional (2D) and three-dimensional (3D) signals produced by incoherent systems [3] by measuring the exponential decay of the autocorrelation function in Equation 16.6.

The noise samples $n(i)$ may be estimated on homogeneous signal segments, in which $f(i)$ is constant, by taking the difference between $g(i)$ and its average $\bar{g}(i)$ on a sliding window of length $2m+1$.

Once the CC of the noise, ρ, and the most homogeneous image pixels have been found by means of robust bivariate regression procedures [3], as described in the next section, the noise samples are estimated in the following way. If Equation 16.3 and Equation 16.6 are utilized to calculate the correlation of the noise affecting g and $\bar{g}$ on a homogeneous window, the estimated noise sample at the ith position is written as

$$\hat{n}(i) = \sqrt{\frac{(2m+1)}{(2m+1) - \left(1 + 2\rho\frac{1-\rho^m}{1-\rho}\right)}} \cdot [g(i) - \bar{g}(i)] \tag{16.7}$$

The resulting set $\{\hat{n}(i)\}$ is made available to find the noise probability density function (PDF), either empirical (histogram) or parametric, via proper modeling.

16.3.2 Estimation of Noise Variance and Correlation

To properly describe the estimation procedure, a 2D notation is adopted in this section, that is, (i, j) identifies the spatial position of the indexed entities. The standard deviation of the noisy observed band $g(i, j)$ is stated in homogeneous areas as

$$\sigma_g(i,j) = \sigma_n \tag{16.8}$$

Therefore, Equation 16.8 yields an estimate of σ_n, namely $\hat{\sigma}_n$, as the y-intercept of the horizontal regression line drawn on the scatterplot of $\hat{\sigma}_g$ versus $\hat{\mu}_g$, in which the symbol ^ denotes *estimated* values, and is calculated only on pixels belonging to homogeneous areas. Although methods based on scatterplots have been devised more than one decade ago for speckle noise assessment [9], the crucial point is the reliable identification of homogeneous areas. To overcome the drawback of a user-supervised method, an automatic procedure was developed [10] on the basis of the fact that each homogeneous area originates a cluster of scatterpoints. All these clusters are aligned along a horizontal straight line having the y-intercept equal to σ_n. Instead, the presence of signal edges and textures originates scatterpoints spread throughout the plot.

The scatterplot relative to the whole band may be regarded as the joint PDF of the *estimated* local standard deviation to the *estimated* local mean. In the absence of any signal textures, the image is made up by uniform noisy patches; by assuming that the noise is stationary, the PDF is given by the superposition of as many unimodal distributions as the patches. Because the noise is independent of the signal, the measured variance does not depend on the underlying mean. Thus, all the above expectations are aligned along a horizontal line.

The presence of textured areas modifies the "flaps" of the PDF, which still exhibit aligned modes, or possibly a watershed. The idea is to threshold the PDF to identify a number of points belonging to the most homogeneous image areas, large enough to yield a statistically consistent estimate and small enough to avoid comprising signal textures that, even if weak, might introduce a bias by excess.

Now let us calculate the space-varying (auto)covariance of unity lag along either of the coordinate directions, say i and j,

$$C_g(i, j;1, 0) \underline{\underline{\Delta}} E\{[g(i, j) - E(g(i, j))] \cdot [g(i+1, j) - E(g(i+1, j))]\} = C_f(i, j;1, 0) + \rho_x \cdot \sigma_n^2 \tag{16.9}$$

The term $C_f(i, j;1, 0)$ on right-hand side of Equation 16.9 is identically zero in homogeneous areas in which $\sigma_g(i, j)$ becomes equal to σ_n. Thus, Equation 16.9 becomes

$$C_g(i, j;1, 0) = \rho_x \cdot \sigma_n^2 = \rho_x \cdot \sigma_g(i, j) \cdot \sigma_g(i+1, j) \tag{16.10}$$

Hence, ρ_x, and analogously ρ_y, is estimated from those points, lying on the covariance-to-variance scatterplots, corresponding to homogeneous areas.

To avoid calculating the PDF, the following procedure was devised:

- Within a $(2m + 1) \times (2m + 1)$ window, sliding over the image, calculate the local statistics of the noisy image to estimate its space-varying ensemble statistics:
  - average $\bar{g}(i, j) \equiv \hat{\mu}_g(i, j)$

$$\bar{g}(i, j) = \frac{1}{(2m+1)^2} \sum_{k=-m}^{m} \sum_{l=-m}^{m} g(i+k, j+l) \tag{16.11}$$

  - RMS deviation from the average, $\hat{\sigma}_g(i, j)$, with

$$\hat{\sigma}_g^2(i, j) = \frac{1}{(2m+1)^2 - 1} \sum_{k=-m}^{m} \sum_{l=-m}^{m} [g(i+k, j+l) - \bar{g}(i, j)]^2 \tag{16.12}$$

  - cross deviation from the average $\hat{C}_g(i, j; 1, 0)$, given by

$$\hat{C}_g(i, j;1, 0) = \frac{1}{(2m+1)^2 - 1} \sum_{k=-m}^{m} \sum_{l=-m}^{m} [g(i+k, j+l) - \bar{g}(i, j)] \times [g(i+k+1, j+l) - \bar{g}(i+1, j)] \tag{16.13}$$

- Draw scatterplots either of $\hat{\sigma}_g(i, j)$ versus $\hat{\mu}_g(i, j)$ to estimate σ_n or of $\hat{C}_g(i, j; 1, 0)$ versus $\hat{\sigma}_g(i, j) \cdot \hat{\sigma}_g(i + 1, j)$ to estimate ρ_x.
- Partition the scatterplot planes into an $L \times L$ array of rectangular blocks.
- Sort and label such blocks by decreasing population, that is, the number of scatter points: if $C(\cdot)$ denotes the cardinality of a set, then $C(B^{(k)}) \geq C(B^{(k+1)})$, $k = 1, \ldots, L^2$.
- For each scatterplot, calculate a succession of horizontal regression lines, that is $\{\hat{\sigma}_n(k)\}$, from the set of scatter points $\{\mathbf{p} \mid \mathbf{p} \in \bigcup_{l=1}^{k} B^{(l)}\}$.

The succession attains a steady value of the parameter after a number of terms that depend on the actual percentage of homogeneous points. The size of the partition and the stop criterion are noncrucial; however, a 100×100 array of blocks and the stop criterion is applied after processing 2–10% of points, depending on the degree of heterogeneity of the scene, are usually setup. The size of the processing window, that is, $(2m + 1) \times (2m + 1)$, is noncrucial if the noise is white. Otherwise, a size 7×7 to 11×11 is recommended, because too small a size will underestimate the covariance.

It is noteworthy that, unlike most of the methods that rely on the assumption of *white* noise, the scatterplot method, which is easily adjustable to deal with signal-dependent noise, can also accurately measure correlated noise, and is thus preferred in this context.

To show an example of the estimation procedure, Figure 16.3a portrays band 25 extracted from the sequence acquired on the *Cuprite Mine* test site in 1997 by the AVIRIS instrument. On this image, homogeneous areas that contribute to the estimation procedure have been automatically extracted (Figure 16.3b). The scatterplot of local standard deviation to local mean is plotted in Figure 16.3c. Homogeneous areas cluster and determine the regression line whose y-intercept is the estimated standard deviation of noise. Eventually, the scatterplot of local one-lag covariance to local variance is reported in Figure 16.3d; the slope of the regression line represents the estimated correlation coefficient.

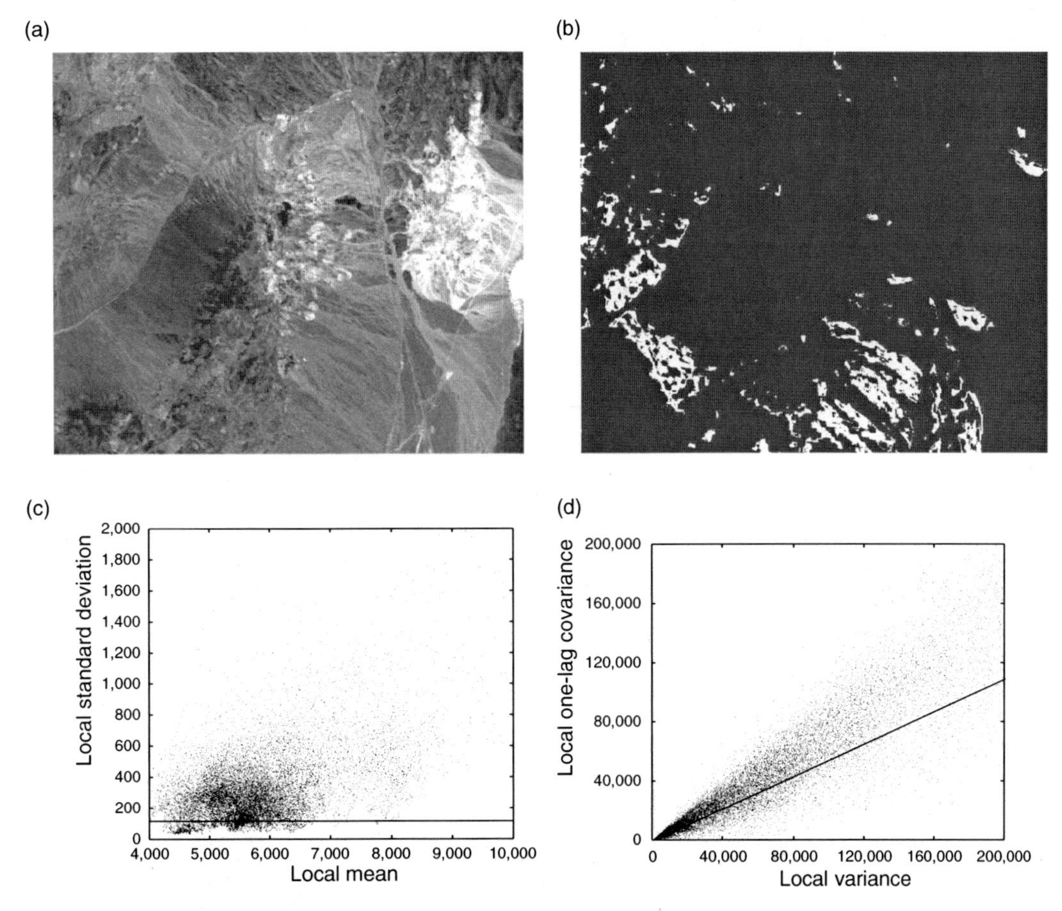

FIGURE 16.3
Estimation of noise parameters: (a) AVIRIS band 25 extracted from the sequence of Cuprite Mine test site acquired in 1997; (b) map of homogeneous areas that contribute to the estimation; (c) scatterplot of local standard deviation to local mean; the y-intercept of the regression line is the estimated standard deviation of noise; (d) scatterplot of local one-lag covariance to local variance; the slope of the regression line is the estimated correlation coefficient.

16.3.3 Source Decorrelation via DPCM

Differential pulse code modulation is usually employed for *reversible* data compression. DPCM basically consists of a prediction stage followed by entropy coding of the resulting prediction errors. For the sake of clarity, we develop the analysis for a 1D *fixed* DPCM and extend its results to the case of 2D and 3D *adaptive* prediction [6,7,11,12].

Let $\hat{g}(i)$ denote the prediction at pixel i obtained as a linear regression of the values of P previous pixels

$$\hat{g}(i) = \sum_{j=1}^{P} \phi(j) \cdot g(i-j) \tag{16.14}$$

in which $\{\phi(j),\ j = 1,\ldots, P\}$ are the coefficients of the linear predictor and constant throughout the image.

By replacing the additive noise model, one obtains

$$\hat{g}(i) = \hat{f}(i) + \sum_{j=1}^{P} \phi(j) \cdot n(i-j) \tag{16.15}$$

in which

$$\hat{f}(i) = \sum_{j=1}^{P} \phi(j) \cdot f(i-j) \tag{16.16}$$

represents the prediction for the noise-free signal expressed from its previous samples. Prediction errors of g are

$$e_g(i) \triangleq g(i) - \hat{g}(i) = e_f(i) + n(i) - \sum_{j=1}^{P} \phi(j) \cdot n(i-j) \tag{16.17}$$

in which $e_f(i) \triangleq f(i) - \hat{f}(i)$ is the error the predictor would produce starting from noise-free data. Both $e_g(i)$ and $e_f(i)$ are zero-mean processes, uncorrelated, and nonstationary. The zero-mean property stems from an assumption of local first-order stationarity, within the $(P + 1)$-pixel window comprising the current pixel and its prediction support.

Equation 16.17 is written as

$$e_g(i) = e_f(i) + e_n(i) \tag{16.18}$$

in which

$$e_n(i) \triangleq n(i) - \hat{n}(i) = n(i) - \sum_{j=1}^{P} \phi(j) \cdot n(i-j) \tag{16.19}$$

is the error produced when the correlated noise is predicted. The term $e_n(i)$ is assumed to be zero mean, stationary, and independent of $e_f(i)$, because f and n are assumed to be independent of each other. Thus, the relationship among the variances of the three types of prediction errors becomes

$$\sigma_{e_g}^2(i) = \sigma_{e_f}^2(i) + \sigma_{e_n}^2 \tag{16.20}$$

From the noise model in Equation 16.3, it is easily noticed that the term $\sigma_{e_n}^2$ is lower bounded by $\sigma_{\varepsilon_n}^2$, which means that $\sigma_{e_n}^2 \geq \sigma_n^2 \cdot (1 - \rho^2)$. The optimum MMSE predictor for a first-order Markov model, like in Equation 16.3, is $\phi(1) = \rho$ and $\phi(j) = 0, j = 2, \ldots, P$; it yields $\sigma_{e_n}^2 = \sigma_n^2 \cdot (1 - \rho^2) = \sigma_{\varepsilon_n}^2$, which can easily be verified. Thus, the residual variance of the noise after decorrelation may be approximated from the estimated variance of the correlated noise, that is, $\hat{\sigma}_n^2$, and from its estimated CC, $\hat{\rho}$, as

$$\sigma_{e_n}^2 \cong \hat{\sigma}_n^2 \cdot (1 - \hat{\rho}^2) \tag{16.21}$$

The approximation is more accurate as the predictor attains the optimal MMSE performance.

16.3.4 Entropy Modeling

Given a stationary memoryless source S, uniquely defined by its PDF, $p(x)$, having zero mean and variance σ^2, linearly quantized with a step size Δ, the minimum bit rate needed to encode one of its symbols is [13]

$$R \cong h(S) - \log_2 \Delta \tag{16.22}$$

in which $h(S)$ is the differential entropy of S defined as

$$h(S) = -\int_{-\infty}^{\infty} p(x) \log_2 p(x)\, dx = \frac{1}{2} \log_2(c \cdot \sigma^2) \tag{16.23}$$

where $0 < c \leq 2\pi e$ is a positive constant accounting for the shape of the PDF and attaining its maximum for a Gaussian function. Such a constant is referred in the following as the *entropy factor*. The approximation in Equation 16.22 holds for $\sigma \gg \Delta$, but is still acceptable for $\sigma > \Delta$ [14].

Now, the minimum average bit rate R_g necessary to reversibly encode an integer-valued sample of g, approximated as in Equation 16.22, in which prediction errors are regarded as an uncorrelated source $G \equiv \{e_g(i)\}$ and are linearly quantized with a step size $\Delta = 1$

$$R_g \cong h(G) = \frac{1}{2} \log_2(c_g \cdot \bar{\sigma}_{e_g}^2) \tag{16.24}$$

in which $\bar{\sigma}_{e_g}^2$ is the *average* variance of $e_g(i)$. By averaging Equation 16.20 and replacing it in Equation 16.24, R_g may be written as

$$R_g = \frac{1}{2} \log_2 [c_g \cdot (\bar{\sigma}_{e_f}^2 + \sigma_{e_n}^2)] \tag{16.25}$$

where $\bar{\sigma}_{e_f}{}^2$ is the *average* variance of $\sigma_{e_f}{}^2(i)$. If $\sigma_{e_f}{}^2 = 0$, then Equation 16.25 reduces to

$$R_g \equiv R_n = \frac{1}{2} \log_2(c_n \cdot \sigma_{e_n}^2) \tag{16.26}$$

in which $c_n = 2\pi e$ is the entropy factor of the PDF of e_n, if n is Gaussian. Analogously, if $\sigma_{e_n}^2 = 0$, then Equation 16.25 becomes

$$R_g \equiv R_f = \frac{1}{2}\log_2(c_f \cdot \bar{\sigma}_{e_f}^2) \tag{16.27}$$

in which $c_f \leq 2\pi e$ is the entropy factor of prediction errors of the noise-free image, which are generally non-Gaussian.

The average entropy of the noise-free signal f in the case of correlated noise is given by replacing Equation 16.20 and Equation 16.21 in Equation 16.27 to yield

$$R_f = \frac{1}{2}\log_2\{c_f \cdot [\bar{\sigma}_{e_g}^2 - (1-\rho^2)\cdot\sigma_n^2]\} \tag{16.28}$$

Since $\bar{\sigma}_{e_g}^2$ can be measured during compression process by averaging $\sigma_{e_g}^2$, c_f is the only unknown parameter whose estimation is crucial for the accuracy of R_f. The determination of c_f can be performed by modeling the PDF of e_f through the PDF of e_g and e_n. The generalized Gaussian density (GGD) model, which can properly represent these PDFs, is described in the next section.

16.3.5 Generalized Gaussian PDF

A model suitable for describing unimodal non-Gaussian amplitude distributions is achieved by varying the parameters ν (shape factor) and σ (standard deviation) of the GGD [15], which is defined as

$$p_{GG}(x) = \left[\frac{\nu \cdot \eta(\nu,\sigma)}{2\cdot\Gamma(1/\nu)}\right]\exp\{-[\eta(\nu,\sigma)\cdot|x|]^\nu\} \tag{16.29}$$

where

$$\eta(\nu,\sigma) = \frac{1}{\sigma}\left[\frac{\Gamma(3/\nu)}{\Gamma(1/\nu)}\right]^{1/2} \tag{16.30}$$

and $\Gamma(\cdot)$ is the well-known Gamma function, that is, $\Gamma(z) = \int_0^\infty t^{z-1}e^{-t}\,dt$, $z > 0$. Since $\Gamma(n) = (n-1)!$, when $\nu = 1$, a Laplacian law is obtained; but, $\nu = 2$ yields a Gaussian distribution. As limit cases, for $\nu \to 0$, $p_{GG}(x)$ becomes an impulse function, yet has extremely heavy tails and thus a nonzero σ^2 variance, whereas for $\nu \to \infty$, $p_{GG}(x)$ approaches a uniform distribution having variance σ^2 as well. The shape parameter ν rules the exponential rate of decay: the larger the ν, the flatter the PDF; the smaller the ν, the more peaked the PDF. Figure 16.4a shows the trend of the GG function for different values of ν.

The matching between a GGD and the empirical data distribution can be obtained following a maximum likelihood (ML) approach [16]. Such an approach has the disadvantage of a cumbersome numerical solution. Some effective methods, which are suitable for real-time applications, have been developed. They are based on fitting a parametric function (*shape function*) of the modeled source to statistics calculated from the observed data [17–19]. In Figure 16.4b, the kurtosis [19], Mallat's [20] and entropy matching [18] shape functions are plotted as a function of the shape factor ν for a unity variance GG. In the experiments carried out in this work, the method briefly introduced by Mallat [20] and then developed in greater detail by Sharifi and Leon-Garcia [17] has been adopted.

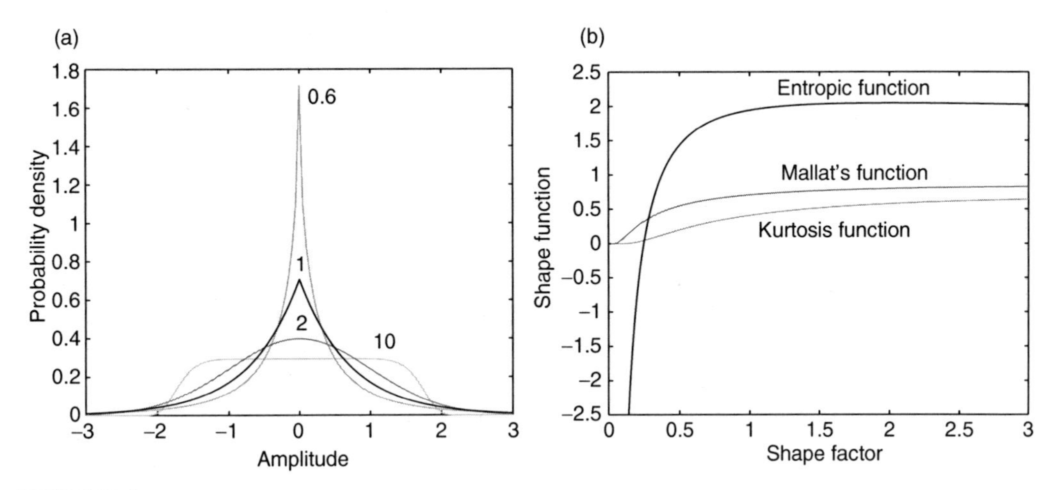

FIGURE 16.4
(a) Unity-variance GG density plotted for several ν's; (b) shape functions of a unity-variance GG PDF as a function of the shape factor ν.

16.3.6 Information Theoretic Assessment

Let us assume that the real-valued $e_g(i)$ may be modeled as a GGD. From Equation 16.24, the entropy function is

$$\frac{1}{2}\log_2(c_g) = R_g - \log_2(\bar{\sigma}_{e_g}) = F_H(\nu_{e_g}) \tag{16.31}$$

in which ν_{e_g} is the shape factor of $e_g(i)$, the average rate of which, R_g, has been set equal to the entropy H of the discrete source. The ν_{e_g} is found by inverting either the entropy function $F_H(\nu_{e_g})$ or any other shape function; in that case, the e_g(i) produced by DPCM, instead of its variance, is directly used. Eventually, the parametric PDF of the uncorrelated observed source e_g(i) is available.

The term $e_g(i)$ is obtained by adding a sample of white non-Gaussian noise of variance $\sigma_{e_n}^2$, approximately equal to $(1-\rho^2)\cdot\sigma_n^2$, to a sample of the noise-free uncorrelated non-Gaussian signal $e_f(i)$. Furthermore, $e_f(i)$ and $e_n(i)$ are independent of each other.

Therefore, the GG PDF of e_g previously found is given by the linear convolution of the unknown $p_{e_f}(x)$ with a GG PDF having variance $\sigma_{e_n}^2$ and shape factor ν_{e_n}. By assuming that the PDF of the noise-free residues, $p_{e_f}(x)$, is GG as well, its shape factor ν_{e_f} can be obtained starting from the forward relationship

$$p_{\mathrm{GG}}[\bar{\sigma}_{e_g}, \nu_{e_g}](x) = p_{\mathrm{GG}}[\sqrt{\bar{\sigma}_{e_g}^2 - \sigma_{e_n}^2}, \nu_{e_f}](x) \otimes p_{\mathrm{GG}}[\sigma_{e_n}, \nu_{e_n}](x) \tag{16.32}$$

by deconvolving the PDF of noise residue from that of noisy signal residue.

In a practical implementation, the estimated value of ν_{e_f} is found such that the direct convolution at the right-hand side of Equation 16.32 yields a GGD, whose shape factor matches ν_{e_g} as much as possible.

Eventually, the estimated shape factor $\hat{\nu}_{e_f}$ is used to determine the entropy function

$$\frac{1}{2}\log_2(c_f) = F_H(\hat{\nu}_{e_f}) \tag{16.33}$$

which is replaced in Equation 16.28 to yield the entropy of the noise-free signal $R_f \equiv H(S)$. Figure 16.2 summarizes the overall procedure. The mutual information is simply given as the difference between the rate of the decorrelated source and the rate of the noise.

The extension of the procedure to 2D and 3D signals, that is, to digital images and sequences of digital images, is straightforward. In the former case, 2D prediction is used to find e_g, and two correlation coefficients, ρ_x and ρ_y, are to be estimated for the noise, since its variance after decorrelation is approximated as $\sigma_n^2(1-\rho_x^2)(1-\rho_y^2)$, by assuming a separable 2D Markov model. Analogously, Equation 16.7 that defines the estimated value of a sample of correlated noise is extended as

$$\hat{n}(i,j) = \sqrt{\frac{N^2}{(N^2 - \left(1+2\rho_x \frac{1-\rho_x^m}{1-\rho_x}\right)\left(1+2\rho_y \frac{1-\rho_y^m}{1-\rho_y}\right)}} \cdot [g(i,j) - \bar{g}(i,j)] \tag{16.34}$$

in which $N = 2m + 1$ is the length of the side of the square window on which the average $\bar{g}(i,j)$ is calculated.

The 3D extension is more critical because a sequence of images may have noise variances and spatial CCs different for each image. Moreover, it is often desirable to estimate entropy and mutual information of the individual images of the sequence. Therefore, each image is decorrelated both spatially and along the third dimension by using a 3D prediction.

16.4 Experimental Results

The results presented in this section refer to AVIRIS and ASTER sensors. Such sensors can be considered as representatives of hyperspectral and superspectral instruments, respectively. Apart from specific design characteristics (e.g., only ASTER has TIR channels), their main difference for users consists in the spectral resolution that is particularly high (10 nm) for AVIRIS, thus yielding the acquisition of a consistent number of spectral bands (224). An obvious question would be to assess to what extent such a high spectral resolution and large number of bands may correspond to an increase in information content. Although it is not possible to answer this and it would require the definition of an ideal experiment in which the same scene would be acquired with two ideal sensors differing in spectral resolution only, some careful consideration is, however, possible by referring to the mutual information measured for the two sensors.

16.4.1 AVIRIS Hyperspectral Data

Hyperspectral imaging sensors provide a thorough description of a scene in a quasicontinuous range of wavelengths [21]. A huge amount of data is produced for each scene: problems may arise for transmission, storage, and processing. In several applications, the whole set of data is redundant and therefore it is difficult to seek the intrinsically embedded information [22]. Hence, a representation of the data that allows essential information to be condensed in a few basic components is desirable to expedite scene analysis [23]. The challenging task of hypervariate data analysis may be alleviated by resorting to information theoretic methods capable of quantifying the information content of the data that are likely to be useful in application contexts [3,24].

The proposed information theoretic procedure was run on a hyperspectral image having 224 spectral bands. The image was acquired in 1997 by the AVIRIS operated by NASA/JPL on the *Cuprite Mine* test site in Nevada.

AVIRIS sequences are constituted by 224 bands recorded at different wavelengths in the range 380–2500 nm, with a spectral separation between two bands of 10 nm nominally. The sensor acquires images pixel by pixel (whisk-broom), recording the spectrum of each pixel. The size of an image is 614 pixels in the across-track direction, while the size in the along-track direction is variable and limited only by the onboard mass storage capacity. The instantaneous field of view (IFOV) is about 1 mrad, which means that, at the operating height of 20 km, the spatial resolution is about 20 m. The sequence was acquired by the sensor with the 12-bit analog-to-digital converter (ADC) introduced in 1995 in substitution of the 10-bit ADC originally designed in 1987. All the data are radiometrically calibrated and expressed as radiance values (i.e., power per surface, solid angle, and wavelength units). The AVIRIS system features four distinct spectrometers capable of operating in the visible (VIS) (380–720 nm), NIR (730–1100 nm), and in the first and second part of the SWIR interval (1110–2500). The three transition regions of the spectrum are covered by two adjacent spectrometers, whose bands are partially overlapped.

The *Cuprite Mine* image for which results are reported is the fourth 614 × 512 portion of the full image composed of four consecutive 614 × 512 frames. A sample of six bands, four acquired in the blue, green, red, and NIR wavelengths, two in the SWIR wavelengths, is shown in Figure 16.5.

The first step of the procedure concerns noise estimation of all 224 bands. Across-track and along-track CCs of the noise, that is, ρ_x and ρ_y, are plotted against band number in

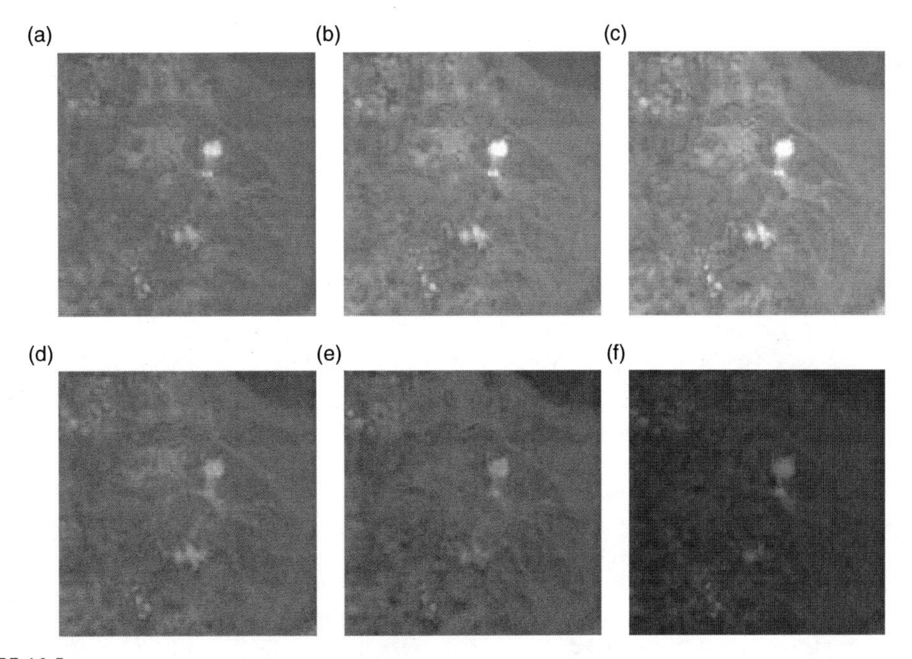

FIGURE 16.5
Sample bands from *Cuprite Mine* AVIRIS hyperspectral image (128 × 128, detail): (a) blue wavelength (band 8); (b) green (band 13); (c) red (band 23); (d) near infrared (band 43); (e) lower short-wave infrared (band 137); (f) upper short-wave infrared (band 181).

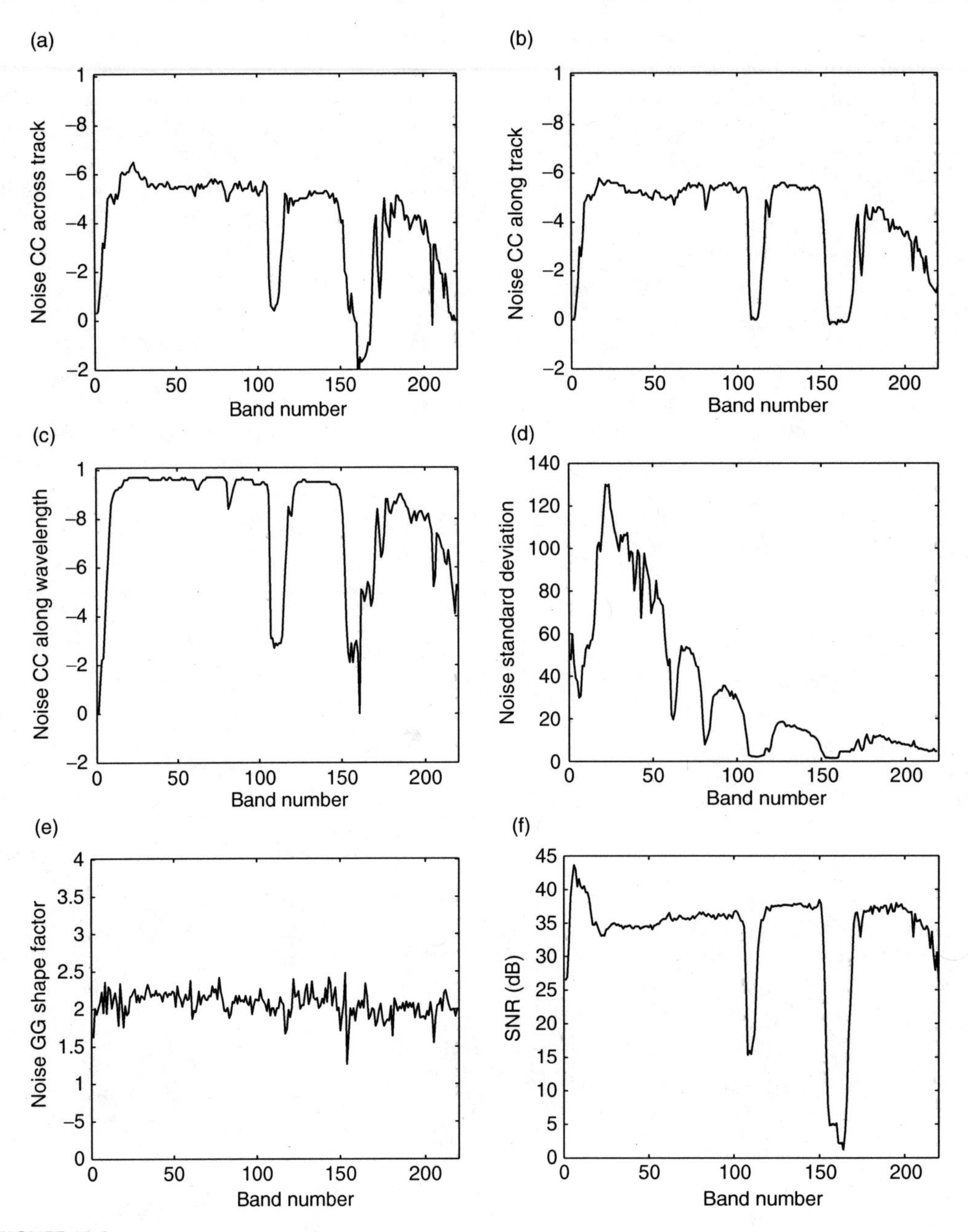

FIGURE 16.6
Noise parameters of test AVIRIS hyperspectral image plotted against band number: (a) CC across track, ρ_x; (b) CC along track, ρ_y; (c) CC along spectral direction, ρ_λ; (d) square root of noise variance, σ_n; (e) shape factor of GG-modeled noise PDF, ν_n (average 2.074); (f) SNR.

Figure 16.6a and Figure 16.6b, respectively. The CCs exhibit similar values (averages of 0.403 versus 0.404), showing that the data have been properly preprocessed to eliminate any striping effect possibly caused by the scanning mechanism that produces the images line by line following the motion of the airborne platform (track). Apart from the transition between the first and the second spectrometers, marked losses of correlation are noticed around changes between the other spectrometers. The spectral CC of the noise

(Figure 16.6c) seems rather singular: apart from the absorption bands, all the spectrometers exhibit extremely high values (average of 0.798), probably originated by preprocessing along the spectral direction aimed at mitigating structured noise patterns occurring in the raw data [3]. The measured standard deviation of correlated noise is drawn in Figure 16.6d; the noise follows the typical spectral distribution provided by JPL as NER [14]; the average value of its standard deviation is 32.86. Figure 16.6e shows that the noise of AVIRIS data is Gaussian with good approximation (average shape factor 2.074), as assumed in earlier works [25,3], with the noise shape factor rather stable around 2. Eventually, Figure 16.6f demonstrates that the SNR is almost uniform, varying with the wavelength, apart from the absorption bands, and close to values greater than 35 dB (average of 33.03 dB). This feature is essential for the analysis of spectral pixels, for which a uniform SNR is desirable.

Once all noise parameters had been estimated, the information theoretic assessment procedure was run on the AVIRIS data set. Figure 16.7 reports plots of information parameters varying with band number (not exactly corresponding to wavelength because every couple of adjacent spectrometers yields duplicate bands at the same wavelengths).

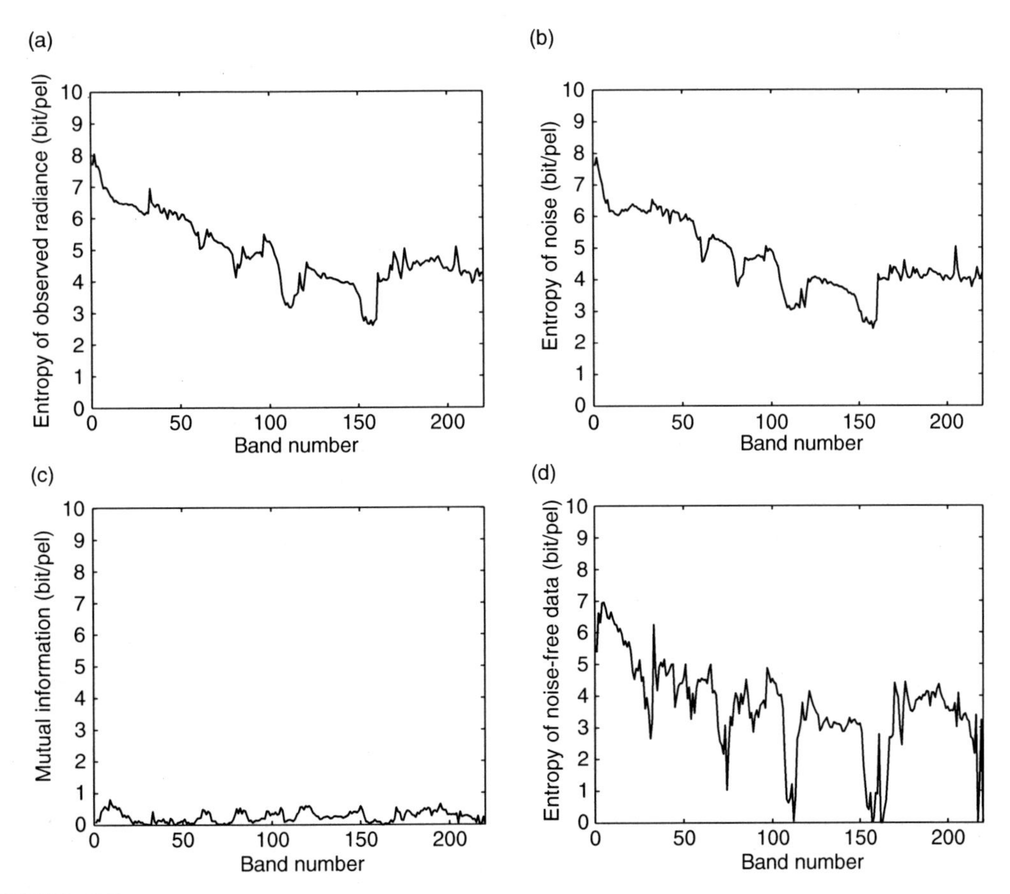

FIGURE 16.7
Information parameters of test AVIRIS hyperspectral image plotted against band number: (a) code rate of noisy radiance, R_g, approximating $H(\hat{S})$; (b) noise entropy, R_n, corresponding to $H(\hat{S}|S)$; (c) mutual information, $I(S;\hat{S})$, given by $R_g - R_n$ (average 0.249); (d) entropy of noise-free radiance, $H(S)$, given by R_f obtained by inverting the parametric source entropy model.

The entropy of noisy radiance was approximated by the code bit rates provided by the RLP 3D encoder [12], working in reversible mode, that is, without quantization (roundoff to integer only) of prediction residues. The entropy of the estimated noise (R_n), plotted in Figure 16.7b, follows a trend similar to that of bit rate (R_g) in Figure 16.7a, which also reflects the trend of noise variance (Figure 16.6d) with the only exception of the first spectrometer. R_g has an average value of 4.896 bit per radiance sample, R_n of 4.647. The mutual information, given by the difference of the two former plots and drawn in Figure 16.7c, reveals that the noisiness of the instrument has destroyed most of noise-free radiance information. Only an average of 0.249 bit survived. Conversely, the radiance information varying with band number would approximately be equal to that shown in Figure 16.7d (about 3.57 bits, in average) if the recording process were ideally noise-free. According to the assumed distortion model of spectral radiance, its entropy must be zero whenever the mutual information is zero as well. This condition occurs only in the absorption bands, where SNR is low, thereby validating the implicit relationship between SNR and information.

16.4.2 ASTER Superspectral Data

Considerations on the huge amount of produced data and the opportunity to characterize the basic components in which information is condensed also apply for *superspectral* sensors like ASTER [26]. In fact, the number of spectral bands is moderate, but the swath width is far wider than that of hyperspectral sensors and therefore the data volume is still huge.

The proposed information theoretic procedure was run on the ASTER superspectral image acquired on the area of Mt. Fuji and made available by Earth Remote Sensing Data Analysis Center (ERSDAC) of Ministry of Economy, Trade, and Industry of Japan (METI). L1B data, that is, georeferenced and radiometrically corrected, were analyzed. ASTER collected data in 14 channels of VNIR, SWIR, and TIR spectral range. Images had different spatial, spectral, and radiometric resolutions. The main characteristics of the ASTER data are reported in Table 16.1. Figure 16.8 shows details of the full size test image. Three out of the 14 ASTER bands are reported as representative of VNIR, SWIR, and TIR spectral intervals. The relative original spatial scale has been maintained between images, while contrast has been enhanced to improve visual appearance.

The first step of the procedure concerns noise estimation of all 14 bands. Across-track and along-track CCs of the noise, that is, ρ_x and ρ_y, are plotted against band number in Figure 16.9a and Figure 16.9b, respectively. In the former, the correlation of noise is lower in VNIR and SWIR bands than in TIR ones, whereas in the latter this behavior is opposite. This can easily be explained since VNIR and SWIR sensors are push-broom imagers that

TABLE 16.1

Spectral, Spatial, and Radiometric Resolution of ASTER Data

VNIR (15 m–8 bit/sample) (μm)	SWIR (30 m–8 bit/sample) (μm)	TIR (90 m–12 bit/sample) (μm)
Band 1: 0.52–0.60	Band 4: 1.600–1.700	Band 10: 8.125–8.475
Band 2: 0.63–0.69	Band 5: 2.145–2.185	Band 11: 8.475–8.825
Band 3: 0.76–0.86	Band 6: 2.185–2.225	Band 12: 8.925–9.275
	Band 7: 2.235–2.285	Band 13: 10.25–10.95
	Band 8: 2.295–2.365	Band 14: 10.95–11.65
	Band 9: 2.360–2.430	

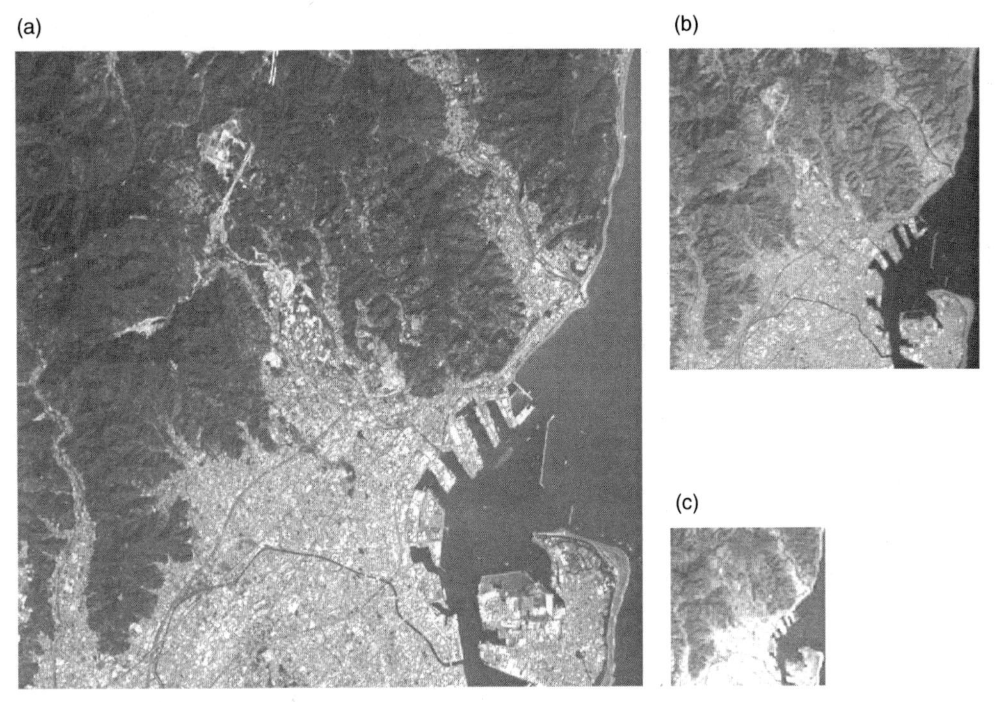

FIGURE 16.8
Detail of ASTER superspectral test image: (a) VNIR band 2, resolution 15 m; (b) SWIR band 6, resolution 30 m; (c) TIR band 12, resolution 90 m.

acquire data along-track, whereas the TIR sensor is a whisk-broom device that scans the image across-track. Therefore, in VNIR and SWIR bands, the same sensor element produces a line in the along-track direction and the noise CC results higher along that direction. The spectral CC of the noise is reported in Figure 16.9c. The CC values are rather low, in contrast to hyperspectral sensors where preprocessing along the spectral direction, aimed at mitigating structured noise patterns occurring in the raw data [3], can introduce significant correlation among bands. The CC values of bands 1, 4, and 10 in Figure 16.9c have been arbitrarily set equal to 0, because a previous reference band is not available for the measurement.

The measured standard deviation of correlated noise is drawn in Figure 16.9d where two curves are plotted to take into account that the ADC of the TIR sensor has 12 bits, while the ADCs of the VNIR and SWIR sensors have 8 bits. The solid line refers to the acquired data and suggests that the VNIR and SWIR data are less noisy than those of TIR. The dashed line has been obtained by rescaling the 8-bit VNIR and SWIR data to the 12-bit dynamic range of TIR sensor; it represents a likely indicator of the noisiness of the VNIR and SWIR data if they were acquired with a 12-bit ADC. Therefore, it is evident that TIR data are less noisy than the others, as shown in the SNR plot reported in Figure 16.9f. In fact, the SNR values of TIR bands are higher than in the other bands because of the 12-bit dynamic range of the sensor and the low value of the standard deviation of the noise, once the data have been properly rescaled. As a further consideration, Figure 16.9d and Figure 16.9f evidence that the standard deviation of the noise of ASTER data in all bands is rather low and, conversely, the SNR is somewhat high when compared to other satellite imaging sensors.

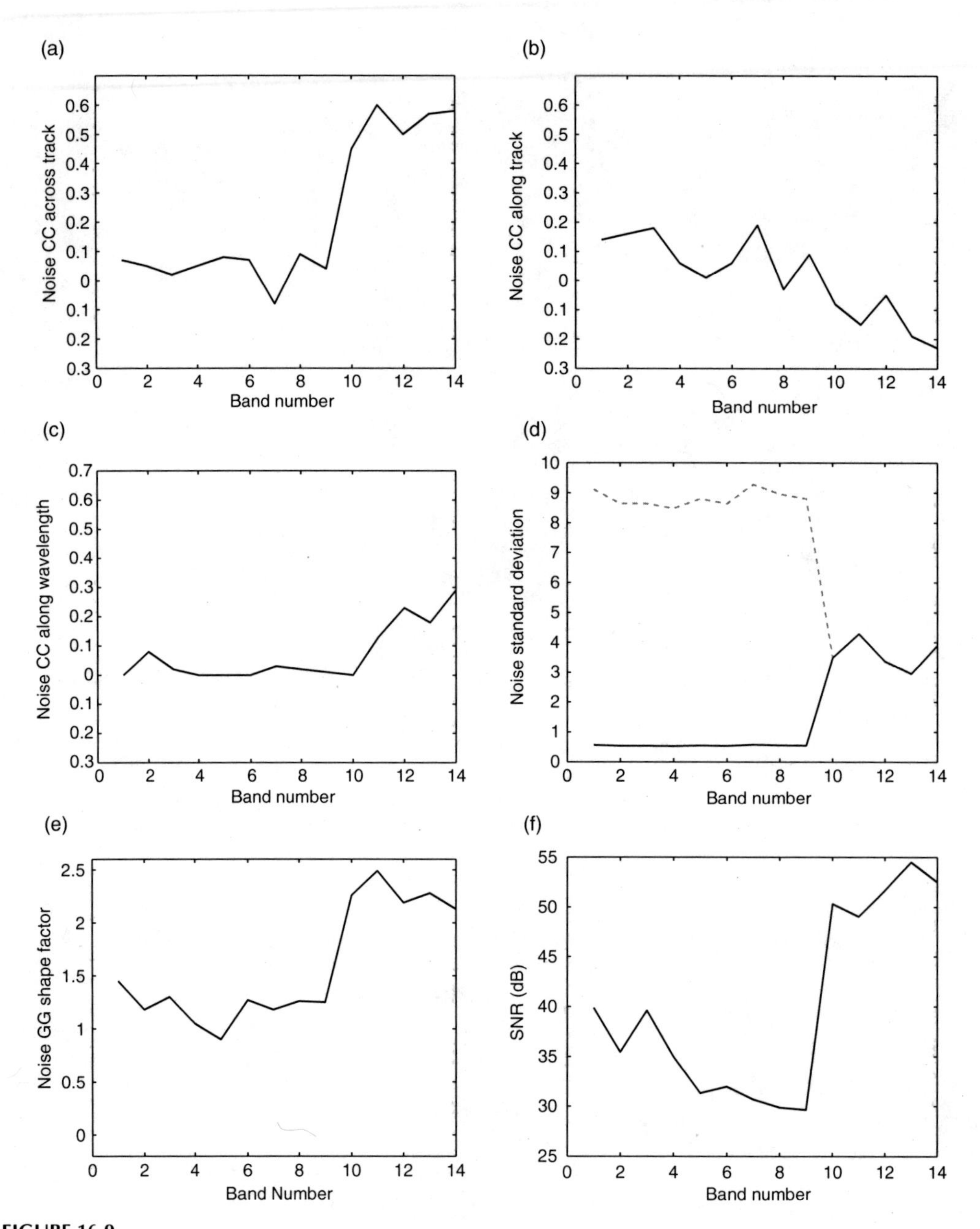

FIGURE 16.9
Noise parameters of test ASTER superspectral image plotted against band number: (a) CC across track, ρ_x; (b) CC along track, ρ_y; (c) CC along spectral direction, ρ_λ; (d) square root of noise variance, σ_n: solid line represents values relative to data with their original dynamic range; dashed line plots values of VNIR and SWIR data rescaled to match the dynamic range of TIR; (e) shape factor of GG-modeled noise PDF, ν_n; (f) SNR.

Eventually, Figure 16.9e plots the noise shape factor. Conversely, from AVIRIS, the noise shape factor is rather different from 2, thereby suggesting that the noise might be non-Gaussian. This apparent discrepancy is due to an inaccuracy in the estimation procedure and probably occurs because the noise level in VNIR and SWIR bands is low (in the 8-bit scale) and hardly separable from weak textures.

Once all noise parameters had been estimated, the information theoretic assessment procedure was run on the ASTER data set. The entropy of observed images was again approximated by the code bit rates provided by the RLP 3D encoder [12], working in reversible mode. Owing to the different spatial scales of ASTER images in VNIR, SWIR, and TIR spectral range, each data set has been processed separately. Therefore, the 3D prediction has been carried out for each spectrometer. As a consequence, the decorrelation of the first image of each data set cannot exploit the correlation with the spectrally preceding band. Thus, the code rate R_g, and consequently R_n and R_f, is inflated for bands 1, 4, 10 (some tenths of bit for bands 1, 4, and roughly 1 bit for band 10).

Figure 16.10 reports plots of information parameters varying with band numbers. Entropies are higher for TIR bands than for the others because of the 12-bit ADC. The bit rate R_g is reported in Figure 16.10a. The *inflation effect* is apparent for band 10 and hardly recognizable for bands 1 and 4. The high value of R_g for band 3 is due to the response of vegetation that introduces a strong variability in the observed scene and makes band 3 strongly uncorrelated with the other VNIR bands. Such variability represents an increase in information that is identified and measured by the entropy values.

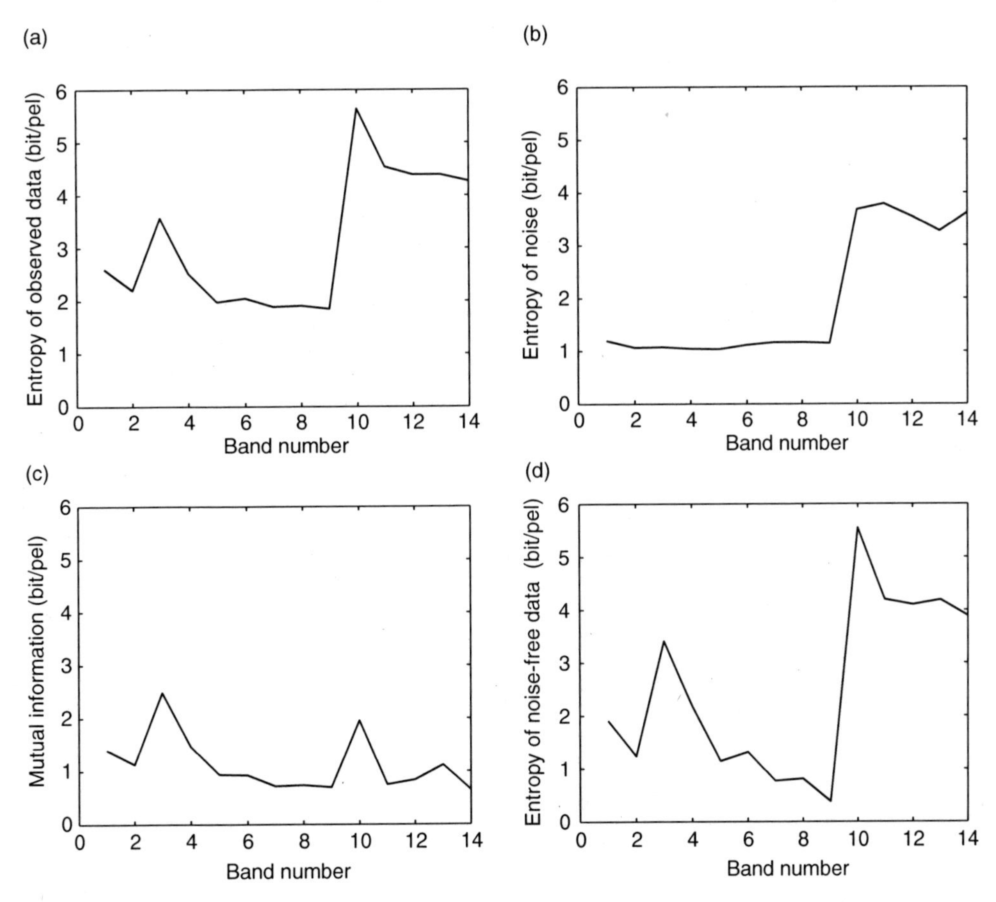

FIGURE 16.10
Information parameters of test ASTER superspectral image plotted against band number: (a) code rate of observed data, R_g, approximating $H(\hat{S})$; (b) noise entropy, R_n, corresponding to $H(\hat{S}|S)$; (c) mutual information, $I(S,\hat{S})$, given by $R_g - R_n$; (d) entropy of noise-free data, $H(S)$, given by R_f obtained by inverting the parametric source entropy model.

The entropy of the estimated noise, plotted in Figure 16.10b, follows a trend similar to that of bit rate in Figure 16.10a, with a noticeable difference in band 3 (NIR), where the information presents a peak value, as already noticed.

The mutual information, $I(S; \hat{S})$, given by the difference of the two former plots and drawn in Figure 16.10c has retained a significant part of noise-free radiance information. In fact, the radiance information varying with the band number is consistent with that shown in Figure 16.10d. This happens because the recording process is weakly affected by the noise. According to the assumed model, when the noise level is low and does not affect the observation significantly, the entropy of the noise-free source must show the same trend of the mutual information and the entropy of the observed data.

16.5 Conclusions

A procedure for information theoretic assessment of multi-dimensional remote-sensing data has been described. It relies on robust noise parameters estimation and advanced lossless compression to calculate the mutual information between the noise-free band-limited analog signal and the acquired digitized signal. Also, because of the parametric entropy modeling of information sources, it is possible to upper bound the amount of information generated by an ideally noise-free process of sampling and digitization. The results on image sequences acquired by AVIRIS and ASTER imaging sensors offer an estimation of the true and hypothetical information contents of each spectral band. In particular, for a single spectral band, mutual information of ASTER data results higher than that of AVIRIS. This is not surprising and is due to the nature of hyperspectral data that are strongly correlated and thus well predictable, apart from the noise. Therefore, a *hyperspectral* band usually exhibits mutual information lower than that of a *superspectral* band. Nevertheless, the sum of the mutual information of the hyperspectral bands on a given spectral interval should be higher than the mutual information of only one band covering the same spectral interval. Interesting considerations should stem from future work devoted to the analysis of data of a same scene acquired at the same time by different sensors.

Acknowledgment

The authors wish to thank NASA/JPL and ERSDAC/METI for providing the test data.

References

1. Huck, F.O., Fales, C.L., Alter-Ganterberg, R., Park, S.K., and Rahman, Z., Information-theoretic assessment of sampled imaging systems, *J. Opt. Eng.*, 38, 742–762, 1999.
2. Park, S.K. and Rahman, Z., Fidelity analysis of sampled imaging systems, *J. Opt. Eng.*, 38, 786–800, 1999.

3. Aiazzi, B., Alparone, L., Barducci, A., Baronti, S., and Pippi, I., Information theoretic assessment of sampled hyperspectral imagers, *IEEE Trans. Geosci. Rem. Sens.*, 39, 1447–1458, 2001.
4. Shannon, C.E. and Weaver, W., *The Mathematical Theory of Communication*, University of Illinois Press, Urbana, IL, 1949.
5. Aiazzi, B., Alparone, L., Barducci, A., Baronti, S., and Pippi, I., Estimating noise and information of multispectral imagery, *J. Opt. Eng.*, 41, 656–668, 2002.
6. Aiazzi, B., Alparone, L., and Baronti, S., Fuzzy logic-based matching pursuits for lossless predictive coding of still images, *IEEE Trans. Fuzzy Syst.*, 10, 473–483, 2002.
7. Aiazzi, B., Alparone, L., and Baronti, S., Near-lossless image compression by relaxation-labelled prediction, *Signal Process.*, 82, 1619–1631, 2002.
8. Blahut, R.E., *Principles and Practice of Information Theory*, Addison-Wesley, Reading, MA, 1987.
9. Lee, J.S. and Hoppel, K., Noise modeling and estimation of remotely sensed images, *Proc. IEEE Int. Geosci. Rem. Sens. Symp.*, 2,1005–1008, 1989.
10. Aiazzi, B., Alparone, L., and Baronti, S., Reliably estimating the speckle noise from SAR data, *Proc. IEEE Int. Geosci. Rem. Sens. Symp.*, 3,1546–1548, 1999.
11. Aiazzi, B., Alba, P., Alparone, L., and Baronti, S., Lossless compression of multi/hyper-spectral imagery based on a 3-D fuzzy prediction, *IEEE Trans. Geosci. Rem. Sens.*, 37, 2287–2294, 1999.
12. Aiazzi, B., Alparone, L., and Baronti, S., Near-lossless compression of 3-D optical data, *IEEE Trans. Geosci. Rem. Sens.*, 39, 2547–2557, 2001.
13. Jayant, N.S. and Noll, P., *Digital Coding of Waveforms: Principles and Applications to Speech and Video*, Prentice Hall, Englewood Cliffs, NJ, 1984.
14. Roger, R.E. and Arnold, J.F., Reversible image compression bounded by noise, *IEEE Trans. Geosci. Rem. Sens.*, 32, 19–24, 1994.
15. Birney, K.A. and Fischer, T.R., On the modeling of DCT and subband image data for compression, *IEEE Trans. Image Process.*, 4, 186–193, 1995.
16. Müller, F., Distribution shape of two-dimensional DCT coefficients of natural images, *Electron. Lett.*, 29, 1935–1936, 1993.
17. Sharifi, K. and Leon-Garcia, A., Estimation of shape parameter for generalized Gaussian distributions in subband decompositions of video, *IEEE Trans. Circuits Syst. Video Technol.*, 5, 52–56, 1995.
18. Aiazzi, B., Alparone, L., and Baronti, S., Estimation based on entropy matching for generalized Gaussian PDF modeling, *IEEE Signal Process. Lett.*, 6, 138–140, 1999.
19. Kokkinakis, K. and Nandi, A.K., Exponent parameter estimation for generalized Gaussian PDF modeling, *Signal Process.*, 85, 1852–1858, 2005.
20. Mallat, S., A theory for multiresolution signal decomposition: the wavelet representation, *IEEE Trans. Pattern Anal. Machine Intell.*, 11, 674–693, 1989.
21. Shaw, G. and Manolakis, D., Signal processing for hyperspectral image exploitation, *IEEE Signal Process. Magazine*, 19, 12–16, 2002.
22. Jimenez, L.O. and Landgrebe, D.A., Hyperspectral data analysis and supervised feature reduction with projection pursuit, *IEEE Trans. Geosci. Rem. Sens.*, 37, 2653–2667, 1999.
23. Landgrebe, D.A., Hyperspectral image data analysis, *IEEE Signal Process. Mag.*, 19, 17–28, 2002.
24. Aiazzi, B., Baronti, S., Santurri, L., Selva, M., and Alparone, L., Information–theoretic assessment of multi-dimensional signals, *Signal Process.*, 85, 903–916, 2005.
25. Roger, R.E. and Arnold, J.F., Reliably estimating the noise in AVIRIS hyperspectral images, *Int. J. Rem. Sens.*, 17, 1951–1962, 1996.
26. Aiazzi, B., Alparone, L., Baronti, S., Santurri, L., and Selva, M., Information theoretic assessment of aster super-spectral imagery. In *Image and Signal Processing for Remote Sensing XI*, Bruzzone, L., ed., Proc. SPIE, Bellingham, Washington, Volume 5982, 2005.

Index